Development and Application of Computer Techniques to Environmental Studies VI

SIXTH INTERNATIONAL CONFERENCE ON DEVELOPMENT AND APPLICATION OF COMPUTER TECHNIQUES TO ENVIRONMENTAL STUDIES
ENVIROSOFT 96

CONFERENCE DIRECTOR

P. Zannetti
Failure Analysis Associates Inc., USA

EXECUTIVE COMMITTEE

C.A. Brebbia
Wessex Institute of Technology, UK

P. Zannetti
Failure Analysis Associates Inc., USA

EXHIBITION CHAIRMAN

G. Guariso
Polytechnic of Milan, Italy

INTERNATIONAL SCIENTIFIC ADVISORY COMMITTEE

D. Al Ajmi
J.M. Baldasano
A. Barbaro
J. Bartnicki
M. Bonazountas
R. Bornstein
C. Dejak
G. Gambolati
J.-M. Giovannoni
G. Guariso
B. Henderson-Sellers
T.V. Hromadka II

G. Ibarra
A.J. Jakeman
A.A. Jennings
D.P. Loucks
T.J. Lyons
A. McMillan
N. Moussiopoulos
D.W. Pepper
J.O. Rumbaugh
R. San Jose
T. Tirabassi
Y.Q. Zhang

Organised by:

Wessex Institute of Technology, UK

Acknowledgement is made to L. Della Ragione et al. for the use of figure 4 on page 725, which appears on the front cover of this book.

Development and Application of Computer Techniques to Environmental Studies VI

EDITORS:

P. Zannetti
Failure Analysis Associates Inc., USA

C.A. Brebbia
Wessex Institute of Technology, UK

Computational Mechanics Publications
Southampton   Boston

P. Zannetti
Failure Analysis Associates Inc.
149 Commonwealth Drive
P.O. Box 3015
Menlo Park
California 94025, USA
e-mail: paolo@cup.portal.com
tel.: (415) 688-6962
fax: (415) 688-7269

C.A. Brebbia
Wessex Institute of Technology
Ashurst Lodge
Ashurst
Southampton SO40 7AA
UK
e-mail: carlos@wessex.ac.uk
tel.: (44) 01703-293223
fax: (44) 01703-292853

Published by

Computational Mechanics Publications
Ashurst Lodge, Ashurst, Southampton SO40 7AA, UK
Tel: 44 (0)1703 293223; Fax: 44 (0)1703 292853
Email: cmp@cmp.co.uk
http://www.cmp.co.uk/

For USA, Canada and Mexico

Computational Mechanics Inc
25 Bridge Street, Billerica, MA 01821, USA
Tel: 508 667 5841; Fax: 508 667 7582
Email: cmina@ix.netcom.com

British Library Cataloguing-in-Publication Data

A Catalogue record for this book is available
from the British Library

ISBN: 1-85312-4117 Computational Mechanics Publications, Southampton

Library of Congress Catalog Card Number 96-083651

*The texts of the various papers in this volume were set
individually by the authors or under their supervision*

PREFACE

ENVIROSOFT 96 is to us a particularly important event for two reasons. First, it is the 10th anniversary of the conference series. The first ENVIROSOFT meeting, in fact, was held in 1986 in Newport Beach, California. Many things have changed since then: the PC revolution and the worldwide spread of the Internet, just to mention the top two. Environmental topics are today even more important than 10 years ago and the emphasis of many R&D studies has expanded from the local region to the global scale of pollution.

We started, 10 years ago, with the intention of providing a forum for discussion to those scientists working at the intersection of two fast-growing, exciting fields: environmental and computer sciences. This intersection has grown dramatically since then. We can say that, today, it is almost unthinkable to work in environmental science without computers.

The future holds surprises and exciting discoveries. Computers will become faster and faster and, more importantly, really user-friendly. People like me, who have always enjoyed working with a Macintosh better than a Cray, know how important user-friendliness is, not just to make things easier but to increase our productivity. With the new generation of fast PCs and Win95/NT operating systems, we should expect a real push toward simplicity and common sense in the next few years. New user interfaces will soon start adopting virtual reality techniques to immerse a large segment of computer users into virtual worlds that will be particularly suitable to environmental applications. The helmet (probably a much more comfortable version of what we have today) will be in the next decade what the mouse has been in the last 10 years. The Internet will give all of us to instant access to information, news, summaries, measurements, and results related to the environment.

The ENVIROSOFT conference will continue to monitor these trends and provide an opportunity to meet, discuss, debate, and sometimes argue environmental issues and the role computers play.

I want to conclude with a sincere thanks to all ENVIROSOFT 96 participants, starting, of course, with those who were with us ten years ago and finishing with a particularly warm welcome to the newest generation of attendees, i.e., those young scientists who are just at the beginning of their long careers as "envirometricians." To them, I have only one suggestion to make. Do not be afraid of using the most recent computer tools, both hardware and software. Do not hesitate to adopt the latest technology (e.g., virtual reality), even when it looks a little weird in the eyes of the older generation. Do try new things. Your productivity can be an order of magnitude higher if you force yourselves to remain at the cutting edge of computer sciences in studying environmental phenomena.

PAOLO ZANNETTI
Menlo Park, California
June 1996

CONTENTS

SECTION 3: ENVIRONMENTAL SCIENCES AND ENGINEERING

SECTION 4: CHEMISTRY, PHYSICS AND BIOLOGY

SECTION 7: FLUID DYNAMICS

SECTION 8: SATELLITE DATA, IMAGE PROCESSING, PATTERN RECOGNITION, REMOTE SENSING

SECTION 9: DATABASES

SECTION 10: ENVIRONMENTAL MANAGEMENT AND DECISION ANALYSIS

SECTION 11: SOFTWARE IMPLEMENTATION

KEYNOTE ADDRESS

Environmental modeling – the next generation

P. Zannetti

Failure Analysis Associates, Inc., 149 Commonwealth Drive, Menlo Park, California 94025, USA

Abstract

This chapter provides a brief introduction to that which will affect the next generation of computer systems for simulating environmental phenomena.

Keywords: Environmental modeling, pollution, computer simulation, numerical modeling, computer sciences, research and development, multimedia modeling, comprehensive modeling systems, virtual reality

1   Introduction

What should we call "environmental sciences"?

I recently proposed a definition of environmental sciences which seems to be acceptable to most of the scientists I interacted with (Zannetti, 1993).

Environmental sciences are defined as the sciences that cover, as their principal subject, anthropogenic pollution: its generation, its transport and fate in different environmental media (air, water, soil, groundwater, and biota), and its adverse effects.

We could use a more general characterization, if you prefer, and define environmental sciences as that concerned with pollution, population control, and resource conservation. But I would resist the temptation of "contaminating" environmental sciences with the inclusion of controversial and ideological topics that better belong to the political and economic sciences and to the agendas of modern-day environmentalism.

One of the main challenges in environmental sciences today is to simulate environmental flows and the behavior of anthropogenic pollutants injected into them. This process is called "environmental modeling" and is performed today mostly by using computer simulation techniques.

Today environmental modeling software is still divided into two major fields. On one side we have relatively simple commercial products, mostly for regulatory applications, i.e., to help the industry in complying with environmental laws and regulations. On the other side, we have complex "research codes," developed at universities and research centers. Research codes are (in theory) more advanced but remain difficult to use. This dichotomy is expected to persist for a while, even though we should expect that within 5 to 10 years computers will become so user-friendly that even the codes developed for research and development (R&D) goals will be available to a larger segment of users.

Environmental phenomena remain a formidable scientific challenge. The simulation of environmental flows and the behavior of pollutants injected into them will remain a difficult task plagued by uncertainties in input parameters and always requiring some form of simplification of the "primitive" equations.

Soon the new generation of computer tools will provide models that will actually quantify the uncertainties of their own calculations. These models will not only supply forecasts and simulations, but they will provide the full probability density function of their outputs. People who use these simulations will receive "uncertainty intervals" that will allow them to better interpret the results.

2   To Model or Not to Model?

As a modeler, my interactions with the rest of the world have been complex and sometimes frustrating. In most cases, I have received gratifying and competent appreciation from my work with models and a fair understanding of what models can do. But in other cases things have been complicated. Often, in certain circles, I have found it difficult to explain what computer modeling is. This is because computer illiteracy in certain segments of the population, including mid-level and high-level business management, is still not uncommon today, despite the clear progress made during the last decade through the personal-computer revolution. In many cases, I faced a profound skepticism for the capabilities of environmental models.

Environmental modeling, like most things on earth, can be done perfectly or can be an example of "garbage in, garbage out." Most applications, of course, lie somewhere between these two extremes. It is difficult to define how a modeling application can be done perfectly, but I have tried to make a list of 10 conditions for what we can call "excellence in modeling":

1. The computer code must be fully tested, both as a whole and in its individual modules.

2. The code should be used by other groups, besides the developers, to assure external peer-review and approval.

3. The code should be fully documented. All equations and assumptions should come from formulations and methods published in peer-reviewed publications.

4. The input data should be reliable and pertinent to the specific application. The spatial and temporal resolution of the input data should be consistent with the time and space scales of the phenomenon that one wants to simulate.

5. The code should apply the best available science, e.g., by using full three-dimensional (3D) representations and avoiding the use of steady-state solutions to simulate time-varying phenomena.

6. The code should be numerically correct. That is, all iterative calculations should reach full convergence and results should not be affected by the size of the grid or the length of the integration time.

7. The code should be fully validated against reliable field data. This validation should include a careful evaluation of the physical significance of both the model outputs and the field measurements. For example, a comparison between model outputs on large spatial grids with point measurements can be phenomenologically incorrect.

8. Any calibration of the code should be done properly. Proper calibration means the tuning of some input parameters within their range of uncertainty in order to maximize the agreement between model outputs and field data. Calibration, however, is not proper when it ends up making the model work for the wrong reason!

This has happened, unfortunately on several occasions in the past. A well-known example is the application of photochemical models to simulate ozone concentrations in urban areas. In the past, reaction rates in these photochemical models were "tuned" to allow the models to agree with the available measurements of ozone in the region. Later, it was discovered that the main model inputs, i.e., the emissions of VOC and NO_x in the region, were incorrect. That means that these models could be tuned to provide "correct" ozone concentrations with highly incorrect emission values of the ozone precursors. Therefore, what people called calibration in that case was just forcing the models to work for the wrong reason. Any practical application of these models, for example to calculate the emission reductions in VOC and NO_x required to achieve a certain predefined reduction of ozone concentrations in the region, was clearly incorrect.

9. The model application should be fully documented in a report presenting all assumptions and a description of all computer runs. The reader should be able to rerun the application on his computer and obtaining the same results in an independent manner.

10. In summary, use your brain, do it right, and do not cheat!

3   Environmental Modeling Tomorrow

This section presents a discussion of expected future developments in environmental modeling. Three factors are discussed: (1) the increasing use of multimedia models, (2) the current design and development efforts toward the creation of comprehensive modeling systems, and (3) the possible incorporation of virtual reality techniques into environmental models.

3.1   Multimedia Modeling

Traditionally, environmental models were designed to cover the dynamics of pollution in one particular medium, e.g., the atmosphere or a body of water. Recently, multimedia models have become necessary (Seigneur, 1993), since pollutants such as pesticides can migrate and be detected in different media. Conceptually, a multimedia model is not necessarily more complex than a model designed for a specific medium. However, multimedia simulations require a more extensive set of input parameters. They must include, for example, intermedia transport terms such as atmospheric deposition, sedimentation, volatilization, and erosion.

3.2   Comprehensive Modeling Systems

In the early 1990s, a few project teams, typically formed by environmental and computer scientists, explored the possibility of developing comprehensive modeling systems (CMS) for environmental sciences. In a nutshell, a CMS should include three major functions (Zannetti, 1993):

1. <u>Education</u>. The education section of a CMS should contain several modules designed for environmental education. These modules could typically cover six topics: the atmosphere, rivers and lakes, seas and oceans, soil and groundwater, human health, and environmental effects.

2. <u>Simulation</u>. The simulation section of a CMS should contain modules that simulate environmental phenomena and adverse effects of pollution. Results could be visualized by still images or animation. The simulation modules could cover ten topics: air, watershed, soil and groundwater, surface water, multimedia, indoor pollution, noise, human health, environmental damage, and environmental engineering.

3. <u>Management</u>. The management section of a CMS should contain several modules to assist the environmental manager and the environmental regulator. These modules could address eight topics: environmental information management, regulatory compliance, risk assessment, emergency response, pollution control, environmental remediation, litigation support, and literature search.

Much progress has been made recently toward the development of a CMS for air pollution (Hansen et al., 1994 and 1995; Dennis et al., 1996; Zannetti et al., 1996). These efforts are expected to be extended to other media in the near future. The air pollution CMS, as envisioned by its designers initially, should provide an infrastructure that helps its users do their jobs better and faster, whether those jobs be regulatory and policy analysis, source impact assessment, understanding atmospheric chemistry and physics, or performing atmospheric research studies. As such, this CMS should provide the following:

- A platform for modeling pollutant emissions, atmospheric physics and chemistry, and the impact of pollution in as scientifically sound a fashion as is desired or possible.

- A readily accessible interface, so that its use is a benefit, not a distraction.

- A powerful set of analysis and decision-support tools, including graphical, visual, economic, and scientific tools, and also including report preparation.

- A method to make maximum use of the available computational resources, including CPU power, disk storage, and communication systems.

The air pollution CMS is being designed in a way that facilitates its continuous evolution with science, computer capabilities, and user needs. This will require that it employ well-acknowledged standards (e.g., computer languages and data protocols).

Perhaps the best way to illustrate the CMS is to provide a realistic description of its future use:

> Palo Alto, 7 April 1999. A user sits in front of an Apple-IBM *Penta III* computer screen. The screen shows a stylish *CMS* logo and several other buttons. The user issues a voice command to the system, or clicks (with a foot mouse) the button *Beginners click here*, and a series of windows is displayed. These windows contain detailed information sections and an animated user's guide to describe the entire system. By clicking the button *Education,* the user brings up a new series of "windows" and "chapters." These sections are connected to CDs, laser disks, and multimedia devices and provide, on the *Penta III* screen, interactive education tools on the subjects of atmospheric sciences, air pollution, laws/regulations, simulation modeling, and databases. A special *Communication* button allows the user to communicate, via user-friendly interfaces, with library databases, meteorological/air quality databases, and other users.

> By clicking the *CMS Regulatory* button, the user accesses a subset of the CMS system in which only models and techniques that have received some regulatory approval are available. The use of these models is "locked," in the sense that they can only be used with computational options that are acceptable to the regulatory agencies. Regulations of different countries (USA, Canada, ECC, Japan, etc.) can be selected, thereby locking the execution of the simulations under different regulatory constraints. By clicking the *CMS full set* button, the user accesses the entire simulation system. Through a password and a voice recognition check, a user-developer is allowed access to the master version of CMS (in remote computer storage) and modify/add/update modules and functions.

> A typical CMS session consists of a CMS-guided computer simulation and "report" preparation. The user defines the computational domain, the simulation period, and other options. CMS assists the user in performing a sequence of simulations and making choices to calculate emission data, meteorological fields, transport and diffusion scenarios, chemical reactions, dry and wet deposition, and some adverse effects of air pollution, such as visibility impairment. Any step can be fully visualized by superimposing input/output data on geographical information using a

GIS and full 3D views (in a fly-through fashion). A special *Real-time* button allows real-time simulation for emergency response of accidental releases, if proper connections are made to access meteorological and other data on-line. At any time, the user can select input/output data and ask CMS to perform special calculations and analyses in different computational environments (such as new versions of *Mathematica, Spyglass, Systat,* etc.) on the *Penta III* screen.

What is described above illustrates the first and by far the most important goal of the CMS – to let the computer do the work. In the crudest sense, this means more complex and more sophisticated computations, but this is only the tip of the iceberg. Far more important is using the computer to compile, cross-reference, and comprehend intentions. The goal of the CMS is as much a manifesto as a plan; it is meant to be revolutionary, not incremental.

3.3   Virtual Reality

In the world of information sciences, two trends seem particularly relevant in the mid-1990s. The first trend is the increasing power of PCs (IBM-compatible and Macintosh-type machines). Internal PC clocks can now run at about 200 MHz providing very substantial computer power.

Many scientists have invested their time and resources heavily in Unix workstations in the last few years. This investment was certainly a smart move at the beginning of the decade, when it became clear that mainframe computers and supercomputers were not cost-effective in comparison with the new generation of Unix-based microcomputers. A similar phenomenon is occurring now. Scientists are discovering the increased power of the new generation of personal computers running under new, powerful operating systems such as IBM OS/2, Microsoft Win95/NT, and Apple OS. These new hardware/software platforms, and especially Microsoft NT, are expected to substantially erode the market for Unix workstations by providing high-speed, advanced features, user-friendliness, and inexpensive software for scientific calculations and simulations.

The second trend is virtual reality (VR). VR may appear frivolous in nature because of its close relation to interactive games and, therefore, its "arcade" connotation. But VR techniques appear to be the most interesting trend in computer software today.

VR has been defined as

". . . a computer-synthesized, three-dimensional environment in which a plurality of human participants, appropriately interfaced, may engage and manipulate simulated physical elements in the environment and, in some forms, may engage and interact with representations of other humans, past, present or fictional, or with invented creatures." (Nugent, 1991, quoted in Larijani, 1993)

and

". . . an interactive computer system so fast and intuitive that *the computer disappears from the mind of the user,* leaving the computer-generated environment as the reality." (Goldfarb, 1991, quoted in Larijani, 1993)

According to Goerbe (1992),

". . . virtual reality enables users to immerse themselves in computer-generated environments that include three-dimensionality through sound, sight, and touch."

Larijani (1994) simply describes VR as ". . . a cartoon world you can get into."

VR techniques cover two basic applications: (1) those in which the user puts on appropriate gadgets and enters an artificial world in which he or she manipulates objects in an interactive manner and (2) those in which the user observes and moves virtual humans operating in a virtual world.

VR techniques differ from multimedia tools. The reader should note that in this section the term "multimedia" has a different connotation. "Multimedia" here indicates the expansion of our communication capabilities and the inclusion of sounds, images, animations, and video. VR techniques employ multimedia tools and expand upon them. Thus, allowing an interaction between the user and "objects" in a virtual world. Examples of VR applications are some of the most sophisticated video games and some training techniques used by the military (for example, jet fighter simulations).

VR techniques are expanding rapidly. According to Martin (1994), the marketing firm Forst & Sullivan reported that the VR market is expected to exceed $1 billion in 1997, following a growth rate in which the market almost tripled from 1991 to 1993 (from $49.7 million to $130 million).

I expect VR techniques to become, within 10 years (and perhaps even 5), the *preferred* way in which any user interacts with any computer. To understand why, one must look at the history of the computer user interface. There we find a logical evolution: from primitive techniques such as punched cards and dumb terminals to today's microcomputer software, developed using object-oriented techniques and available to users through GUIs (graphical user interfaces). The next step is the use of 3D icons and, almost inevitably, the total immersion of the user in a 3D world – a VR world.

Clearly, the environmental modeling community will be affected, at least indirectly, by this trend. Certainly, the use and the development of environmental models will be influenced by the increasing availability of VR-based productivity tools. But the real question is how this trend will *directly* affect environmental scientists (Zannetti, 1995).

As discussed in Section 3.2, an important trend in environmental modeling today is the ongoing development effort directed toward "comprehensive" systems. These new air quality modeling systems are expected to be scientifically advanced, comprehensive (e.g., covering local, regional, and continental scales), easy to access and use, self-consistent, and computationally efficient. For example, a comprehensive modeling system (CMS), such as that described by Hansen et al. (1995), is expected to change the way people use air pollution models by allowing both scientists and nontechnical people to operate complex atmospheric simulations and run air pollution scenarios almost anywhere.

I would argue that the ultimate step in the development of environmental models – and probably the only step that can assure full user-friendliness and accessibility to nontechnical people – is the incorporation of VR techniques. After all, what is environmental modeling? Environmental modeling can be seen as an organized and interactive series of computational modules that transform one set of databases (emissions, human activities, meteorology, geography, etc.) into another set (concentrations, depositions, health damage, etc.) The more comprehensive environmental modeling systems become, the more users will be able to apply simulation techniques as black boxes, i.e., without necessarily understanding the details of the computations. Users, therefore, will be able to focus on exploring the relationships among the databases (inputs and outputs) and experimenting with changes without worrying about the computational details.

In less than 10 years, the use of environmental models may very well become a simple and entertaining VR exploration into databases. It may look like a video game, but it will incorporate the best of science. Manipulations of input databases (e.g., emissions) will automatically generate related changes in output databases (e.g., concentrations). Inverse calculations (e.g., the emission controls required to achieve predefined concentration standards) could be simply triggered by imposing a constraint on an object in the output database. It is not science fiction to envision a user wearing a VR device such as a helmet, entering a virtual database, walking through objects and their functional relationships, and performing direct and inverse simulations, analyses, and optimizations by simple manual operations on 3D objects. It will be like walking through a gigantic spreadsheet in which a single cell may contain the same number of calculations as one of today's air pollution models (e.g., 10^5 to 10^6 lines of FORTRAN code).

3.4   Conclusions

Clearly, at this point one can hardly guess to what extent CMSs and VR will figure into future environmental modeling studies. To forecast the evolution of computer systems and user needs is problematical, and one may easily be fooled by fashionable trends with no future. However, let me conclude this chapter with two pieces of advice to the reader. First, this is probably not the best time to make a big Unix investment – wait till 1997 if you can, and do consider NT as an alternative. Second, next time your kids go to the arcade and start playing with helmets and other gadgets, join them; you may develop very useful skills for the computer work that is awaiting you just a few years ahead.

Disclaimer

The opinions presented herein are those of the author alone and should not be interpreted as necessarily those of Failure Analysis Associates, Inc.

References

Anderson, M.P. and W. W. Woessner (1992). *Applied Groundwater Modeling–Simulation of Flow and Advective Transport.* San Diego: Academic Press, 1992.

Dennis, R.L., D.W. Byun, J.H. Novak, K.J. Galluppi, C.J. Coats, and M.A. Vouk (1996). "The Next Generation of Integrated Air Quality Modeling: EPA's Models-3." *Atmospheric Environment*, **30**(12), pp. 1925-1938.

Goerbe, C. (1992). "Visionary Marketers Hope for Concrete Gains from the Fantasy of Virtual Reality." *Marketing News,* December 7.

Goldfarb, N. (1991). "Virtual Reality: The State of the Art." *MicroTimes,* October 14, p 62.

Hansen, D.A., et al. (1994). "The Quest for an Advanced Regional Air Quality Model." *Environ. Sci. Technol.*, **28**,(2), pp. 71A-77A.

Hansen, D.A., P. Zannetti, and J.M. Hales (1995). "Design of a Framework for the Next Generation of Air Quality Modeling System." Proceedings of AIR POLLUTION 95, Porto Carras, Greece. Computational Mechanics Publications, Southampton, UK.

Larijani, L.C. (1994). *The Virtual Reality Primer*. McGraw-Hill.

Martin, C. (1994). "Imagine what you could do with it. (Virtual Reality)." *Computer Weekly*, June 9.

Nugent, W.R. (1991). "Virtual Reality: Advanced Imaging Special Effects Let You Roam in Cyberspace." *J. Am. So. for Information Science*. September.

Seigneur, C. (1993). "Multimedia Modeling." Chapter 3 of *Environmental Modeling* Volume I, (Zannetti, Ed.), Computational Mechanics Publications and Elsevier Applied Science.

Zannetti, P. (1990). *Air Pollution Modeling – Theories, Computational Methods and Available Software*. New York: Van Nostrand Reinhold.

Zannetti, P. (1993). "Introduction and Overview." Chapter 1 of *Environmental Modeling*, Volume I(Zannetti, Ed.), Computational Mechanics Publications and Elsevier Applied Science.

Zannetti, P., B. Bruegge, D.H. Hansen, N. Lincoln, W.A. Lyons, D.A. Moon, R.E. Morris, and A.G. Russell (1996). "Framework Design – Design and Development of a Comprehensive Modeling System (CMS) for Air Pollution." FaAA Report SF-R-96-02-21, February. Submitted to the Electric Power Research Institute.

Zannetti, P. (1995). "Is Virtual Reality the Future of Air Pollution Modeling?" (Keynote Address). AIR POLLUTION 95, Porto Carras, Greece. Computational Mechanics Publications, South Hampton.

SECTION 1:
POLLUTION: AIR, WATER AND SOIL

Severe Nuclear Accident Program (SNAP) - an operational dispersion model

J. Bartnicki, A. Foss, J. Saltbones
Norwegian Meteorological Institute, P.O.Box 43 Blindern, N-0313 Oslo, Norway

Abstract

SNAP model has been developed at the Norwegian Meteorological Institute (DNMI) to provide decision makers and Government officials with real-time tool for simulating large accidental releases of radioactivity to the atmosphere from nuclear power plants or other sources. SNAP is developed in the Lagrangian framework in which atmospheric transport of radioactive pollutants is simulated by emitting a large number of particles from the source. After release each particle carries a given mass of radioactive nuclides which can be in the form of gas or aerosol. During the transport, the radioactive mass of each particle is reduced by dry and wet deposition. SNAP can be used to predict the transport and deposition of a radioactive cloud in the future (up to 48 hours, in the present version) or to analyse the behaviour of the cloud in the past. SNAP is now fully operational at DNMI, which means the meteorologist on duty can start a simulation at any time - day and night. It was evaluated during the European Tracer EXperiment (ETEX) showing good agreement with the measurements. SNAP has also been used for simulating the Chernobyl Accident.

1 Introduction

In the days following the Chernobyl accident in 1986, a real-time dispersion model was not available for the decision makers responsible for handling the emergency situation. Since then, real-time dispersion models have been developed in many countries and have been implemented for operational use in national weather services. Discussion of the requirements for a real-time disper-

sion model for operational use in Norway is given by Saltbones [11]. Such a
model, SNAP, is now available at DNMI.

The first version of SNAP [11], [12], [13], [14] was developed based on
the same theoretical assumptions and basic architecture as the NAME model,
[10], designed and implemented at the UK Meteorological Office. Although a
first version of SNAP was in operational use at DNMI already in 1994, the fully
operational version was implemented in december 1995, as part of DNMI's
contribution to the Norwegian Application of the Major Emergency Management (MEMbrain) Project, which is a part of the EUREKA (EU-904) activity.
MEMbrain is a computer based decision support system for emergency management in case of large accidents. A/S Quasar Consultants has been project
leader for the Norwegian part of MEMbrain, with five Norwegian research
institutions as sub-contracters to the system under development. The user of the
fully developed system is the Norwegian Radiation Protection Authority
(NRPA) in its function as secretariat for the Norwegian Nuclear Emergency
Preparedness Organization. The development of the system have been given
financial support from EKSPOMIL and Norwegian State Authorities.

The main objective of SNAP is to provide decision makers and Government officials with a real-time tool for simulation of large accidental releases of
radioactivity to the atmosphere in case of a nuclear accident. SNAP is now fully
operational at DNMI, which means that the meteorologist on duty, on request
from the NRPA, can start a simulation at any time - day and night.

We can list models of similar capability as SNAP in operational use in
EUROPE: NAME of the UK Meteorological Office [10]; MATCH of the Swed-
ish Meteorological and Hydrological Institute (SMHI) [9]; Lagrangian Particle
Model of the German Weather Service [3]. Many of these models, including
SNAP, have been tested in the European Tracer Experiment (ETEX) [2], [6],
[7]. In this real-time modeling exercise, participants from 28 research institutions in 21 countries in Europe, North America and Japan could compare their
model results with air samples of the released tracer PFC, perfluoromethylcyk-
lohexane.

In this paper, we describe the operational version of SNAP.

2 Description of SNAP

SNAP is a Lagrangian particle transport model. The main advantage of the
Lagrangian approach, compared to Eulerian, is a possibility of a better description of the advection process. A disadvantage is that large computer memory is
necessary to run the model.

2.1 Basic model assumptions

In SNAP, the emitted mass of radioactive debris is distributed among a large
number of particles. After the release, each particle carries a given mass of

radioactive nuclides which can be in the form of gas or aerosol. A 'particle', in this approach, is given an abstract mathematical definition, rather than a physical air parcel containing different pollutants. It is used in SNAP as a vehicle to carry the information about radionuclides emitted from the source point. The particle is not given a definite size and can not be subdivided or split into parts. On the other hand, the mass carried by the particle can be subdivided and partly removed under certain circumstances. In this sense the particle can be interpreted as a vehicle or 'a truck', dumping off its 'load' during transport.

2.2 Numerical structure of the model

A simple diagram illustrating the information flow in the SNAP model computations is shown in Figure 1. The key information about the nuclear accident and the type of simulation to be performed by SNAP, is specified in two input files: *run_snap* and *snap.input*.

The first file *run_snap* starts the execution of SNAP and in addition, includes the following information: start time and period of the simulation, period of release and geographical location of the source, vertical release profile and time release profile, number of particles per release and intensity of release.

The second input file *snap.input* includes more information about model parameters, variables and model operation.

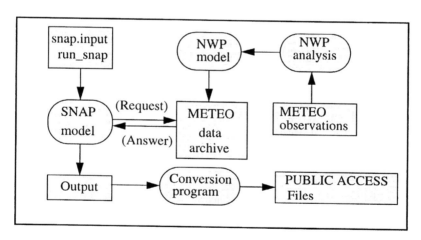

Figure 1: Information flow in the SNAP model computations.

2.3 Model domain and meteorological input

Simulation with the SNAP model can be performed in different grid systems. The system most frequently used is shown in Figure 2. This is the grid system used by the model HIRLAM, currently DNMI's main operational numerical weather prediction (NWP) model.

The vertical structure of SNAP is made compatible with HIRLAM, using data from a subset (14 layers, 7 layers below 1800 m) of the vertical layers available. Concentration and deposition fields computed by SNAP are stored and presented in the same grid.

There are two classes of meteorological input data for SNAP. The first is produced directly by HIRLAM and includes: wind, temperature, surface pressure and precipitation. The second is the mixing height, which defines the height of the Atmospheric Boundary Layer (ABL). The mixing height is also computed from NWP model output (HIRLAM), but it is in a class of its own; both due to its importance for SNAP as input, but also because the procedure to determine the mixing height is so sensitive for the outcome of the simulation.

HIRLAM operates in η terrain following vertical coordinates system. For SNAP use, horizontal wind, vertical velocity and temperature are available on-line from HIRLAM at 14 η-layers.

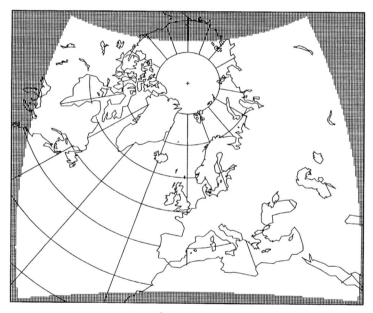

Figure 2: Numerical grid system used for the SNAP computations.

The procedure to identify and calculate the mixing height is based on the 'Critical Richardson Number' concept, first formulated by L.F. Richardson in 1925. The results from SNAP are sensitively dependent on the values of the mixing height. An outline of the ideas behind the chosen procedure is presented in [10] and [15].

2.4 Source-term specification in SNAP

Three groups of substances are taken into account in the source-term specification in SNAP: Aerosols (A), Noble Gases (NG) and other gases (G). All groups are assumed to be emitted into the atmospheric boundary layer (ABL). Only in case of an explosion, a certain fraction of the emission may reach the atmosphere above the ABL as a direct injection in the emission process.

During the model simulation we use unit emission rate ($1\ g\ s^{-1}$) for each group. Since the basic model equations of SNAP are linear in relation to emission, concentration and deposition (mass), the results of a 'run' using '*unit emission rate*' can easily be scaled by multiplying output concentration and deposition according to 'corrected' emission rates. In this way, unit output fields are very convenient for the user, because the transport of arbitrary isotopes can be taken into account by scaling the results as a post-processing procedure. An additional advantage of this approach gives the possibility of performing "off-line" modeling of the radioactive decay. These features give the user more flexibility in testing different emission scenarios for different radioactive species. However, we have to mention that there are certain restrictions for scaling SNAP results, as discussed in more detail in [15].

2.5 Advection and diffusion of particles

The displacement of each particle is calculated for each model time step. For this calculation, three-dimensional velocity is interpolated to particle position from the eight nearest nodes in the model grid. Bilinear interpolation in space is applied to horizontal components of the velocity field and linear interpolation for vertical component. In addition, linear interpolation in time is applied between six-hourly meteorological input fields. The displacement of each particle in one time step is calculated according to the following equation:

$$\vec{x}'_{t+\Delta t} = \vec{x}_t + \vec{v}\left(\vec{x}_t\right)\Delta t \tag{1}$$

where $\vec{x}_t = (x, y, \eta)$ is the position of the particle (vector) and $\vec{v}_t = (u, v, \dot{\eta})$ is the velocity field (vector) at time t. The intermediate position of the particle after advection procedure is denoted by the vector $\vec{x}'_{t+\Delta t}$.

In the model computations, the processes - advection and diffusion - are separated so that advection is first performed, followed by the diffusion process in a stepwise manner. A '*random walk*' approach is used to parameterize horizontal and vertical diffusion. The sequence of steps for the parameterization of the process is described in detail ref. [10] and [15]. Slightly different parameterization is used for particles located within the ABL and for those above, but it can be described by the same equations:

$$x'' = x' + r_x l$$

$$y'' = y' + r_y l \tag{2}$$

$$\eta'' = \eta' + r_\eta l_\eta$$

where $\vec{x}''_{t+\Delta t} = (x'', y'', \eta'')$ is the vector of particle position for $t + \Delta t$ after application of the horizontal diffusion operator, r_x, r_y and r_η are the randomly sampled values from the range <-0.5, +0.5>, taken from a uniform distribution; l and l_η are the length scale parameters for horizontal and vertical turbulent motion.

In the case of vertical diffusion, parameterization is different within the ABL and above. The time step of 15 minutes (900 s) corresponds to the turn-over time of a large eddy within the ABL: $\Delta t \approx z_i / w_c \approx 900 s$, where z_i is the ABL height and w_c is a characteristic convective velocity. This means that after approximately 15 min., a particle does not 'remember' its previous position within the ABL. Therefore, vertical diffusion in this region is parameterized as a random relocation of particles between surface and z_i, applied after each time step. To allow a vertical exchange of particles between the ABL and a free atmosphere, the mixing zone is extended above the ABL from z_i to $z_i + \Delta z_i$.

2.6 Dry and wet deposition

These processes are described in detail in [15]. A particle carry an initial mass - $M_p(t_0)$ - from the source. Only particles in the boundary layer, are subject to dry deposition. The particle is randomly 'assigned' a number - n_p - from the number set $<z_b>$, where z_b is the number of 'meters' of the boundary layer thickness. If $n_p > n_{crit}$, no mass is lost by this particle. If $n_p \leq n_{crit}$, a certain fraction of the particle's mass - ΔM_{pd} - is deposited onto the ground. n_{crit} is prescribed as an element in the lower part of the number set $<z_p>$.

The 'adjustable' probabilistic parameters for the process, are given numerical values corresponding to a rate of efficiency which can be compared with the more commonly used parameterization, for example, 'dry deposition velocity'.

In the parameterization of the wet deposition process, we use 'probability curves', defined as follows: If we choose an arbitrary point (x,y,t) in a 'grid-volume' ($\Delta x, \Delta y, \Delta t$), a 'probability curve' will tell us: "What is the probability for 'real' precipitation to occur in this specific point in space and time, - given we have access to the amount of precipitation (Prec) predicted by DNMI's operational NWP model". These ideas and concepts have been developed and used in models describing transport/deposition of acidifying substances [1].

For Prec > 0, we define a two-step process. First step: We read from the 'probability curve' for the precipitation amount in question, what the probability (P_w) is for 'real' precipitation to occur at the 'space-time' point of the particle. Next step: We randomly assign '0' or '1' to each particle. The '1' value is chosen with a probability P_w, and '0' value with a probability $(1 - P_w)$.

The 'adjustable' probabilistic parameters for the process, are given numerical values corresponding to a rate of efficiency which can be compared with, for example, 'scavenging ratio'. The 'probability curve' used in SNAP, has been worked out by P.E. Haga [4].

3 Applications of SNAP

The main application of SNAP is real-time simulation of transport of radionuclides released during a nuclear accident (operational application). In addition SNAP can be used to simulate events from the past, for which meteorological data are available in archives, and special events related to model intercomparisons and tests. For more details, see ref. [15].

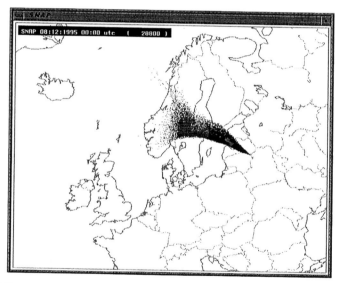

Figure 3: On-line display of the particles in SNAP simulation.

3.1 Operational applications.

During a model run, the user can graphically display (on-line and off-line) the following fields:
- particles effected by precipitation - individual locations
- particles not effected by precipitation - individual locations
- particles in the ABL - individual locations
- particles above the ABL - individual locations
- all particles in the atmosphere - individual locations

- precipitation amount for the last three hours - isolines
- concentration of mass/activity in the ABL - isolines
- accumulated wet deposition - isolines
- accumulated dry deposition - isolines
- accumulated total (wet + dry) deposition - isolines

Examples of on-line and off-line graphics are shown in Figure 3 and Figure 4.

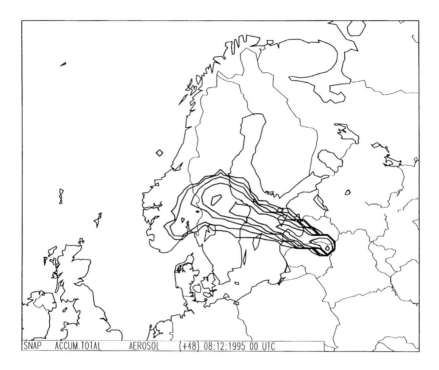

Figure 4: Total deposition after 48 hrs simulation of a hypothetical release from the Ignalina nuclear power plant in Lithuenia.

3.2 Other applications

Besides the operational applications, SNAP is often used for special tasks. For example, SNAP was used in ETEX [15], in which model results were compared with the measurements, as well as with the results produced by other models. Also, the Chernobyl accident in April/May 1986 has been simulated using SNAP. The results from this simulation have been taken out on video cassette. The results from SNAP have also been compared with other Scandinavian models in the frame of Nordic collaboration in functional exercises [11]. However, despite these additional activities, the operational application of the model is the most important. The future development of SNAP, will mainly be directed towards these operational aspects.

4 Conclusions

The SNAP model in its present stage of development is fully operational and can be used for the simulation of nuclear accidents anywhere in Europe, in real time. SNAP has been tested and evaluated against measurements in the ETEX experiment. The results from this evaluation indicate that SNAP is among the best fifth of the nearly 30 real-time models tested.

However, the model is still under development with the main focus on improving of the parameterization of the physical processes during release and transport of radioactive material. The main effort in the future model development will go towards:

- better parameterization of the source-term
- improved description of wet deposition
- more user-friendly set-up for operational services at DNMI

References

[1] Barrett, K., Ø. Seland, A. Foss, S. Mylona, H. Sandness, H. Styve and L. Tarrason (1995). European Transboundary Acidifying Air Pollution. Ten years calculated fields and budgets to the end of the first Sulphur Protocol. EMEP/MSC-W Report 1/95. Meteorological Synthesizing Centre - West, Oslo.

[2] ETEX (1994). S.P.I. 94.46. (Information about the project, jointly sent out by IAEA, WMO and ECE/JRC).

[3] Fay B., H. Glaab and I. Jacobsen (1991). A Lagrangian particle model for long-range simulation of accidental releases of radioactivity at the German Weather Service. In: *Air Pollution Modelling and Its Application VIII* (H. van Dop and D. G. Steyn, eds), Plenum Press.

[4] Haga P.E. (1991) Hvordan influerer nedbørprosessers tids- og romskala på langtransport av svoveldioksyd og partikulært sulfat? (in Norwegian). Thesis, Oslo University.

[5] Haugen J.E. and K.H. Midtbø (1995) Det operasjonelle HIRLAM systemet ved DNMI. In Norwegian (Available from Jan Erik Haugen, The Norwegian Meteorological Institute, Po. O. Box 47, N-0313 Oslo, Norway).

[6] Klug W., Graziani G., Grippa G., and Tassone C. (1992) Evaluation of Long Range Atmospheric Transport Models Using Environmental Radioactivity Data from the Chernobyl Accident, The ATMES Report, Elsevier Scientific Publishing Company, London.

[7] Klug W., G. Graziani, S. Mosca, F. Kroonenberg, G. Archer, K. Nododp, A. Stingele (1995). Real Time Long Range Dispersion Model Evaluation; ETEX first experiment, Draft report. (Prepared for the ETEX Modeller's Meeting, Prague, October 1995).

[8] Kristjansson J.E. and X.-Y. Huang (1990) Implementation of a consistent scheme for condensation and clouds in HIRLAM. HIRLAM Techn. Rep. No. 7 (Available from the Research Division, DNMI).

[9] Langer J., C. Persson and L. Robertson (1994). MATCH: A Modelling System for Mesoscale Atmospheric Dispersion. Application to Air Pollution Studies in Sweden. Preprints for the XIX Nordic Meteorologists Meeting, (6-10)/6-1994, Kristiansand, Norway.

[10] Maryon R.H., J.B. Smith, B.J. Conway and D.M. Goddard (1991). The United Kingdom Nuclear Accident Model. *Prog. Nucl. Energy*, **26**: 85-104.

[11] Saltbones, J. (1995). Real-Time Dispersion Model Calculation as a Part of NORMEM - WP19. *Safety Science*, **20**: 51 - 59.

[12] Saltbones J. and A. Foss. (1994). Real-Time Dispersion Model Calculations of Radioactive Pollutants at DNMI. Part of Norwegian Preparedness Against Nuclear Accidents. Preprints for the XIX Nordic Meteorologists Meeting, (6-10)/6-1994, Kristiansand, Norway.

[13] Saltbones J., Foss A. and J. Bartnicki (1995a). Severe Nuclear Accident program (SNAP) - A Real-Time Dispersion Model. In: *Proceedings for Oslo Conference on International Aspects of Emergency Management and Environmental Technology* (K.H. Drager, ed.), pp. 175-184. A/s Quasar Consultants, Oslo, Norway.

[14] Saltbones J., Foss A. and J. Bartnicki (1995c). Severe Nuclear Accident program (SNAP) - A Real-Time Dispersion Model. In: *Proceedings of the 21st NATO/CCMS International Technical Meeting on Air Pollution and Its Application* (S.E. Gryning and F.A. Schiermeier, eds.), pp. 333-340. American Meteorological Society, USA.

[15] Saltbones J., Foss A. and J. Bartnicki (1995d). SNAP: Severe Nuclear Accident program. A Real-Time Dispersion Model for Major Emergency Management. Final Report for the MEMbrain project available at the Norwegian Meteorological Institute in Oslo.

XDF system for graphical monitoring of weather and pollution

E.W. García, E.B. González, E.G. Ortiz, M.M. González
Instituto Mexicano del Petróleo, Lazaro Cardenas 152, Mexico, D.F.

The XDF v3.1. system is a powerful tool which has allowed Mexican researchers and authorities the ability to follow in a graphical manner the evolution and impact air pollution has had on different areas of the Valley of Mexico. It also permits the simultaneous monitoring of wind vector fields, wind flows temperature distributions, relative humidity distributions, and probability of rain. It is fully three dimensional and permits viewing from any angle desired. Designed to run on a variety of platforms and hardwares, it has the ability to switch to a two dimensional representation should the graphical interphase of the user not be able to handle the PEX protocol. The system is constructed with an object oriented approach, but makes use of no other than C ANSI language to enhance compatibility with other systems. This paper includes a general description of the system, which contains the basic system requirements and components. That is followed by a description of the methodology where the image superposition tecnique is detailed and the application of such method is explained. Then a description of the computational modules is done, and, finally, conclusions are drawn.

Introduction

Purpose

Mexico City, one of the largest metropolitan areas in the world, and also one of the most polluted, had seen little research with the noxious substances in the atmosphere, until Petróleos Mexicanos (PEMEX) and the National Laboratory at Los Alamos, New Mexico undertook to finance a joint project with the Instituto Mexicano del Petróleo (IMP).

The objective in the creation of the XDF system was to have a tool to evaluate the air quality in the metropolitan area of Mexico City (MAMC), applying the methodology of impact evaluation on health and the environment. This system should be capable of visualizing in graphical and expedite form the information related to the air quality with the purpose of making quick and trustworthy diagnostics of the information that is collected at the monitoring centers every hour.

Background

Due to the fast and uncontrolled growth of the MAMC and the geographical conditions of the Anahuac Valley, air pollution has grown in the last few years, impelling the authorities to implement different measures. Within these measures were the financing of several research projects, the graphical XDF system is among one of the results.

The investigation project has amongst its main objetives to study the global air quality within the MAMC to inform the authorities on the tangible effect that the diverse measures taken, such as the reduction of sulphur content in diesel fuel, the implementation of catalytic converters and unleaded fuel in automobiles, and the closing down of the Azcapotzalco oil refinery. The study of the effects of the geographical and meteorological variables on the concentration and dispersion of the pollutants, and how this varies during the different seasons of the year is also important. The XDF system is one of the tools developed for this purpose.

Initially the system was developed in a parallel manner using two computers for different aspects. The database and numerical calculation were done on a Digital Microvax 3800, while the three dimensional graphical interphase was constructed in a Stardent Titan. As soon as a Dec 3000 AXP was available, however, both aspects of the software being developed were put together in a joint effort by the two groups.

General Description of the System

The system, initially, was programmed using Fortran and Pascal for the database, but this was replaced by the more efficient C language, which allowed for further improvements. It must be noted that ANSI C was used to ensure compatibility across different platforms. The graphical interfase began with DORE imbeded C code, which was later traded in for PHIGS, and finally brought down to basic PEX instructions, making the system independent of individual computer manufacturers software licence requirements and completely transportable to suitable hardware. In likewise fashion, basic XLIB instructions were preferred over the easier to use object oriented implementations, again to insure that the system has the ability to outlive the computer on which it is currently installed.

The system can be divided into three groups of programs : the front end section, the database manipulation, and the number cruncher.

The front end is transparent for the user. Once the program has been invoked, the user can forget about the keyboard and navigate through as many images as she requires with the sole use of the keyboard. The independent development of the SOMOD (Sistema Objeto Multiramal Orientado a Disco) database system at the IMP, allows for practically instantaneous access to more than a million records, allowing the generation of any image with practically no wait time.

The front end is composed of three windows. The first is the control panel on the left part of the screen where the different menus appear. The second window occupies the rest of the screen (but can be resized to any scale according to the needs of the user) and contains the graphical images. Finally, the third window remains hidden until summoned forth by some event, such as an alteration in the images displayed, such as dates, intervals and viewing angle.

System requirements

The XDF v.3.1 (as all those based on X-Windows) may be executed on a variety of platforms. The output unit is a graphical screen which may be a workstation display, an X-terminal, or even a personal computer wired to the network. The input unit is the mouse of the workstation, X terminal, or personal computer. In view of the fact that most terminals and PC's are not capable of processing the PEX protocol, the program automatically detects the availability of PEX compatibility and switches to a two dimensional graphical output accordingly. In this manner the XDF system can be adapted to a variety of platforms, and even a combination of them, when reduced budgets obligate such configurations.

System Components

The XDF v.3.1 system is composed of 20 files, seventeen of which contain the code in ANSI C, with imbedded XLIB and PEX, responsible for the generation of the frontend, database management, and number crunching. The remaining three files contain global variables and definitions and macros for the system.

Methodology

Ever since the introduction of the *Evaluation of Environmental Impact* (EEI) in the United States in 1970,[1] the tecnique has had a great number of applications which have helped to predict the impact produced by a certain industry within a given region. For this to be done, all factors have to be considered, such as introducing telephone networks, water, sewage, power lines, and transportation facilities.

The EEI has several branches, and one of the most important is the EEIH, which deals specifically with the impact on health. The primordial goal that is sought for with a EEIH is the identification of all environmental aspects, from the degradation of a physical resource, until the effects have consecuences on living organisms.[2] The atmosphere is a natural resource that receives potential impacts on air quality, generated by mankind during the development of agriculture, industry, mining, transportation, urbanization, and the generation of electric energy, amongst other activities. The atmosphere of the MAMC is no exception and is highly impacted with respect to air quality due to the pollution produced by industrial activity and the transportation requirements of the constantly growing population.

The air pollution in Mexico City is a problem that has been growing for several decades, and in 1986 the *Secretaría de Desarrollo Urbano y Ecología* (SEDUE) now known as the *Instituto Nacional de Ecología* (INE) organized an automatic monitoring network in twenty five points of the Metropolitan Area (later augmented to 33) with the purpose of keeping tabs on air quality. These points were distributed in five zones : northeast, northwest, southeast, southwest, and center. In these stations the parameters that best represent the atmospheric situation are monitored. These parameters are :

- Ozone concentration (ppm)
- Nitrogen oxides concentration (ppm)
- Nitrogen dioxide concentration (ppm)
- Non-methane hidrocarbons (ppm)
- Wind direction (degrees)
- Wind speed (m/s)
- Dry bulb temperature (°C)
- Relative moisture (%)
- Total suspended particles ($\mu g/m^3$)
- Carbon monoxide concentration (ppm)
- Sulphur dioxide concentration (ppm)
- Hydrogen sulfide concentration (ppm)

With these parameters it is possible to determine air quality and its evolution with time, but the analysis has not been simple, due to the quantity of information generated since 1986 to the present. This is the reason why highly efficient computer programs had to be developed.

Image Superposition Technique

The image superposition tecnique, also called cartographic system,[3] is used in the location of impact on the environment and they are an excellent means of communication with the public.

The sequence for its utilization is as follows :

1. It is necesary to have a scaled map of the area under study. Above this map a transparency is laid and is divided by means of horizontal and vertical lines in order to obtain a division of the area in a series of geographical units.

2. In each unit a set of environmental factors are studied applying *Environmental Impact Indicators* (EII) previously determined.
3. The EII are marked separately on other transparencies above the map acording to the factor evaluated, i.e., an equal number of transparencies results from a given number of EII.
4. The transparencies are laid one over another on the map to obtain the final results. A system of grey tones is recommended, in this manner each unit is darkened succesively as the negative impact increases. Upon comparison of the maps with gray scales, illustrating the natural tendencies of the terrain, the zones with greater factibility for the location of industrial areas, housing units, and commercial zones can be identified. The map superposition can be done by hand, but the use of a computer greatly the work.

To obtain an evaluation of global impact, all the transparencies should be laid on each other, in this manner the effects of different pollutants is combined. It is true that an area may not be impacted by a particular pollutant when analyzed individually, but upon a combined analysis the situation may be very different. In particular, Mexico City is highly impacted with respect to ozone in the southwest, but the global impact of all air pollutants is far worse in the northeast.

Application of the Technique

To apply the preceeding technique in the XDF v.3.1 software the following steps are taken :
1. The outline of the Distrito Federal is drawn, together with the outline of the metropolitan area.
2. The monitoring stations are located according to their geographical position, in computer coordinates.
3. The unit cells are logically drawn in the area covered by the screen, where each unit corresponds to 250 square meters (50m x 50m).
4. The EII are calculated and interpolated for every unit cell. The interpolation is based on Davis' modified formula, which we shall hence explain:

Suppose we have 2 points, x_1 and x_2, where $x_1, x_2 \in \Re^n$, and two respective measurements y_1 and y_2, where $y_1, y_2 \in \Re^+$. We wish to know the value y for every element of the interval $x \in (x_1, x_2)$. To do such, Davis implicitly supposed the existance of an adimensional number D_x, that depends on x, such that:

$$D_x = \frac{\|x_1 - x\|}{\|x_2 - x\|} = \frac{|y_1 - y|}{|y_2 - y|}.$$

In other words,, if we consider x at the origin, the variation of y with respect to y1 and y2 is linear, and the slope is the ratio of the norms of x1 and x2.

Without loss of generality, we can suppose $y_1 \leq y \leq y_2$. Then:

$$\frac{\|x_1 - x\|}{\|x_2 - x\|} = \frac{y - y_1}{y_2 - y},$$

of which results:

$$y\left(\frac{1}{\|x_1 - x\|} + \frac{1}{\|x_2 - x\|}\right) = \frac{y_1}{\|x_1 - x\|} + \frac{y_2}{\|x_2 - x\|}.$$

Generalizing to a n-dimensional system this leads us to Davis' generalized formula:

$$y = \frac{\sum\limits_{i=1}^{n} \dfrac{y_i}{\|x_i - x\|}}{\sum\limits_{i=1}^{n} \dfrac{1}{\|x_i - x\|}}.$$

The space in which Davis worked was a common geological space, where his unknown, y, was the topography, and therefore the most natural norm was the euclidean distance. But in out case the unknown we are dealing with is the diffusion of a gaseous substance. Let us remember the diffusion equation:

$$\iint_{C.S.} \rho(\vec{v} \cdot \vec{n}) dA + \iiint_{C.V.} \rho dV = 0.$$

From which, with the aid of the kinetic theory of gases and the Lennard Jones potential, we can obtain the diffusivity of A through B:

$$D_{AB} = \frac{0.001858 T^{3/2}\left(\dfrac{1}{M_A} + \dfrac{1}{M_B}\right)^{1/2}}{P \sigma_{AB}^2 \Omega_D}.$$

Where T is the temperature in Kelvin, M_A, M_B are molecular weights, P is the absolute pressure in atmospheres, and Ω_D is an adimensional function that depends on temperature and the intermolecular potential field between A and B; σ_{AB} is the collision diameter, a Lennard Jones parameter in Angstroms. From this equation, considering the characteristics of the atmosphere, it is clear that diffusivity is inversely proportional to σ_{AB}^2. By dimensional analysis we find that the most appropriate mathematical norm to use is the distance squared, which leads us to Davis' modified formula:

$$y = \frac{\sum\limits_{i=1}^{n} \dfrac{y_i}{d^2(x_i - x)}}{\sum\limits_{i=1}^{n} \dfrac{1}{d^2(x_i - x)}}.$$

Extensive numerical experimentation was done using this norm and other common mathematical expressions with the same properties. The norm which represented the statistical information most accurately was the one used in Davis' modified formula.

5. The plane superposition is done transforming EII according to the modifications mentioned above and mathematically combining the matrices into the final image.

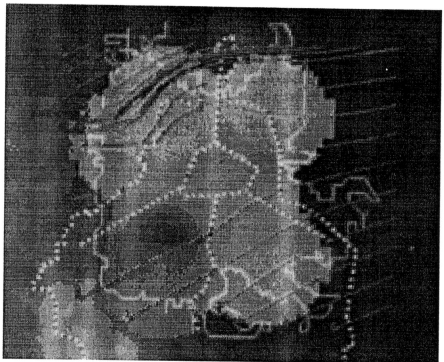

figure 1

Graphical Modules

The XDF v.3.1 is composed of several modules. Amongst them we can count the following :

1. The first module creates a two dimensional image of the geographical topology of the Valley of Anahuac where the altitudes above sea level can be observed. It also draws the political division of the Federal District, the urban sprawl, green areas, water bodies, and the main traffic arteries of Mexico City. This image serves as a background for the visualization of the air pollution impact maps as well as the metereological phenomena such as rain and wind vector fields.

2. This module, as a complement to the first, is in charge of creating a three dimensional surface to represent the geographical topology of the Valley of

Mexico. This module is used to give the user the view from any angle she desires. It is composed, as before, from an array of 123x123 kilometer digitalization centered at Mexico City. Included is the political division of the Federal District, the urban sprawl, green areas, water bodies, and the main traffic arteries of Mexico City.

3. The wind is dealt with in this module. In it, the wind velocity and direction are drawn as vectors over the points of measurement. It also constructs a vector field over the area under study. The third possibility contemplated in this module is the drawing of wind flows. With this ability is is possibly to follow a particular point as the wind carries it away. The vectors or wind flows can be seen in hourly representations or as a daily, weekly, monthly or seasonal hourly average. With this module it has been possible to explain the appearance of pollution pockets that form when the wind forms whirlwind patterns. Figure 1 shows one manner in which these wind flows are represented. The options of this module, as all others that follow, may be seen in a two dimensional or three dimensional representation, whichever the user chooses, or the hardware supports.

4. The next module is the one that realizes the impact maps, with the methodology that was detailed beforehand. This module plots the impact in a continuous scale of color, beginning with green (no impact) going through yellow, orange, and red until reaching magenta (dangerous impact). With the use of the SIMEDA system, independently developed at the *Instituto Mexicano del Petróleo*, it is not only possible to visualize hourly images, but also daily, weekly, monthly or seasonal averages as well. This module has allowed to visualize the concrete effects the introduction of new gasolines has had on the air quality in the city. Figure 1 above also includes the representation generated by this module.

5. At times a pollution reduction is observed that cannot be explained by shifting wind patterns. But rain may be the cause. This was the motivation to develop this module in which the probability that rain has ocurred is calculated. It does so by calculating changes in environmental factors between the hour under study and the previous and next hours. By suggestion of the researchers at the *Instituto de Ciencias de la Atmósfera* the following variables were used in order to estimate the probability of rain: temperature, relative moisture, and concentration of NO_x, SO_2 and CO. This was done based on the following suppositions ; when it rains, temperature drops and relative moisture rises, also, the concentration in the atmosphere of the molecules of NO_x, SO_2 and CO drop (because of their acidity they undergo a certain solubility in water). In order to correlate these changes in variables to the actual facts, the statistical information of the rain measuring centers of the *Dirección General de Operación Hidraulica del D.D.F.* was correlated to obtain the most accurate formulas. In this manner a reliable routine was created that has helped to explain why

in certain occasions pollution levels have dropped in certain parts of the city.

6. To help explain the formation of secondary pollutants, such as ozone, it was necessary to have a module that could draw over the impact maps, isotherm and isohumidity curves. With these curves it has been possible to explain why certain pollutants, which depend on chemical reactions for their formation, form the patterns of distribution they do, depending on

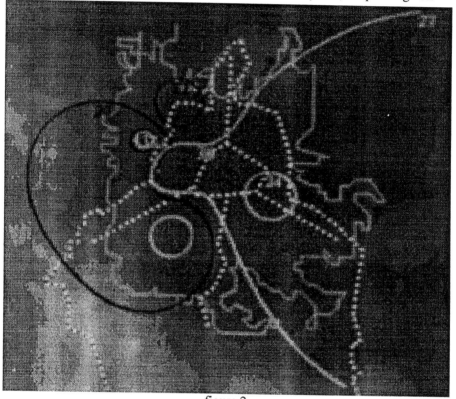

figure 2

temperature and humidity factors. A figure that depicts these curves is shown on figure 2.

Conclusions

The XDF v.3.1 system was developed for the graphical monitoring of air quality in Mexico City, and is a valuable and trustworthy tool that saves many hours of work by accessing megabytes of information in extremely short times. With this tool it is possible to have all the information collected in Mexico City at your finger tips with practically instant access. The generality with which it was constructed allows for implementation on a variety of platforms and the applicability to practically any city which has a network of automatic monitoring stations. All that is necessary is to count with the geographical location of the stations, access to their readings, and, if a geographical topology is desired, a digitized array of altitudes. This is available for all cities in Mexico by means of the *Instituto Nacional de Geografía, Informatica e Historia.*

References

[1]Estevan-Bolea, M.T., *Las evaluaciones de impacto ambiental.* Cuadernos del Centro Internacional de Formación en Ciencias Ambientales, Madrid, 1980.

[2]ECO (Centro Panamericano de Ecología Humana y Salud). *Manual básico de evaluación del impacto en el ambiente y la salud de proyectos de desarrollo* ; version preliminar, Henyk Weitzenfeld. ECO. Metepec, México. 1990.

[3]McHarg Y. *Design with Nature,* 1969.

[4]Davis,J.C. *Statistics and Data Analysis in Geology,* John Wiley and Sons, USA, 1982

[5]Comisión Metropolitana para la prevención y control de la contaminación atmosférica : *Programa integral contra la contaminación atmosférica, un compromiso común.* Mexico, 1990

[6]Straw, W., Mainwaring,S.J. *Contaminación del aire, causas, efectos y soluciones.* Trillas, Mexico, 1990

[7]Abelson,S. *Structure and Interpretation of Computer Programs,* MIT. Cambridge, 1985

[8]Gaskins Tom, *Pexlib Programming Manual.* O'Reilly, 1992

[9]Reis L., *Aplique X Window,* McGraw Hill, Mexico, 1993

[10]Welty, J.R., Wicks, C.E., Wilson, R.E., *Fundamentos de transferencia de calor, momento y masa,* Editorial Limusa, Mexico 1982.

[11]Haaser,N.B.,Sullivan,J.A.,*Real Analysis,* Dover Pub., New York, 1991.

[12]Tijonv,A.N.,Samarsky,A.A., *Ecuaciones de la física matemática,* MIR, Moscow, 1983.

Numerical simulation of pollution in a groundwater basin in Austria

G. Kammerer, W. Loiskandl, G. Zelezo
Institute of Hydraulics and Rural Water Management, Universität für Bodenkultur Vienna, A-1190 Wien, Nussdorfer Lände 11, Austria

Abstract

Groundwater contamination was recently encountered near the Danube river. There is evidence that an accidental input of pollutants in that area occurred some years ago. By simulating groundwater flow and solute transport with the computer model MOC - Method Of Characteristics - the spread of the contamination plume was evaluated. For our application the program had to be extended to approximate unconfined flow and time-dependent boundary conditions. With reasonable assumptions and inverse modeling of the unknown parameters of the scenario of the accident estimated with the extended MOC, agreement was achieved between the simulation and the observed concentration distribution.

1 Introduction

A routine measurement in a water-supply well in eastern Austria (discharge well 1 in figure 2), upstream on the Danube near Vienna, indicated a concentration of HOX exceeding the limits defined by Austrian laws. Although measures for remediation were immediately taken, a considerable amount of the substance remained in the aquifer, documented by downstream measurements one year later. The evidence of an accident (see input point in figure 2) raised the question if this single pulse of pollutant input was responsible for concentrations higher than the limits in the groundwater basin.

With the groundwater basin being bounded by the Danube, a very heterogeneous subsoil is formed geologically by the influence of the river. Numerous hydrogeological exploration data and well-documented observations of the groundwater levels and hydrographs of the Danube are available. With this extensive data base and the determination of the simulation domain the transient groundwater flow could be modeled.

The Danube to the south and an impermeable rock formation (range of hills) to the north and northeast parallel to the stream enclose the simulation domain on two opposite sides. Groundwater observation wells were used for defining the boundary conditions upstream and at the so-called "Korneuburger" gate downstream.

Data characterizing the properties of the subsoil related to contaminant transport were much more poorly documented than those for hydrology. Only a few analyses of HOX-pollutants in the region (probably long after the accident) were available. The simulation is for a period of 6 years and begins a short time before the accident presumably happened.

2 Model

The investigated aquifer demanded a model capable of simulating the groundwater flow for the heterogeneous subsoil subject to constraints of known discharges and groundwater heads in observation wells within the domain as well as describing solute transport taking into account convection, dispersion and adsorption was required. Although the pollutants are highly volatile and the solution-behavior depends on several (unknown) parameters, multiphase flow was neglected owing to the importance of convection for the given scale and the fact that there was a dearth of corresponding data. Because horizontal flow predominates, the problem was reduced to two dimensions with vertical components of precipitation and evapotranspiration being considered. The simulation model MOC, based on the method of characteristics, developed by Bredehoeft and Konikow [1] for the US. Geological Survey, was selected by reason of its flexibility to easily accommodate alterations by programmers (source code is at our disposal). This 2D single phase model based on steady-state boundary conditions, allows transient conditions due to pumping within the domain. The mixing of fluids, adsorption and chemical degradation are described with the assumption that density, viscosity and temperature gradients do not act on the flow regime. The groundwater flow is solved using the finite difference method with the solute transport being described by the method of characteristics.

Because the boundary values (observation wells and particularly the Danube as recipient) change with time and to calculate unconfined groundwater flow, the main program was written with the MOC-code included as a subroutine (figure 1). For the case of a phreatic aquifer the iterative calculation of transmissivities for every time step is required. To keep computing time within reasonable limits, the transmissivities and the boundary conditions were kept constant for one month. The main program "KORNEU" calculates the new transmissivity matrix as the product of hydraulic conductivity times groundwater head minus impermeable layer surface. At the end of each time step the differences between the calculated groundwater heads at the boundary and the given values for the boundary condition at the same time cause an error in mass balance. Therefore, the time step had to be subdivided into several internal time steps for MOC. With the first time step lasting 4 hours and 20 minutes and every successive time step increasing by 10 percent, the mass balance error was reduced to less than two percent.

To improve the handling of input data, the main program governs the input data management of variable parameters and serves as a pre-processing unit for the graphical output. Constant parameters are in the input-file for MOC. Since the accompanying parts for MOC for pre and postprocessing could not be adjusted to the requirements of the given case, extra programs had to be used and in- and output had to be modified in MOC.

The groundwater model was developed in two steps with groundwater flow being a prerequisite for the calculation of solute transport. After the model was calibrated and the hydraulic validation with known groundwater levels was satisfactorily performed, the unsettled questions of contaminant transport were clarified in accordance with the measured HOX-concentrations.

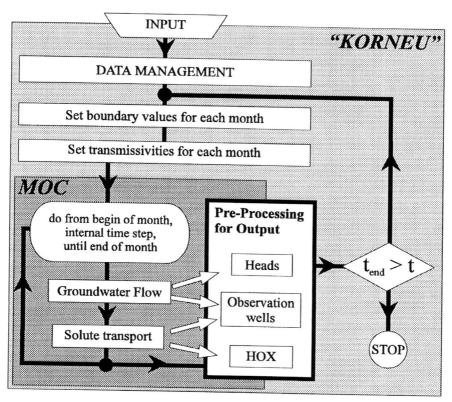

Figure 1: Simplified flow diagram for the model "Korneuburger Bucht"

First simulations were performed on a PC. When the grid was refined to 148 x 104 (= 15392) cells (size of one cell: 50 x 50 meter) and the number of particles per cell was increased from 4 to 16, the model was necessarily transferred to a mainframe computer.

3 Input Parameters

The quality of simulation results is directly related to the accuracy of the input parameters. They are divided in hydro/geological data and data for the description of the solute transport. These values and their availability are summarized in table 1. The basic data for the layout of the model domain are supplied by geological surveys (Grill [2]) and previous studies in that area (Schuch [7]). For each time step the Danube and the observation wells define constant head boundaries. In MOC this is achieved by setting the leakage factor to a high value (e. g. $1.0 \cdot s^{-1}$). The boundary to the range of hills is defined by a constant flux, established in the model by cells with a determined diffuse recharge rate. Within the domain some impermeable "islands" break through the groundwater-leading layer. The time dependent hydrological data and the pumping data on hand for the discharge wells for water supply are provided as monthly values (means), while the discharge rate for the remediation well is considered daily.

Parameter	known	representative value
depth of impermeable layer	Schuch [7]	rel. 0 - 12 m
hydraulic conductivity	Schuch [7], modified	$0.15 - 4.8 \cdot 10^{-2}$ m/s
hydrograph of Danube	Donaukraft	fluctuation max. 3.6 m
groundwater observation points	Hydrographisches Zentralbüro [3]	fluctuation max. 3.0 m
precipitation	Hydrographisches Zentralbüro [3]	4 - 113 mm/month
discharge wells	major pumping rates	about 10 - 150 l/s
pollutant	HOX	retardation factor 1.13; half-life 10 a
date of input	uncertain	
amount of input	unknown	inverse modeled
longitudinal dispersivity		$\alpha_L(s) = 0.25 \cdot s^{0.7}$
transversal dispersivity		$\alpha_T = 0.04 \cdot \alpha_L$

Table 1: Parameters for the groundwater model "Korneuburger Bucht"

The initial conditions for the groundwater flow are determined by a steady state calculation using all observation wells along the boundary and inside the flow domain as constant-head cells. In the absence of *in-situ* experiments for description of the pollutant behaviour in the soil, we made simple assumptions based on bibliographical references for adsorption (linear adsorption isotherm) and chemical degradation (estimated half life: 10 years). Instead of a constant longitudinal dispersion coefficient, a power function related to the distance from the source was used in the calculations. During the simulation runs, the observation wells inside the domain served as control points for verification of the results. As seen from Table 1, the initial conditions for the solute transport are not well defined. With the first measured concentrations in discharge well 1 and 2 being known only some time after the accident, the missing data for the determination of the pollutant source are evaluated by inverse calculations. The date of the beginning and the ending of the contamination input relevant for the modeling of the drifted plume in the downstream part could be fixed exactly, since the solute could leave the drawdown cone of discharge wells 1 and 2 only when the pumps were turned off (June 1987 - April 1988 and after February 1990).

4 Computed results

Considering all of the monthly generated groundwater heads, a pattern from January 1991 was chosen (figure 2) to represent the groundwater flowing parallel to the Danube in the same direction. Three drawdown cones of wells with larger discharge rates (5, 7 and 9) and two small weirs in the ditch (dashed line in figure 2) for the regulation of the water supply for the riparian forest have only local influence.

Groundwater hydrographs of two different observation wells, approximately representing a cross section in the middle of the domain, are shown in figure 3. Well K situated nearer the Danube has obvious fluctuating groundwater heads. Because these fluctuations are almost a third of the total distance between the top of the impermeable layer and the free water surface, it was not possible to simply treat the groundwater basin as a confined aquifer. A flood of the Danube in August 1991 is documented by the plot of observation well K. Indeed, with the top of the impermeable layer being at 159 m, the flood raised the water table about three meters from 163 to 166 m. With measured and calculated groundwater heads being nearly identical, the accuracy of the groundwater simulation is sufficient for the subsequent simulation of convective contaminant transport.

The four plots of figure 4 allow a comparison between calculated and measured concentrations in two locations. Because the measured concentrations in discharge well 1 were used to calibrate the pollution source (distance: 250 m), the measured and calculated values of this well are almost the same. The agreement between the measured and calculated values at well K, 3.5 km downstream, shows the reliability of the performed simulation.

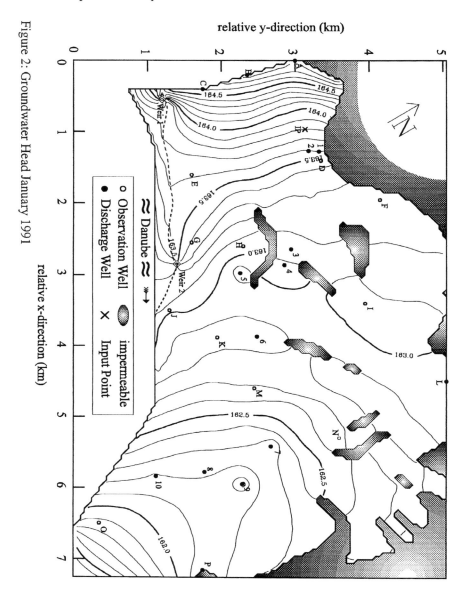

Figure 2: Groundwater Head January 1991

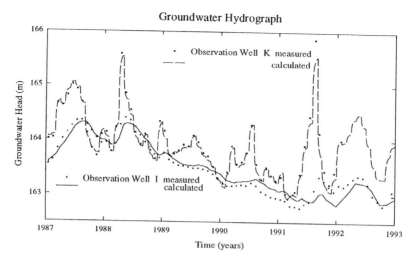

Figure 3: Measured and calculated groundwater hydrograph for two observation wells

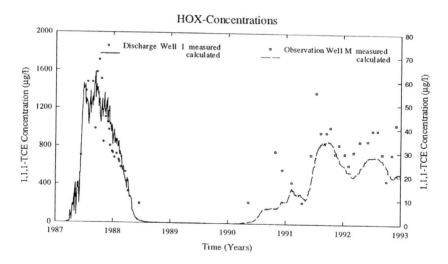

Figure 4: Measured and calculated concentrations

Concentration Distribution September 1987

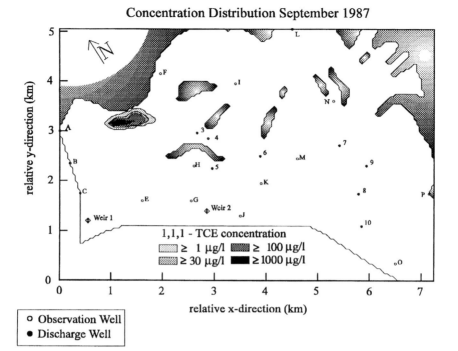

Concentration Distribution January 1991

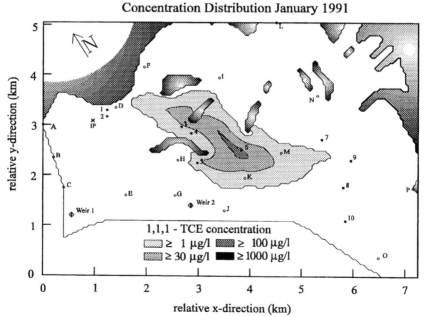

Figures 5 and 6:   Calculated distribution of the contaminant at September 1987
and January 1991

The upper pattern of the two concentration distributions (figure 5) shows the results for September 1987. Three months earlier continuous pumping in discharge well 1 had been stopped. As long as the pumps were in operation, almost the entire plume had been attracted to the well. Subsequently the pollutant was able to escape from the drawdown cone and began to move in the main direction of groundwater flow.

Three and a half years later (January 1991, figure 6) the centre of the plume has travelled about three kilometres downstream. It has been guided by the shapes and the positions of the impermeable inclusions of the groundwater layer. Although the plume has been diluted and considerably widened by dispersion, its concentration still remains too high for drinking water.

Fortunately the "island" near discharge well 5 (pumping rates up to 150 l/s) diverted the plume sufficiently away from well 5 to allow its continued use as a drinking water supply.

The implemented function for the dispersivity (see table 1) appears to adequately describe the real situation in the investigated alluvial aquifer. The chosen ratio between α_T and α_L (1:25) leads to a large widening of the plume and is probably too high.

5 Conclusions

Based on the groundwater flow calculations for the contaminant a plausible scenario was achieved. The moment, when the contamination plume started to move with the main groundwater flow, could be evaluated and it was shown that the presumed amount of pollutant input by the accident is not sufficient for the measured concentration distribution downstream. By inverse calculation an estimation of the necessary input could be given.

The proposed modification of MOC to approximate the phreatic aquifer and time-dependent boundary conditions proved to be satisfactory for the considered two-dimensional groundwater flow problem.

Combining a well-known computer code - MOC with extensions for the respective application case, appears to be an efficient way for developing a regional groundwater flow model, particularly with regard to the flexibility and the independence on the computer level of MOC.

References

1. Konikow. L.F. & Bredehoeft, J.D, Techniques of Water Resources Investigations of the United States Geological Survey, Chapter 2, *Computer Model of two-dimensional Solute Transport and Dispersion in Ground Water*, United States Government Printing Office, Washington, 1978
2. Grill, R., *Erläuterungen zur Geologischen Karte der Umgebung von Korneuburg und Stockerau*, Geologische Bundesanstalt, Wien, 1962

3. Hydrographisches Zentralbüro, *Hydrographisches Jahrbuch von Österreich*, BM f. Land- u. Forstwirtschaft, Wien,

4. Konikow, L.F. & Bredehoeft, J.D., Computer model of two-dimensional solute transport and dispersion in ground water - *Techniques of water-resources investigations of the United States Geological Survey*, Virginia, 1984

5. Nachtnebel, H.P. et al, *Expertensystem Grundwassermodell Östlicher Donauraum*, Institut für Wasserwirtschaft, Hydrologie und konstruktiven Wasserbau, Wien, 1991

6. Österreichische Bodenkartierung, *Erläuterung zur Bodenkarte, Kartierungsbereich Korneuburg*, NÖ, BM f. Land- u. Forstwirtschaft, Wien, 1985

7. Schuch, M.F., *Bericht über die Ergebnisse der hydrogeologischen Untersuchung in der Korneuburger Bucht*, Wien, 1980

8. Umweltbundesamt, *Grundwassergüte Tullnerfeld-Pilotstudie*, Beitrag zum österreichischen Grundwasserkataster, Wien, 1992

9. Uniconsult, *Ermittlung des Gefährdungspotentials Korneuburger Bucht Schongebiet Bisamberg*, Wien, 1990

Soil pollution: environmental integrated systems and prediction mapping

M. Kanevsky,[a] R. Arutyunyan,[a] L. Bolshov,[a] V. Demyanov,[a]
S. Kabalevskiy,[a] E. Kanevskaya,[a] V. Kiselev,[a] N. Koptelova,[a]
I. Linge,[a] E. Martynenko,[a] E. Savelieva,[a] A. Serov,[a] S. Chernova,[a]
S. Chernov,[a] M. Maignan[b]

[a]*Institute of Nuclear Safety, B. Tulskaya 52, 113191, Moscow, Russia*
[b]*Lausanne University, Switzerland*

Abstract

The work deals with development of modern Environmental Integrated Systems (EIS) and prediction 3D-mapping for analysis of late off-site consequences of severe nuclear accidents and analysis, processing and presentation of spatially distributed radioecological data on soil contamination by Chernobyl radionuclides. The main modules of the EIS are: Data bases and data base management systems, Bank of models, Information systems, Special Geographical Information System (SGIS), User's interfaces/dialogue modules including data processing, post processing, presentation and visualisation.

1. Introduction

The goal of the present work is to design and develop a concept of integrated system - a computer integrated environment to support decision making in complex radioecological projects, as well as to work out a prototype of such system using advanced informational technologies. The developed prototype has been applied to the consequences of the accident at Chernobyl nuclear power plant (NPP) in 1986, which is considered as an example of a severe nuclear accident.

Analysing the environmental effect from such a complex technological object as an NPP is, there have been proposed three steps of analysis: 1) pre-accident modelling (probability analysis of safety); 2) real time modelling of the accident: 3) analysis of the accident consequences, setting up intervention levels, countermeasures and their efficiency, social, economical and psychological consequences, data base analysis, as well as prediction mapping including maps of probability and uncertainty.

Integrated systems should be flexible enough due to development and improvement reasons. Modern informational technologies, like data bases and data banks, systems for their management (DBMS), geographic information systems (GIS), decision support systems (DSS), computer models possessing friendly user interface, expert systems, informational support systems, are to be used in computer realisation of an integrated system. It should be noted that both architecture of computer environments and their development and realisation depend on the final objectives and the decisions to be taken.

2. Environmental Decision Support Systems

Classical Decision Support System (DSS) include Data bases, Bank of models and User's interfaces [1]. At present time DSS is essential part of any good ecological project. Intelligent DSS also includes expert systems or knowledge bases. With the help of DSS it is possible to manage both data and models in order to prepare support for decision making process. Environmental Integrated System (EIS) deals with spatially/geographically distributed data. It means that EIS has to be connected with cartographic (mapping) systems or Geographical Information Systems (GIS).

The main modules of Environmental Decision Support Systems are as follows (see figure 1):

- Data bases and data base management systems. Data bases contain information concerning environment, demography, contamination, economics and others necessary for the description of the region of interest and decision-making process. Data bases describing spatially distributed information have to be compatible with GIS. The data bases can be presented on digitised maps of 10 regions in European part of Russia.
- Bank of Models and Model Base management system. Bank of Models includes computer models on different aspects of radionuclides behaviour in environment and their impact on people: atmospheric dispersion models, migration in soils, food chains, dose effect and risk models, etc. An essential part of the Model Bank is a module with models for spatial data analysis including, statistics, geostatistics, fractal interpolations and simulations. For each subject (e.g. migration in food chains) there are several models of different complexity. Input and output of models dealing with spatially distributed information are compatible with GIS. All coefficients and parameters of models are organised as a model-dependent data bases. In this case uncertainty/sensitivity analysis could be done very easily and efficiently. At present time Model bank consists of a number of models on different subjects: short and long range atmospheric dispersion models, compartment and advection-diffusion models on radionuclides migration in soils, compartment and dynamic models on radionuclides migration in food chains, dosimetric models, dose-effect models, risk models, etc. An important phase

of model development is models' verification ands validation. These were done for majority of models by using different kinds of data, also data on Chernobyl accident consequences. There have been included computer models on radionuclides' migration in soil. *RAMIS* program for Windows perform convective/diffussion transport model. Another important part of Model bank deals with spatially distributed information (data of monitoring networks and sampling campaigns, etc.) and consists of a variety of deterministic and stochastic tools and models for the spatial data analysis, processing and presentation. In more details this is described below.

- Information support systems. They include different kind of information concerning computer science, radioecological modelling, decision-making process, intervention levels and countermeasures, etc. and are organised as a glossaries, electronic books, help systems, hypertexts.
- User's interfaces or dialogue modules. This is an important part of any DSS and helps user to prepare, process, present, and understand results.

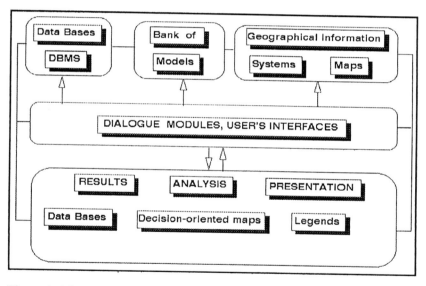

Figure 1: Block scheme of an EIS.

The off-site analysis includes different subjects: atmosphere, hydrosphere, soil, food chains, dose calculations and dose-effect analysis, etc. First step is to prepare scenario (chain of processes and models) compatible with available information between input and output/(objectives of the analysis). For example: source term-atmospheric dispersion-migration of radionuclides in food chains-dose calculations-countermeasures. As it was mentioned above each subject is described with the help of models of different complexity. During scenario preparing we select appropriate models. Models are processors. Data bases are the integrators (flow of information) of EIS. Data base management systems controls these flows between models. Processors/models are exchanged by information (data bases) and results. Such "subject oriented approach" is like an object-oriented programming. Moreover, all coefficients and parameters of models are organised as a model-dependent data bases. In this case uncertainty/sensitivity analysis could be done very easily and effectively.

The main results of the EIS 'GeneSYS", specially developed in Nuclear Safety Institute for analysing and mapping radiological data, are decision-oriented maps, legends, analysis (reports and recommendations) (figure 2).

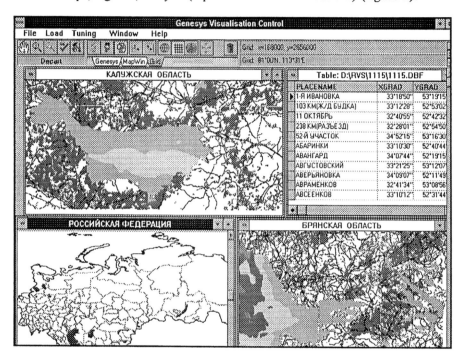

Figure 2: Prediction mapping of soil contamination by using EIS "GeneSYS".

3. Description of data

The EIS prototype has been applied to the consequences of the Chernobyl accident. The data on radioactive surface contamination and other relevant information have been compiled in to special data banks and data bases. The Central Data Bank (CDB) integrates and includes a number of data bases related to the decision making and analysis of Chernobyl accident consequences. In this data bank the information has been accumulated for about 10,000 populated areas belonging to 15 regions of Russia. The total number of data types (fields) is about 600. However for a specific region it usually doesn't exceed 100. Only very brief description of CDB is presented below. One of the most important software developed is Comprehensive data base management system for the CDB [2].

3.1 Radioactive-contaminated territories.

Main data groups:
- populated area name, administrative code and geographical co-ordinates
- social and economic development data
- data on types of protection measures used
- data on soil and food products contamination by radionuclides
- data on measured population internal and external exposure doses
- estimates of expected population irradiation doses.

Administrative and economic characteristics
- administrative belonging (state, region, district, village soviet)
- populated area type (town, settlement, village) and name
- information about participation in a collective or state farm
- quantity of houses (for country-side populated areas)
- size of population
- indication of belonging to a zone (e.g., the zone of relocation or the zone of favourable socio-economic status, etc.)
- indication whether the populated area is concerned with the program of urgent (long-term) countermeasures.

Geographic data: geographical co-ordinates, prevailing building type.

3.2 Data on radioactive-contaminated territories

- Cs-137 soil contamination (minimum, maximum, and average Cs-137 soil contamination density in the populated area according to data of the State Committee on Hydrometeorology)
- data on Strontium-90 soil contamination (minimum, maximum, and average Sr-90 soil contamination density in the populated area according to the data of the State Committee on Hydrometeorology)
- data on radionuclides of Cs and Sr content in milk (average and quantile value during the current year pasture period) on the basis of data of sanitary or veterinary inspection; otherwise the Cs and Sr content in milk can be estimated using regional migration coefficients.

3.3 Data on doses

Estimates of actual irradiation doses of the whole body for the period from 1986 to preceding the current year performed by the research institutes or calculated by using the methods adopted by these institutes; predicted external and internal irradiation doses, evaluated by research institutes or calculated using methods adopted by them.

3.4 Data preparation

Further data processing includes extracting necessary data on one or another region of study, on types of radionuclides for consideration, types of soil in a certain site, in different co-ordinate systems, etc. Data bases prepared for specific purposes are used in further analysis and spatial predictions. The related software handles data in various formats: dBase, PARADOX, ASCII, Geo-EAS (special format widely used in geostatistical software), bitmap and raster format for images, binary formats for GIS links. There have been developed converters for these formats as well as co-ordinate system converters e.g. from geographical to Lambert projection. *DataMan* program for Windows is used as a general pre processor to prepare data bases for further analysis in accordance to immediate decision maker's requirements and expert opinion. Such software is essential for raw data bases analysis. Usually there are a lot of processing to be done using expert opinion to prepare data base for the spatial prediction.

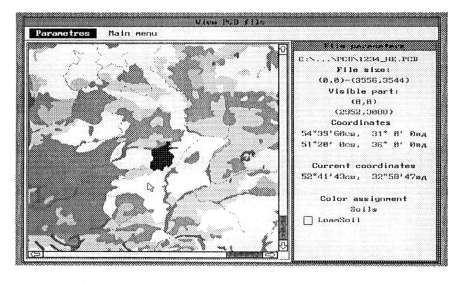

Figure 3: Soil type variability map prepared by *PCD* program, Bryansk region, Russia.

4 Analysis of spatially distributed radioecological data

Environmental data, as a rule, are discrete, spatially distributed and time dependent. Before using data have to be analysed and prepared, correspondingly. The results of analysis are strongly dependent on both input data (how do they represent reality) and data processing methods used. Preliminary spatial data analysis is extremally important because of highly variable data originating conditions: soil types (see figure 3), weather conditions, orography, source of contamination.

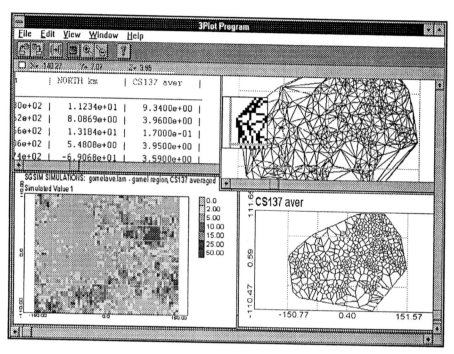

Figure 4: 3Plot program saample interface including data file (Cs137 in Gomel region), triangulation, polygons and mosaic map (stochastic simulation) windows.

4.1 Monitoring network analysis

One of the first and important question is: what is a spatial and dimensional resolution of the network and which phenomena can be detected by the given network? These problems are related to the qualitative and quantitative description of monitoring network and its clustering (spatial inhomogeneity). Clustering of network (preferential measurements of some regions) leads to wrong calculations of mean value, variances, histograms that is essential in many applications. We have used several indicators in order to describe described network clustering and its spatial resolution: The most frequently used are Dirichlet cells/Voronoi polygons/Thiessen tesselations (area-of-influence polygons) and Morishita index [7]. Voronoi polygons along with Delauney triangulation can be used also for network structure visualisation. Within EIS there have been developed a modules for visualisation monitoring network structure. The *3Plot* program for Windows (figure 4) builds Delauney triangulation and polygons (Dirichlet cells) as well as simple post plots of samples on arbitrary grids.

4.2 Spatial predictions

The are several approaches for the spatial data analysis and processing: statistics, interpolations, geostatistics, fractal dimension analysis and modelling. All the methods can be considered as deterministic (ordinary interpolation techniques) or statistical (geostatistics and some kinds of modelling). They differ in the degree of preliminary data preparing, underlying mathematics and in the interpretation of models and results. Statistical methods need more deep preliminary analysis of spatial/temporal data correlation (including anisotropy) and they estimate also errors of prediction. Geostatistical models can be and are used for the monitoring network optimisation and redesign.

Geostatistics offers a number of techniques which allow to model spatial data structures and make spatial predictions including maps of estimates and maps of estimation errors. The main assumption which distinguishes geostatistics from traditional spatial interpolation approach is in concerning data as a result of a random process. This implies characterising the distribution and spatial correlation by statistical moments of first and second order.

At present time more and more linear and non-linear geostatistical models (different kinds of kriging: simple, ordinary, disjunctive, indicator) are used for the analysis of ecological data. In the works [4,6] several modern methods/models were used and compared. It was shown that different methods give different both quantitatively and qualitatively decision-oriented maps. Only modern geostatistical models give us information about errors of mapping (maps of errors). With the help of such methods we can improve and optimise monitoring networks. This problem is also essential for remediation and restoration of contaminated sites. We are using both data from Central Data Bank and REM BANK Data [8] during validation and verification of our software for spatial data analysis.

Described above methods were used for estimating soil contamination by Chernobyl radionuclides. Traditional (deterministic) interpolators (like inverse distance squared, multi-quadric equations, ect.) have been implemented in to *Interpol* program for Windows. These methods provide quick results and perform fairly enough in case of well understood simple spatial data structure. Progrms Geostatistical Software Library (GSLIB) [5] was used to calculate geostatistical predictions. There have been developed a friendly user shell for Windows (*WinGSLIB*) to work with the programs. Geostatistical predictors require significant preliminary structural analysis and modelling of spatial correlation.

5. Prediction Mapping

Decision support in multi scale spatial accident heavily depends on prediction maps developed by using various models. Radioactive contamination originated after a release from NPP can be distributed on a large territory under changing weather, sedimentation, orographic conditions, thus there should be used advanced models to built maps of predicted spatial contamination distribution.

The calculation results should be post processed and visualised. The *Upfile+* program for Windows have been designed to present prediction results as colour images (mosaic maps), to process them and to export the results to other presentation tools: "GeneSYS", Surfer software.

Geostatistical methods mentioned above are used to built maps of predictions accompanied by maps of errors, maps of probability (p-maps), maps exceeding of a certain probability level, maps of equiprobable realisations as result of stochastic modelling, maps of estimation uncertainty characterised by thick estimation isolines. One of probability surfaces accompanied by a contour map is shown on figure 5.

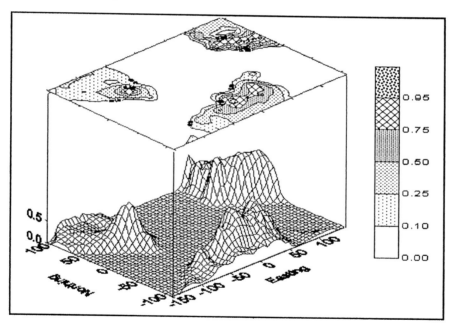

Figure 5: Cs137 contamination in Gomel region (Belarus), probability of exceeding the level of 10 Ci/sq.km.

6. Conclusions

The present work deals with a concept of Environmental Integrated System and includes description of the prototype of such system applied to analysis of Chernobyl accident consequences. The system prototype contains various compatible program modules which are used in corresponding steps of the analysis (scenario). The programmes have friendly user interfaces. Some programmes were developed for Windows. The software has been developed using different programming languages: FORTRAN, C, C++, Visual Basic, Clipper, Delphi. The most advanced object-oriented approach in programming has been applied to generalise and integrate the existing and developing software. Different modules are connected with one another through import/export data file formats. At present time the developed EIS and its separate modules are widely used for decision oriented and prediction mapping of spatially distributed environmental and ecological data.

7. References

1. M.Kanevsky, V.Kiselev, P.Fache, J.Touche. Decision-oriented mapping in emergency and post-accident situations. Radioprotection, Special Issue, February 1993, p. 441-445.
2. M.F. Kanevsky, I.E. Krayushkin, N.A. Koptelova, E.D. Martynenko, V.A. Vorob'ev, S.Yu. Chernov, E.A. Savel'eva. Development of applied integrated systems in radioecology. In: Safety problems of nuclear power. Moscow: "Nauka", p.165-194, 1993.
3. S.A.Kabalevsky, V.P.Kiselev, A.N.Serov. Development of applied software for information of radioecological monitoring. In: Safety problems of nuclear power. Moscow:"Nauka", p.211-220, 1993.
4. M. Kanevsky, R. Arutyunyan, L. Bolshov, V. Demyanov, T. Haas, L. Savel'eva. Environmental Data Analysis. Case Study: Chernobyl Fallout. 10th International Congress on Mathematical and Computer Modelling and Scientific Computing. Boston, July 5-9, 1995.
5. Deutsch C. V., and Journel A.G. (1992) *GSLIB Geostatistical Software Library and User's Guide.* Oxford University Press, New York, Oxford.
6. Kanevsky M., Arutyunyan R., Bolshov L., Demyanov V., Savelieva E., and Haas T. (1995a) Geostatistical approach to the analysis of Chernobyl fallout *Izvestija Akademii Nauk. Energetika* No.3, 34-46 (in Russian).
7. Kanevsky M., and Savelieva E. (1995) *Environmental monitoring networks and quantitative description of clustering.* IAMG'95 Annual Conference. Abstracts for Technical Programs. 31-32.
8. M. de Cort, G. Graziani, F. Raes, D. Stanners, G. Grippa, I. Ricapito. Radioactivity Measurements in Europe after the Chernobyl Accident. EUR-12800 EN, 1990, 165 pp.

SECTION 2:
MATHEMATICAL MODELLING

AIR AIR & NOISE air pollutants emissions and noise from airports

C. Trozzi, R. Vaccaro

TECHNE s.r.l., Via Nicola Zabaglia 3, 00153 Roma, Italy

Abstract

The work describes the methodology, the content and the aim of the computer program AIR AIR & NOISE produced by our company. The program is developed on Personal Computer in Windows environment with an object-oriented language.

The program is composed by two integrated modules, Air e Noise, which have a common specific airport data base. AIR module allows the estimate of air pollutants emissions and NOISE module the evaluation of the impact of the noise from aircrafts operation.

AIR AIR & NOISE can be used for two main purposes: the characterization of air pollutant emissions and the noise generation; the simulation of measures for their reduction.

The module AIR allows the estimate of the air pollutants emissions produced by aircraft traffic within the airport area. The emissions are calculated from emission factors for single LTO (Landing TakeOff) cycle and number of LTO cycles.

The module NOISE supplies the impact of the noise generated by the aircrafts in the airport and in its neighbourhood. The impact of the noise is supplied through maps of the isolines of the noise through a raster representation of the airport area. These isolines can be visualized in a geographical information system. The module NOISE is based on the application of the model INM (Integrated Noise Model) dell'U.S. Department of Transportation, Federal Aviation Administration.

A simple case study, the Naples airport in Italy, is reported either for the air emissions and for the noise evaluation.

1 The program

The AIR AIR & NOISE package runs in Windows environment and has been developed with Visual Basic object oriented language.

The package contains several integrated module (Figure 1): data entry (DATA), air pollutant emission estimate (AIR), noise evaluation (NOISE), mapping (MAP).

The following actions are also planned:

- integration of a simplified version of APEX [1], for the management of emissions inventory from fixed (power plant, incinerators, waste water treatment, ecc.) and mobile sources (road, traffic);
- integration of SETS [2] model for the estimate of emissions from road traffic;
- integration of a dispersion model (DIFF).

AIR and NOISE modules share the section of input of the data related to aircraft movimentation. AIR allows the estimate of air pollutant emissions [3] and NOISE the evaluation of the impact of the noise generated by aircraft within the airport and its neighbourhood.

Moreover, it's currently provided a cartographic reproduction of noise isolines on MapInfo Geographic Informative System [4].

2 Data entry module

The package requires as input data:
- airport altitude above sea level and average daily airport temperature;
- runway configuration (name, heading, starting and ending coordinates);
- effective tracks related to landings and takeoffs divided in legs and, for each leg:
  - the type (STRAIGHT, LEFT, RIGHT),
  - the bending angle (0÷360),
  - length or radius [length if STRAIGHT, radius if LEFT or RIGHT];
- number of landings and takeoffs operations by:
  - runway,
  - type of aircraft,
  - class of length of travel (associated with a takeoff weight which represents a typical passenger load factor and fuel required for such a trip),
  - profile of approach used (for landings only),
  - period of the day (day [7-19], evening [19-22], night [22-7]);
- specific noise profiles by aircraft model (different from the INM default ones);
- Sound Exposure Level (SEL) and Effective Perceived Noise Level (EPNL) per different thrust levels and at different distances;
- specific profiles of approach by aircraft model (in particular in the INM model are presented two default approaches at 3 and 5 degrees);
- specific takeoff profiles by aircraft model (different from the INM default profiles).

3 NOISE module

The NOISE module provides the impact of the noise generated by aircrafts within the airport and its neighbourhood. The impact of the noise is provided showing maps of the isolines (lines of equal exposition to noise) for several noise level indicators. The NOISE module is based on the application of the INM (Integrated Noise Model) model of the U.S. Department of Transportation, Federal Aviation Administration.

The first version of the INM (Integrated Noise Model) model has been released on January 1978 by the Federal Aviation Administration (FAA) of the U.S. Department of Transportation [5, 6].

The model was originally developed to provide specialists and airport planners with an instrument to evaluate the impact of the flying over noise in the neighbourhood of the airports. Since its first public release the model has been used in different studies on airport of big or small size in the United States and Europe. The model has become the standard instrument for the analysis of the noise in the neighbourhood of the airports.

The model has been subsequently updated in the course of the years up to version 5. The impact of the noise is provided showing maps of the isolines (lines of equal exposition to noise) for various indicators of measure of noise.

The indicators analysed are the following:

- Noise Exposure Forecast (NEF). The NEF index was developed in 1967 and is based on the effective perceived noise decibel (EPNdB) as the unit of aircraft noise. All aircraft operation during the period 10 p.m. to 7 a.m. are weighted by a factor of 16.7 per one operation.

- Equivalent Sound Level (Leq). The equivalent sound level constitutes an index of the disturbance or damage level determined by a sequence of noises detected within a determinate measure interval ; that is, it corresponds to the constant and continuous noise level that within the above mentioned time interval has the same average energetic level of the analysed noise.

- Equivalent Sound Level - day (Leq-day). Equivalent sound level for all operations carried out between 7 a.m. and 10 p.m.

- Equivalent Sound Level - night (Leq-night). Equivalent sound level for all operations carried out between 10 p.m. and 7 a.m.

- Day-Night Average Sound Level (Ldn). The Ldn index was developed in 1973-74 by the U.S. EPA (Environmental Protection Agency) and keeps into account the different importance of nocturnal noise. In particular the Ldn index is based upon the Leq index with the aircraft operations during the period 10 p.m. to 7 a.m. weighted by a 10 decibel penalty.

- Community Noise Equivalent Level (CNEL). The CNEL index keeps into account the different importance of evening and night noise. In particular the CNEL index is based on the Leq index with the aircraft operations during the period 10 p.m. to 7 a.m. weighted by a 10 decibel penalty and the aircraft operations during the period 7 p.m. to 10 p.m. weighted by a 3 decibel penalty.

- Weighted Equivalent Continuos Perceived Noisiness Level (WECPNL). The WECPNL index keeps into account the different importance of evening and night noise. In particular the WECPNL index uses as noise measurement unit the equivalent sound level weighted in decibel (ECPNL) with the aircraft operations during the period 10 p.m. to 7 a.m. weighted by a 10 decibel penalty and the aircraft operations during the period 7 p.m. to 10 p.m. weighted by a 5 decibel penalty. Moreover, the index introduces a seasonal correction factor as a function of the air temperature.

- Time above a specified threshold of A-Weighted Sound (TA). The TA index indicates the time in minutes that a dBA level is exceeded during a 24-hour period.

For the calculation of the indexes the package is based on the database contained in the INM model and updated on the base of its application to Italy. The package supplies at output the isolines of the noise in the airport area.

4 AIR module

The AIR module supplies the estimate of the emissions of the main air pollutants produced within the airport sites by aircraft movement. The emissions produced by other airport activities (boilers, generator sets, ground vehicles, ecc.) are not taken into account. The pollutants taken into account are: Carbon Monoxide (CO), Volatile Organic Compounds (VOC), Carbon Dioxide (CO_2), Nitrogen Oxides (NOx), Total Suspended Particles (PST) and Sulphur Oxides (SOx).

The emissions are calculated by aircraft model and single LTO (landing-takeoff) cycle. An LTO cycle includes all aircraft operations in flight and on the ground. In particular are considered: descent and approach from an height of about 3000 feet (915 mt.) above ground level, touchdown, landing run, taxi in, idle and shutdown, startup and idle, checkout, taxi out, takeoff and climbout to 3000 ft. above ground.

To the means of the emission estimates four phases can be divided: approach, taxi/idle, takeoff and climbout. Every aircraft class has its typical LTO cycles, meant as sets of typical operation times. In the calculation model several aircraft classes have been individuated: long range jet, commercial turboprop, medium range jet, business turboprop, jumbo jet, piston. For each of these classes and for the four phases defined above it's necessary to insert the typical operation times. Once inserted the typical times the program calculates, through the data contained inside its database, the emission factors by single aircraft and LTO cycle. Moreover, starting with the number of LTO cycles by aircraft model the program calculates the emissions by type of aircraft.

The internal database contains:
-    a set of engines commonly used on aircrafts and the respective consumptions and hourly emission factors;
-    a set of aircrafts with the respective model and number of engines.

5 Cartographic module

The package supplies automatically the noise maps in Mapinfo format and allows the full integration of the maps within said software. The maps can be personalized inserting a raster image of the airport.

6 Napoli Capodichino case study

The airport of Napoli Capodichino is 7 km away from the urban centre, has an height of 88 m, classification ICAO "A". From the analysis of the historical series of the temperatures of the airport Napoli Capodichino is obtained a temperature of 19°C, annual average related to the hours 7 a.m.-10 p.m..

The airport consists of one runway only, the RWY 06-24, wide 45 m and long 2640. Within the Master Plan is provided a lengthening of the runway of 300 m at heading 24.

The runway 06-24 is mainly used in an unidirectional sense; in particular about the 96% of the takeoffs takes place from heading 06 and the 4% from heading 24, and the 97% of the landings takes place from threshold 24 and the 3% from threshold 06. In regards to the landing, at first approximation, is assumed the standard landing profile at 3° for all the landings; the profile is enough well comparable with the one in use by the airport at 3.33°.

In Table 1 are reported the previsions of the Master Plan, in thousands of movements, related to the passenger aircraft divided by aircraft size category.

Category	Aircraft type	1995	2000	2005	2010	2015
Large	L1011, A300, 767, A310	0,7	2,3	4,5	7,2	10,2
Medium	A320, 757, MD80/90, 737, 146, DC9	24,9	31,0	37,0	43,8	53,0
Small	ART, CRJ, Emb	3,6	5,1	5,8	6,9	7,6
Total		29,2	38,4	47,3	57,9	70,8

Table 1 - Traffic forecast of the Master Plan by aircraft size

To obtain a forecast of the noise the following hypotheses and assumptions will be done:
- some standard aircrafts by class of size, according to the existing data for the 1995;
- within the type of aircraft, the distribution between trip length class within aircraft type related to 1995 year;
- within the type of aircraft, the distribution between periods of the day (7am-7pm; 7pm-10pm; 10pm-7am) relative to the 1995;
- for the forecast of aircraft types, the development hypothesized in the master plan is more oriented towards scheduled traffic than charter traffic;
- in the selection of the standard vehicles will be taken into account the subdivision of the vehicles as a function of the noise and in particular the vehicles of older conception (DC9 and BAC 111) will eliminated since year 2000.

In Table 2 is reported the forecast by type of aircraft.

flights/day	1995	2000	2005	2010	2015
A300	1,88	6,18	12,10	19,36	27,43
MD82	45,05	64,81	79,61	96,01	117,21
737300	17,12	20,74	24,75	29,30	35,45
DC950	6,16	0,00	0,00	0,00	0,00
BAE146	1,89	2,29	2,74	3,24	3,92
DHC8	7,69	10,90	12,39	14,75	16,24
CL601	0,97	1,37	1,56	1,85	2,04
SF340	1,16	1,65	1,87	2,23	2,45
F28MK2	0,28	0,00	0,00	0,00	0,00
Total	82,22	107,94	135,02	166,73	204,74

Table 2 - Average day 1995-2015 by type of aircraft

In Figure 2 are reported the estimated values of the air pollutants emissions under the hypotheses of the Master Plan.

In Figure 3 are shown the areas (in square miles) exposed at the different levels of noise expressed in W_{ECPNL}.

From the Figure 3 is noticed the immediate effect of the withdrawal of the older aircraft models (in particular DC9) hypothesized within five years, before year 2000. Such an effect, even if the number of flights of said aircraft is low, is very considerable. The effect of the lengthening of the runway, planned for 2010, calculated in the hypothesis of no following modification of the takeoff and landing tracks and the respective frequencies in both directions, is practically null.

In Figure 4 are shown the isolines of the noise expressed in W_{ECPNL} for the year 2015. In the terms of the index W_{ECPNL} the area exposed at noise levels above 88 dBA remains very limited in the course of the time (from 0,18 square miles in 1995 it drops to 0,15 square miles in 2000 to rise to 0,31 square miles in 2015). This area exposed remains substantially the airport area closer to the runways. The area included in the 75-88 range drops from 3,13 square miles in 1995 to 2,78 square miles in 2000 to rise to 4,99 square miles in 2015. The town of Casalnuovo di Napoli is very close to this area (from the 2010).

We nevertheless believe that the most relevant result of the simulation is the effect of the replacement of the aircrafts. Already in 2000 the replacement of the noisiest aircrafts allows a sharp reduction of noise in spite of substantial increase of the traffic. Since 2010 a progressive introduction of low noise vehicles (of the kind, for example, of the BAe146) can be expected with a further reduction of the noise.

Acknowledgements

The case study has been realized under contract of GESAC spa - Capodichino Airport - Napoli (Italy)

References

1. Trozzi C., Vaccaro R., Nicolò L., Trobbiani R., Valentini L., Di Giovandomenico P., APEX - Air Pollutants Emissions inventory Computer System - In: Software per l'ambiente. A cura di G. Guarisio, A. Rizzoli. Patron Editore, Bologna, 1995. pagg. 215-220Azienda Autonoma Assistenza al Volo (1995)

2. Trozzi C., Vaccaro R., Di Giovandomenico P., Crocetti S., SETS - Estimate of air pollutants emissions from road traffic - In: Software per l'ambiente. A cura di G. Guarisio, A. Rizzoli. Patron Editore, Bologna, 1995. pagg. 221-226

3. Trozzi C., Vaccaro R., Nicolò L., Trobbiani R., Martinelli A., AIR AIR - Estimate of air pollutants emissions from airport - In: Software per l'ambiente. A cura di G. Guarisio, A. Rizzoli. Patron Editore, Bologna, 1995. pagg. 209-214

4. MapInfo User's guide. MapInfo Corporation, Troy, New York, 1994

5. U.S. Department of Transportation - Federal Aviation Administration - INM - Integrated Noise Model, Version 3, User guide - Revision 1 DOT/FAA/EE/92-02, 1992

6. U.S. Department of Transportation - Federal Aviation Administration - INM - Integrated Noise Model, Version 4.11, User guide - Supplement DOT/FAA/EE/93-03, 1993

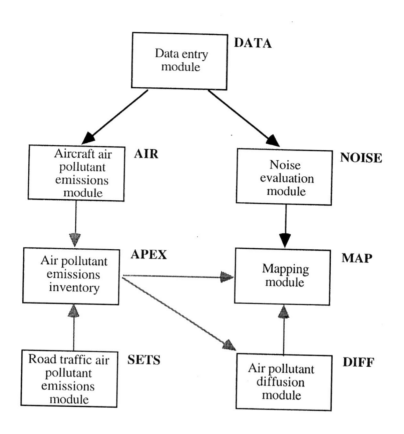

Figure 1: AIR AIR & NOISE package structure

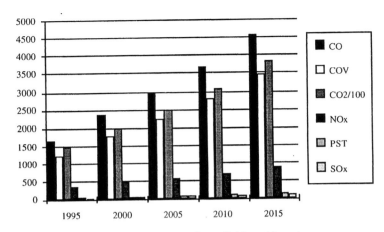

Figure 2 - Capodichino Airport
Estimate of the emissions of the main air pollutants in kg/day and
of CO_2 in hundreds kg/day in the hypotheses of the Master Plan

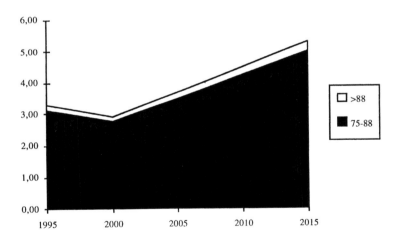

Figure 3 - Capodichino Airport
Area (square miles) affected by the different levels of noise expressed in
W_{ECPNL}

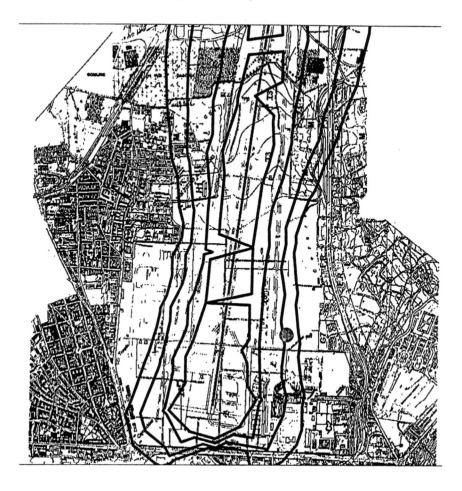

Figure 4 - Capodichino Airport, 2015 Prevision year
Isolines of the noise espressed in W_{ECPNL} a 75, 78, 83, 88 dBA

Models based on a statistical-thermodynamics analogy for the computation of varying diffusivity tensors and integration of the diffusion equation

C. Dejak, R. Pastres, G. Pecenik, I. Polenghi, C. Solidoro

Department of Physical Chemistry, Section of Ecological Physical Chemistry, University of Venice, Dorsoduro 2137, 30121 Venice, Italy

Abstract

The possibility is discussed of estimating space and time varying eddy diffusivities tensors based on numerically computed velocity fields of a waterbody. Valid results, are obtained with an approach derived from statistical thermodynamics analogies, which make to correspond the mean free path to the dimension of tidal vortexes.

Through the computation of the Lagrangian time scale, the possibility is also shown, to integrate the shear stress with a Fickian equation. Shear stress is evaluated through a modified von Kàrmàn formulation.

Additional approaches are also prospected in the determination of diffusivity tensors components.

Introduction

Models of complex waterbody ecosystems with chemical and biological purposes, demand the modelling of transport phenomena such as advection and diffusion. This need is determined by the influence of dilution on the relative kinetics which are effected not just through a time reversible advection mechanism, that can not lead to a steady state, but basically through an irreversible diffusion process. Steady states are similar to equilibrium states as, all intensive variables (such as temperature, chemical concentrations, biological densities) are constant in time, but they differ from these because they vary in space. In fact continuos source and open boundary conditions assure constant gradients, in the waterbody, of these intensive variables.

Fluxes of the conjugated extensive variables (heat, chemical and biological mass) are derived from gradients from Fick and Fourier laws according to eddy diffusivities. This dilution and the concomitant time irreversible processes, produce entropy, which is still minimum at steady state, and it is removed from the ecosystem contemporaneously with mass and heat fluxes.

In the real environment, however, and particularly in water bodies of complex topography, vector fluxes are not collinear with the corresponding gradients, and in such instances, the diffusivities constitute a transformation matrix related to the spatial coordinates and hence a tensor. Such conditions clearly refer to turbulent diffusivities strictly linked to the filed properties and not to isotropic molecular phenomena quantities. In respect to the former, however, molecular diffusivities are several order of magnitude lower and their effect is felt only in restricted and particular space such as the thermohaloclyne.

In all environmental interdisciplinary models the greater uncertainty is associated to the description of processes related to the trophic chain and it increases with the growing up of complexity: this occurs beginning with nutrients, photosynthesis, primary and secondary production until the simulation is possible with continuous distribution.

For instance, in the extended simulation of seasonal phenomena all descriptive formulations require an increasing number of parameters, to account for variation of rate processes with irradiance and temperature and so the uncertainty increases further. Because of this situation, a balance between uncertainty sources is requested, it is left arbitrary the decision to where, the greater accuracy should be required for description of a phenomenon in respect to another.

In this view, even a diffusion process may be reproduced through a less accurate constant diffusivity instead of an approach involving a tensorial analysis. This practice has been successfully applied so far in the majority of applicative model, such as the eutrophication-diffusion model of the Venice lagoon[1], Florianopolis bay (Brasil) and others.

However, all efforts presently devoted to more accurately estimate trophic processes parameters, by sensitivity analysis and calibration of each single parameter[2], also basing on time series of structural environmental data and their accurate statistical treatments[3], induce to a more detailed analysis of transport mechanisms. This demands for introduction of time and space varying diffusivity tensors whose average should be comparable to the constant value so far adopted in transport models.

Basing on dimensional analysis, several semi-empirical formulations have been elaborated in order to estimate eddy diffusivities, but they are not capable to render the complexity of eddy diffusivities in real situations. A different approach is necessary, deriving from the application of thermodynamic-statistical analogies on velocity fields computed through numerical programs, now widely available also in 3D.

The simplest analogy[4] leads to a molecular diffusivity which is proportional to the product of a particle mean velocity between two collisions and the mean free path: the first obtained on a short time interval, and the second, dependent only on the density, time averaged on a longer period. The most attainable procedure to represent the turbulent motion is the statistical one, but this would demand a very large number of measurements carried out in a long series of repeated similar experiments, that are not available at all. In fact only one single experiment is usually considered, which should last for a time long enough to calculate a reliable time averaged mean property. Then it is needed to find out how close this empirical mean quantity lies to the theoretical probability mean value. This conception is completely analogous to that in ordinary statistical thermodynamics, where the theoretical "mean over all possible states of the systems" (called "ensemble mean"), may also be replaced by the directly observed time-mean. Although this averaging interval seems to be tolerable, still the ergodic hypothesis must be verified, i.e. all the possible states of the system must be actually reached in that time interval, which is shown to be acceptable. The central limit theorem in statistics generates, in these cases, gaussian distributions and this may be considered as a posterior verification, instead of a prior cause.

The use of statistical-thermodynamical analogies in this aim, is so completely justified even though the velocity fields applied to theoretical formulations, are computed from numerical programs and not from experimental data.

If the water body to model is governed by periodical forcing functions as the tide, the tidal velocities exhibit a negligible mean value if averages during a proper time interval, and therefore in the transport phenomenon, turbulent diffusion plays the most important role.

In this case the analogy with the ultrasimplified theory of molecular diffusion suggests to substitute the particle velocity between two collisions with a mean velocity averaged in a time interval (1 hour), small compared to the tidal period (12 hours).

Using this velocity field to obtain the particle displacements, since residual currents are negligible, it is observed that almost all the different trajectories (starting at different tidal moments, from every point) are closed curves. Therefore they may be statistically described with a dispersion ellipse that includes 95% of all the trajectories points. In this way it is possible to determine the main direction of the flux vector in the Lagrangian reference system. Each ellipse is in fact characterized by the length of the major and minor axis, by the angle between the major axis and the ordinate axis of the Cartesian reference system (N-S direction) and by the eccentricity. The distribution of these four parameter in the modelled area (the central part of the Venice lagoon) is reported in Figure 1a,b,c,d. As one can see, in the majority of the points, the eccentricity is markedly different from 0 and this substantiates the need of a tensor reproducing the concentration patterns in all the different zones of the modelled area. Carrying on with the analogy, the eddy diffusivities along the canonical axes of each ellipse are determined by the products between the two ellipse semiaxes (analogous to the mean free path) and the projections of the hourly averaged velocities on them. The Lagrangian diffusivity tensor so computed, is represented by a diagonal matrix, having, in the principal diagonal, the two components of this vector and, outside it , zero values. To obtain the Eulerian tensor, the Lagrangian one is rotated to the orientation of the fixed Cartesian axes by applying a linear transformation and its inverse, which lead to a non diagonal but still symmetrical space and time varying diffusivity tensor with the following properties:

$K_{xx} > 0$, $K_{yy} > 0$, $K_{xy} = K_{yx}$ and $K_{xx} \cdot K_{yy} - K_{xy}^2 > 0$, as requested by the second law of thermodynamics.

Results are summarised in Figure 2a,b,c that show respectively the daily averaged diagonal components, K_{xx} and K_{yy} and the cross term, K_{xy} of the tensor. For a comparison, in Figure 2d it is shown the bathymetry of the central part of the lagoon characterized by a complex network of shallow areas and deep channels joined to the Adriatic sea through three inlets (maximum depth: 20 m).

The numerical integration of the diffusion equation in presence of a continuos point source, has been performed with a finite difference explicit method for each horizontal layer and with an implicit Laasonen scheme along each vertical column using space variable diffusivities to simulate the stratification[5]. The explicit method adopted to solve the horizontal diffusion second order partial differential equation with space and time varying diffusivity tensors, must assure the respect of the conservativity, the convergence, consistency, and stability of the solutions. To guarantee the conservativity, the most important property for chemical and biological applications, the scheme adopted is forward in time and centred in space except for the term of the scalar product of the diffused property gradient with the divergence of a vector with components K_{xx} and K_{yy}, that is forward in space. The stability condition, on the other hand, requires a quite long running time but still acceptable. Initial instabilities, though, are inevitably caused by the cross term (especially in the few grid points where K_{xy} is greater than a tenth of any of the diagonal components), but later smoothed out as the numerical integration proceeds. The results are presented in Figure 3, where the spatial distribution of a persistent pollutant is shown after 40 days of simulation. From the comparison with Figure 2d it is observed that all the deepest channels are realistically rendered also using only diffusivity tensors without any advective direct contribution.

Analysing, now, the ergodicity of the process, it must be observed that the points of the trajectories within the ellipses, do not have a 2D Gaussian distribution, but a toroidal one. This is unavoidable as the resolution of minor vortexes, originating from the turbulent energy cascade of bigger eddies, is not possible to be simulated with the presently available advective models, and so it is not possible to smooth out the toroidal distribution set of curves. Nevertheless, the solution adopted theoretically guarantees the statistical-thermodynamic ergodicity, since the whole reachable phase space is really reached for both the position and momentum subspace, and only the frequencies of reachability are different. For this purpose it is possible to resort to theoretical treatments, capable of taking into account the proximity of big vortexes along the main direction of the flux vector and the smaller ones present in the surroundings. However, they must strongly interact to realistically render the distributions: to this aim shear stress fluxes orthogonal to the main ones, are added, and they start to overwhelm the molecular diffusion only in an intermediate time, after which the process goes back to a Fickian law but with eddy diffusivities. During this time period, the area covered by the vortex (or of the variance σ^2)[6] is proportional not to the time but to the cube of it. The relative determining parameter is the Lagrangian integral time scale T_L, which is computed from the ratio between the autocovariance and the variance[7] and in the present case, it results lower than a tidal period[8], while the stationary state requires a longer time intervals to be reached. It is so legitimized to add the shear stress component to the tensor one in a normal integration with the Fick equation for diffusion and not with a faster telegraph equation.

The classical method to obtain the shear stress for systems with an irregular geometry, is due to von Kàrmàn theory[9]: the first and second velocity derivatives are calculated in every grid point in respect to its orthogonal direction, and for dimensional reasons, the rate between the cube of the first derivative and the square of the second one is performed and multiplied by the square of the well known von Kàrmàn "universal" constant, k=0.36÷0.4.

Another way of evaluating this contribute of the flux is due to Aris[10] and Saffman[11], who based their analysis on the direct calculation of the zero, first and second moments of the distribution function along the main direction through their dependence on the diffusivity in the orthogonal direction, but the solution holds in non-stationary conditions of flow, and with a semi-empirical parametrization.

A further method was proposed by Carter and Okubo[12] who based their analysis on the integration of the advective diffusion equation, modifying the velocity vector components in the main direction. Both relationships seemed to be tested experimentally by Elder[13] and Okubo and Karweit[14] respectively.

The computation of the shear stress component of the eddy diffusivity, was first carried out with the Von Kàrmàn analysis, but the distribution of the results appears to be spatially almost random, Figure 4a, according to $R_1 = (\partial v / \partial n)^3 / (\partial^2 v / \partial n^2)^2$ and the frequencies distribution shows a very high variability. It is very difficult to reduce with the removal of distonormal $(R_1 \pm 3\sigma_{R1})$ values which are distributed in a range between -1.4×10^{14} and 1.6 10^{14}. Studying separately the numerator from the denominator the situation improves, but only slightly. Cutting all the values that detach from the mean more than, after few iterations, 40% of the points are excluded and about 29% are in a tight range around the mean value.

Aware of the real difficulties encountered in the application of the method based on second derivatives to numerical computations, von Kàrmàn dimensional logic was applied not to velocities but to the Lagrangian main components of the tensor, as made by Aris, but with a completely different theoretical approach. In this case the dimensional analysis is much easier and does not utilize neither second derivatives nor high powers. In fact it leads to this new ratio: $R_2 = (\partial K / \partial n)^2 / (\partial v / \partial n)$. In this ratio two cases have been distincted for the denominator: in the first one the velocity is considered along its own direction, and in the second projected along the main tensor direction. While the first approach gives good results as the original, but still quite scattered, the second case shows reliable results, Figure 4b and its frequency distribution is less scattered but with a behaviour still similar to R_1, as it is multiplied by a corrective factor obtained by the statistical comparison of R_2 and R_1.

Adding the two shear stress components to the already computed Lagrangian eddy diffusivity tensors, and transforming in a new Eulerian matrix, the integration of the diffusion equation could be arranged with the scheme described above. Some initial instabilities appear and now they require a long computing time to disappear. To overtake this problem, super computers with high parallelism and suitable optimization will be used.

In conclusion the method here proposed is suitable to evaluate the effect of a diffusive transport with space and time eddy diffusivity tensors including also shear stress, and the results of the integration is completely conservative and also applicable to chemical and biological modelling.

References

1. Dejak, C. & Pecenik, G. *Ecological Modelling*, 1987, **37**, 21-404.
2. Pastres, R., Franco, D., Pecenik, G., Solidoro, C., & Dejak, C. First order sensitivity analysis of a distributed parameter ecological model.*Proceedings of the International Symposium SAMO 95*, Belgirate, 1995.
3. Dejak, C., Franco, D., Pastres, R., Pecenik, G., & Solidoro, C.An informational approach to model time series of environmental data though negentropy estimation. *Ecological Modelling*, 1993, **67**, 199-220.
4. Hirshfelder, J., Curtiss, C.F. & Bird, R.B. *Molecular theory of gases and liquids.* J. Wiley&Sons Inc. N.Y. Edition 3rd, 1966.
5. Dejak, C., Franco, D., Pastres, R., Pecenik, G. & Solidoro, C. Thermal exchanges at air-water interfaces and reproduction of temperature vertical profiles in water columns. *Journal of Marine Systems*, 1991, **13**, 465-476.
6. Csanady, G.T. *Turbulence diffusion in the environment.* D. Reidel Publishing Company, II ed., 1994.
7. Monin, A. S.& Yaglom, A.M. *Statistical fluid mechanics,* J.L. Lumley ed., 1975.

8. Dejak, C., Pastres, R., Polenghi, I., Righetto, S. & Solidoro, C. *Conditions for the application of transport models to problems of environmental chemistry*. J. of Analitical and Environment Chemistry, in press, 1996.
9. von Kàrmàn, T. Mechanische Ähnlichkeit und Turbolenz. *Nach. Ges. Wiss., Göttingen, Math. Phys. Klasse*, 1930, **58**.
10. Aris, R. *Proc. Roy. Soc. London* A235, 1956, **67**.
11. Saffman, P.G. *Quart. J. Roy. Meteorol. Soc*, 1962, **88**, 382.
12. Carter , H.H. & Okubo, A. A Study of the physical processes of movement and dispersion in the Cape Kennedy Area, Cheasepeake Bay Institute, Johnson Hopkins Univ. 1965 Ref **65**-2, 150.
13. Elder, J.W. *Fluid Mechanics*, 1959, **5**, 544.
14. Okubo, A. & Karweit M.J. *Limnol. Oceanog.*, 1969, **14**, 514.

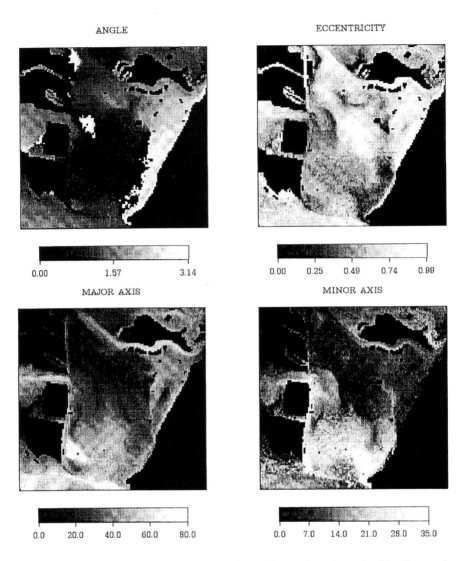

Figure 1:Parameters of the dispersion ellipse. a) major axis; b) minor axis; c) angle; d) eccentricity.

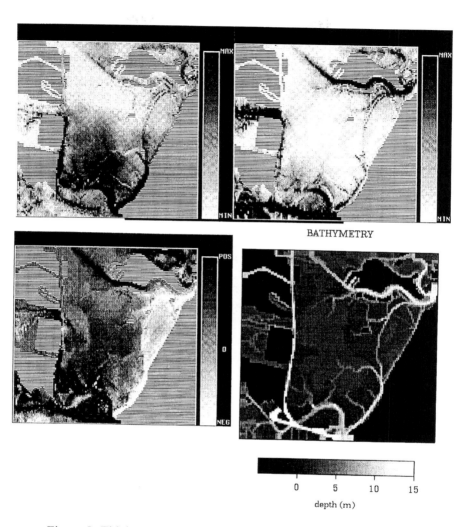

Figure 2: Tidal averaged Eulerian diffusivity tensors components. A) N-S component K_{xx}; b) E-W component K_{yy}; c) diagonal component K_{xy}; d) bathymetry of the Venice Lagoon.

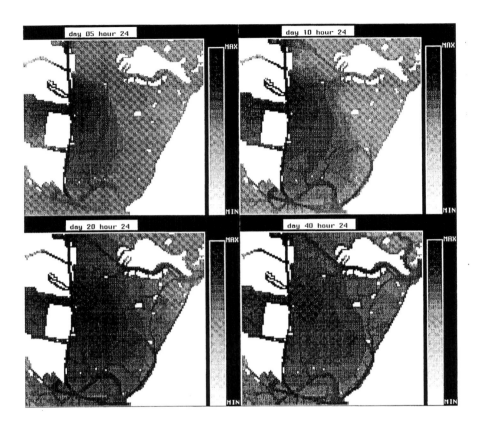

Figure 3: Integration results showing the distribution pattern of a passive tracer continuously released from a source point after: a) 5 days; b) 10 days; c) 20 days; d) 40 days.

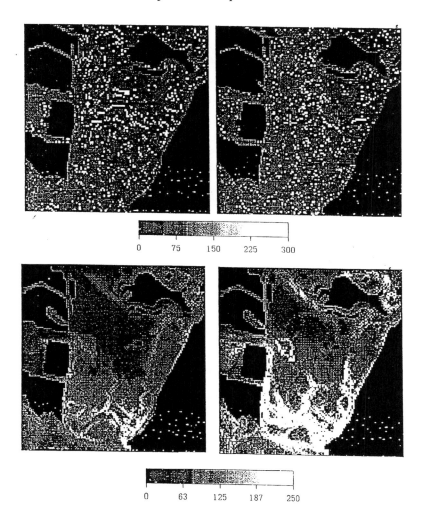

Figure 4: Distribution of the results of von Karman analysis applied a) to the velocity field, b) to the Eulerian components of the diffusivity tensor.

A hybrid deterministic/ nonlinear stochastic model of the activated sludge process

W. Ward,[a] D.A. Vaccari,[a] D. McMahon,[a] S. Rivera,[a] E. Wojciechowski[b]

[a]*Stevens Institute of Technology, Hoboken, NJ, USA*
[b]*ITT Corp., Nutley, NJ, USA*

Abstract

The activated sludge process (ASP) is the most common method of treating wastewater. Failures in the ASP are usually related to the discharge of total suspended solids (TSS) in the effluent. A recent case involves a 70 million gallon per day regional wastewater treatment plant. As a result of this plant's violating its permitted discharge limitations for TSS, a construction ban was imposed, resulting in significant local economic problems. The remedy for this problem involved construction of new treatment units to increase capacity.

Many plants have similar problems, or will have to face them as regional growth increases their loading. However, it is likely that capacity could be increased without new capital construction by using control techniques to operate the current units more efficiently. Few plants, if any, currently apply modern process control techniques to optimize effluent performance.

The development of modern controllers, such as the model predictive controller [1]requires considerable experimental data for the system identification. Control techniques may be tested by applying them to process simulators, if the simulators are realistic enough in their behavior.

Most of the activated sludge models available in the literature lack important behaviors necessary for their use as "experimental animals" in controller development. They often focus on deterministic behavior such as substrate uptake and microbial growth. Few are available which can predict the stochastic behaviors such as variability in sludge volume index (SVI), and the resulting effect on sedimentation. Most importantly, it is rare to see a model for predicting effluent suspended solids, although this is the most important performance parameter in the activated sludge process. This paper describes a hybrid model which combines fundamental deterministic models with empirical stochastic ones to produce a highly realistic model incorporating the behaviors mentioned above, as well as numerous others. The empirical model is the nonlinear PARX model.

The Hybrid System Model

The activated sludge process has many complex and interacting behaviors. Most published models only describe a few of those behaviors at a time. This work is an attempt to combine a variety of models which together can be used to simulate the process in as full a degree of complexity as could be detected by our usual sensor or laboratory measurements (short of using a microscope or our noses).

In summary, a nonlinear stochastic model, called the PARX model, will be used to generate influent B.O.D. and suspended solids data; the modified IAWQ ASP Model No. 1 [2] will be used to compute solids growth and substrate uptake; the linear stochastic model of Christodoulatos and Vaccari [3] together with the Wahlberg and Keinath [4] relationship will be used to predict changes in sludge thickening properties; the compressive thickening model of Vaccari and Uchrin [5] will be used to model thickening, and finally, a combination of short-term and daily nonlinear stochastic models will be used for effluent suspended solids.

Model	Modeled behaviors
PARX	Stochastic influent loading
IAWQ Model No. 1	Microbial kinetics and oxygen uptake
Busby and Andrews	Substrate mass transfer kinetics
Film theory	Oxygen mass transfer kinetics
Vaccari and Wojciechowski	Sludge volume index
Wahlberg and Keinath	Sedimentation parameters vs. SVI
Vaccari and Uchrin	Thickening with compression
PARX	Influent flow and COD
PARX	Daily average effluent TSS
PARX	Short-term TSS variations

Each of these models will be described here, mostly in brief. The PARX models are less familiar and will be described in more detail below. All of these models are coupled by material balance considerations, which are not described here.

Fundamental Deterministic Models

Microbial growth kinetics

The core of the model will be a deterministic model for microbial reactions in the aeration tank and for thickening in the secondary settler. Microbial reactions will be modeled using the IAWQ ASP Model No. 1 [2]. Only carbonaceous reactions will be included (that is, nitrogen reactions will be ignored). Table 1 summarizes the process.

Table 1. Process kinetics and stoichiometry for partial IAWQ model [6].

Components i		1	2	3	4	5	Rates, ρ_j (ML^{-3}T^{-1})
j	Process	X_B	X_p	X_s	X_I	S_s	
1	Growth	1				-1/Y	$\mu \dfrac{S_s}{K_s+S_s}\dfrac{S_o}{K_o+S_o}X_B$
2	Decay	-1	f	1-f			$b\,X_B$
3	Hydrolysis			-1		1	$K_H \dfrac{X_s}{K_X+(X_s/X_B)}\dfrac{S_o}{K_o+S_o}$

where X_B is the concentration of biomass; X_p is the particulate products of decay; X_S is particulate substrate; X_I is inert particles; and S_S is the soluble COD. The other parameters are kinetic coefficients.

Mass transfer of substrate and oxygen

The IAWQ model distinguishes between dissolved and particulate substrate and assimilation of products of microbial decomposition. However, the IAWQ model will be modified to include one process it lacks, which the Busby and Andrews model [7] includes: substrate mass transfer. The absorption of soluble and particulate substrate can be modeled by standard mass transfer kinetics. This can become important during conditions of short aeration tank retention times, whereupon some substrate can appear in the effluent without having been absorbed. This can occur during storm flow events or if flow is diverted to the last tank of a step-feed version of the ASP. In this model two additional processes are modeled -- mass transfer of X_S and mass transfer of S_S. Although this has not yet been implemented, proposed models for these rates can be based on Busby and Andrews formulation for particulate matter, and partitioning behavior for soluble substrate:

$$\rho_{Xa} = K_L a_{Xa} \cdot X_T \left(f \frac{X_S}{K_{Xa}+X_S} - \frac{X_S}{X_T} \right) \qquad [1]$$

$$\rho_{Sa} = K_L a_{Sa} \left(k_{Sa} \cdot S_a - S_S \right) \qquad [2]$$

where X_a and S_a are the aqueous-phase concentrations of particulate and dissolved substrate, respectively (as distinguished from X_S and S_S , which are floc-phase concentrations; the K_{La} are mass transfer coefficients, k is a partition coefficient, and X_T is the total floc-phase particulate concentration.

Oxygen mass transfer will also be modeled using standard mass transfer kinetics. In this process an additional component is added to Table 1 with S_o, the oxygen concentration. Its stoichiometric coefficient is 1.0, and the rate is:

$$\rho_{So} = K_L a_o (S_{os} - S_o) \qquad [3]$$

where K_{La_o} is the mass transfer coefficient for oxygen and S_{os} is the solubility of oxygen.

Dynamic process parameters

Dynamic sludge age: An improved state parameter for the process was developed called the dynamic sludge age (*DSA*) [8]. The *DSA* parameter is a theoretically correct measure of the average age of a culture of growing and wasting particles. The *DSA* parameter is independant of the *F/M*, and reflects the loading history of the process, leaving the *F/M* ratio to indicate current operating conditions. The *DSA* is computed using the net growth, *K*, and the solids production rate, *P*, are averaged over the sampling or calculation period:

$$K = (M - M_o)/t \qquad [4]$$

$$P = K + W \qquad [5]$$

where M = mass at time t, M_o = mass at last time period, t = length of sampling period; W = average waste rate over the sampling period. There are four different scenarios that might arise; one of the following equations is used to compute the new *DSA*:

Case 1: $W=0$; No sludge was wasted or lost over the period t:

$$DSA_t = (DSA_{t-1} + t)M_o/M + (1 - M_o/M)/2 \qquad [6]$$

Case 2: $K=0$; There was no net sludge production ($W = P$)

$$DSA_t = M_o/W + (DSA_{t-1} - M_o/W)(-W_o/M_o) \qquad [7]$$

Case 3: $W=2 \cdot P$; Exactly twice as much sludge is wasted as was produced:

$$DSA_t = DSA_{t-1}(M/M_o) = (t + M_o/K) \cdot \ln (M_o/M) \qquad [8]$$

Case 4: None of the above conditions hold.

$$DSA_t = \left(DSA_{t-1} - \frac{M_o}{P+K} \right) \left(\frac{M}{M_o} \right)^{(-P/K)} + \frac{M}{P+K} \qquad [9]$$

Use of the MCRT can produce misleading results for unsteady-state processes. For example, if a plant initially at steady state is perturbed by cutting the waste sludge flow rate in half, the traditional calculation predicts an immediate doubling of age. Smoothing techniques do not eliminate the theoretical problem.

Specific sludge production rate: Another dynamic state parameter was developed based on an overall mass balance on the process. This is the specific sludge production rate, or solids production-to-mass (*P/M*) ratio. Although conceptually simple, it is not computed by real plants because no plants routinely accurately measure the biomass in the system, including that stored in the clarifiers. Ignoring clarifier biomass can lead to gross errors in biomass measurement as solids shift to and from the basins from day to day. Solids waste rate (*W*) was shown to have the same theoretical relationship to *P/M* for unsteady-state plants as the food-to-microorganism ratio has to the inverse of the MCRT for the steady-state [9]:

$$W/M = (P/M)/Y - k_d \qquad [10]$$

where Y is the microbial yield; k_d is the decay rate coefficient; M is the mass of biomass in the system; W is the mass waste rate; P is the rate of solids production. In the dynamic formulation of the mass balance, the ratio W/M is equivalent to $1/MCRT$ (mean cell residence time) and P/M corresponds to the food-to-microorganism ratio.

Compressive thickening

The function of the secondary settler in the ASP can be separated into thickening (production of a high concentration underflow) and clarification (production of a low concentration overflow). These can be modeled somewhat distinct from each other, although, of course, there are interactions.

The standard model for thickening based on a balance of forces between viscous resistance and gravitation. It has proved quite satisfactory for explaining some of the most important behaviors of the process, especially the formation and growth of the sludge blanket in an overloaded settler. However, it is not capable of describing some other behaviors which are also important, such as blanket height in an underloaded thickener.

A model was developed which has these additional capabilities. It adds a third force, interparticle compression, to the balance of forces [10, 5]. This model directly predicts sludge blanket heights even for underloaded settlers. This is important for two reasons. First, the dynamics of sludge return depend upon the storage of solids in the clarifier; secondly, effluent suspended solids may be correlated to blanket height. The compressive thickening model can be calibrated using a single batch settling test in the laboratory [11]. In the compressive model, settling velocity depend upon solids concentration gradient as well as concentration:

$$v_s = v_m \cdot \left(1 - \frac{\partial C / \partial x}{K(C)}\right); \quad C > C_c \tag{11}$$

$$K(C) = K_c \cdot \sqrt{\frac{C - C_{max}}{C_c - C_{max}}} \tag{12}$$

where v_s is the settling velocity; v_m is the maximum settling velocity in the absense of a concentration gradient. The other parameters relate to solids compression forces and are based on the concentration distribution which would be found in a column of completely settled sludge. C_c is the compression point concentration (the concentration at which interparticle forces become significant), found at the top of the settled column; C_{max} is the maximum solids concentration, such as would be found at the bottom; K_c is the concentration gradient found at the top where the concentration is C_c. Example values of the model coefficients used are based on measurements made with a non-bulking sludge with an SVI of 112 mL/g [5]. The values are $C_C = 2.6$ g/L, $C_{max} = 11.77$ g/L, and $K_C = 0.168$ g/L/cm. Ultimately, these parameters will be correlated to SVI.

Nonlinear stochastic models

Sludge volume index and settling velocity

No fundamental model is capable of predicting how settling properties depend upon process variables. Thus, this must be determined empirically. *SVI* was simulated using an autoregressive (AR) model [12, 13]. The best model developed for *SVI* was purely autoregressive [3]:

$$SVI_t = 13.6 + 0.594 \, SVI_{t-1} + 0.255 \, SVI_{t-2}; \quad R^2 = 0.68 \tag{13}$$

In turn, the *SVI* was used to estimate the v_m as a function of total suspended solids concentration, X [4]:

$$v_m = v_0 \exp(-K_v X) \tag{14}$$

$$v_0 = (10.86 + 0.1854 \, SVI) \exp(-SVI/62.5) \quad \text{(meters/hr)} \tag{15}$$

$$K_v = 0.16 + SVI/370. \tag{16}$$

The PARX Model

Polynomial Autoregressive with Exogenous variables (PARX) models are similar to multiple linear regression (MLR) models with added terms for nonlinearity and linear and nonlinear interactions, and for previous values of any variable [14]. MLR models relate N independent variables to a dependent variable. Lagged values (values from previous intervals in a time series) can be included as independent variables, resulting in autoregression models. The coefficients of the models can be determined using least squares regression.

Nonlinear and nonlinear interaction terms could also be added to the model. As an example, consider a dependent variable, Y, which depends upon three predictors, Q, R and S. In general:

$$Y = \sum a_i \cdot Q^{b_i} \cdot R^{c_i} \cdot S^{d_i} \qquad ; \quad 0 < b_i, c_i, d_i < n \qquad\qquad [17]$$

An example would be:

$$Y = a_0 + a_1 Q + a_2 R^2 + a_{12} Q^2 \cdot R \qquad\qquad [18]$$

Typically, the exponents a, b, and c will be integers from zero to n, where n is the degree of the model. The number of terms increases rapidly with n and the number of independent variables. However, the model can be made parsimonious by including only the terms which have coefficients which are significantly different from zero. In practice, this greatly reduces the number of terms in a model. A stepwise algorithm has been developed to select terms for the model [15]. To begin with, all models with only one term are tested. The best (if any), according to some criteria, is selected as the starting model. Then, the model is changed by one term at a time in successive steps by testing each possible one-term change to the model. That is, all models with one more term are tested, as well as each model with one term removed. The best of all of these becomes the new model. Ultimately, no further improvement is possible by a change in a single term. This procedure selects for a sub-optimum model. It uses the data itself to determine the form of the model.

The PARX model was compared with both the Box-Jenkins type model and the artificial neural network (ANN) model for fitting wastewater treatment plant performance data, and found to be superior to both of them [16]. In comparison to the ANN, the PARX model produced better predictions with fewer model coefficients.

Influent Flow and COD

PARX models were developed for influent flow and COD based on data from [17].

Flow Model: Influent flow was modeled by first subtracting out a polynomial function for flow versus time of day, then additional terms were added for an autoregressive component for flow, allowing interaction between flow and time. where. Logarithmic transformation of flow was found necessary to prevent negative predictions.

$$\log_{10}\left(\frac{Q_{t+1}}{Q_{avg}}\right) = 1.1604 \cdot \log_{10} q - 6.5064e - 5 \cdot t^3 \qquad [19]$$

$$+2.1683e - 3 \cdot t^2 - 0.013901 \cdot t - 0.004532 * \log q \cdot t^2 + a_t$$

where t is hour of the day; Q_t is the flow at hour t; $q = Q_t/Q_{avg}$. The noise has $\sigma = 0.082$

COD Model: Influent COD depended on flow and previous COD:

$$S_t / S_{avg} = \left(\frac{1.3368 \times Q_{t-1}}{Q_t}\right) - \left(\frac{0.32381}{(Q_{t-1} / Q_{avg}) \cdot (S_{t-1} / S_{avg})}\right) + a_t \qquad [20]$$

where a_t is a random normally distributed variable with variance σ^2. The performance was $\sigma = 0.0782$; $R^2 = 0.62$. These models were stochastically simulated using a_t as uniformly-distributed between $-a$ and $+a$, computed from the mean square error of the model (*MSE*) following [18].

$$-a < a_t < a ; \qquad a = \sqrt{MSE} / 3 \qquad [21]$$

Effluent suspended solids

Much research into optimum control schemes for the ASP use effluent soluble B.O.D. as an objective to minimize (e.g. [6]. However, when a plant has problems with its performance, the problem is almost always in the form of high effluent suspended solids, and rarely due to soluble B.O.D. This practical consideration was not studied because no predictive model for effluent TSS was available. Traditional deterministic approaches were not useful because effluent TSS was a noisy parameter which depended upon many other parameters in a way which could not be predicted by theory.

Suspended solids are in the effluent due to two causes. Should there be a failure in the thickening function of the settler, whether due to overloading or to deterioration in settling properties, then the sludge blanket may accumulate to the point of overflowing the weirs into the effluent. Independantly, solids (called flocs) appear in the effluent even during normal operation, because clarification is an imperfect process. Clarification depends upon numerous hydrodynamic and floc structure factors. These are influenced by flow rates, changes in flow, oxygen and substrate loading in the aeration tank, aeration tank turbulence, the types of microorganisms in the flocs (especially the number and type of filamentous organisms), the average age of the solids, etc. For the most part there is no theoretical relationship which can predict effluent suspended solids.

The PARX model is ideally suited to describe this type of nonlinear multivariate relationship. This was done in two parts: The first predicts the daily average VSS, and will take into account process parameters such as *DSA* and BOD loading rate. It was developed using data from a large, regional wastewater treatment plant. A second model was developed to model the minute-to-minute variations as they depend mainly upon hydraulic effects. The short-term model will be developed using data which was obtained from [19]. The short term predictions will then be superimposed upon the daily average predictions. The resulting model to predict the daily average effluent volatile suspended solids from a real plant is [20]:

$$X_{e,avg} = 52.5 \cdot DSA \cdot PM^2 - 68.2 \cdot DSA \cdot PM^3 \cdot TSA$$

$$+0.0406 \cdot BODIN \cdot BH \cdot DSA^3 + 0.226 \cdot DSA^3 \cdot TSA$$

$$-2.19 \cdot 10^{-14} \cdot DSA^3 \cdot M^3 \cdot U + 57.6 \cdot PM$$

$$-1.23 \cdot DSA^3 \cdot PM \cdot TSA + 1.66 \cdot DSA^3 \cdot PM^3 \cdot TSA$$

[22]

where: BODIN is influent BOD (mg/L); BH is the height of the sludge blanket in the secondary settlers (ft); M is total mass of volatile suspended solids in system (1000s of pounds); PM is the specific sludge production rate (d^{-1}); TSA is the traditional sludge age computed by the steady-state calculation (d); DSA is the dynamic sludge age (d); and U is the substrate utilization rate (d^{-1}). $R^2 = 0.756$

The short-term model predicts effluent total suspended solids (X_e) at one-minute intervals. It is based on an average TSS, $Xe,avg = 3.7$ g TSS/m^3. Then X_e, was then converted to g VSS/m^3 by an assumed factor of 0.80.

$$\log_{10}\left(\frac{Xe}{Xe,avg}\right) = \frac{-0.20912 \cdot q_{r,t}^3}{q_t} + 5.7363 \cdot e$$

$$-2q_{t-10}^3 \cdot x_{r,t-30}^3 + 0.1063 \cdot q_{r,t}^3 \cdot q_{r,t-10}^3 + a_t$$

[23]

where $q = Q/Q_{avg}$; $q_r = Q_r/Q_{r,avg}$; $x_{r,t} = X_{r,t}/X_{r,t,avg}$;   and $\sigma = 0.1064$

Summary

The hybrid model described here describes a much wider range of behaviors of the activated sludge process than previous models are designed to do. A model of this type is necessary to simulate the activated sludge process with enough realism for use in development and testing of practical control schemes. Previous models in the literature have not been comprehensive enough for this purpose. Most importantly, little research has addressed the prediction of effluent suspended solids, which are the major cause of activated sludge process failure. Nonlinear stochastic models show promise for filling the gap in predicting phenomena which cannot be modeled fundamentally.

References

1.   Rivera, S.L., D.A. Vaccari, D. McMahon, AIChE Annual Meeting, Miami, FL (November, 1995).

2.   Henze, M., C.P.L. Grady, W. Gujer, G.v.R. Marais, T. Matsuo, IAWQ, London, UK (1986).

3.   Christodoulatos, C., D.A. Vaccari, *Water Res.*, v27, n1, pp51-62 (1993).

4.   Wahlberg, E.J. and T.M. Keinath, *JWPCF* **60**, 12, 2095-2100 (1988).

5.   Vaccari, D.A., and C.G. Uchrin, *J. Environ. Sci. Health*, A24(6), 645-674 (1989).

6.   Kabouris, J.C. and A.P. Georgakakos, *Water Research*, v24, n10, pp1197-1208 (1990).

7.   Busby, J.B. and J.F. Andrews, *JWPCF*, v47, n5, pp1055-1080 (1975).

8.   Vaccari, D.A., A. Cooper, and C. Christodoulatos, *JWPCF*, v60, n11, p 1979-1985, (1988).

9.   Vaccari, D.A., and C. Christodoulatos, IAWPRC Conf. on Instr. and Control, Yokohama, Japan (August 1990).

10.   Kos, P. Ph.D. Thesis, U. of Massachusetts (1968).

11.   Cacossa, K. and D.A. Vaccari, *Water Science and Technology*, (1995).

12.   Box, G.E.P. and G.M. Jenkins, "Time Series Analysis, Forecasting and Control", (Holden-Day, 1976).

13.   Bertheouex, Paul M., *et al*, JEED ASCE, pp127-138 (1975).

14.   Chen, S. and Billings, S.A., *Int. J. Control*, v49, n3, 1013-1032 (1989).

15.   Draper, N.R. and H. Smith, "Applied Regression Analysis", John Wiley Pub. Co., New York (1966).

16.   Vaccari, D.A., and C. Christodoulatos, *Instr. Soc. of Am. Transactions*, v31, n1, pp97-102 (1992).

17.   Wallace, A.T., and Zollman, D.M., *Journal of the San. Eng. Div.*, ASCE v97, SA3, Proc. Paper 8167, (1971)

18.   Papoulis, A., "Prob., Random Variables, and Stochastic Processes", McGraw-Hill, Inc., New York (1965).

19.   Hill, Robert D., Ph.D. Thesis, Rice University, Houston, Texas (1985).

20.   Vaccari, D.A. and E. Wojciechowski, Minn., MN (June 25-28, 1995).

A numerical model of hydrocarbon contaminant flow in soil and aquifers

Ph. Tardy,[a,b] M. Quintard,[a] P. Le Thiez,[b] E. Seva[c]

[a]L.E.P.T.-ENSAM (URA-CNRS), Esplanade des Arts et Métiers, 33405 Talence, France
[b]Institut Français du Pétrole, 1&4 Avenue de Bois Préau, BP 311, 92500 Rueil-Malmaison, France
[c]S.N.A.M., P.O.Box 12060, Milano 20120, Italy

Abstract

We present a numerical model which can be used to design remediation strategies for cleaning up soils and aquifers in both the saturated and unsaturated zones. This three-phase (air, water and pollutants) flow model takes into account the advective and the dispersive mass-transport, and includes local non-equilibrium mass transfer.

The simulator is based on a finite-volume method. Several innovative techniques have been introduced to improve the efficiency and accuracy of the numerical model. In particular, for each time step the splitting method is used to solve the equations in two stages: (i) the first stage deals with transport and dispersion, (ii) the second stage deals with the mass exchange between the phases. The validity of the model has been tested against experimental studies in the following two cases

    1. A two-dimensional physical model representing the flow in an aquifer with pollutant injection,

    2. A history matching of dissolved hydrocarbon in core samples.

The results of the one- and two-dimensional simulations agree very well with the experimental data.

1. Introduction

Hydrocarbon infiltration is among the most important causes of soil and aquifer contamination. Such infiltration can result from the spill of hydrocarbon during the production, transport, and storage processes, and can damage the groundwater quality for long periods of time. Once in the soil, chemicals from the NAPL (non-aqueous phase liquid) can dissolve in the groundwater with subsequent convective and dispersive contamination in the aquifer. Transport of the contaminant in the water in the presence of the dissolving NAPL will be referred to as active dispersion. Most of the transport models (Abriola & Pinder [1], Corapcioglu & Baehr [6]) assume local mass equilibrium between water and NAPL. This assumption is supported by some laboratory experimental studies (Hunt et al. [11]), while the non local equilibrium is supported by some other researchers (Geller & Hunt [9], Le Thiez & Ducreux [12]).

For a comprehensive discussion of active dispersion models, i.e., derivation of the macroscopic equations and calculations of the transport properties from the pore-scale transport problem, we refer the reader to the study developed by Quintard & Whitaker [13]. Our objective in this paper is the implementation of a discretized model solving the three-phase flow equations with or without an assumption of local equilibrium between water and NAPL. The technique is based on the finite volume method and uses a two-stage splitting. The first stage deals with the transport and dispersion. It is solved with high order schemes for systems of conservation laws. The second stage solves a system of differential equations modeling the NAPL dissolution into the water phase. As the different characteristic times associated with interphase transfer, advection and dispersion are very different, we have to be careful that the numerical scheme is not limited by too small time steps. Borden & Piwoni [5] have developed numerical models for two-phase flows with or without local-equilibrium condition, and they compared their performed a comparison with experimental data. They found numerical difficulties due to small time steps in the non-equilibrium case, which are related to the fact that the mass transfer model is a stiff problem. We payed much attention to this problem, and we propose in Section 4 to solve the second stage with an unconditionally stable scheme. Hence, it is possible to increase the speed of the simulations.

The validity of this model has been tested against many experimental studies performed at the Institut Français du Pétrole. Results of two of these comparisons are presented in Section 5. They confirm that the physical assumptions, as well as the mathematical and numerical approaches used, are sufficiently accurate to correctly represent the behavior of realistic systems.

2. Mathematical model

In the following, we use subscripts η and λ = o,w or g to represent the oil-phase, the NAPL or the gaz-phase.

Compositional three-phase flows are mathematically modeled by postulating the mass conservation of each of the n_c chemical species present in the soil.

$$\partial_t (\phi \rho_\eta s_\eta c_\eta^i) + \text{div}(\rho_\eta c_\eta^i \mathbf{V}_\eta) + \text{div}(\mathbf{D}_\eta^i . \nabla c_\eta^i) = \sum_{\lambda \neq \eta} T_{\eta\lambda}^i \tag{1}$$

Filtration velocities are given by Darcy's equation of the form

$$\mathbf{V}_\eta = -\mathbf{K} \cdot \frac{kr_\eta}{\mu_\eta} (\nabla p_\eta - \rho_\eta g) \quad \text{with} \quad \begin{cases} p_w = p_g - p_{c_{gw}} (s_w, s_g) \\ p_o = p_w + p_{c_{ow}} (s_w, s_o) \end{cases} \tag{2}$$

We assume that the air phase is compressible. If there is local non-equilibrium, we represent the mass transfer with a first order kinetic, and a linear interphase equilibrium relation.

$$T_{\eta\lambda}^i = -\alpha_{\eta\lambda}^i (c_\eta^i - k_{\eta\lambda}^i c_\lambda^i) \tag{3}$$

where

$$\alpha_{\eta\lambda}^i = \alpha_{\eta\lambda}^i (s_\eta, s_\lambda, V_\eta, V_\lambda) \tag{4}$$

The local equilibrium condition is equivalent to

$$\alpha_{\eta\lambda}^i = +\infty \quad c_\lambda^i = k_\lambda^i c_\eta^i \tag{5}$$

We only deal with the NAPL dissolution and ignore volatilization. In the case of motionless gas-phase, it is proven in [7] that system (1) admits a weak solution under usual assumptions.

3. Splitting

3.1. Choice of the principal variables

To describe the flow, we seek a phase pressure, two saturations and $2n_c$ mass fractions. However, since

$$\sum_{i=1,n_c} c_\eta^i = 1 \tag{6}$$

we only need $2n_c-2$ mass fractions. As it is not very convenient to define a mass fraction in a phase whose saturation is none, we prefer to consider the volume ratio

$$u_\eta^i = c_\eta^i s_\eta \tag{7}$$

as the unknown function. The system of Equations (1) is re-written with these functions. In [16], it is shown that it is not necessary to define concentrations in a phase which is absent, i.e., at zero saturation.

In order to use numerical methods appropriate to each driving phenomenon, we split the system in two stages, as it is shown in the following sections.

3.2. First stage : transport and dispersion

We deal first with the transport and dispersion of the various components at time t. After simple linear combinations, we find :

$$\begin{cases} \partial_t(\phi\rho_w s_w) + \mathrm{div}(\rho_w \mathbf{V}_w) = 0 \\ \partial_t(\phi\rho_o s_o) + \mathrm{div}(\rho_o \mathbf{V}_o) = 0 \\ \partial_t(\phi\rho_g s_g) + \mathrm{div}(\rho_g \mathbf{V}_g) = 0 \end{cases} \tag{8}$$

$$\begin{cases} \partial_t(\phi\rho_w u_w^i) + \mathrm{div}(\rho_w \frac{u_w^i}{s_w} \mathbf{V}_w) + \mathrm{div}(\mathbf{D}_w^i \cdot \nabla \frac{u_w^i}{s_w}) = 0 \\ \partial_t(\phi\rho_o u_o^i) + \mathrm{div}(\rho_o \frac{u_o^i}{s_o} \mathbf{V}_o) + \mathrm{div}(\mathbf{D}_o^i \cdot \nabla \frac{u_o^i}{s_o}) = 0 \end{cases} \quad i = 1, n_c - 1 \tag{9}$$

with initial data at time t=0. The system consisting of Equations (8) has already been studied theoretically in the case of incompressible flows [8], and it was proven that a weak solution exists under reasonable assumptions.

3.3. Second stage : mass transfer

After Stage 1, we consider the transfer of each component between oil and water at the same time t :

$$
\begin{cases}
\partial_t(\phi\rho_w u_w^i) = T_{wo}^i \\
\partial_t(\phi\rho_o u_o^i) = T_{ow}^i = -T_{wo}^i
\end{cases}
\tag{10}
$$

The initial data used for the system (10) are the solutions of (8)-(9) calculated at time t. A theoretical study of this system of ordinary differential equation is given in [16]. It gives qualitative behaviors of the solution according to the regularity of α. In particular, it is proven that, if oil saturation is small enough and if water is not in equilibrium, NAPL can vanish in finite time when oil exists as residual films in the porous medium. In all other cases concentrations and saturations have asymptotic behavior and converge to the equilibrium state.

4. Discretization

4.1. First stage

We give here the time-discretization of Eqs. (8) through (9) at time t^n on cell mesh number j. (More complete discretization is described in [16]).

$$
\begin{cases}
\dfrac{\left(\phi\rho_w s_w\right)_j^{n+\frac{1}{2}} - \left(\phi\rho_w s_w\right)_j^n}{\delta t^n} + \operatorname{div}(\rho_w \mathbf{V}_w)_j^{n,n+\frac{1}{2}} = 0 \\[2ex]
\dfrac{\left(\phi\rho_o s_o\right)_j^{n+\frac{1}{2}} - \left(\phi\rho_o s_o\right)_j^n}{\delta t^n} + \operatorname{div}(\rho_o \mathbf{V}_o)_j^{n,n+\frac{1}{2}} = 0 \\[2ex]
\dfrac{\left(\phi\rho_g s_g\right)_j^{n+\frac{1}{2}} - \left(\phi\rho_g s_g\right)_j^n}{\delta t^n} + \operatorname{div}(\rho_g \mathbf{V}_g)_j^{n,n+\frac{1}{2}} = 0
\end{cases}
\tag{11}
$$

$$
\begin{cases}
\dfrac{\left(\phi\rho_w u_w^i\right)_j^{n+\frac{1}{2}} - \left(\phi\rho_w u_w^i\right)_j^n}{\delta t^n} + \operatorname{div}(\rho_w \tfrac{u_w^i}{s_w}\mathbf{V}_w)_j^{n,n+\frac{1}{2}} = -\operatorname{div}(\mathbf{D}_w^i \cdot \nabla \tfrac{u_w^i}{s_w})_j^{n,n+\frac{1}{2}} \\[2ex]
\dfrac{\left(\phi\rho_w u_w^i\right)_j^{n+\frac{1}{2}} - \left(\phi\rho_w u_w^i\right)_j^n}{\delta t^n} + \operatorname{div}(\rho_o \tfrac{u_o^i}{s_o}\mathbf{V}_o)_j^{n,n+\frac{1}{2}} = -\operatorname{div}(\mathbf{D}_o^i \cdot \nabla \tfrac{u_o^i}{s_o})_j^{n,n+\frac{1}{2}}
\end{cases}
\tag{12}
$$

Here, we note f_j an average of the function f on the cell number j. We also note $f_{j+1/2}$ an evaluation of f at the interface between cells number j and j+1. The superscripts denote the discrete times t^n, $t^{n+1/2}$ used for the evaluations.

The discretisation of Eqs. (8)-(9) has been widely investigated in reservoir engineering, for example by Allen [3], Aziz & Settari [4]. Here, we use first- and second-order finite-volume schemes to solve the saturations and the gas pressure in (11). The general framework is based on Roe's scheme for conservation systems [14] with correction of the numerical diffusion by Harten's technique [10] and flux limiting process [15]. Then, we use upstream schemes for linear advection terms and centered schemes for diffusion terms in (12). Figure 1 illustrates the low sensitivity of Harten's high-order scheme (on the left) to the grid orientation. This figure corresponds to oil injected in a large horizontal square saturated with 50% water and 50% gas. In this case, isosaturations should be circular. This feature is correctly obtained with Harten's scheme (Figure on the left), while the solution simulated with the well-known first-order upstream scheme (Figure on the right) shows a significant grid orientation effect.

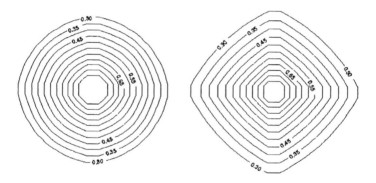

Figure 1 : Sensitivity to grid orientation effect.

4.2. Second stage

To avoid time step limitations, we use a backward Euler-scheme to solve Eqs. (10), which is written as

$$
\begin{cases}
\text{for } i = 1, n_c - 1 \\[2mm]
\dfrac{\left(\phi\rho_w u_w^i\right)_j^{n+1} - \left(\phi\rho_w u_w^i\right)_j^{n+\frac{1}{2}}}{\delta t^n} = \left(T_{wo}^i\right)_j^{n+1} = -\alpha_{wo}^i{}^{n+1}\left(\dfrac{u_w^i}{s_w} - k_{wo}^i\dfrac{u_o^i}{s_o}\right)_j^{n+1} \\[4mm]
\dfrac{\left(\phi\rho_o u_o^i\right)_j^{n+1} - \left(\phi\rho_o u_o^i\right)_j^{n+\frac{1}{2}}}{\delta t^n} = \left(T_{ow}^i\right)_j^{n+1} = -\left(T_{wo}^i\right)_j^{n+1}
\end{cases}
\tag{13}
$$

It is proven in [16] that with this technique the numerical solution remains close to the exact solution. Moreover, the numerical solution is stable for any choice of the time step δt^n.

Results with backward and forward Euler-schemes are shown in Figure 2. This figure illustrates the very strong limitation on time steps when using a forward Euler-scheme. On the contrary, good behavior and unconditional stability is obtained with the proposed scheme. In fact, we can verify that the solution of (11)-(13) verifies the following equations

$$
\begin{cases}
\dfrac{\left(\phi \rho_\eta s_\eta c_\eta^i\right)_j^{n+1} - \left(\phi \rho_\eta s_\eta c_\eta^i\right)_j^n}{\delta t^n} + \operatorname{div}(\rho_\eta c_\eta^i \mathbf{V}_\eta)_j^{n,n+1} + \operatorname{div}(\mathbf{D}_\eta^i \cdot \nabla c_\eta^i)_j^n = \sum_{\lambda \neq \eta}\left(T_{\eta\lambda}^i\right)_j^{n+1} \\[4mm]
\dfrac{\left(\phi \rho_g s_g\right)_j^{n+1} - \left(\phi \rho_g s_g\right)_j^n}{\delta t^n} + \operatorname{div}(\rho_g \mathbf{V}_g)_j^{n,n+1} = 0
\end{cases}
\tag{14}
$$

System (14) is the direct discretization of (1) without splitting and using implicit transfer and IMPES (IMplicit Pressure, Explicit Saturations) transport. This proves incidentally that the splitting is consistent with the entire system.

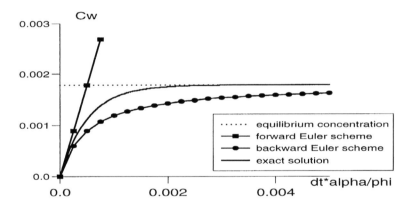

Figure 2 : Loss of stability with an explicit scheme

5. Comparison with experimental data

In order to validate the model, some physical experiments have been performed at the Institut Français du Pétrole. The first were designed to test the ability of the system (11) to simulate two-dimensional non-compositional three-phase flows. A laboratory cell of 2.25 meter long, 0.46 meter high, and 0.10 meter wide was filled with clean sand with 38% porosity, and a permeability of 200 Darcy. A total volume of 1.84 m³ of diesel oil (viscosity 4 mPa.s) was injected over a 51 hour period; the injections were made during the following periods: the first 7 hours, 24 to 31 hours, and 48 to 51 hours. Water was injected at the left bottom and produced at the right bottom to simulate the aquifer flow. The visual monitoring of the oil migration was made possible since the physical model was built in Plexiglas.

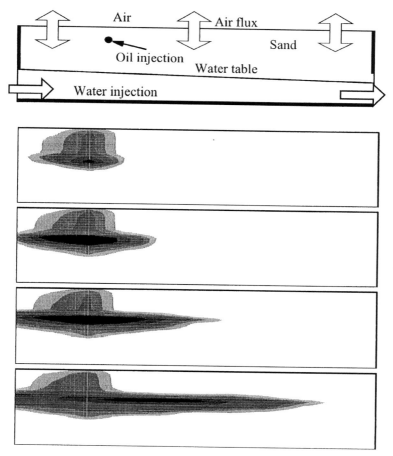

Figure 3 : Oil saturations at t = 3h, 7h, 1d 7h, 3d (d=day, h=hour)

	above	0.400
	0.350	0.400
	0.250	0.350
	0.150	0.250
	0.100	0.150
	0.050	0.100
	below	0.050

The experiment setup is described in Figure 3 with the numerical results.

We obtain a very good agreement between the experiment and the numerical solution, especially for the following items:

- The spreading of oil in the unsaturated zone under the injection point,
- the accumulation of oil in the capillary fringe above the water table,
- the spreading in the capillary fringe in the direction of the aquifer gradient,
- the final shape of the contaminated zone after 3 days.

The second experiment deals with the validation of the non-equilibrium assumption between oil and water, and the capability of the model (11)-(13) to accurately and quickly simulate such compositional water-oil flows. One-dimensional physical experiments were performed on different core samples. First, they were saturated with oil containing benzene and toluene. Then, water was injected at one side. At the other side, mass fractions of dissolved benzene and toluene were recorded. Results are plotted in Figure 4 together with simulations corresponding to the local equilibrium model. These simulations clearly show that the equilibrium assumption is not valid for this experiment. On the other hand, by fitting the relation between α and the saturation with a polynomial such as those proposed in [13], it was possible to accurately simulate the history of the dissolved hydrocarbons. In addition, we found that if it is necessary to consider that oil and water are in equilibrium, the local non-equilibrium model can simulate satisfactorily equilibrium conditions provided α is taken as a large constant. In that case dissolution is instantaneous, and Figure 5 shows that it is possible to simulate equilibrium as expected. The interesting thing is that with the proposed algorithm computational costs are not prohibitive.

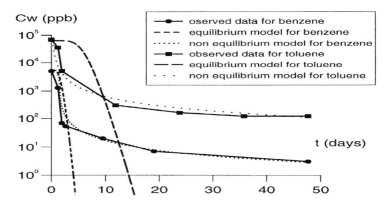

Figure 4 : History matching of dissolved hydrocarbon

6. Conclusion

It seems that for many real cases, the dissolution of hydrocarbons into water invalidate the assumption of local mass equilibrium. This is mainly due to the heterogeneity of the porous medium, or high filtration velocities. It might happen in a real aquifer that different zones exist in which equilibrium conditions are not the same. The example of the benzene and toluene dissolution in the sample cores presented in this paper shows that the aquifer contamination may take longer than predicted by a local-equilibrium model. Thus, it is important to design good numerical models that can work with non-equilibrium conditions. The proposed model allowed us to simulate accurately multiphase flow and dissolution with or without equilibrium. Computational costs were maintained acceptable by a proper choice of the numerical schemes. The validity of the simulations clearly relies on an accurate determination of the effective properties, in particular the mass transfer terms. These quantities can be approached from a change of scale point of view as emphasized in [13].

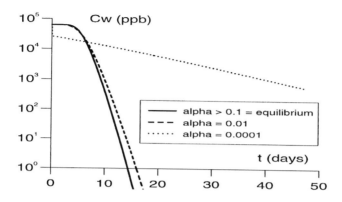

Figure 5 : Simulation of equilibrium with kinetic models

Nomenclature

In the following, letters η and λ denote the oil, water or gas phase.

Greek Letters

$\alpha^i_{\eta\lambda}$ mass exchange coefficient for species i between the η-phase and the λ-phase, kg m^{-3} s^{-1},

ϕ porosity,

ρ_η density of the η-phase, kg m^{-3},

μ_η viscosity of the η-phase, Pa s,

Roman Letters

c^i_η mass fraction of species i in the η-phase,

dt^n time step = $t^{n+1} - t^n$, s,

D^i_η dispersion tensor of species i in the η-phase, m^{5}s^{-1}kg^{-1},

$\mathbf{g}$ gravity vector,

k^i_η equilibrium concentration for species i in the η-phase,

$\mathbf{K}$ permeability tensor of the porous medium, m^2,

kr_η relative permeability of the η-phase,

n_c number of species present in the three phases,

p_η η-phase pressure, N m^{-2},

$p_{c\eta\lambda}$ capillary pressure = $p_\eta - p_\lambda$,

s_η saturation of the η-phase,

$T^i_{\eta\lambda}$ mass transfer for i between the η- and the λ-phase, kg m^{-3} s^{-1},

u^i_η volume ratio of species i in the η-phase,

$\mathbf{V}_\eta$ filtration velocity of the η-phase, m s^{-1},

References

[1] **Abriola L.** and **Pinder G.F.** (1985) *"A multiphase approach to the modeling of porous media contaminated by organic compounds, 1, Equation development"*, Water Resources Research, 21, pp. 19-26

[2] **Aigueperse A.** and **Quintard M.** (1994) *"Macroscopic models for active dispersion in aquifers contaminated by non-aqueous phase liquids"*, in Computer Techniques in Environmental Studies V Vol. 1: Pollution Modelling, ed. by P. Zanetti, Computational Mechanics Publications, Boston, 157-164.

[3]**Allen M.B.** (1987) *"Numerical modeling of multiphase flow in porous media"*, Advances in Transport Phenomena in Porous Media, pp. 849-920.

[4] **Aziz K.** and **Settari A.** (1979) *"Petroleum reservoir simulation"*, Applied Science Publisher Ltd., London.

[5] **Borden R.C.** and **Piwoni M.D.** (1992) *"Hydrocarbon dissolution and transport: a comparison of equilibrium and kinetic models"*, Journal of Contaminant Hydrology", 10, pp. 309-323.

[6] **Corapciaglu M.Y.** and **Baehr A.L.** (1987) *"A compositional multiphase model for groundwater contamination by petroleum products 1: Theoretical considerations"*, Water Resources Research, 23, pp. 191-200.

[7] **Fabrie P., Le Thiez P.** and **Tardy Ph.** (1996) *"On a system of nonlinear elliptic and parabolic degenerate parabolic equations describing compositional water-oil flows in porous media"*, in press, NonLin. Analysis.

[8] **Fabrie P.** and **Saad M.** (1993) *"Existence de solutions faibles pour un modèle d'écoulements triphasiques"*, Annales de la Faculté des Sciences de Toulouse, 2, N.3, pp. 337-371.

[9] **Geller J.T.** and **Hunt J.R.** (1993) "Mass transfer from non-aqueous phase organic liquids in water-saturated porous media", Water Resou. Res., 29, pp. 833-845.

[10] **Harten A.** (1983) *"High resolution schemes for hyperbolic conservation laws"*, Journal of Computational Physics, 49, pp. 357-393.

[11] **Hunt J.R., Sitar N.** and **Udell K.S.** (1988) "Nonaqueous phase transport and cleanup 1: Analysis of mechanisms", Water Resources Research, 24, pp. 1247-1258.

[12] **Le Thiez P.** and **Ducreux J.** (1994) *"A 3-D numerical model for analyzing hydrocarbon migration into soils and aquifers"*, Proceeding of the Eighth International Conference on Computer Methods and Advances in Geomechanics (Morgantown), Balkema, Rotterdam.

[13] **Quintard M.** and **Whitaker S.** (1994) *"Convection, dispersion and interfacial transport of contaminant: homogeneous porous media"*, Advances in Water Resources, 17, pp 221-239.

[14] **Roe P.L.** (1981) *"Approximate Riemann solvers, parameter vectors, and difference schemes"*, Journal of Computational Physics, 43, pp. 357-372.

[15] **Sweeby P.K.** (1984) *"High resolution schemes using flux limiters for hyperbolic conservation laws"*, SIAM Journal on Numerical Analysis, 21, pp. 995-1011.

[16] **Tardy Ph.** (1995) *"Modélisation de la pollution des sols et des nappes par des hydrocarbures"*, Ph.D. thesis, Université Bordeaux I

An industrial 3-D numerical model for simulating organic pollutants migration and site remediation

P. Le Thiez,[a] J.M. Côme,[b] E. Seva,[c] G. Pottecher[d]

[a]Institut Français du Pétrole, 1&4 Avenue de Bois Préau, BP 311, 92500 Rueil-Malmaison, France

[b]Burgeap, 27 Rue de Vanves, 92100 Boulogne, France

[c]S.N.A.M., P.O.Box 12060, Milano 20120, Italy

[d]Anjou Recherche, 13 Villa de la Croix Nivert, 75015 Paris, France

Abstract

This paper presents a general numerical model able to simulate both organic pollutants migration (3-phase compositional flows, mass transfer, transport) in soils and aquifers and decontamination techniques such as pumping, in situ stripping, venting, hot venting, steam injection, surfactant injection and biodegradation.

To validate the simulator, a 3-D experiment in a large pilot was carried out. A total of 475 liters of diesel oil was injected in the pilot and numerous in-situ measurements were performed to determine pollutants location and concentrations within the vadose and saturated zones. Prior to the pilot test, a predictive run computed the extent of the contaminated zone and the oil saturations. Numerical results showed good agreement between experiment and simulation.

To demonstrate the simulator abilities in modeling field cases, two examples are given with comparison with on-site observations: (1) soluble hydrocarbon transfer and transport at regional scale; (2) free-oil pumping.

1 Introduction

Among the hazards for soil and aquifer pollution, the infiltration of hydrocarbons and chlorinated solvents represents one of the most serious threats. This infiltration resulting from accidental spills during production, transport or storage of these compounds can deteriorate underground water quality for long periods of time. Contamination by a nonaqueous phase liquid (NAPL) remains a complex problem. Understanding the evolution of NAPL in soils and aquifers requires knowledge of various phenomena:

- vertical migration of NAPL from surface to water table, then spreading in the capillary fringe and creating a contaminated zone at residual oil saturation;
- evaporation of volatile compounds creating a gazeous envelope of oil vapor;

- dissolution and transport of soluble compounds in water phase leading to a "plume" of contaminated water in the aquifer;
- interactions between pollutants and the porous matrix (adsorption);
- natural biodegradation of both dissolved pollutants and residual NAPL trapped in the pores.

Good comprehension of the above mechanisms can be obtained with the help of numerical simulators as reviewed by Pinder and Abriola [1], Corapcioglu and Panday [2] and Panday et al. [3].

Numerical models aim also at simulating decontamination processes such as pumping, air stripping, venting, steam injection, biodegradation, surfactant injection, ... While these various methods have been studied at the laboratory scale and have been used in field conditions, the design of an operation is often empirical. Use of a comprehensive simulator may significantly increase our knowledge about the complex phenomena associated with the above mentioned processes and our capability in optimizing such reclamation operations.

The purpose of the present study is to describe a general simulator, SIMUSCOPP, able to simulate most of the organic pollutants (hydrocarbons, chlorinated solvents) behavior and remediation processes discussed above. Then, 3 cases are presented to compare computed results and observations: (1) migration of diesel oil into a large-scale basin, (2) description of an accidental spill of gasoline 30 years ago, transfer of soluble products within the aquifer and transport at regional scale, and (3) site remediation by free-gasoline pumping.

2 Model description

The SIMUSCOPP model is a general 3-D simulator, dedicated to pollution/re-mediation problems, able to simulate most of the phenomena and remediation techniques listed above, including biodegradation, steam injection and surfactant injection. It solves problems of 1, 2 or 3-phase flow, compositional or not, from the laboratory scale to the site scale. Geometry can be described by cartesian coordinates or radial coordinates, with possible local mesh refinement. Its code is portable between most common computers: PC, Unix workstations and also vector computers.

The mathematical description of compositional multiphase flow in porous media is based on equation of mass conservation for each component and momentum conservation for each phase. The general mass conservation equation for a component i existing in the three fluid phases (p) and adsorbed on the matrix (r) is:

$$\frac{\partial}{\partial t}\left[\phi\left(\sum_{p=1}^{N_p}\rho_p \cdot S_p \cdot x_i^p\right)+(1-\phi)\cdot\rho_r \cdot q_i^r\right]+Q_i+R_i$$

$$+\mathrm{div}\left(\sum_{p=1}^{N_p}\rho_p \cdot x_i^p \cdot \alpha_i^p \cdot \vec{u}_p + \vec{J}_i^p\right)=0 \qquad (1)$$

In this equation the flux modifiers, α_i^p, have been introduced to limit transfer between phases. They have been tested in a previous study to describe nonequilibrium dissolution of hydrocarbon in water [4].

The momentum equation for each fluid phase is reduced to Darcy's generalized law.

In order to close the system of equations, constitutive equations must be written for phase densities and viscosities, equilibrium partition constants, molecular diffusion and hydrodynamic dispersion, adsorption and biodegradation rates for both dissolved and residual organic products. These various relationships are not described in the present paper.

3. Diesel oil migration into a large-scale basin

The simulation results discussed here were performed by simulating 3-D diesel oil infiltration on a pilot scale. This pilot (SCERES basin), located at the CNRS-Campus of Strasbourg (France), is a large impervious concrete basin (25 m long x 12 m wide x 4 m deep) packed with quartz sand. The porosity of the sand is 40%, and its hydraulic conductivity is 8.10^{-4} m/s (intrinsic permeability = 80 darcy). This pilot was build to test some decontamination processes such as surfactant-aided recovery and in-situ bioremediation [5,6]. During the summer of 1994, the basin was contaminated with 475 liters of diesel oil through a limited area of 1.8 m^2, 3 m above the water table. After a few months of stabilization, core-drill investigations were performed in order to determine the shape and extent of the residual contaminated zone and pollutant concentrations (i.e., oil-phase saturations) in the various parts of this contaminated zone. The diameter of the oil-impregnation body in the unsaturated zone was about 1.85 m. The horizontal extent of the oil phase in the capillary fringe was in the range of 5.60 m to 5.90 m, with a mean thickness of 10 to 15 cm. Residual oil saturations were close to 5% of pore volume in the unsaturated zone and in the range of 21 to 24% in the capillary fringe [7].

In a second stage, an elevation of 0.5 m of the water table was performed in order to vertically displace the oil phase initially trapped in the capillary fringe. New measurements gave a value of 10% for the residual oil saturation in the saturated zone after the rise of the water table.

The purpose of this numerical modeling was to reproduce the migration of diesel oil in the SCERES basin and the shape and final extent of the contamination. For this, we performed runs with three-phase non-compositional flow. Only three components were taken into account: oil, water and air, without any mass transfer. The SCERES basin was discretized with 12000 grid-blocks (25 x 24 x 20). Three-phase relative permeabilities for the oil, water and air phases were obtained by laboratory experiments. The water-air capillary pressure curve was obtained by inversion of the vertical water saturation profile measured in the basin by a neutron probe before the oil spill. Water-oil and air-oil capillary pressure curves were then obtained by scaling the water-air curve according to their respective interfacial tensions. The main data used for the runs are given in Table 1. Four boundary conditions were used for this run:

1. aquifer influx: pressure in the water phase was fixed at 1 atmosphere at the top of the water level, and water was injected between -3 and -4 m;
2. aquifer outflux: pressure was fixed at 1 atmosphere at -3.125 m, corresponding to the slope of the water table (5‰),and water was produced between -3.125 and -4 m;
3. pollution infiltration: area at the top of the model ($z = 0$) where diesel oil was injected at a constant rate of 9 liters per minute (475 liters injected);

4. surface of the model: pressure in the gas phase was fixed at 1 atmosphere, allowing air to get out of the system when oil was infiltrating.

oil density	= 838 kg/m^3
oil viscosity	= 3.91 mPa.s
irrdeducible water saturation	= 0.2
residual oil saturation (air-oil flow)	= 0.05
residual oil saturation (water-oil flow)	= 0.1
max. air-water capillary pressure	= 5000 Pa
max. water-oil capillary pressure	= 1700 Pa
max. air-oil capillary pressure	= 1200 Pa

Table 1. - Data for SCERES basin simulation

Figure 1 gives the shape of the contaminated zone and the oil saturations on a cross-section for two times: at the end of the diesel-oil injection (1 hour), and 2 months later, and an areal view at the top of the water table after 2 months. As shown in Fig. 1, the horizontal extent of the oil lens calculated by the simulator is close to 5.50 m, which is in very good agreement with the value observed. Regarding the oil spreading in the unsaturated zone, the calculated extent of the contamination is greater than the observed one. As shown in Fig. 1, The main cylinder of residual oil centered under the injection zone has a diameter of 2 m, which is in good agreement with the observations. Oil saturation in this zone is 5%. Around this main cylinder, the numerical model computed very low oil saturations, which have not been observed in the pilot. This is probably due to numerical dispersion in this zone or to excessive air-oil capillary forces leading to a lateral spreading of the contamination zone. Fig. 2 gives the vertical profile of oil saturations under the surface of pollution injection after 1 hour (end of injection), after 2 months and after the 0.5 m rise of the water table. As shown in Fig. 2, injected diesel-oil migrated vertically through the unsaturated zone and spread within the water-air capillary fringe. A vertical equilibrium was then reached, where the maximum of oil saturation was 28% (an average value of 24% was measured in the basin). The rise of the water table displaced this oil lens vertically and trapped the pollutant at its residual saturation of 10%. Calculated saturation profiles before and after the rise of the water table effectively reproduce the observations of the pilot.

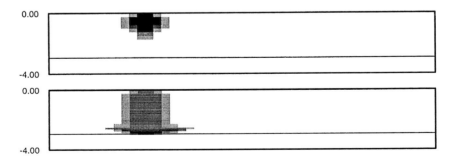

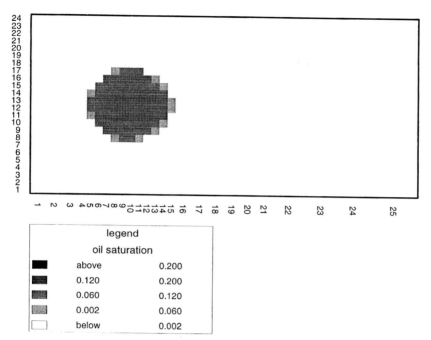

Figure 1: Oil saturations in the SCERES basin

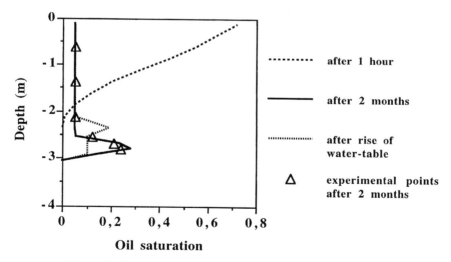

Oil saturation

Figure 2: Computed oil saturations in the SCERES basin

4. Site 1: transfer and transport at regional scale

The simulation results given here concern a site of hydrocarbon storage contaminated by gasoline. The spill of gasoline occurred in 1963 and the volume of infiltration was not known. A diagnosis of the site was carried out in 1994 to locate the residual contamination under this storage area and to measure the concentration of dissolved hydrocarbon present in various pumping wells located near the storage area. A first run was performed to calibrate the piezometry on the aquifer, taking into account bounding rivers, rain and water-pumping wells. Then, knowing the regional flow and more precisely the direction of the aquifer flow under the storage area, a second set of runs was done in order to describe the spill of gasoline. Several scenarios were built with spill volumes ranging between 1 m^3 to 100 m^3 of gasoline. The comparison between simulations and on-site observations demonstrated that the spill that occurred in 1963 could be close to 15 m^3. In this latter case the computed extent of the residual contamination was compatible with the observed one. Moreover, analyses performed in 1994 on the pollutants showed clearly that the initial product have been slightly modified during 30 years: minor dissolution of aromatics and no significant biodegradation and vaporization. Knowing approximately the volume of gasoline infiltrated, a third set of runs were carried out to describe the dissolution of aromatics and their transport at large scale in the aquifer. Measurements carried out in 1994 showed that the water under the storage area contained 15 ppm of BTEX (benzene, toluene, ethylbenzene, xylene). In an observation well located 2 km downstream the storage, hydrocarbon concentrations in water were between 20 to 50 ppb.

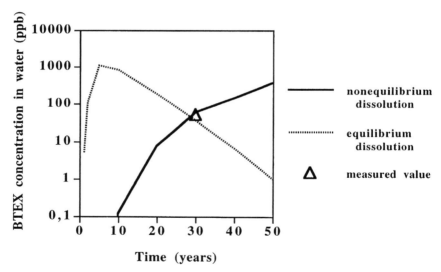

Figure 3: Computed BTEX concentrations in the aquifer under the storage

A first calculation assumed thermodynamical equilibrium for dissolution. As shown in Fig. 3, the computed BTEX concentration in the observation well after 30 years is in good agreement with the value observed, but concentrations under the storage (Fig. 4) are close to zero, which is not compatible with the 15 ppm observed in 1994. In order to limit the transfer of hydrocarbon from the residual oil to the aquifer, a second run was carried out, using flow limitors which allow to reproduce nonequilibrium dissolution effects [4]. A calibration run was then performed to reproduce both the concentrations under the storage and in the observation well, as shown in Figs. 3 and 4. Then, the calculation was extended in order to forecast the evolution of contaminant transport in the aquifer for the next 20 years. Simulation results obtained with the nonequilibrium assumption show clearly that BTEX concentrations in water are still increasing and that the aquifer will be contaminated for long time.

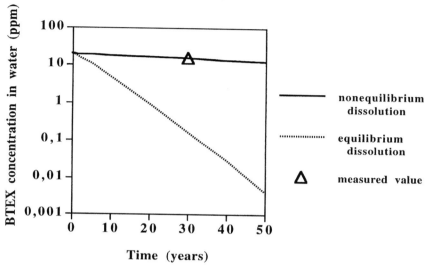

Figure 4: Computed BTEX concentrations in the aquifer under the storage

These results show the importance of taking into account nonequilibrium dissolution of hydrocarbon in simulations to ensure predictive calculations of pollutants propagation within the aquifer. The dissolution rates can be obtained by history matching and calibration of the model. After that phase, long-term forecasts can be handled.

5 Site 2: remediation by free-oil pumping

This last case consists of a site located in San Diego's area (California) where a layer of free-phase hydrocarbon was discovered in the groundwater, 15 to 18 feet below the ground surface. A pilot was then implemented on the site in order to recover the free-phase hydrocarbon by pumping groundwater together with the oil phase. Simulation discussed here describes a 30-day pumping test of one of the pilot wells. The porous medium consists of a sand with 10^{-3} m/s of hydraulic conductivity, polluted by gasoline. Three-phase relative permeabilities and capillary pressures were obtained via the Van Genuchten parameters measured on sand samples. Simulation were performed with a 2-D (rz) radial grid with 19 layers describing the unsaturated zone, the oil zone above the water table (thickness 2.9 ft) and the aquifer. The well test performed on-site can be summarized as follows: 15 days of water pumping at 1440 US gallon/day, 8 days at 2160 gallon/day and 7 days at 2880 gallon/day. Figs. 5 and 6 show the comparison between observation and calculation for two parameters: oil recovery rate and drawdowns. These results show fair agreement between observed and computed results.

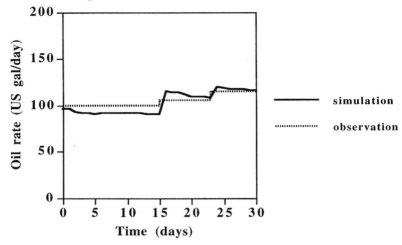

Figure 5: Comparison between observed and computed gasoline recoveries

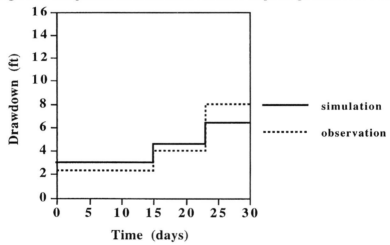

Figure 6: Comparison between observed and computed drawdowns

6 Conclusions

A general numerical model was developed to simulate organic pollutants migration in soils and aquifers, and most in-situ remediation techniques such as pumping, stripping, venting, hot venting, steam injection, surfactant injection and biodegradation. Three cases were simulated here, describing some of the various aspects encountered in pollution/remediation problems: (1) diesel-oil migration at pilot scale by means of three-phase flow in saturated and unsaturated zones, (2) dissolution of hydrocarbon under nonequilibrium conditions and transport in the aquifer, and (3) free-oil recovery by water pumping on a site contaminated by gasoline. A good agreement between simulated results and observations was obtained for the three cases. This study shows that the SIMUSCOPP model can be an efficient and useful tool, capable of handling complex problems of risk assessment and remediation of sites contaminated by organic pollutants.

Nomenclature

$\vec{J}_i^p$ diffusion-dispersion flow of i in phase p, mol/s

P phase pressure, Pa

q_i^r mole fraction of i adsorbed on the solid phase

Q_i component i injection/production source term, mol/s

R_i component i biodegradation source term, mol/s

S_p phase p saturation

t time

$\vec{u}_p$ darcy velocity of phase p, m/s

x_i^p molar fraction of i in phase p

α_i^p flux modifier of i in phase p

ϕ porosity

ρ_p phase p molar density, mol/m^3

Subscripts

g gas phase

i component

o oil phase

p fluid phase (g, o, w)

r rock phase

w water phase

Acknowledgments

This work has been carried out as part of the franco-italian EUREKA project "RESCOPP". Partners for the development of the simulator are Institut Français du Pétrole, SNAM, Burgeap and Italgas. The authors want to acknowledge GRS and IFARE for providing data, and Beicip-Franlab and TEMA for their participation in this modeling project.

References

[1] Pinder, G.F., and Abriola, L.M.:"On the simulation of nonaqueous phase organic compounds in the subsurface", *Water Resour. Res.*, **22(9)**, 109S-119S, 1986.

[2] Corapcioglu, M.Y., and Panday, S.: "Compositional multiphase flow models", *Adv. Porous Media*, **1**, 1-59, 1991.

[3]   Panday, S., Forsyth, P.A., Falta, R.W., Wu, Y., and Huyakorn, P.S.: "Considerations for robust compositional simulations of subsurface nonaqueous phase liquid contamination and remediation", *Water Resour. Res.*, **31(5)**, 1273-1289, 1995.

[4]   Le Thiez, P.A., and Ducreux, J.: "A 3-D numerical model for analyzing hydrocarbon migration into soils and aquifers", *Proceedings of the Eighth International Conference on Computer Methods and Advances in Geomechanics* ,Morgantown, 1984, Balkema, Rotterdam, 1994.

[5]   Ducreux, J., Bavière, M., Seabra, P., Razakarizoa, O., Schäfer, G, and Arnaud, C.: "Surfactant-Aided Recovery / In Situ Bioremediation for Oil-Contaminated Sites", *Proceedings of the International Symposium In-Situ and On-Site Bioreclamation*, San Diego, April 24-27, 1995.

[6]   Ducreux, J., Bavière, M., Bocard, C., Monin, N., Muntzer, P., Razakarizoa, O., and Arnaud, C.: "Key parameters of oil contaminated soil cleanup by in-situ surfactant-aided drainage", *5th Annual Symposium on Groundwater and Soil Remediation*, Toronto, 2-6 Oct., 1995.

[7]   Razakarizoa, O., Muntzer, P., Ott, Schafer, G, Zilliox, L., and Ducreux, J.: "Evaluation of the degree and extent of aquifer contamination by diesel oil in an experimental controlled site", *Proceedings of the International Conference on Restoration and Protection of the Environment*, Patras, Greece, 24-26 Aug. 1994.

Flowpath-simulation in potential hazard studies for contaminants in groundwater

E. Holzbecher

Technical University Berlin, Sekr. W, P.O.Box 100 320, 10563 Berlin, Germany

Abstract

Studies on the potential hazard of a contamination often use flowpath simulation as a technique applied in combination with a flow model. Some numerical aspects and alternatives are presented and evaluated. The time-integrated approach turns out to be the method of choice. For the mentioned purpose the application of particle tracking should be used with care, because it is not conservative. The danger may be underestimated in areas transverse from the pathlines downstream from a contaminated site.

1 Introduction

In hydro-engineering practice it has become a widespread technique to use particle tracking as post-processing tool to a flow model. The visualisation of the flow pattern by pathlines resp. streamlines is used for different purposes. Reilly & Pollock [10,11] study areas contributing recharge to wells and their seasonal and long-term changes. Delineation of contributing areas is the problem tackled by Barlow [1] with the very technique. Within the international HYDROCOIN [6] project tracking algorithms have been studied for safety analyses for nuclear waste repositories. Johnson, Ravi & Rumery [7] estimate solute concentrations using a pathline counting method.

   In this paper flowpath simulations are studied that are applied as a tool to estimate the potential hazard caused by contaminations in groundwater. Using the method flowpaths, here used synonymous with pathlines, are traced downstream from a contaminated site or upstream from a vulnerable region. In that way regions of influence can be determined. They provide informations to make a decision about the danger resulting from the transport of pollutants. Mostly the criterion is quite simple: there is no danger if one out of the two following situations is given

- there is no well within the region of influence from a contaminated site
- there is no contamination within the regions of influence for the wells

There are some advantages, which may explain the popularity of the method and which makes it the method of choice in comparison to transport modeling techniques. The main advantage is that there are no additional hydrogeological, physical or chemical parameters needed. Thus all those difficulties connected with the determination of parameters becomes obsolete. No need for calibration runs with a numerical model on a computer in order to determine aquifer dispersivities. No need for tracer experiments. No need for column or batch experiments in the laboratory in order to get Langmuir- or Freundlich-sorption isotherms (the determination of linear sorption parameters may be a reasonable extension of the method, as will be lined out below).

Nevertheless it will be shown that, what seems the advantage of the method, in fact is the main disadvantage as well. In cases where the neglected processes are relevant, the flowpath technique may not be able to provide good predictions. The danger resulting from transport of contaminants may be overestimated in parts of the environment and underestimated in others.

2 Numerical methods

2.1 General description
In order to simulate advective transport in a flow field various 'particle tracking' algorithms can be used. Starting at user-defined origin the path of a particle can be traced forward until the boundary of the region or a sink is reached. Tracing is likewise possible backward in time: from start the path is traced until boundary or a source are reached.

Codes are different in the form, in which the flow field needs to be given. Bear & Verruijt [2] present cases, in which velocities are given by explicit analytical expressions. Making a choice for the timestep Δt (positive for tracing in future, negative for tracing in the past) the path is traced by using the *Euler-*method:

$$\Delta x = u_x \Delta t \qquad \Delta y = u_y \Delta t \qquad \Delta z = u_z \Delta t \qquad (1)$$

or:
$$\mathbf{r} = \mathbf{r}_0 + \Delta t \cdot \mathbf{u}(\mathbf{r}_0, t_0) \qquad (2)$$

Velocity components u_x, u_y and u_z are estimated using the location, where the particle stayed at time t at the beginning of the timestep. This procedure is only applicable, if velocities are known at each location in the region. This is not possible, if the flow field comes out of a numerical model.

Tracking-algorithms, which can be applied in connection with a flow model need to take into account the discrete form of the velocity field. Values are given at certain locations only: at block-medium points, at volume boundaries or element edges of a discrete grid.

Using the code STLINE (Ward & Harrover & Vincent & Lester [13]) it is assumed, that velocities (given in the block centers of a rectangular grid) change linearly along the space in between. The distances $(\Delta x, \Delta y, \Delta z)$ covered in a timeperiod Δt can be calculated according to (1) or (2), if velocity components are taken from the nearest block centers. An example: in x-direction the linear formula $u_x = a + bx$ is used, where x represents the local variable in the block. The coefficients are then given by:

$$a = u_{x1} \qquad\qquad b = (u_{x2} - u_{x1}) / \Delta x \qquad (3)$$

(u_{x1} resp. u_{x2} denote velocity components at block centers). Then holds:

$$\Delta x = \frac{(a + bx)\Delta t}{1 - b\Delta t / 2} \qquad (4)$$

Proceeding like this for all coordinate directions the location of the particle at the end of the timeperiod can be determined. A problem with the method is, that the final location should not be outside the block from which the velocities have been taken. STLINE overcomes the problem partially by using smaller timesteps. Nevertheless this does not prevent the method, to calculate a final location of the block another time. On the other hand a repeated use of the procedure reduces the distance of overshooting and thus the errors resulting from a false velocity estimation.

Some other aspects should be mentioned briefly. Velocity fields determined by linear interpolation as described above, fullfill the usual differential equation for groundwater flow. The field is rotation free, for u_x is a function of x and u_y is a function of y. But divergence may not necessarily vanish in steady state. Treating both directions as described in the 2-dimensional case the following is obtained:

$$\operatorname{div} \mathbf{u} = \frac{\partial u_x}{\partial x} + \frac{\partial u_y}{\partial y} = \frac{u_{x2} - u_{x1}}{\Delta x} + \frac{u_{y2} - u_{y1}}{\Delta y} =$$

$$= \frac{1}{\varphi S \Delta x \Delta y} (v_{x2}\Delta y - v_{x1}\Delta y + v_{y2}\Delta x - v_{y1}\Delta x) = 0 \qquad (5)$$

In equation (5) particle velocities (u) are related to darcy-velocities (v), porosities (φ) and saturations (S). Finite volume discretisations generally fullfill the last equation in the block interior, because they are based on the finite formulation given in the brackets. At block edges there may be jumps of u_x resp. u_y, when there are different parameter values for φS in the adjacent blocks. Furtheron at edges with constant y the u_x-component and at edges with constant x the u_y-componente may have discontinuities themselves. In order to avoid this inconsistency Goode [4] recommends bilinear interpolation. Nevertheless these have the disadvantage, that they do not fullfill the differential equation in the interior of the blocks.

For the numerical integration in time approaches of higher order can be used, replacing the simple Euler method given in equation (1). Methods of Runge-Kutta type are often described in publications. Sauter & Beusen [12] apply the 2.order Heun-method. In their algorithm the Euler-method is used to determine a first approximation for the particle location **r** at the end of the timestep. In a second step the velocity at position **r** is taken to determine the second and final location:

$$\tilde{\mathbf{r}} = \mathbf{r}_0 + \Delta t \cdot \mathbf{u}(\mathbf{r}_0)$$

$$\mathbf{r} = \mathbf{r}_0 + \frac{\Delta t}{2} \cdot (\mathbf{u}(\mathbf{r}_0) + \mathbf{u}(\tilde{\mathbf{r}})) \qquad (6)$$

There are various other Runge-Kutta type approaches, which can be used alternatively. I want to mention just one more, which starts with an Euler step with timestep $\Delta t/2$ gemacht. The position from that calculation is then used to determine an improved velocity approximation for the second step starting again at the initial location:

$$\tilde{\mathbf{r}} = \mathbf{r}_0 + \frac{\Delta t}{2} \cdot \mathbf{u}(\mathbf{r}_0)$$

$$\mathbf{r} = \mathbf{r}_0 + \Delta t \cdot \mathbf{u}(\tilde{\mathbf{r}}) \qquad (7)$$

All these approaches can be generalised for 'particle tracking' in transient flow fields. With higher order schemes the problems, which occur when a block boundary is approched, cannot be solved. This can be achieved by a time-integration, which is based on the analytical solution, that exists under certain conditions. Δt looses it's characteristic as a user-defined control parameter for the accuracy of the solution.

2.2 Time-integrated approach

As mentioned above the time-stepping approach leads to problems when block resp. element boundaries are passed in the considered timeperiod Δt. In order to overcome this problem, a time-integration approach can be used. The method is decribed as implemented in the FASTpath code, which was developed by the author (Holzbecher [5]). The program is designed as a post-processor for steady-state flow fields, which are calculated by block-centered finite difference or finite volume models. Nevertheless the method can be applied to general cases.

In FASTpath the problem is avoided by giving a different role to timesteps. Δt is no longer a user-defined guess to influence the lengths of discrete steps in the tracking procedure. By time-integration an analytical solution is determined, from which the exact locations can be calculated where a flowpath enters and leaves a block. The method starting at one origin then connects block-exit points.

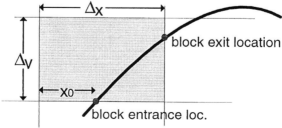

Figure 1: Distances and denotations in a rectangular grid block

The FASTpath algorithm is based on the assumption, that velocity components are given normal to the block edges. As in other codes the method is based on the assumption, that within a block velocities change linearly. The time which the particle takes to move from one block edge to the other can be determined analytically.

If u_{x1} resp. u_{x2} denote velocity components at block edges, a and b from equation (3) can be used to interpolate u for the block interior. The traveltime for a particle between coordinates x_0 and x can be calculated by:

$$\Delta t_x = \int_0^{\Delta t_x} dt = \int_{x_0}^x \frac{ds}{u_x(s)} = \int_{x_0}^x \frac{ds}{a+bs} = \frac{1}{b}\ln(a+bs)\Big|_{x_0}^x$$

$$= \frac{1}{b}[\ln(a+bx) - \ln(a+bx_0)] = \frac{1}{b}\ln\left(\frac{a+bx}{a+bx_0}\right)$$

(8)

In figure 1 the situation is shown for x_0 as the value of the local variable at the block entrance point. The formula holds for all x-values that fall inside the block. Especially the maximum traveltime within the block if the maximum value for x is taken ($\Delta x - x_0$ if flow is in positive x-direction, 0 if flow is in negative x-direction):

$$\Delta t_x^{max} = \begin{cases} \dfrac{1}{b}\ln\left(\dfrac{a+b(\Delta x - x_0)}{a+bx_0}\right) & \text{for positive velocity} \\[3mm] \dfrac{1}{b}\ln\left(\dfrac{a}{a+bx_0}\right) & \text{for negative velocity} \end{cases}$$

(9)

The traveltime from equation is the real one, if the flowpath reaches one of both edges x=0 or x=Δx. In a moredimensional case there are other possibilities. Thus in the same way the other coordinate axes have to be treated and the actual traveltime in the block (in 3-dimensional flow) is the minimum of the ones determined for the directions:

$$\Delta t = \min\left\{\Delta t_x^{max}, \Delta t_y^{max}, \Delta t_z^{max}\right\}$$

(10)

After the time period Δt the flowpath reaches the boundary of the block. Using Δt the exact location of the block-exit-point can be calculated easily using equation (2). Solving for the local variable x provides:

$$x = \frac{1}{b}[(a+bx_0)\exp(-b\Delta t) - a]$$

(11)

Similar equations hold for the other coordinate directions. The transformations leading to equation (11) are valid, if b≠0 and if a and x_0 do not vanish simultaneously. If b=0 then there is a constant velocity in x-direction and the maximum traveltime results from the simple formula:

$$\Delta t_x = \frac{1}{a}(x - x_0) \tag{12}$$

If a=0 holds, then there a stagnation point at the concerning block edge. If with x_0=0 the block is reached at this edge, the particle does not move further into the block interior. In the algorithm the maximum traveltime can be formally set to the maximum machine constant (equivalent ∞) and the position is x=0.

Other exeptions have to be formulated for source- resp. sinkblocks. These are characterised by the fact that there is are outflow resp. inflow at all edges. Sinkblocks thus have no exit in forward particle tracking, sourceblocks have no entrance in backward tracking. Traveltimes within these blocks are highly speculative, because the very local nature of flow in the vicinity of infiltration resp. pumping wells is reduced to a mass balance relation, when a computer model for groundwater flow is used.

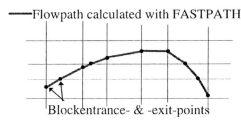

Figure 2: A flowpath as connection of entrance- resp. exit-points

An equivalent formulation of the algorithm implemented in the FASTpath code was published by Pollock [9]. A generalisation for non-rectangular grids can be found in Cordes & Kinzelbach [3].

The method has been described for steady state flow fields here. Nevertheless the extension to transient cases is straight-forward. It has additionally been taken into account, that there is a moment, when the current velocity field needs to be replaced by a new one. It is intended to improve the FASTpath code in that way.

3 Application

Flow fields can be visualised by particle tracking. In two-dimensional steady state cases the flow pattern can be presented in a single plot. Figure 3 shows the flow towards a pumping well, which is located on the boundary of the modelled area; i.e. only one half of a horizontal cross-section is considered. The other half can be obtained by symmetry. Approximately half of the water entering at the boundary on the right hand side reach the well (5 out of 10 pathlines are trced into the well).

Figure 4 shows the same system, where a second well with a lower pumping rate was installed in order to catch polluted fluid entering from the right hand side. The limited effect of the protection well is clearly visible in the figure: only those pollutants are withdrawn, which are transported with the flow field from the the small area (A). Moreover the first well pumps water from the far side of the system: the upper 4 pathlines only reach and cross the outflow boundary on the left side. If the contaminated area extends into that region (B), the installation of the second well is even contraproductive: the quality of pumped water worsens.

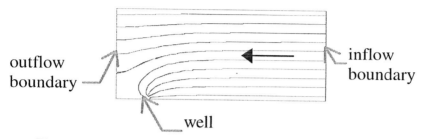

Figure 3: Visualisation by flowpaths: flow towards a pumping well

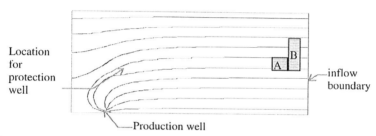

Figure 4: Flow towards a pumping and a designed protection well

The example demonstrates the application of particle tracking. In order to determine, if water quality in a pumping well is worsened from a contaminated site, it needs to be recognized, if well and polluted area are combined by at least one pathline. Practically this is done by an interactive use of the flowpath software: startpoints for pathlines are specified by the user just on the basis of the visual impression from the flow graph.

Backward and forward tracing in principle are both useful to determine possible connections. In both cases regions of influence are of interest. The region of influence downstream from a contaminated zone is reached by advectively transported pollutants. For a single well or a well gallery the pumped water originates from the region of influence. In 2-dimensional horizontal flow tracking backward in time leads to catchment areas.

Particle tracking simulates advective transport as the only process. All other processes which influence the distribution of pollutants in the subsurface are neglected. In the remaining part of this section will be discussed, how the results from the flowpath technique as described above, can be judged in relation to results from a common transport model, which considers other processes. Especially the question will be answered, if the method is conservative.

3.1 Neglection of longitudinal dispersion

Along the flowpath polluted fluids are subject to mixing processes as diffusion and longitudinal dispersion. Peak concentrations are reduced. On the other hand values increase in low concentrations regions. When the flowpath method is used as described above - by determination of influence regions - the method always leads to an overestimation of risk. In principle the technique is based on the conservative assumption, that there is no mixing along the pathline and thus concentrations are kept high. Nevertheless in real systems mixing always is a substantial process.

From the aspect of mixing along a pathline the use of the flowpath technique always leads to conservative results.

3.2 Neglection of transverse dispersion

In directions transverse to the flowpath mixing processes occur as well. Diffusion and transverse dispersion are responsable, that a plume of pollutants not only spreads in the direction of flow. Though these moments are neglected by using flowpath tracing as described above, the potential danger is underestimated in regions which are not hit by pathlines which pass a contaminated region.

From the aspect of mixing transverse to a pathline the use of the flowpath technique does *not* lead to conservative results.

3.3 Neglection of decay resp. degradation processes

If any decay or degradation are relevant processes the result is a decrease of pollutant concentrations relative to another component which is not degraded. The flowpath technique does not account for these processes and thus predicts a higher risk for decaying contaminants. If the pollutant reaches a vulnerable region, it may be mostly degraded. From the aspect of decay the use of the flowpath technique always leads to conservative results.

It is generally not possible to estimate peak maximum concentrations considering the traveltime along the flowpath. In real systems, mixing and decay interact in a way, which cannot easily be predicted, even if initial concentrations are known. A more specific statement can be made, when analytical solutions are studied - which is beyond the scope of this contribution. It is worth mentioning, that under certain conditions the decrease of maximum concentrations along a pathline can be estimated. If there is a single peak for the pollutant concentration as initial condition, the peak of the n-dimensional analytical solution declines with time according to (Kinzelbach [8]) :

$$c(.,t) \cong \frac{\exp(-\lambda t)}{\left(\sqrt{t}\right)^n} \tag{13}$$

Nevertheless in practical cases it is difficult to check if the conditions for the application of these formulae are fullfilled.

3.4 Considering linear sorption as retardation

If the interaction between fluid and porous medium concerning the concentrations of a component can be described by a constant linear isotherm, the description of the processes can be simplified by the introduction of a retardation factor. This does not change anything, if the decision about the potential hazard of a contaminated site is made following the above mentioned criterion. The condition depends on the connection of pathlines only and not on the timescale of the flow.

Retardation can be considered with the flowpath technique. It should be noted, that the problem of transverse mixing is less relevant for retardated contaminants, because the timescale for dispersion processes is reduced as well.

3.5 Neglection of non-linear sorption processes

It is not possible to do the same analysis as in the preceding section, if sorption acts obeying certain non-linear relationsships, as the Langmuir- or the general Freundlich-isotherm. Retardation factors can be introduced in that case as well, but these are concentration dependent and thus change along the flowpath spatially and temporally. The resulting distribution patterns are thus more complicated.

The same holds for non-equilibrium sorption. It is beyond the scope of this paper to attempt a general statement concerning over- resp. underestimation by the flowpath technique for generalised sorption behaviour.

4 Conclusional remarks

The technique of flowpath simulation to study the potential hazard of contaminations should be applied with caution. The danger resulting from a pollution may be underestimated in parts of the environment, but overestimated in others. The modeler should keep in mind, that advection is the only transport process, which is visualized by flowpath tracing. Effects from dispersion, diffusion, decay and degradation cannot be taken into account. Linear equilibrium sorption can be considered, as long as it acts as a constant retardation. The movement of a concentration peak can thus be predicted but not the peak's height.

Particle tracking combined with a decision based on pathline connections, is not a conservative method. Pollutants in real systems partially move transverse to flowpaths and there may increase concentrations above reasonable limits. To overcome this obstacle the criterion based on the connected pathlines could be altered. It could be required, that there is a certain distance between a pathline for polluted fluids and the catchment of a well. Nevertheless the quantification of that spacing can only be based on an estimation of transverse dispersivity.

References

1. Barlow P.M., Two- and three-dimensional pathline analysis of contributing areas to public-supply wells of Cape Cod Massachussetts, *Groundwater*, Vol.32, No.3, 1994, 399-410
2. Bear J./ Verruijt A., *Modeling groundwater flow and pollution*, D. Reidel Publ., Dordrecht, 1987
3. Cordes C./ Kinzelbach W., Continuous velocity fields and pathlines in finite element groundwater flow models, in: Comp. Meth. in Water Res. IX (ed Russell T.F./ Ewing R.E./ Brebbia C.A./ Gray W.G./ Pinder G.F.), *Proceedings Vol.1*, pp. 247-254, 1992
4. Goode J., Particle interpolation in block-centered finite differences groundwater flow models, *Water Res. Res.*, Vol.26, No.5, 1992, 925-940
5. Holzbecher E., *Modellierung dynamischer Prozesse in der Hydrologie: Grundwasser und ungesättigte Zone*, Springer Verlag, Heidelberg, 1996
6. HYDROCOIN - *an international project for studying groundwater hydrology modelling strategies*, OECD, Paris, level 3
7. Johnson J.A./ Ravi V./ Rumery J.K., Estimation of solute concentrations using the pathline counting method, *Groundwater*, Vol.32, No.5, 1994, 719-726
8. Kinzelbach W., *Numerische Methoden zur Modellierung des Transports von Schadstoffen im Grundwasser*, Oldenbourg, München, 1987
9. Pollock D.W., Semianalytical computation of path lines for finite difference models, *Groundwater*, Vol.26, No.6, 1988, 743-750
10. Reilly T.E./ Pollock D.W., Factors affecting areas contributing recharge to wells in shallow aquifers, *U.S.-Geol. Survey Water-Supply Paper* 2412, 1993
11. Reilly T.E./ Pollock D.W., Effects of seasonal and long-term changes in stress on sources of water to wells, *U.S.-Geol. Survey Water-Supply Paper* 2445, 1995
12. Sauter F.J./ Beusen A.H.W., Streamline calculations using continuous and discontinuous velocity fields and several time integration methods, in: Groundwater Quality Management GQM 93 (ed Kovar K./ Soveri J.), *Proceedings*, Tallinn, IAHS Publ., No.220, pp. 347-355, 1994
13. Ward D.S./ Harrover A.L./ Vincent A.H./ Lester B.H., Data input guide for SWIFT/486: *The SANDIA Waste-Isolation Flow and Transport Model for Fractured Media*, GeoTrans, 1993
14. Zheng C., Analysis of particle tracking errors associated with spatial discretisation, *Groundwater*, Vol.32, No.5, 1994, 821-828

Pollutant dispersion analysis in porous media using dual reciprocity boundary element method

T.I. Eldho
Institute of Hydromechanics, University of Karlsruhe, D-76128, Karlsruhe, Germany

Abstract

Here dual reciprocity boundary element method is used for the analysis of pollutant dispersion in porous media. In the governing dispersion equation, the time dependent term and convective terms are approximated using dual reciprocity method and Green's theorem is used to form boundary integral equations. Linear elements are used for boundary discretization. The model is verified with one dimensional analytical solution and found to be satisfactory. The model is applied for the solution of some two dimensional pollutant dispersion problems in homogeneous isotropic porous media.

Introduction

In any development and management of groundwater resource system, a major problem of interest is that of the quality of the water being extracted. For the assessment of groundwater quality, it is necessary to identify the source of pollution and predict the movement of the pollutants in the groundwater environment using effective tools. As very few analytical solutions are available, prediction of the movement of pollutants is done using numerical methods like finite difference (FDM), finite elements (FEM) and boundary elements (BEM) [1]. FDM and FEM are domain oriented methods and hence tedious in discretization and data preparation. BEM is more suitable since it is boundary oriented and hence very easier in discretization and data preparation for very large groundwater basins.

Here a BEM model based on dual reciprocity principle [2] is used in the prediction of pollutant dispersion in porous media. As the BEM reduces the computational dimension of the problem by one, two dimensional problems are solved in one dimension. The dual reciprocity BEM model is validated with analytical solution and applied in the prediction of some two dimensional pollutant dispersion problems in porous media.

Formulation of the problem

Governing equation and boundary conditions
The basic differential equation in two dimensions, describing the process of solute transport in saturated porous media, considering the effects of convection and hydrodynamic dispersion and neglecting sources, decay etc., can be expressed as [1]:

$$D_x \frac{\partial^2 C}{\partial x^2} + D_y \frac{\partial^2 C}{\partial y^2} = \frac{\partial C}{\partial t} + \frac{1}{n_e}\left(U_x \frac{\partial C}{\partial x} + U_y \frac{\partial C}{\partial y}\right) \qquad (1)$$

where C is solute concentration, D is dispersion coefficient, U is Darcy's velocity of flow in the porous media, n_e is the effective porosity of the media, and t is the time. A dimensionless form of eqn.(1) can be achieved by substituting,

$$c = \frac{C\, n_e}{C_0} \; ; \; X = \frac{x}{L'} \; ; \; Y = \frac{\beta y}{L'} \; ;$$
$$u_x = \frac{U_x L'}{D_x} \; ; \; u_y = \frac{\beta\, U_y L'}{D_x} \; ; \; t' = \frac{D_x\, t}{L'^2} \qquad (2)$$

in which C_0 and L' are some reference concentration and length and $\beta = (D_x/D_y)^{1/2}$.

For pollutant dispersion modelling, the appropriate initial and boundary conditions should be prescribed. The initial condition, usually concentration distribution at some initial time say t=0, is described throughout the considered domain Ω. Two boundary conditions are commonly used in the solution of solute transport problems. One is boundary of prescribed concentration and another one is boundary of prescribed flux (gradient of concentration) as given below:

$$C(x,y,t) = g_1(x,y,t) \; on \; \Gamma_1 \; ; \; C_n = \frac{\partial C}{\partial n} = g_2(x,y,t) \quad on \; \Gamma_2 \qquad (3)$$

where n is the unit outward normal to the boundary.

Boundary element formulation
For a two dimensional dispersion problem, assuming $D_x = D_y = D_T$, $V_x = U_x/n_e$ and $V_y = U_y/n_e$, then eqn.(1) can be written as:

$$\nabla^2 C = \frac{1}{D_T} \left(V_x \frac{\partial C}{\partial x} + V_y \frac{\partial C}{\partial y} + \frac{\partial C}{\partial t} \right) = \frac{b}{D_T} \qquad (4)$$

In the dual reciprocity method, the time dependent term and convective terms are approximated [3] and eqn.(4) is represented as:

$$\nabla^2 C = \frac{1}{D_T} \sum_{j=1}^{N+L} \alpha_j \left(\nabla^2 \hat{C}_j \right) = \frac{1}{D_T} \sum_{j=1}^{N+L} f_j(x,y) \, \alpha_j(t) \qquad (5)$$

where $\hat{C}_j$ is a series of particular solutions, N- number of boundary nodes, L- number of internal nodes, D_T- dispersion coefficient, α_j- a set of initially unknown coefficients and f_j- approximating functions. Number of $\hat{C}_j$ is equal to total number of nodes N+L. Here the right hand side of eqn.(4) has been replaced by a summation of products of coefficients α_j and the Laplacian operating on the particular solution $\hat{C}_j$.

Now using reciprocity principle on both sides of eqn.(5), the boundary integral equations for each source node i can be written as [4],

$$g_i \, C_i + \int_\Gamma \frac{\partial C^*}{\partial n} C \, d\Gamma - \int_\Gamma C^* \frac{\partial C}{\partial n} \, d\Gamma$$

$$= \frac{1}{D_T} \sum_{j=1}^{N+L} \alpha_j \left(g_i \, \hat{C}_{ij} + \int_\Gamma \frac{\partial C^*}{\partial n} \hat{C}_j \, d\Gamma - \int_\Gamma C^* \frac{\partial \hat{C}_j}{\partial n} \, d\Gamma \right) \qquad (6)$$

where C^* is the fundamental solution of Laplace equation ($C^* = (1/2\pi)\ln(1/r)$, where r is distance from the point i of application of the source or sink to any other point under consideration), n is the unit outward normal and g_i is Green's constant. $g_i = 1$ for a point inside the domain Ω, $g_i = 0$ for a point out side the domain and $g_i = \theta/2\pi$ for a point on the boundary, in which θ is the internal angle at point i in radians.

Matrix formulation and assembly of equations

From eqns.(4) and (5), by taking the value of b at (N+L) different points, a set of equations in the following matrix form can be written:

$$b = F \alpha \; ; \; \alpha = F^{-1} b = F^{-1} \left[V_x \frac{\partial C}{\partial x} + V_y \frac{\partial C}{\partial y} + \frac{\partial C}{\partial t} \right] \qquad (7)$$

In eqn.(6), $\hat{C}$ and $\partial \hat{C}/\partial n$ are not necessarily be approximated over the boundary using interpolation functions as they are known once the 'f' function is chosen. But doing so will reduce the necessary boundary integration (eventhough the accuracy is slightly affected) and the same matrices may be used on both sides. Hence discretizing the boundary into N linear elements and using L internal nodes, introducing the interpolation functions and integrating over each boundary elements, eqn.(6) can be written as:

$$g_i C_i + \sum_{k=1}^{N} H_{ik} C_k - \sum_{k=1}^{2N} G_{ik} \frac{\partial C}{\partial n}_k =$$

$$\frac{1}{D_T} \sum_{j=1}^{N+L} \alpha_j \left(g_i \hat{C}_{ij} + \sum_{k=1}^{N} H_{ik} \hat{C}_{kj} - \sum_{k=1}^{2N} G_{ik} \frac{\partial \hat{C}}{\partial n}_{kj} \right)$$

(8)

Index k is used for the boundary nodes which are the field points. After application to all boundary nodes using a collocation technique and using eqn.(7), eqn.(8) can be represented as:

$$H C - G \left[\frac{\partial C}{\partial n} \right] = \frac{1}{D_T} \left[H \hat{C} - G \left[\frac{\partial \hat{C}}{\partial n} \right] \right] F^{-1}$$

$$\left[V_x \frac{\partial C}{\partial x} + V_y \frac{\partial C}{\partial y} + \frac{\partial C}{\partial t} \right]$$

(9)

Here the approximating function 'f' can be a function of 'r' (r is the distance from the point i of application of the source or sink to any other point under consideration) and generally used 'f' function is '1+r'. Correspondingly [3]:

$$\hat{C} = \frac{r^2}{4} + \frac{r^3}{9} \; ; \; \frac{\partial \hat{C}}{\partial n} = \left[r_x \frac{\partial x}{\partial n} + r_y \frac{\partial y}{\partial n} \right] \left[\frac{1}{2} + \frac{r}{3} \right]$$

(10)

A mechanism must now be established to relate the nodal values of C to the nodal values of its derivatives $\partial C/\partial x$ and $\partial C/\partial y$. Similar to eqn.(7), putting C=Fβ (where β≠α) and differentiating,

$$\frac{\partial C}{\partial x} = \frac{\partial F}{\partial x} \beta \; ; \; \frac{\partial C}{\partial y} = \frac{\partial F}{\partial y} \beta$$

(11)

Rewriting $\beta = F^{-1}C$, then

$$\frac{\partial C}{\partial x} = \frac{\partial F}{\partial x} F^{-1} C \; ; \; \frac{\partial C}{\partial y} = \frac{\partial F}{\partial y} F^{-1} C$$

(12)

Corresponding to the function 'f=1+r' [3]:

$$\frac{\partial f}{\partial x} = \frac{\partial f}{\partial r} \frac{\partial r}{\partial x} = \frac{\partial f}{\partial r} \frac{r_x}{r} = \frac{r_x}{r} \; ; \; \frac{\partial f}{\partial y} = \frac{\partial f}{\partial r} \frac{\partial r}{\partial y} = \frac{\partial f}{\partial r} \frac{r_y}{r} = \frac{r_y}{r}$$

(13)

Eqn.(9) can be now expressed as:

$$H C - G \left(\frac{\partial C}{\partial n} \right) = \frac{1}{D_T} \left(H \hat{C} - G \left(\frac{\partial \hat{C}}{\partial n} \right) \right) F^{-1}$$
$$\left[\left(V_x \frac{\partial F}{\partial x} + V_y \frac{\partial F}{\partial y} \right) F^{-1} C + \frac{\partial C}{\partial t} \right] \tag{14}$$

Therefore, eqn.(14) can be written as,

$$P \frac{\partial C}{\partial t} + (H - R) C = G \left(\frac{\partial C}{\partial n} \right) \tag{15}$$

where

$$S = \left(H \hat{C} - G \frac{\partial \hat{C}}{\partial n} \right) F^{-1} \; ; \; P = -\frac{1}{D_T} S \; ; \; R = \frac{S}{D_T} \left(V_x \frac{\partial F}{\partial x} + V_y \frac{\partial F}{\partial y} \right) F^{-1}$$

For the sake of simplicity, a two level time integration scheme will be employed here. Within each time step, proposing a linear approximation for the variation of C and $\partial C/\partial n$,

$$C = (1 - \theta_u) C^m + \theta_u C^{m+1}$$
$$\frac{\partial C}{\partial n} = (1 - \theta_q) \frac{\partial C^m}{\partial n} + \theta_q \frac{\partial C^{m+1}}{\partial n} \tag{16}$$
$$\frac{\partial C}{\partial t} = \frac{1}{\Delta t} \left(C^{m+1} - C^m \right)$$

where θ_u and θ_q are parameters which positions the values of C and $\partial C/\partial n$, respectively between time levels m and m+1.

Substituting these approximations into eqn.(15) gives:

$$\left[\frac{1}{\Delta t} P + \theta_u (H - R) \right] C^{m+1} - \theta_q G \left[\frac{\partial C}{\partial n} \right]^{m+1}$$
$$= \left[\frac{1}{\Delta t} P - (1 - \theta_u) (H - R) \right] C^m + (1 - \theta_q) G \left[\frac{\partial C}{\partial n} \right]^m \tag{17}$$

The right hand side of eqn.(17) is known at time $(m+1)\Delta t$, since it involves values which have been specified as initial conditions or calculated at previous time step. Introducing the boundary conditions at time $(m+1)\Delta t$, one can rearrange the left side of eqn.(17) and write as:

$$A \; X \; = \; F \tag{18}$$

where X is unknown matrix of C and $\partial C/\partial n$, F is a known matrix and A is the coefficient matrix. Solution of eqn(18) give the unknown values of C or $\partial C/\partial n$.

Validation of the model

The DRBEM dispersion model is compared with some analytical solutions given by Marino [5] out of which one is presented here. The configuration of the porous media considered is shown in Fig.1. It is 600 m X 100 m size and is homogeneous, isotropic and fully saturated. The dispersion coefficient of the media (D_T) is assumed to be 7.616 m^2/day and seepage velocity is 1.219 m/day. The initial and boundary conditions are,

$$C(x,y,0) \; = \; 0.0$$
$$C(0,y,t) \; = \; C_0 = 1.0 \; ; \; C(600,y,t) \; = 0.0 \tag{19}$$
$$\frac{\partial C}{\partial n} \; (x,0,t) \; = \; \frac{\partial C}{\partial n} \; (x,100,t) \; = \; 0.0$$

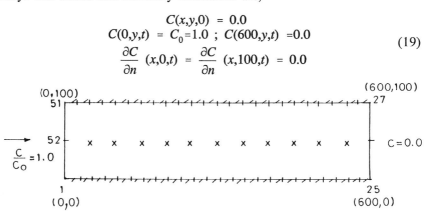

Figure 1. Solution region for validation model

The analytical solution to this problem, considering as a one dimensional problem given by Marino [5] is:

$$\frac{C}{C_0} \; (x,t) \; = \; \frac{1}{2} \left[erfc \left[\frac{(x-Vt)}{2\sqrt{D_t t}} \right] \; + \; \exp \left[\frac{V_x}{D_T} \right] \; erfc \left[\frac{(x+Vt)}{2\sqrt{D_t t}} \right] \right] \tag{20}$$

In the dual reciprocity boundary element dispersion model, the boundary of the domain is discretized into 52 linear elements and 11 internal nodes are considered as in Fig.1. Time step $\Delta t = 1$ day; $\theta_u = 0.5$, $\theta_q = 1.0$ and f=1+r. The dispersion analysis is carried out for 300 time steps and concentration distribution is found out.

The concentration distributions along the domain at various times are plotted in Fig.2. using analytical solution and dual reciprocity BEM solution.

Good agreement is observed between both solutions.

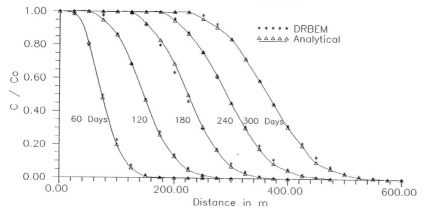

Figure 2. Concentration distribution: Model validation

Case studies

Here two case studies are presented to show the effectiveness of dual reciprocity BEM in the solution of pollutant dispersion problems.

Case study 1

Here the two dimensional dispersion in a homogeneous isotropic porous media of size 600 X 300 m, with a symmetric and uniform point source located at (0,150) with a relative concentration (C/C_0) of one is considered. Due to symmetry in the input concentration, only one half of the full domain is considered as in Fig. 3. The flow is considered along the longitudinal direction

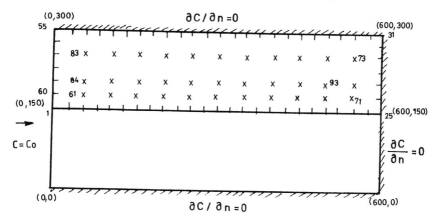

Figure 3. Solution region for Case study 1

with a seepage velocity of 1.219 m/day. The longitudinal dispersion coefficient (D_{Tx}) is 7.616 m²/day and lateral dispersion coefficient (D_{Ty}) is 0.7616 m²/day. In the DRBEM model, θ_u=0.5, θ_q=1.0, f=1+r, time step Δt=2 day and analysis is done for 360 days. As in Fig. 3, 60 linear elements and 33 internal nodes are used for discretization. The contours of relative concentration distribution for the time instance of 360 days are shown in Fig. 4. The concentration movement with time at three nodes 7 (150,150), 13 (300,150) and 19 (450,150) are shown in Fig. 5.

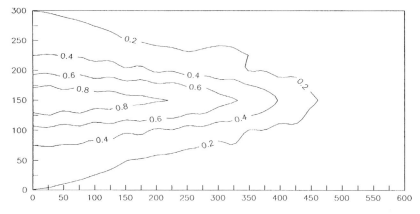

Figure 4. Contours of concentration distribution - 360 days

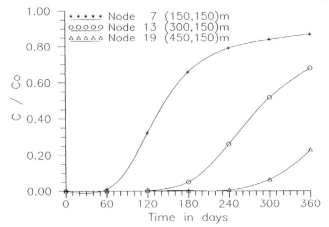

Figure 5. Concentration distribution with time: Case study 1

Case study 2

Here the two dimensional dispersion in a homogeneous isotropic porous media of size 600 X 300 m with unsymmetric input concentration is considered. The input concentration varies along the y- axis by the following equation:

$$C(0,y,t) = C_0 \exp\left[\frac{-(y-125.0)^2}{3140}\right] \qquad (21)$$

where C_0 is the relative value of concentration and assumed to be unity. The dispersion is considered in unidirectional velocity field with a velocity of 1.22 m/day. The dispersion coefficient in x- direction (D_{Tx}) is 7.622 m²/day and in y- direction (D_{Ty}) is 0.4795 m²/day. As in Fig. 6, the boundary of the domain

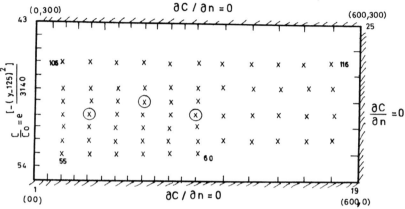

Figure 6. Solution region for Case study 2

is discretized into 54 linear elements and 62 internal nodes are considered. In the DRBEM model, θ_u=0.5, θ_q=1.0, f=1+r, time step Δt=1 day and the analysis is carried out for 200 days. The contours of concentration distribution for 200 days are plotted in Fig. 7. The concentration movement with time at three nodes, 79 (100,125), 92 (200,150) and 83 (300,125) are shown in Fig. 8.

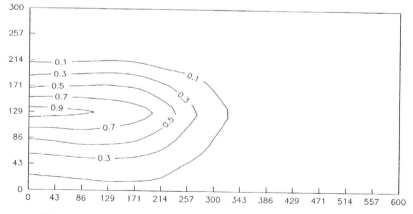

Figure 7. Contours of concentration distribution - 200 days

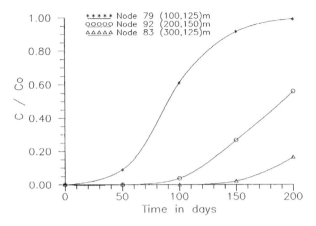

Figure 8. Concentration distribution with time: Case study 2

Conclusions

Here dual reciprocity BEM is used in the prediction of pollutant dispersion in porous media. The convective terms and time dependent term in the governing equation is approximated using dual reciprocity method so that a boundary only solution is achieved. Application of the model to various case studies demonstrates the effectiveness of the model in dispersion analysis in porous media. The model utilizes all basic advantages of BEM like reduction in computational dimension, easiness in data handling and less numerical dispersion compared to other numerical methods.

Index:

Pollutant Dispersion, Dual Reciprocity Method ,Boundary Elements

References

1. Bear, J. & Verruijt, A., *Modeling Groundwater Flow and Pollution*, D.Riedel Co., Dordrecht, 1987.
2. Eldho, T.I., Transient flow and dispersion analysis in porous media using boundary element method, *Ph.D Thesis*, Dept. of Civil Engg., Indian Institute of Technology, Bombay, 1995.
3. Patridge, P.W., Brebbia, C.A, & Wrobel, L.C, *The Dual Reciprocity Boundary Element Method,* Computational Mechanics Publications, Southampton & Elsevier Science Publishers, New York, 1992.
4. Brebbia, C.A., *The Boundary Element Method for Engineers,* Pentech Press, London, 1978.
5. Marino, M.A., Longitudinal dispersion in saturated porous media, *J. of the Hydraulics Engg., ASCE*, 1974,100 (1), 151-157.

Evaluation of CO dispersion models applied to a Milan city database

G. Manini,[a] E. Angelino,[b] R. Gualdi[b]

aEnvironmental Consultant - Italtel Telesis, Italy
bAzienda U.S.S.L. AT.n. 38 - P.M.I.P. (Health Protection Local Office) - U.O. Fisica e Tutela Ambiente di Milano, Italy

Abstract

This paper describes the principal results obtained in evaluating the performance of three modelling techniques (CALINE4, CPB-3, and Johnson's formula) in simulating CO concentrations in three streets of Milan with different facility types and location. The models have been applied to compute 1-hour CO concentrations along a period of two months. Moreover the application of a lagrangian model (GEM) to one of the three streets is illustrated.

1 Introduction

Traffic is certainly one of the main sources of atmospheric pollution: in urban area it is the prevalent source of carbon monoxide and is responsible for the presence of other pollutants, as nitrogen oxides, benzene and particulate.

In the frame of the CNR (Research National Council) research project entitled "Traffic planning and air pollution in urban areas" the PMIP of Milan collaborating with Italtel-Telesis and CNR has carried out a study in order to describe the dispersion of a pollutant inside a street canyon. The first phase of the project regards the intercomparison of different dispersion models in order to evaluate the most adequate to be applied during the field experiment in Milan as provided in the second phase of the study.

2 The database

The modeling techniques have been applied using the traffic data (hourly traffic volumes counted for each lane) measured by the Urban Police Force during November and December '89 at Viale Marche, Via Monreale and Via Senato. These three streets have been chosen because represent different facility types and locations and are characterised by urbanistic structures typical in Milan.

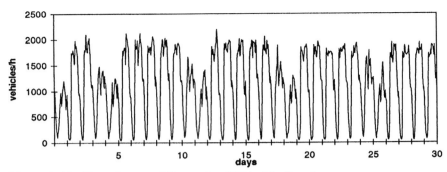

Figure 1: Traffic volumes of November '89 in Via Senato.

Milan has a concentric structure with radial penetration axes joining Milan with motorways and neighbouring towns. The three streets considered are briefly described below:
— Viale Marche is a thorough-fare which is part of the Milan ring-road;
— Via Monreale is a low-traffic side street;
— Via Senato is a street canyon with high traffic intensity.
It is evident for all the three sites that different temporal trends characterise working-days and holidays, even if daily traffic volumes depend on the street (Manini[1]). From traffic analysis for the two months under study it can be noticed that:

• In Viale Marche the vehicles number is very elevated, also in the night-time, and keeps on similar values on the different days of the week: only festivities cause evident changes of the daily trends.

• In Via Monreale traffic volumes are medium-low during working-days. Vehicles number decreases during holidays excepting Sundays in correspondence with the soccer-match played in the near stadium.

• Via Senato is situated in the centre of the city, so traffic volumes are high in the morning of the working-days. Already for Saturday vehicles number tends to decrease and reaches the minimum value during holidays.

As an example, Figure 1 shows the hourly traffic volumes recorded in Via Senato during November '89.

The hourly traffic volumes have been used to calculate traffic CO emissions relating to each street, as described afterwards.

With regard to meteorological parameters (temperature, wind speed and direction, stability class, solar radiation), it has tried to use for each site the values observed by the monitoring station situated on the spot (belonging to the Air Quality Monitoring Network of the Province of Milan). The data of Viale Marche, owing to the lack of data for several hours, have been integrated with the values measured by the monitoring station situated in Via Monreale. Moreover the monitoring station of Via Senato is not equipped with meteorological analysers; so have been used the data measured in Viale Marche. In fact this station, in spite of some missing data, is the more adequate to

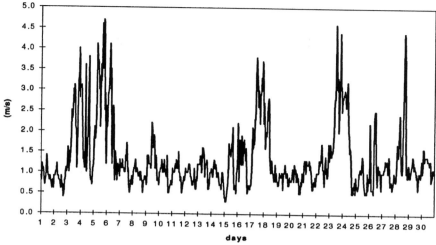

Figure 2: Wind speed trend in Milan during November '89.

describe the meteorological situation in Via Senato, because it is the nearest monitoring station working in the considered period.

Analysing the months of November and December '89, the following meteorological considerations can be briefly noted (Tebaldi[2]; Tebaldi[3]). The temperature has kept mild during the first ten days in November and the second ten days in December, whereas harsh temperatures have been recorded during the last week in November, at the beginning and particularly at the end of December. A high atmospheric pressure field has concerned the North of Italy during the second ten days in November and a another one from the end of the month to the beginning of December. The rainfalls are almost completely concentrated in the first week of November. Regarding the anemological conditions in Milan during the period under study the wind speed has kept low values (about 1 m/s) except some isolated episode. Only during short periods we have recorded values above 2.5 m/s and sometimes wind speed reached even 4.5 m/s. Figure 2 shows, as an example, the temporal trend of the wind speed in Milan during November '89.

3 The modelling techniques

To simulate the spatial distribution of pollutant concentration inside a street it is necessary to use a microscale dispersion model. The choice of the modelling technique depends on the particular topography considered since there is no model suitable for all situations. So the results obtained by different models have been compared for each street.

In this study four models have been considered: CALINE4 (Benson[4]), CPB-3 (Yamartino[5]; Yamartino[6]), Johnson's formula (Johnson[7]), GEM (Lanzani[8]; Lanzani[9]). The first one is a line source Gaussian model; the second one is a

Plume-Box model; the third one is an empirical box model and the last one is a Lagrangian particle model. These models require four different kinds of data:

- topographical characteristics of the street (street width, buildings height, number and width of the lanes);
- meteorological parameters (wind direction and speed, wind direction standard deviation, stability class, temperature, solar radiation);
- traffic characteristics (vehicle number and speed per lane);
- composite emission factors per lane.

Moreover the background concentration of the pollutant is required.

In this study meteorological parameters and CO concentrations have been extracted from the databases of the Province of Milan Air Quality Monitoring Network while traffic data from measurements made by Urban Police Force of Milan. The model FATEMVE (Tamponi[10]) has been used to calculate the composite emission factors. This module is based on the CORINAIR (Co-ORdination Information AIR)-COPERT'90 method of the European Economic Community (Bocola[11]; Eggleston[12]) adapted for the Italian situation.

An appropriate processor, MSP (Milan Street Processor), applies at user's request one or more models described above, extracting the hourly input data from traffic and meteorological datasets sequentially for several days. To obviate the circumstance in which, as in the present one, no information about the vehicle travelling speed is available, user can choose one of five empirical curves relating speed with traffic volumes for five different kinds of street. Since this choice is difficult, it is opportune to test several options for the same street. At last all the dispersion models require the background concentration of the pollutant. As a starting point for estimating the background the minimum among the 4:00, 5:00 and 6:00 a.m. measured concentration values for every day is used. If this datum is not available, the background is evaluated from the average daily minimum reached in the month.

4 Results

The dispersion models CALINE4, CPB-3 and Johnson's formula have been applied over a period of two months: November and December 1989. Figure 3 shows the concentrations predicted by the three models as a function of the concentrations observed in Viale Marche, Via Monreale and Via Senato during the two months. From these graphs, it is evident that CPB-3 is not adequate to describe the situation of Viale Marche and CALINE4 is not suitable to simulate the concentrations of Via Monreale and Via Senato.

Comparing the observed CO concentrations time series over the period of study to the results of the three models, it can be deduced that:

- in the case of Viale Marche, CALINE4 is the model which best describes the experimental data. Figure 4 shows, as an example, the CO concentrations measured during November '89 and the values predicted by CALINE4. It can be noticed that the model tends to underestimate particularly when the wind speed is high (≥ 2.5 m/s) and the wind direction is parallel to the street

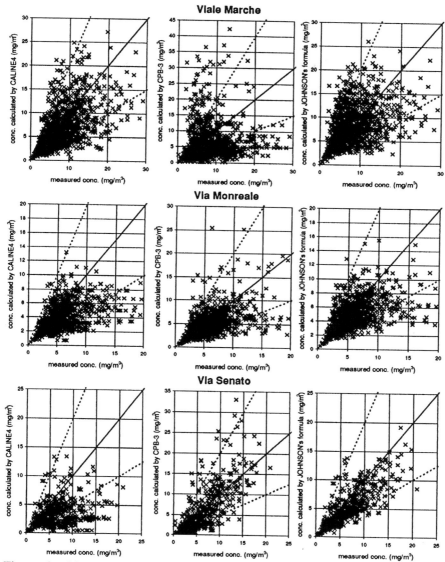

Figure 3: CO concentrations of November-December '89 in the three sites considered: values predicted by CALINE4, CPB-3 and Johnson's formula vs. observed values. The solid line (bisector) is the perfect fit; the dotted lines delimit the factor-of-two region.

axis.
- with regard to Via Monreale, the best results are obtained by CPB-3 model. As an example Figure 5 shows the CO concentrations measured during

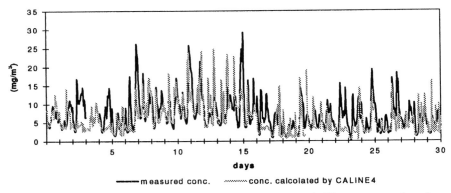

Figure 4: Comparison between observed and computed CO concentration time series of November '89 in Viale Marche.

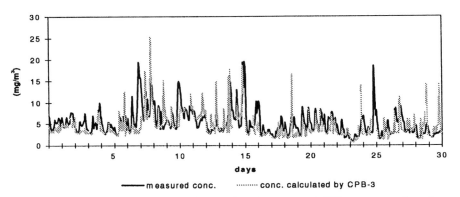

Figure 5: Comparison between observed and computed CO concentration time series of November '89 in Via Monreale.

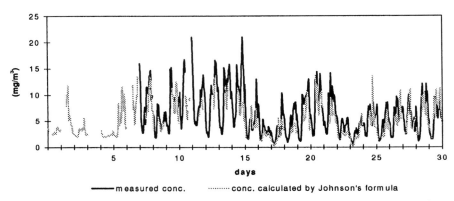

Figure 6: Comparison between observed and computed CO concentration time series of November '89 in Via Senato.

November '89 and calculated by CPB-3.

- in the case of Via Senato the best model is Johnson's formula, even if it tends to lightly underestimate the concentrations. For example the concentrations observed during November '89 and predicted by Johnson's formula are reported in Figure 6.

Model GEM has been applied for Via Senato only: in fact the typical canyon structure of this street speeds up the elaboration time of the model. Nevertheless the elaboration can last for several hours if the stationary condition is not reached. To obviate this problem, only the data referring to wind speed with component perpendicular to the street axis greater than 1 m/s have been processed. Moreover, to reduce computation time, whole traffic volume has been concentrated in the middle lane, since concentrations on pavement at man's height

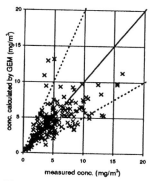

Figure 7: CO concentrations of November-December '89 in Via Senato: values predicted by GEM vs. observed values (wind speed > 1 m/s).

are similar to those obtained dividing the traffic into the lanes. Figure 7 shows the calculated values as a function of the observed concentrations.

5 Statistical parameters

A statistical evaluation was performed for each model and for each street using some statistical parameters recommended by the U.S. Environmental Protection Agency (EPA[13]; DiCristofaro[14]). Among these we can number NMSE, BIAS, FB (Fractional BIAS) and CPM (Composite Performance Measure), defined by:

$$NMSE = \frac{\overline{(C_C - C_M)^2}}{\overline{C_M} \cdot \overline{C_C}}$$

$$BIAS = \frac{\overline{C_C} - \overline{C_M}}{\overline{C_M}}$$

$$FB = \frac{2 \cdot (\overline{C_C} - \overline{C_M})}{\overline{C_M} + \overline{C_C}}$$

$$CPM = \frac{1}{2} AVG(AFB(i)) + \frac{1}{2} AFB(1)$$

where C_M is the measured concentration, C_C is the calculated concentration, the line above the concentrations means the average value, $AFB(1)$ is the absolute value of the Fractional BIAS for the one-hour averages, $AFB(i)$ is the absolute value of the Fractional BIAS for each diagnostic category i and $AVG(AFB(i))$ is the average of $AFB(i)$ weighted for each category.

To calculate CPM the dataset has been divided in three subsets referring to the following categories:

- wind speed ≤ 2 m/s and stability class neutral or stable;
- wind speed ≤ 2 m/s and stability class unstable;
- wind speed > 2 m/s and all stability classes.

The wind speed (v) ≤ 2 m/s and neutral/stable category is weighted more than the other two categories because of the importance of this category for regulatory modelling purposes. Thus, the average of $AFB(i)$ is

$$AVG(AFB(i)) = 0.5\ AFB(v \leq 2\ m/s,\ Neutral/Stable)\ +$$
$$+ 0.25\ AFB(v \leq 2\ m/s,\ Unstable)\ +$$
$$+ 0.25\ AFB(v > 2\ m/s,\ All\ stabilities)$$

Table 1 shows the values of the correlation coefficient r and statistical parameters $NMSE$, $BIAS$, FB and CPM calculated for each model and for each street considering the whole dataset (November-December '89). Besides the conclusions already drawn previously, it can be deduced that all the models tend to underestimate on the average the CO concentrations ($BIAS$ and FB values always negative).

With regards to the application of GEM to Via Senato (only using cases with perpendicular component of wind speed > 1 m/s), the values of the statistical parameters are similar to those obtained for CPB-3 model and Johnson's formula infact $r = 0.68$, $NMSE = 0.26$, $BIAS = -0.10$ and $FB = -0.11$.

To assess differences in model performance under different model regimes, for each site we have computed the values of r, $NMSE$, $BIAS$ and FB concerning three subsets related to the three meteorological categories considered. It can be noticed that, when wind speed is greater than 2 m/s, for

Tab. 1. Statistical parameters computed referring to observed and predicted concentrations of November and December '89 at each site.

Street	Model	r	NMSE	BIAS	FB	CPM
V.le Marche	CALINE4	0.58	0.35	-0.15	-0.16	0.19
	CPB-3	0.13	1.00	-0.24	-0.27	0.38
	JOHNSON	0.47	0.33	-0.01	-0.01	0.12
V. Monreale	CALINE4	0.53	0.46	-0.32	-0.38	0.35
	CPB-3	0.56	0.31	-0.16	-0.18	0.17
	JOHNSON	0.59	0.29	-0.19	-0.21	0.20
V. Senato	CALINE4	0.30	1.19	-0.51	-0.68	0.67
	CPB-3	0.74	0.34	-0.09	-0.10	0.13
	JOHNSON	0.79	0.22	-0.22	-0.25	0.25

every street *NMSE*, *BIAS* and *FB* values approaching more zero are those referring to Johnson's formula.

The performance of each model as regards the peak concentrations has been put in evidence for every site calculating *FB* and *CPM* for the 25-highest concentrations. It can be remarked that for the wind speed ≤ 2 m/s and neutral/stable category CPB-3 overestimates at all three sites. After a careful examination of the time series of the observed and predicted concentrations and of the meteorological parameters, it can be deduced that CPB-3 tends to overestimate particularly with atmospheric stability when wind speed is low (< 0.7 m/s) and wind direction is parallel to the street axis.

6 Conclusions

This study certainly is not exhaustive but could be extended in the future: models for example could be applied to several kinds of streets with more detailed information and an updated database.

In any case, the results point out that each of the three modelling techniques evaluated shows very different performances in simulating concentration of CO at the three streets under study. This is comprenhensible because every street represents a typical case in terms of emissions and urbanistic characteristics.

Evaluating the obtained results, it must bear in mind that the lack of some input data has caused some approximations. In fact the composite emission factors have been calculated for the vehicle fleet composition of the year 1991 and for estimated vehicle speeds. Moreover in some cases the meteorological parameters are measured not on the spot, but in the closest meteorological station. In spite of these approximations, we can draw the following conclusions:

- CALINE4 is the model which best describes the experimental data in the case of the broader road, Viale Marche, even if generally tends to underestimate the concentrations;
- CPB-3 model and Johnson's formula show the best outcomes for Via Monreale and Via Senato respectively, giving similar results.

It can be noticed that these specific urban canyon modelling techniques simulate quite well CO levels in Via Monreale even if this is not a typical street canyon like via Senato, but a street with buildings on both sides.

The results of the application of GEM to Via Senato are very interesting and encourages to go on with tests. Plans for the future could be to extend evaluation of this model to a more detailed database.

Acknowledgements

This research was supported by the CNR (Research National Council). We thank Urban Police Force and ATM (Azienda Municipale Trasporti) of Milan for useful informations. In particular we are grateful to Dr. G. Lanzani for providing very precious suggestions about application of model GEM.

References

1. Manini, G. & Angelino, E. *Modelli di dispersione del monossido di carbonio: applicazione ad alcune strade di Milano,* U.O.F.T.A.-P.M.I.P. di Milano, U.S.S.L. A.T. 38, Internal Report, Dec.1995.
2. Tebaldi, G. *Inquinamento atmosferico - novembre 1989,* Centro Operativo Provinciale (U.O.F.T.A.- P.M.I.P. di Milano), Provincia di Milano, 1989.
3. Tebaldi, G. *Inquinamento atmosferico - dicembre 1989,* Centro Operativo Provinciale (U.O.F.T.A.- P.M.I.P. di Milano), Provincia di Milano, 1990.
4. Benson, P.E. *CALINE4 - A Dispersion Model for Predicting Air Pollutant Concentrations near Roadways,* California Department of Transportation, Sacramento, California, Report No. FHWA/CA/TL-84/15, 1984.
5. Yamartino, R.J. & Wiegand, G. Development and evaluation of simple models for the flow, turbulence and pollutant concentration fields within an urban street canyon, *Atmospheric Environment,* 1986, **20**, No. 11.
6. Yamartino, R.J., Strimaitis, D.G. & Messier, T.A. *Modification of Highway Air Pollution Models for Complex Site Geometries,* DRAFT, Report No. FHWA-RD-89-112, 1989.
7. Johnson, W.B., Dabberdt, W.F., Ludwig, F.L. & Allen, R.J. *Field Study for Initial Evaluation of an Urban Diffusion Model for Carbon Monoxide,* Comprehensive Report CRC and Environ. Protect. Agency (EPA) Contract CAPA-3-68 (1-69), 1971.
8. Lanzani, G.G. *Analisi fisico-modellistica della dispersione nella bassa atmosfera del monossido di carbonio in area urbana,* thesis in Physics, Università degli Studi, Milano, academic year 1991-1992.
9. Lanzani, G. & Tamponi, M. A microscale lagrangian particle model for the dispersion of primary pollutants in an urban topography. Sensitivity analysis and first validation trials within a street canyon, *Atmospheric Environment,* 1995, **29**, 3465-3475.
10. Tamponi, M., Angelino, E., Tebaldi, G. & Belloni, E. *Studio modellistico sull'inquinamento atmosferico da traffico e da riscaldamento nell'area metropolitana milanese,* Provincia di Milano, 1992.
11. Bocola, W., Cirillo, M., Gaudioso, D., Trozzi, C., Vaccaro, R. & Napolitano, C. *Progetto CORINAIR: Inventario delle emissioni di inquinanti dell'aria in Italia nell'anno 1985,* RTI/Studi VASA (89)8, ENEA, Roma, 1989.
12. Eggleston, H.S., Gaudioso, D., Gorissen, N., Joumard, R., Rijkeboer, R.C., Samaras, Z. & Zierock, K.-H. *CORINAIR Working Group on Emission Factors for Calculating 1990 Emissions from Road Traffic,* DRAFT Final Report, December 1991.
13. EPA. *Evaluation of CO Intersection Modeling Techniques Using a New York City Database,* EPA-454/R-92-004, August 1992.
14. DiCristofaro, D., Strimaitis, D., Braverman, T. & Cox, W. The use of simultaneous confidence intervals to evaluate CO intersection models, pp. 503-511, *Air Pollution Modeling and its Application X,* Nato Challenges of Modern Society, Plenum Press, New York, 1994.

An introductory mathematical model and numerical analysis of the evolutionary movement of oil slicks in coastal seas: a case study[1]

J.F.C.A. Meyer, R.F. Cantão

Department of Applied Mathematics, UNICAMP - State University at Campinas, Campinas, Brazil

Abstract

A cooperative effort is being developed presently in São Paulo State to prepare both the State Environmental Agency (CETESB) and the National Petroleum Company (Petrobrás) for more efficient clean-up strategies and actions. Two state universities (USP and UNICAMP) participate in this work. It is one of the subprojects of a greater effort for the creation of an official manual for organizing protection and cleaning activities, in the case of accidental oil spills. A mathematical model is presented, justified and a numerical scheme for obtaining and exhibiting approximate solutions is presented. Two simulations in the São Sebastião Channel region, in São Paulo, Brazil are discussed.

1 Introduction

Over the last twenty two years, more than two hundred oil spills occurred in the São Sebastião Channel region, at a local oil terminal (Petrobrás/DTCS). This terminal processes about 55% of all brazilian transported oil. Efforts for environmental protection and preservation, for clean-up activities and for attributing responsibilities have not always been uniform, partially due to the many agencies involved. Only recently has a joint cooperation been set up. This cooperation effort is being developed by two state universities (UNICAMP and USP), the national Petroleum Company (Petrobrás) and the state environmental protection agency (CETESB), and includes many other aspects, ranging from the knowledge of pre-spill assessment to the post-spill attribution of responsibilities, besides the precise description of the regional circulatory dynamics. A part of this project is the transformation of convenient

---

[1]This work was sponsored by FAPESP

mathematical models into numerical schemes for qualitative prediction of oil slick movements in the region. This program, or programs must be available for in-site operations, and must run on whichever hardware is at hand at the moment. A deterministic model was chosen, in which a Diffusive-Advective Partial Differential Equation was chosen, due to the phase in which the oil spill is studied (the *diffusive-advective* phase, see Fay [6]). The numerical algorithm for approximating solutions of the original mathematical model should be able to qualitatively predict oil trajectories in several given scenarios, creating a manual to be used in helping local teams in making decisions for the allocation of resources and equipment in clean-up processes. The program can be used during the oil spills, but only if circulatory data, besides all other information, is available in a sufficiently precise format.

2 The Mathematical and the Discretized Models

2.1 The Mathematical Model

Since the height of an oil spill (5 cm or less) is so much smaller than the surface dimensions of the oil slick that merit the use of the program described below (hundreds of square meters, or more), we have adopted what most authors use: a diffusive-advective evolutionary model in two space dimensions, for an oil concentration indicated by $u = u(x,y;t)$, given by:

$$\frac{\partial u}{\partial t} + \text{div}(-\alpha\nabla u) + \text{div}(\mathbf{V}u) + \sigma u = f, \quad (x,y) \in \Omega \subset \mathbf{R}^2, \quad t \in [0,T]$$

$$u(x,y;0) = u_0(x,y), \quad (x,y) \in \Omega \quad \text{and}$$

$$-\alpha\frac{\partial u}{\partial \eta} + ku = g(x,y;t), \quad (x,y) \in \partial\Omega, \quad \text{for} \tag{1}$$

$$u, f:\Omega\times[0,T] \to \mathbf{R}, \quad g:\partial\Omega\times[0,T] \to \mathbf{R} \quad \text{and} \quad u_0:\Omega \to \mathbf{R}$$

where $\quad = \alpha(x,y;t,u)$ is the diffusivity, $\mathbf{V}(x,y;t)$ identifies the vector field of the local circulation effects, $\quad$ indicates the several degradation processes, k is a constant which depends upon the type of oil, and the chosen boundaries (coast and open seas) and a polluting source f.

Besides the given initial condition u_0, we will have two distinct behaviors on different parts of the boundary: the boundary in contact with land as Γ_1 (roughly, both "sides" of the channel, in one case, or mainland and island coasts, in the other) and Γ_2 for the sea boundaries. Reasonable simplifications lead to the following assumptions:

- no oil "crosses" the land boundaries, i.e., there is no flow across Γ_1:

$$\frac{\partial u}{\partial \eta}\bigg|_{\Gamma_1} = 0 \text{ and}$$

- flow across open sea boundaries is proportional to the oil concentration: $-\alpha \,\partial u / \partial \eta = ku$, along Γ_2, where k is given by $c(\mathbf{V} . \eta)$, where $\mathbf{V}$ is the current resultant and η is the external normal vector along Γ_2.

Some authors consider giving the oil evaporation a special treatment, different from other degradation processes. A possible model could be:

$$\frac{\partial \sigma}{\partial t} = \begin{cases} \kappa \dfrac{(1-\sigma)e^{-12\sigma}}{u} & \text{if } u > 0 \\ 0 & \text{if } u = 0 \end{cases}$$

but this does present a decay which happens in the very beginning of the second phase, and, in the period in which accidents are identified, most of the oil evaporation has already occurred, justifying the use of a general function for modelling all decay in (1). Due to the generality of the given parameters in the case of accidents, at best a qualitative prediction can be produced. Historical descriptions of previous oil spills can be used for validation of the models, permitting even further simplification of some of the involved phenomena.

2.2 The Simplified Model

We will consider α to be constant. This is not actually the case, but, for shorter periods of time, this assumption is acceptable (see Benqué *et al* [1], and Meyer [9]). Furthermore, due to the enormous difficulty in obtaining suitable data, $\mathbf{V} = (V_x, V_y)$ has been to chosen constant in certain parts of the channel, coinciding with generally accepted circulation models (Furtado [7]). Nevertheless, the program can accept precise information on circulation dynamics. For similar reasons, σ, the decay component is considered constant along the whole second phase period, describing the several different decays (Meyer [10]).

With these assumptions, system (1) becomes:

$$\frac{\partial u}{\partial t} - \alpha \Delta u + \mathbf{V} . \nabla u + (\sigma + \operatorname{div} \mathbf{V}) u = f, \quad (x,y) \in \Omega \subset \mathbf{R}^2, \quad t \in [0, T]$$

$$u(x, y; 0) = u_0(x, y)$$

$$-\alpha \frac{\partial u}{\partial \eta} + ku = 0 \qquad\qquad (2)$$

for $u, f : \Omega \times [0, T] \to \mathbf{R}$ and $u_0 : \Omega \to \mathbf{R}$

2.3 The Discretized Model

Our choice for approximating system (1) and (2) with Galerkin's Method for Second-order Finite Elements in space dimensions and Crank-Nicolson "half-steps" for time leads us naturally to a variational formulation. An additional advantage of this approach is the fact of being able to model phenomena with functions which do in fact lack continuity in space and time variables.

With usual notation, the variational formulation of (2) becomes:

$$\left(\frac{\partial u}{\partial t}\Big|v\right)_\Omega + \alpha\,(\nabla u|\nabla v)_\Omega + V_x\left(\frac{\partial u}{\partial x}\Big|v\right)_\Omega + V_y\left(\frac{\partial u}{\partial y}\Big|v\right)_\Omega + \left\langle\frac{\partial u}{\partial \eta}\Big|v\right\rangle_{\partial\Omega} +$$

$$+ (\sigma + \mathrm{div}\mathbf{V})(u|v)_\Omega = (f|v)_\Omega \tag{3}$$

for $\forall v \in \mathbf{S}$, a conveniently chosen space, where functions and their first derivatives can be integrated in the sense of Lebesgue.

This system is expressed in what is a natural choice for the above-mentioned discretizations, so a solution of equation (2) will be sought in a finite-dimensional subspace $\mathbf{S}_h$ of $\mathbf{S}$, the base of which is given by $\beta = \{\varphi_1,\ldots,\varphi_n\}$, the second-order finite elements. Each of these is a second-order polynomial in x and y on each triangle of the triangularization of the domain. Consequently,

$$u_h = \sum_{j=1}^{n} c_j(t)\,\varphi_j(x,y), \qquad \varphi_j \in \beta \subset \mathbf{S}_h \tag{4}$$

approximates the solution u of (2), with coefficients $c_j(t)$ that will determine the approximate solution in each time-step. This discretization procedure involves a separation of space $\varphi_j(x,y)$ and time $c_j(t)$ variables. Equation (3), for the chosen subspace, will therefore become a system of ordinary differential equations given by:

$$\sum_{j=1}^{n}\frac{dc_j}{dt}(\varphi_j|\varphi_i)_\Omega + \alpha\sum_{j=1}^{n}c_j(\nabla\varphi_j|\nabla\varphi_i)_\Omega + k\sum_{j=1}^{n}c_j\left\langle\varphi_j|\varphi_i\right\rangle_{\partial\Omega\backslash\Gamma_1} +$$

$$+ V_x\sum_{j=1}^{n}c_j\left(\frac{\partial\varphi_j}{\partial x}\Big|\varphi_i\right)_\Omega + V_y\sum_{j=1}^{n}c_j\left(\frac{\partial\varphi_j}{\partial y}\Big|\varphi_i\right)_\Omega + (\sigma + \mathrm{div}\mathbf{V})\sum_{j=1}^{n}c_j(\varphi_j|\varphi_i)_\Omega = (f|\varphi_i) \tag{5}$$

for $i = 1,2,\ldots,n$.

In addition, Crank-Nicolson's approximation will lead us to the expression given by:

$$\sum_{j=1}^{n} c_j^{(n+1)} \left[\left(1+(\sigma+\mathrm{div}\mathbf{V})\frac{\Delta t}{2}\right)(\varphi_j|\varphi_i)_\Omega + \alpha\,\frac{\Delta t}{2}\,(\nabla\varphi_j|\nabla\varphi_i)_\Omega + k\frac{\Delta t}{2}\left\langle\varphi_j|\varphi_i\right\rangle_{\partial\Omega\backslash\Gamma_1} + \right.$$

$$\left. +V_x\,\frac{\Delta t}{2}\left(\frac{\partial\varphi_j}{\partial x}\Big|\varphi_i\right)_\Omega + V_y\,\frac{\Delta t}{2}\left(\frac{\partial\varphi_j}{\partial y}\Big|\varphi_i\right)_\Omega \right] =$$

$$= \sum_{j=1}^{n} c_j^{(n)} \left[\left(1-(\sigma+\mathrm{div}\mathbf{V})\frac{\Delta t}{2}\right)(\varphi_j|\varphi_i)_\Omega - \alpha\,\frac{\Delta t}{2}\,(\nabla\varphi_j|\nabla\varphi_i)_\Omega - k\frac{\Delta t}{2}\left\langle\varphi_j|\varphi_i\right\rangle_{\partial\Omega\backslash\Gamma_1} \right.$$

$$\left. -V_x\,\frac{\Delta t}{2}\left(\frac{\partial\varphi_j}{\partial x}\Big|\varphi_i\right)_\Omega - V_y\,\frac{\Delta t}{2}\left(\frac{\partial\varphi_j}{\partial y}\Big|\varphi_i\right)_\Omega \right] + \Delta t (f^{(n+\frac{1}{2})}|\varphi_i)_\Omega \qquad (6)$$

for $i = 1, 2, \ldots, n$, and for:

$$\mathrm{div}\mathbf{V} = \frac{\partial\mathbf{V}_x}{\partial x} + \frac{\partial\mathbf{V}_y}{\partial y} = \sum_{l=1}^{N} V_x^{(l)}\frac{\partial\varphi_l}{\partial x} + \sum_{k=1}^{N} V_y^{(k)}\frac{\partial\varphi_k}{\partial y}$$

where N corresponds to the number of nodes in the discretized domain. We should note that the current-wind vector field is therefore also approximated by an interpolation of the second-order finite elements.

In matrix notation, appropriate choices for matrices $\mathbf{A}$ and $\mathbf{B}$ and for vector $\mathbf{d}$ lead to the linear system given by:

$$\mathbf{A}\mathbf{c}^{(n+1)} = \mathbf{B}\mathbf{c}^{(n)} + \mathbf{d}, \qquad (7)$$

to be solved in each time-step. For those parameters that vary with time, an update will be necessary at each step, using $\mathbf{A}^{(n)}$, $\mathbf{B}^{(n)}$ and $\mathbf{d}^{(n)}$.

This discretization scheme leads us to $O(\Delta x^2)$ and $O(\Delta y^2)$ approximations in space, besides the previously mentioned $O(\Delta t^2)$ in time. Special attention is given to the rates between velocity components, space discretization dimensions, and diffusivity, so that the Peclet kernel is respected (see Brooks *et al* [2]), since no other techniques are adopted at present.

3 Computational Aspects

From the point of view of the developed software, specialized and interrelated C++ classes were used to manipulate three main data groups:

3.1 The Discretized Domain

The choice for the present triangularization of the domain was that of available software in the computer equipment of the Applied Mathematics Department at UNICAMP: *PDEase2*. The results were adapted to fit a convenient file format. Although the present program will accept quite an extensive refinement, our choice has been to stop at an appropriate level (200 nodes or less, for the channel, or roughly 2000 nodes for the total region) of visual possibilities so qualitative information of reliable value is preferred to the usual numerical analyst's choice of greater refinement producing greater precision (program structure will accept and manipulate over 15000 nodes if necessary). In figure 1, we have one of the discretized domains.

3.2 Matrix Dimensions, Difficulties and Sparsity

The linear system indicated in (7) has matrices and vectors of a dimension equivalent to the number of nodes N in the chosen discretization of the domain. In order to be able to create a numerical scheme which could, on the one hand, be used in on-site executions, and, on the other, accept mesh refinements of a reasonable order, a sparsity structure has been developed. This is due to the fact that the mentioned matrices have about 7 non-null elements (and never more that 10) per row/line. This does not eliminate all problems in running the program for large values of N, since data structures grow with the necessary dimensions.

3.3 Visualization of Results

There is a difficulty inherent to the exhibition of the results: on the one hand, the finer the mesh, the better the numerical results, although they do demand a greater period to be obtained. On the other hand, however, large numbers of nodes tend to create images of difficult interpretation, the visualization of which is difficult even with colored outputs. Original choices for a mesh in the channel region with about 1000 nodes had to be abandoned for this reason, and that is exactly why the channel figures presented in this work do not exceed 275 nodes: the need for in-site interpretations. The greater domain, which includes the whole São Sebastião Island, however, illustrates the visualization difficulty: it uses about 1800 nodes. The use of the program, due to the need for minimum precision in a qualitative sense, may demand many more nodes, and has reasonably processed more than 3000 nodes over 1000 time steps in experimental executions.

4 Numerical Simulation

The domain and the discretization are indicated below, in figure 1. Due to the lack of sufficiently precise data in some of the important aspects of the model, coherent - although symbolic - values were used in the following examples. The prevailing currents (in accordance with Furtado [7]) are described by a vector constant by regions (although the program is designed to accept very general data, from circulatory dynamics data banks, for instance). The concentration of oil corresponds to the vertical height, the horizontal variables are pertaining to the local geography. The time-step in which the graph was sketched is indicated in each caption.

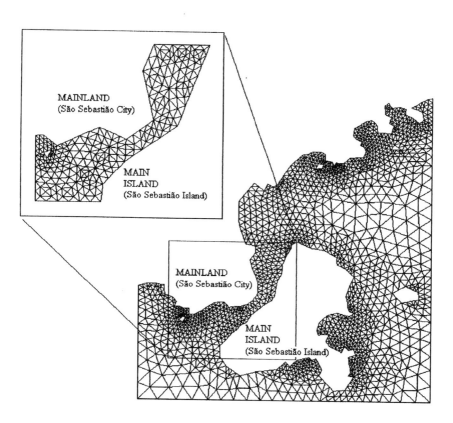

Figure 1: Discretized domain - São Sebastião channel - São Paulo state.

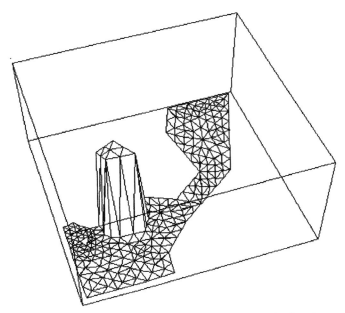

Figure 2a: Initial condition - Experiment using only the channel domain.

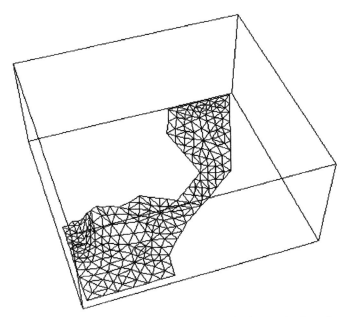

Figure 2b: Solution after 2000 time-steps illustrating diffusion and advection.

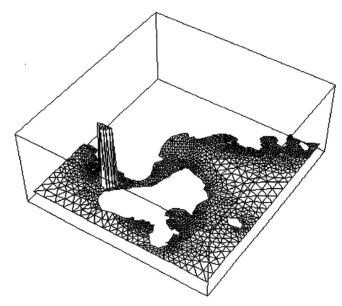

Figure 3a: Initial condition - Experiment using the whole island domain.

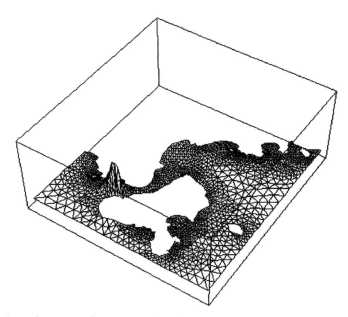

Figure 3b: Solution after 2000 time-steps: the oil slick can be seen to move up the channel.

Bibliographical References

1. Benqué, J. P., Hauguel, A. & Viollet, P. L. in *Engineering Applications of Computational Hydraulics II*, Pitman Advanced Publ. Programme, 1980.

2. Brooks, A. & Hughes, T. J. R. Streamline Upwind/Petrov-Galerkin Formulations for Convection Dominated Flows with Particular Emphasis on the Incompressible Navier-Stokes Equations, *Computer Methods in Applied Mathematics and Engineering*, 1982, **32**, 199-259.

3. Carbonel, C. Modelling of Coastal Ocean Response Induced by Wind Fields, *Laboratório Nacional de Computação Científica*, 1995.

4. Cuesta, I., Grau, F. X. & Giralt, F. Numerical Simulation of Oil Spills in a Generalized Domain, *Oil & Chemical Pollution*, 1990, 7, 143-159.

5. De Leon, D. A. S. Modelo Numerico para la Prediccion del Transporte y la Dispersión de Petroleo en el Oceano. *Revista del Instituto Mexicano del Petroleo*, 1986, 74-80.

6. Fay, J. A. The Spread of Oil Slicks on a Calm Sea. *Oil in the Sea*, 1969, Plenum Press, 53-63.

7. Furtado, V. V. Contribuição ao Estudo da Sedimentação Atual do Canal de São Sebastião - Estado de São Paulo. PhD Thesis, IG-USP, 1978.

8. Mackay, D., Buist, I., Mascarenhas, R., Paterson, S. Oil Spill Processes and Models, 1980, *Report of Environmental Canada, Research and Development Division*, Ottawa.

9. Meyer, J. F. C. A. Derrames de Petróleo em Águas Costeiras: Modelagem Matemática e Simulação Numérica, *in III° Simpósio de Ecossistemas da Costa Brasileira*, pp. 238-247, Serra Negra, São Paulo, Brasil, 1993.

10. Meyer, J. F. C. A., Monte, M. S. & Guimarães, A .F.: Modelagem e Simulação Numérica do Derramamento de Petróleo em Águas Costeiras, *Biomatemática*, **2**, 1992, 156-172.

11. Psaraftis, H.M. & Ziogas, B. O.: A Tactical Decision Algorithm for Optimal Dispatching of Oil Spill Clean-up Equipment. *Management Science*, December 1985, **31**, 1475-1491.

12. Stolzenbach, K. D., Madsen, O. S., Adams, E. E., Pollack, A. M. & Cooper, C.K. A Review and Evaluation of Basic Techniques for Predicting the Behaviour of Surface Oil Slicks. M.I.T. *Sea Grant Report*, 1977, **222**, Mass. Inst. of Technol.

13. Valencia, M.J.: East Asian Seas: Hypothetical Oil Spill Trajectories. *Mar. Pollut. Bull.*, **23**, 1991.

SECTION 3:
ENVIRONMENTAL SCIENCES
AND ENGINEERING

ADM: a model for water treatment in an anaerobic biological reactor

A. Elias,[a] G. Ibarra,[b] J. Ormazabal,[a] I. Murgia,[a] P. Zugazti[a]

[a]Dpto. I.Q. y Medio Ambiente, Alda, Urkijo s/n, 48013 Bilbao, Spain

[b]Dpto. Mecánica de Fluidos, Alda, Urkijo s/n, 48013 Bilbao, Spain

ABSTRACT

The analysis and control of a biological reactor is usually carried out inadequately, owing to lack of resources or apparent complexity of the whole process. The present work is based on the description of the algorithms and the use of the ADM model; this model, developed after empirical work in an anaerobic reactor at our laboratory, allows description of the reactor behaviour with a minimum number of variables. It represents a helpful tool for that working with anaerobic biological reactors in order to optimize production of methane and the quality of the treated wastewaters.

1. INTRODUCTION

The waste water treatment by biological methods has been an alternative to unjustly discredited purification. The starter of an anaerobic reactor means the adaptation of biomass to the waste waters. If this is fulfilled with controlled increase of flow and organic load (COD) or, (what amounts to the same), organic load rate (g COD/lr.d), the reactor will be ready to degrade the wastewater, and, likewise, it will be able to endure breakdowns, pollution overloads, etc. The starter of a treatment plant is made without thinking at this stage of adaptation on numerous occasions, it is a question of setting out to achieve high organic load rates by means of unsuitable sludges for them.

In the treatment plants, however austere they may be, they are equipped with measure systems of control parameters: flow, produced gas, and COD (organic load). These data are collected without analysis by the operator of the plant, because the own plant maintenance takes up most of the time. Nevertheless, if one resorts to striking a balance of COD (g COD/h), gaps are encountered because the real flow values are different from the indicated ones by the measuring teams due to the formation of biofilms which block pipes. Moreover, COD values lower than the fixed ones in the laboratory are introduced. This is due to the inevitable degradations that the wastewater suffers from the entrance to the plant or to difficulties of homogeneization in the taking of the sample. Therefore, this originates an unknowledge of the state of the reactor and, finally, blighted operation of the purification stage and the abandoning of the plant in many cases.

2. OBJECTIVES

The present work was performed with *Universidad del País Vasco* funding. Its objectives were to analyse the influence of acidified and non-acidified substrata in the adaptation and operation with UASB reactors. Subsequently, a shrewd approaching of the reactors behaviour (the aim of this paper) was carried out, endeavouring to facilitate the starter and operation control of an anaerobic digestion plant by means of the conventional parameters through the ADM model. This model will allow us to obtain essential data for the plant. The operator will be able to know if he/she must increase or decrease the organic load rate. Finally, this means in practice: the adaptation and maintenance of biomass, good performance in view of incidentals which arrive at the plant, breakdowns, etc.

3.METHODOLOGY

The system was installed two years ago. The adaptation of the granular sludge, and finally, the modelling was based on two UASB biological reactors (Up-flow anaerobic sludge bed), whose operation scheme can be seen in fig.1.

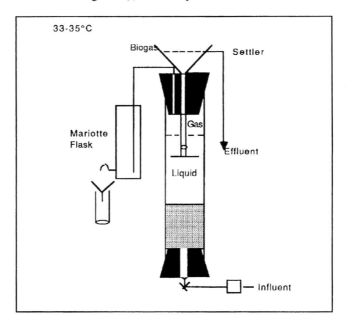

Figure 1.Diagram showing experimental laboratory equipment with the UASB reactor

The starter of the reactors was carried out inoculating sludge coming from industrial degradation plant, a starch plant. The selection of the sludge was performed among sampled sludges in alcohol, sugar and starch plants. The substrata fed to two independent reactors consisted of dissolution of glucose, (totally degradable substrate) and mixtures of volatile fatty acids.

These dissolutions simulated wastewaters, with known and controlled composition. The use of two different substrata originated the adaptation of the biomasses to influents, in one case acidified, and in the other one acidificable. The quality of the influent was guaranteed storing it at 5 $^\circ$C. The temperature of reaction was always between 32-34 $^\circ$C.

3.1. CONTINUOUS OPERATION

In addition to consecutive rises in organic load rate, the reaction systems were adapted to the substrata. The control parameters evaluated during the experimentation were, COD, flow, pH, temperature, VFA content, containment in solids, sludge volume index, % C, H, N among others.

The determination COD was carried out on filtered samples of effluent. The answer of the reactor to the degradation of organic load which was supplied was the production of biogas and an effluent which was believed to contain not more than 15% of the introduced COD. During all the starting and operation time of the reactors, the control of both of them was carried out by the COD balance in the system. This balance the distribution of the COD which comes into the reactor in the working flow. This organic load (COD/h) will be transformed into waste organic load in the influent, in biogas and a small percentage will be consumed in of biomass creation. The balance applied to both systems of reaction showed that the percentage of cells did not surpass 15% in either of the two reactors with eliminations of organic load of 90-95% and known the generated methane. The two reactors behaviour is reaching the state was different. The reactor adapted to glucose generated a very unsteady reaction system of biomass, owing to its biofilms and with frequent problems in overloading states. The adaptation of biomass with the mixture of acetic, propionic and butiric led to a steady reaction system, with dense biomass and with a good capacity of in case of overloading.

3.2. VERIFICATION OF EXPERIMENTAL DATA

The database which is going to be studied consisted of experimental data of influent and effluent organic load (CODi, CODe: g/l), flow (Q: l/d), biogas production (g $COD-CH_4$/l.d) and pH which were measured daily during months of operation. The values in pseudo stationary regimen (%elim. COD = 90-95) were considered as significative data for a given organic loading data. Verification was carried out by the COD balance in the system (cause-effect). The obtained results, for the reactor fed with glucose as well as for the reactor fed with VFA, showed valid correlations.

4. ADM APPLICATIONS

From the verified data, the ADM model allows us to predict the answer of the reactor as well as to determine the reaction state.

With regard to precautionary measure, we can introduce the flow value and COD and the pH feeding to the reactor, analyzed and measured in plant. After the ADM model, the plant operator can determine the COD value which is achieved in the effluent, as weel as the biogas production. This application is interesting when it is noted that an overloaded waste arrives at the plant, a situation in which the reactor needs a lower feeding flow in order to avoid toxic situations and to maintain the same capacity of degradation and production of gas in the reactor. Another unexpected situation opposite to the already afore mentioned can be the load decreasing in the influent. The model ADM will provide the operator with the flow increase in order to keep the same organic load rate in the reactor.

Concernig operation control, with the same data, we could examine whether the methane production of the reactor is of a steady state of operation, or whether its value is much lower than the one forecasted, having to give time for the reaction system to be adapted before performing an organic loading rate increase. Likewise when combining the rest of the parameters.

This control avoids the uncontrolled loading rate increasing. Therefore, the biomass consequently does not find itself in contact with high toxic contents of degradation products (VFA), and that the reactor keeps a constant biomass content in the same one, without being dragged by the excessive flow.

5. OPERATIONAL LIMITS

The use of this model has been valid for the experimented sludge type and in a UASB reaction system. The VSS contained in the reactor in the starter, must come into the minima based on our experiments and collected in the bibliography. Inevitably, a rector has a maximum vco in operation continuous, that is to say, a feeding value of organic load from which the ADM model will show us clear desastibilization of the system, and thus the order of decreasing vco through the flow or the COD value.

At present, anaerobic reactors are being extracted with different inoculants; a stage which is inevitably slow. However, once they are adapted, the reaction systems will be able to allow the verification of the ADM model, to anaerobic sludges from another. Besides, the influence of the reaction temperature would be tested, performing the digestions at lower temperatures, which implies a reduction of the process on scale plants. The application of this study, in the process of completion, would be the verification in large scale plants

6. MATHEMATICAL STEPS.

The method followed to model reactors RM and RT was based upon an stochastic approach. This kind of methods have been widely used in many fields of science when a need for modelling appears. It is based on the principle of the black box. This means that no attempt is made to predict the output of the system, (in this case RM and RT UASB reactors) with previously known chemical reaction and formulas which will describe step by step the behaviour of the influent inside the reactor.

In our case, the input data of the influent employed were its pH, flow rate and DQO and the output was its pH, DQO and produced methane.

Therefore 6 variables were used for the study as can be seen in Table I.

TABLE I. Variables used for this work in both types of reactors.

INPUT VARIABLE NAME	OUTPUT VARIABLE NAME	REACTOR RM/RT.
DQOAL	DQOSAL	DQO
PHAL	PHFUNC	PH
QAL	-----------------------	FLOW RATE
-----------------------	CH$_4$SAL	CH$_4$

After a work of years and after disregarding measured data in which not all variables were present, 129 valid cases with six measures each were available for RM reactor and 121 for RM reactor. With the help of a statistical package (SPSS) both group of data were found to fit best to the following expressions:

RM

CH4SAL = EXP(-13.65 + 1.24LN(QAL)-0.194LN(PHAL) + 1.22LN(DQOAL) + 0.125LN(DQOSAL))

DQOSAL = (EXP(-102.72 + 4.771LN(QAL)-12.972LN(PHAL) + 10.182LN(DQOAL) - 2.183LN(CH4SAL)))^0.25

RT

CH4SAL = EXP(-8.47 + 0.846LN(QAL)-2.158LN(PHAL) + 1.32LN(DQOAL) - 0.193LN(DQOSAL))

DQOSAL = (EXP(-91.407-5.126LN(QAL) + 43.049LN(PHAL) + 6.55LN(DQOAL) - 0.56LN(CH4SAL)))^0.25

These analytical expressions are the base for the ADM model which also includes a subroutine to take into account the volume of the reactor, so that ADM can be used in different anaerobic UASB reactors.
In the above pair of expressions for each reactor, the methane output (CH4SAL) not only depends on the input variables but also on the other output DQOSAL and, inversely DQOSAL also depends on CH4SAL. To cope with this, ADM uses a first estimate of CH4 with data only from input variables, and after an iterative proccess, true values for DQOSAL and CH4SAL are reached. In figure number 1 to 8 measured data in the reactor are plotted against results from the model for reactors type RM and RT.

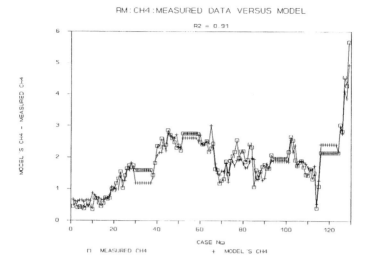

Figure 2. RM reactor. CH4 measured versus simulated. Case by case.

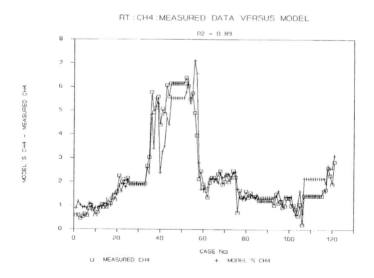

. . Figure 3. RT reactor. CH4 measured versus simulated. Case by case.

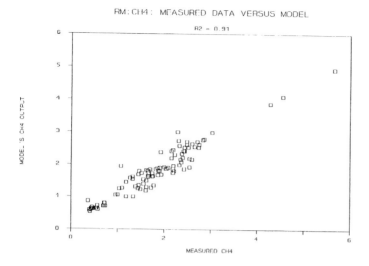

Figure 4. RM reactor. CH4 measured versus simulated.

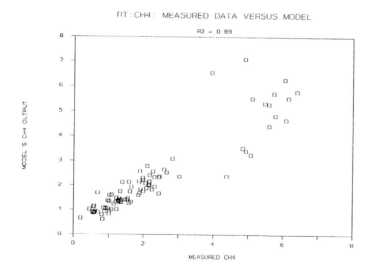

Figure 5. RT reactor. CH4 measured versus simulated.

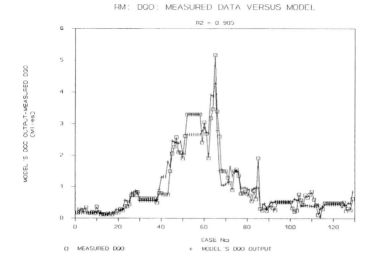

Figure 6. RM reactor. DQO measured versus simulated. Case by case.

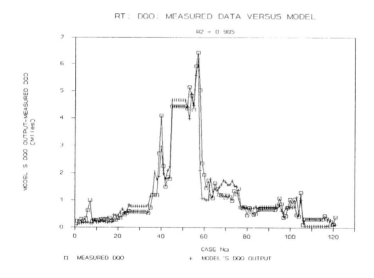

Figure 7. RT reactor. DQO measured versus simulated. Case by case.

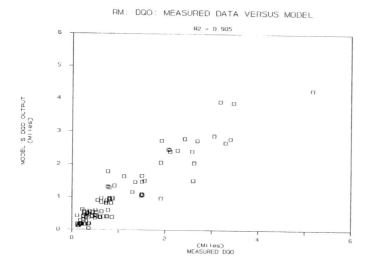

Figure 8. RM reactor. DQO measured versus simulated.

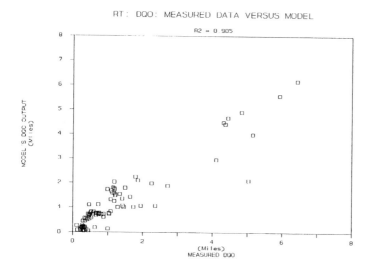

Figure 9. RT reactor. DQO measured versus simulated.

7. CONCLUSIONS.

The parameter used to test the goodnes of fit was the typical used in these cases: Correlation coefficient (R^2)

Table II. R^2. Goodness of fit indicator.

	R^2
RM CH4	0.91
RM DQO	0.905
RT CH4	0.89
RT DQO	0.905

The indicator shows a very good fit of the model to experimental data. The results show a better agreement between model and reality for RM reactor and CH4, being the "worst" one for RT reactor and DQO. However it can be concluded that for the four output variables, ADM model can succesfully predict reactor behaviour out of input variables' data only.

REFERENCES

1.   **Dold, P.L. and Marais, G.V.R (1986).***Evaluations of the general activated sludge model proposed by the IAWPRC task group.*Water Science and Technology 18 (6). pp. 63-89.

2.   **Costello, D.J., Greenfield, P.F. and Lee, P.L. (1991).***Dynamic modelling of a single step high rate anaerobic reactor. I Model Derivation.*Water Res. 25 . pp. 847-858.

3.   **Keller, J., Rosuli, M. and Lee, P.L. (1993).***Dynamic modelling simulation and verification of two stage high rate anaerobic treatment process with recycle.*Water Science and Technology. 28 (11-12). pp. 197-207.

4.   **Gunn, H.G., Rehberg, G.M., Rippon D.W. (1980).***Operator training and plant start-up for a regional water reclamation project.* Journal WPCE. 52.0 pp. 2351-2366.

5.   **Metzger, M. (1994).***Modeling and simulation of time-delay element via advective equation.* Proceedings of the European Simulation Multiconference ESM 94. Barcelona 1-3 June 1994. Simulation Councils Inc.

State space modelling of horizontal contaminant transport processes by extending discrete vortex simulations

M. Nonaka,[a] N.H. Thomas[b]

[a]*Field of Geosystems Engineering, Graduate School, The University of Tokyo, Tokyo 113, Japan*
[b]*School of Chemical Engineering, The University of Birmingham, Birmingham B15 2TT, UK*

Abstract

To diagnose and predict the contaminant transport process in large scale geosystems using a non-linear filtering strategy the state space model has been derived from simulating horizontal contaminant dispersion by extending the discrete vortex method. The dispersion area S and the dispersion rate $\dot{S}$ have been taken into account for evolving the state space model of the horizontal contaminant transport process. Eventually, $\dot{S}/S$ has been found to be represented by a function composed of exponential decay terms, which is transformed to the autonomous state equation and serves as the state space model accompanied with the observation equation. The state equation is extended to involve a noise process corresponding to disturbances such as viscous diffusion and inhomogeneity of the vortex strength as observed in real systems. The parameters involved in the state equation are adaptively estimated from the current state observations using a non-linear filtering algorithm. The state space model has been verified by a series of numerical demonstrations carried out using the discrete vortex method.

1. Introduction

It should be essential for the environmental protection to diagnose the current state and to predict the future state of a contaminant transport system. Contaminant transport processes have been numerically analysed by using the advection-diffusion model[1-5], Lagrangian modelling[6] or Monte Carlo simulation[7,8]. Stochastic equations have also been used for modelling pollutant diffusion processes[9-11]. Moreover, the time series models have been applied to predict the contaminant transport behaviour in the atmosphere[12] and in the lake[13]. Although these methodologies may be useful to evaluate the pollution state in a localised area, the more sophisticated model has to be developed to diagnose and predict the contaminant transport behaviour in a large scale geosystem such as the ocean or the atmosphere.

The discrete vortex method[14] has been used to simulate the complicated flow behaviour with large Reynolds number for saving the calculation time instead of conventional large code simulations. The discrete vortex simulation should be useful to diagnose and predict contaminant transport in a large scale flow system. However, it is generally impossible to know all of the initial conditions before starting the simulation and to catch information on all vortices on the way of calculation. Only when the space co-ordinates and/or velocities of the marked vortices are observable the discrete vortex method can directly take part in state space modelling, which is, however, unrealistic in any large scale geosystems. Therefore, we have to build up a state space model to be applied to estimate the state variables by adapting the parameters referring the current state observations in time variant systems and under unknown initial conditions, which seems to be realisable by extending the discrete vortex simulation.

Although contaminant transport processes have to be evaluated generally in three dimensional systems it is also significant to discuss on the horizontal dispersion processes to simply search for the capability of the discrete vortex method. In this paper we focus on the dispersion area and the dispersion rate in the horizontal contaminant transport process to evolve the state space model. In the second section we offer the discrete vortex system defined in a two dimensional flow field, which is mathematically analysed and extended to build up the state space model. In the third section the state equation derived from the theoretical discussion is verified by a series of numerical demonstrations. Finally, the concluding remarks are summarised.

2. State Space Modelling

To introduce the discrete vortex method for describing contaminant dispersion we first regard the minimum scale of contaminants as a vortex, i.e. the i-th element of contaminants corresponds to the i-th element of vortices. The dispersion area of contaminants starting from the points around the origin is defined by

$$S(t) = \frac{\pi}{4} \sum_{i=1}^{M} \left[\{x_i(t) - x_o(t)\}^2 + \{y_i(t) - y_o(t)\}^2 \right] \qquad (1)$$

in a horizontal dispersion system where $S(t)$ is the dispersion area at time t, $x_i(t)$ and $y_i(t)$ are the space co-ordinates of the i-th element of contaminants, $x_o(t)$ and $y_o(t)$ are the space co-ordinates of the origin translated with the field flow and M is the number of contaminant elements. Differentiating Equation (1) the dispersion rate is written as

$$\dot{S}(t) = \frac{\pi}{2} \sum_{i=1}^{M} \left[\{x_i(t) - x_o(t)\}\{\dot{x}_i(t) - \dot{x}_o(t)\} + \{y_i(t) - y_o(t)\}\{\dot{y}_i(t) - \dot{y}_o(t)\} \right] \quad (2)$$

The vortices are generated in the flow system with velocity, density and viscosity distributions, which are transported along with the flow of the field. We can approximately regard the flow as the perfect fluid with a constant density and without viscosity when the Reynolds number is large enough to disregard viscous diffusion of the vortices. According to the Kelvin-Helmholz's theorem[15] the distribution of vortices in the perfect fluid is replaced by the discrete distribution of vortex filaments. Therefore, it is possible to simulate the flow by analysing the Lagrangian behaviour of the vortex filaments. The discrete vortex method has been developed based on these considerations and applied to evaluate various kinds of flow problems[14,16-18] for saving the calculation time. Adopting the two dimensional discrete vortex formulation the velocities of the i-th vortex in the x- and y-directions are given by

$$\dot{x}_i(t) = U(x_i, y_i) + \frac{1}{2\pi} \sum_{\substack{j=1 \\ j \neq i}}^{N} \Delta\Gamma_j \cdot \frac{y_i(t) - y_j(t)}{\{x_i(t) - x_j(t)\}^2 + \{y_i(t) - y_j(t)\}^2} \quad (3)$$

$$\dot{y}_i(t) = V(x_i, y_i) - \frac{1}{2\pi} \sum_{\substack{j=1 \\ j \neq i}}^{N} \Delta\Gamma_j \cdot \frac{x_i(t) - x_j(t)}{\{x_i(t) - x_j(t)\}^2 + \{y_i(t) - y_j(t)\}^2} \quad (4)$$

where $U(x_i, y_i)$ and $V(x_i, y_i)$ are the velocities of the field flow in the x- and y-directions at the point where the i-th vortex exists, $\Delta\Gamma_j$ is the strength of the j-th vortex and N is the total number of vortices in the flow field. The origin also moves along with the field flow and the moving velocities are given by

$$\dot{x}_o(t) = U(x_o, y_o) \quad (5)$$
$$\dot{y}_o(t) = V(x_o, y_o) \quad (6)$$

Regarding $[x_i(t) - x_o(t)]$ and $[y_i(t) - y_o(t)]$ as the new variables $x_i(t)$ and $y_i(t)$, respectively Equations (3) and (4) are written as

$$\dot{x}_i(t) = U(x_i', y_i') - U(x_o', y_o') + F(x_i, y_i) \equiv f(x_i, y_i) \quad (7)$$
$$\dot{y}_i(t) = V(x_i', y_i') - V(x_o', y_o') + G(x_i, y_i) \equiv g(x_i, y_i) \quad (8)$$

where the variables with prime denote the space co-ordinates before transformation. The Jacobian matrix of the non-linear system described by Equations (7) and (8) is given by

$$J = \begin{bmatrix} \dfrac{\partial f(x_i, y_i)}{\partial x_i} & \dfrac{\partial f(x_i, y_i)}{\partial y_i} \\ \dfrac{\partial g(x_i, y_i)}{\partial x_i} & \dfrac{\partial g(x_i, y_i)}{\partial y_i} \end{bmatrix} \quad (9)$$

The Jacobian matrix defined at $x_i = <x_i>$ and $y_i = <y_i>$ $(t = <t>)$ is diagonalised by using the eigenvalues $\lambda_{1,i}$ and $\lambda_{2,i}$ and is written as

$$J\Big|_{x_i = <x_i>, y_i = <y_i>} = \begin{bmatrix} \lambda_{1,i} & 0 \\ 0 & \lambda_{2,i} \end{bmatrix} \quad (10)$$

Therefore, Equations (7) and (8) are linearised around $x_i = <x_i>$ and $y_i = <y_i>$ as given by

$$\dot{x}_i(t) = \lambda_{1,i}\left\{x_i(t) - <x_i>\right\} + f(<x_i>,<y_i>) \tag{11}$$

$$\dot{y}_i(t) = \lambda_{2,i}\left\{y_i(t) - <y_i>\right\} + g(<x_i>,<y_i>) \tag{12}$$

Equations (11) and (12) can be analytically solved in the vicinity of $x_i = <x_i>$ and $y_i = <y_i>$. The solutions are written as

$$x_i(t) = \alpha_i \exp(\lambda_{1,i}\,t) + \gamma_i \tag{13}$$

$$y_i(t) = \beta_i \exp(\lambda_{2,i}\,t) + \delta_i \tag{14}$$

where $\alpha_i, \beta_i, \gamma_i$ and δ_i are the time variant parameters and the following relationships are given:

$$\alpha_i = \frac{<x_i> - \gamma_i}{\exp(\lambda_{1,i}<t>)} \quad ; \quad \gamma_i = <x_i> - \frac{f(<x_i>,<y_i>)}{\lambda_{1,i}} \tag{15}$$

$$\beta_i = \frac{<y_i> - \delta_i}{\exp(\lambda_{2,i}<t>)} \quad ; \quad \delta_i = <y_i> - \frac{g(<x_i>,<y_i>)}{\lambda_{2,i}} \tag{16}$$

Regarding $\alpha_i, \beta_i, \gamma_i, \delta_i, \lambda_{1,i}$ and $\lambda_{2,i}$ at $(n-1)\Delta t \le t < n\Delta t$ as $\alpha_i\{(n-1)\Delta t\}, \beta_i\{(n-1)\Delta t\}, \gamma_i\{(n-1)\Delta t\}, \delta_i\{(n-1)\Delta t\}, \lambda_{1,i}\{(n-1)\Delta t\}$ and $\lambda_{2,i}\{(n-1)\Delta t\}$, respectively, Equations (13) and (14) are rewritten as

$$x_i(n\Delta t) = \left[x_i\{(n-1)\Delta t\} - \gamma_i\{(n-1)\Delta t\}\right]\exp\left(\lambda_{1,i}\Delta t\right) + \gamma_i\{(n-1)\Delta t\} \tag{17}$$

$$y_i(n\Delta t) = \left[y_i\{(n-1)\Delta t\} - \delta_i\{(n-1)\Delta t\}\right]\exp\left(\lambda_{2,i}\Delta t\right) + \delta_i\{(n-1)\Delta t\} \tag{18}$$

at $(n-1)\Delta t \le t < n\Delta t$.

Using the transformed space co-ordinates the dispersion area and the dispersion rate are given by

$$S(t) = \frac{\pi}{4}\sum_{i-1}^{M}\left\{x_i^2(t) + y_i^2(t)\right\} \tag{19}$$

$$\dot{S}(t) = \frac{\pi}{2}\sum_{i-1}^{M}\left\{x_i(t)\dot{x}_i(t) + y_i(t)\dot{y}_i(t)\right\} \tag{20}$$

Taking into account the function $\dot{S}(t)/S(t)$ it is discretely written as

$$\frac{\dot{S}(n\Delta t)}{S(n\Delta t)} = \frac{2\sum_{i-1}^{M}\left\{x_i(n\Delta t)\dot{x}_i(n\Delta t) + y_i(n\Delta t)\dot{y}_i(n\Delta t)\right\}}{\sum_{i-1}^{M}\left\{x_i^2(n\Delta t) + y_i^2(n\Delta t)\right\}}$$

$$= \frac{\sum_{i-1}^{M}\sum_{j-1}^{n-1}\eta_j^i\exp\left(\xi_j^i\,j\Delta t\right)}{\sum_{i-1}^{M}\sum_{j-1}^{n-1}\left\{\varsigma_j^i\exp\left(\theta_j^i\,j\Delta t\right) + \omega_j^i\right\}} \tag{21}$$

We can recognise

$$\lim_{n \to \infty} \frac{\dot{S}(n\Delta t)}{S(n\Delta t)} = 0 \quad ; \quad \lim_{n \to 0} \frac{\dot{S}(n\Delta t)}{S(n\Delta t)} = c \quad (> 0) \tag{22}$$

When the numerator and the denominator in Equation (21) are divided by the most dominant terms in the denominator and the denominator is evolved by the Taylor expansion we can finally expand the function of Equation (21) as a sum of exponential terms. Furthermore, Equation (21) should be non-singular at $t = j\Delta t \ (j = 1, \cdots, n)$. Eventually, Equation (23) is simply written as

$$\frac{\dot{S}(t)}{S(t)} = \sum_{k=1}^{L} a_k \exp(-b_k t) \tag{23}$$

where and a_k and b_k are the parameters to be adaptively identified and L is the number of exponential terms, which are determined by the horizontal dispersion characteristics. We can also derive

$$\ln \frac{S(t)}{S(0)} = \sum_{k=1}^{L} \left[\frac{a_k}{b_k}\right] \left[1 - \exp(-b_k t)\right] \tag{24}$$

from Equation (23) where $S(0)$ is the initial dispersion area.

Although Equation (23) constitutes a state space equation it might be convenient for estimation to drive the autonomous form from Equation (23). Now putting

$$X_1 = S(t), \ X_2 = \dot{S}(t), \ X_3 = S_1(t), \ \cdots, \ X_{L+2} = S_L(t)$$
$$S_k(t) = a_k \exp(-b_k t) \quad (k = 1, \cdots, L) \tag{25}$$

the autonomous state equation is formulated by

$$\begin{bmatrix} \dot{X}_1 \\ \dot{X}_2 \\ \dot{X}_3 \\ \cdot \\ \dot{X}_{L+1} \\ \dot{X}_{L+2} \end{bmatrix} = \begin{bmatrix} 0 & 1 & 0 & \cdot & 0 & 0 \\ \sum_{k=1}^{L} -b_k X_{k+2} & \sum_{k=1}^{L} X_{k+2} & 0 & \cdot & 0 & 0 \\ 0 & 0 & -b_1 & \cdot & 0 & 0 \\ \cdot & \cdot & \cdot & \cdot & \cdot & \cdot \\ 0 & 0 & 0 & \cdot & -b_{L-1} & 0 \\ 0 & 0 & 0 & \cdot & 0 & -b_L \end{bmatrix} \begin{bmatrix} X_1 \\ X_2 \\ X_3 \\ \cdot \\ X_{L+1} \\ X_{L+2} \end{bmatrix} \tag{26}$$

where $\begin{bmatrix} X_1 & X_2 & \cdots & X_{L+2} \end{bmatrix}^T$ denotes the state vector.

In order to characterise real horizontal dispersion processes in large scale geosystems viscous diffusion should be taken into account in the discrete vortex simulation. Moreover, we have to account for inhomogeneity of the vortex strength in the real flow field. Therefore, the discrete vortex simulation has to be extended so as to account for these disturbances. We regard the disturbances as a noise process embedded within Equation (23) or Equation (26). The available observation variables are $S(t)$ and/or $\dot{S}(t)$, i.e. X_1 and/or X_2 in Equation (26). However, the parameters have to be adaptively identified in any non-linear estimation algorithm and hence, the observation equation is written as

$$\begin{bmatrix} Z_1 & Z_2 \end{bmatrix}^T = \begin{bmatrix} H \end{bmatrix} \begin{bmatrix} X_1 & X_2 & \cdot & X_{L+2} \end{bmatrix}^T \tag{27}$$

where $\begin{bmatrix} Z_1 & Z_2 \end{bmatrix}^T$ is the observation vector and $\begin{bmatrix} H \end{bmatrix}$ is the 2 x (L+2) observation matrix. The observation noise is accompanied with Equation (27) when the observation environment is noisy.

Equations (26) and (27) accompanied with respective noise process constitute the state space model of the horizontal contaminant dispersion process. Consequently, it is possible to apply any adaptive estimation algorithm to the state space model, which contributes to diagnose the current state and to predict the future state of the horizontal dispersion process. The parameters involved in Equation (26) are identified by referring the current observation variables, i.e. the dispersion area and/or the dispersion rate. However, it takes a long calculation time to identify the parameter L from the current observations. Therefore, it is reasonable in real systems to identify the parameter a_k and b_k assuming an appropriate L.

3. Verification of the State Space Model

In order to verify the theoretical derivation of the state space model the contaminant transport processes dominated by horizontal dispersion have been simulated by using the discrete vortex method. The characteristic function $\ln[S(t)/S(0)]$ has been calculated by regarding a part of vortices as contaminants and compared with Equation (24). As a demonstration here we suppose two models of contaminant dispersion processes, one of which is that the contaminant behaviour is determined by the vortices uniformly distributed around the contaminants, called small eddy dispersion. Another is that the contaminants are dispersed by the dominant large eddies composed of vortices and distributed around the contaminant eddy, called large eddy dispersion. The vortex strength is assumed to be unity and the initial space co-ordinates of vortices are determined by using uniform random numbers generated in a given range.

In small eddy dispersion we focus on the system where the contaminants composed of 20 vortices are surrounded by other 60 vortices. On the other hand, in the large eddy system the contaminants composed of 20 vortices are included in a large eddy which is surrounded by 10 large eddies with 10 vortices, respectively. The trajectories of all of the vortices have been calculated by applying the Runge-Kutta-Gill method to Equations (3) and (4). The dispersion process has been simulated in the systems where disturbances are embedded within the uniform flow field, which is regarded as the Gaussian white noise and generated by the Gaussian random numbers with the variance of 100. The results are shown in Figures 1 and 2. The non-linear least squares method proposed by Marquardt[19] has been applied to estimate the parameters of the characteristic function. Although L in Equation (24) should be optimally decided so as to minimise the estimation error we simply assume $L=2$ here. The dispersion trajectories seem to be characterised by Equation (24) with noise. Assumed four parameters in Equation (24) have been estimated as given by $a_1=7.20$, $b_1=6.67$, $a_2=0.425$ and $b_2=0.323$ in Figure 1 and $a_1=5.38$, $b_1=3.96$, $a_2=0.395$ and $b_2=0.235$ in Figure 2.

The contaminant dispersion processes have also been simulated in the flow field with the velocity distribution and disturbances. The dispersion processes evaluated in the flow with the uniform velocity gradient of -0.5 in the y-direction are shown in Figures 3 and 4 where the same initial conditions and disturbances are given as those in Figures 1 and 2, respectively.. As can be seen, although the dispersion patterns are different from Figures 1(b) and 2(b) both dispersion processes are characterised by Equation (24) accompanied with noise processes. The Marquardt's method has also been applied to estimate four parameters in Equation (24), in which $a_1=5.82$, $b_1=5.07$, $a_2=0.342$ and $b_2=0.114$ in Figure 3 and $a_1=9.68$, $b_1=15.8$, $a_2=1.60$ and $b_2=0.816$ in Figure 4 are identified.

We regard this demonstration as convincing evidence of the capability of the discrete vortex method to state space modelling of horizontal contaminant dispersion processes and the validity of the theoretically derived state space model as an useful tool for diagnosing and predicting the horizontal contaminant dispersion states in large scale geosystems instead of complicated conventional large code simulations. The state space model is easily incorporated within the non-linear estimation algorithms[20], in which especially the SEEK algorithm[21,22] or the SEEK-FIND algorithm[23] plays a role in accurately estimating the non-linear dispersion processes.

4. Conclusions

It should be essential for the geoenvironmental protection to diagnose the current state and to predict the future state of the pollution process. The discrete vortex method has recently been applied to analyse complicated flows with high Reynolds numbers for saving calculation time. In order to search for the capability of the discrete vortex method to state space modelling of contaminant transport processes. the horizontal dispersion processes have been theoretically evaluated by the two dimensional vortex element analysis and the state space model has been derived from extending the discrete vortex simulation. Two state variables, i.e. the dispersion area $S(t)$ and the dispersion rate $\dot{S}(t)$ have been taken for evolving the state space model of the horizontal dispersion process. Eventually, $\dot{S}(t)/S(t)$ has been found to be written as a function composed of exponential decay terms involving parameters adaptively identified by referring the current observations. The state space model is extended to account for disturbances corresponding to viscous diffusion and inhomogeneity of the vortex strength, which is achieved by incorporating a noise process. The state space model has been verified by a series of numerical demonstrations carried out by using the discrete vortex method and the simulation results have supported well the theoretically derived state space modelling. This demonstration is regarded as convincing evidence of the capability of the discrete vortex method to state space modelling of the horizontal dispersion process and the validity of the state space modelling as an useful tool for diagnosing and predicting the horizontal contaminant dispersion state in large scale geosystems instead of complicated conventional large code simulations. The state space model is easily incorporated within the non-linear estimation algorithms, in which especially the SEEK and SEEK-FIND algorithms plays a role in adaptively estimating horizontal contaminant dispersion processes.

Acknowledgement

A part of this work is aided by The Steel Industry Foundation for the Advancement of Environmental Protection Technology. The authors are grateful to The Foundation for its financial support.

REFERENCES

1. Lomen, D. O. : Modeling, Identification and Control in Environmental System (Vansteenkiste, G. C., ed.), p. 361-378, (1978), North-Holland, Amsterdam
2. Zlatev, Z., Berkowicz, R. and Prahm, L. P. :   Atmospheric Environment, Vol. 17, p. 491-499, (1983)
3. Pozlewicz, A. : Hydraulic and Environmental Modelling of Coastal, Estuarine and River Waters (Falconer, R. A., Goodwin, P. and Matthew, R. G. S. eds.), Co., , p. 381-389, (1989), Gower Publishing, Aldershot, England
4. Ozmidov, R. V. : Diffusion of Contaminants in the Ocean, (1990), Kluwer Academic publishers, Dordrecht, The Netherlands
5. Nihoul, J. C. J. : Environmental Modeling, Vol. 1, (Zannetti, P. ed.), p. 75-140, (1993), Computational Mechanics Publications, Southampton, England
6. Muller, K. H. : Applied Mathematical Modelling, Vol. 11, p. 104-109, (1987)
7. Horie, T. : Technical Note of The Port & Harbour, Research Institute of Japan, No. 360, p. 1-222, (1980)
8. Brusasca, G., Tinarelli, G., Moussafir, J., Biscay, P., Zannetti, P. and Anfossi, D. : Computer Techniques in Environmental Studies, , p. 431-450, (1988), Computational Mechanics Publications, Southampton, England
9. Monin, A. S. and Yaglom, A. M. : Statistical Fluid Mechanics, Vol. 1,2, (1975), MIT Press, Cambridge, MA
10. Seinfeld, J. H. : Atmospheric Chemistry and Physics of Air Pollution, Chap. 13,15, (1986), Wiley Interscience, New York.
11. Omatsu, S., Seinfeld, J. H., Soeda, T. and Sawaragi, Y. : Automatica, Vol. 24, p. 19-29, (1988)
12. Sawaragi, Y., Soeda, T., Tamura, H., Yoshimura, T., Ohe, S., Chujo, Y. and Ishihara, H. : Automatica, Vol. 15, p. 441-451, (1979)
13. Tanaka, M. and Katayama, T. : Transactions of the Institute of Systems, Control and Information Engineers, Vol. 1, p. 117-126, (1988)
14. Lewis, R. I. : Vortex Element Methods for Fluid Dynamic Analysis of Engineering Systems, (1991), Cambridge University Press, Cambridge, England.
15. Milne-Thomson, L. M. : Theoretical Hydrodynamics, 5th Edition, p. 84-86, (1968), The Macmillan Press, London
16. Porthouse, D. T. C. and Lewis, R. I. : J. Mechanical Engineering Science, Vol. 23, p. 157-167, (1981)
17. Inoue, O. : AIAA Journal, Vol. 23, p. 365-373, (1985)
18. Chein, R. and Chung, J. N. : Multiphase flow, Vol. 13, p. 785-802, (1987)
19. Marquardt, D. W. : J. Soc. Indust. Appl. Math., Vol. 11, p. 431-441, (1963)
20. Maybeck, P. S. : Stochastic Models, Estimation, and Control, Vol. 2, p. 212-271, (1982), Academic Press, New York
21. Nonaka, M. and Thomas, N. H. : Proceedings of the Sixth IAHR International Symposium on Stochastic Hydraulics, Taipei, p. 647-654, (1992)

22. Nonaka, M. and Thomas, N. H. : Computer Simulation, Air Pollution II (Baldasano, C. A. et. al. eds.), Vol.2, p. 105-112, (1994), Computational Mechanics Publications, Southampton, England
23. Nonaka, M. and Thomas, N. H. : J. MMIJ, Vol. 112, p. 89-93, (1996)

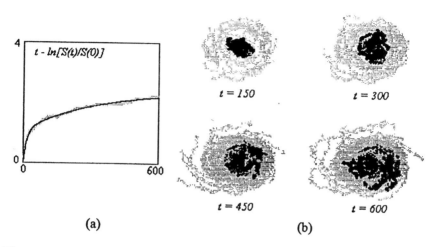

Figure 1: Dispersion behaviour of contaminants in the small eddy system with disturbances in the uniform flow ; (a): t vs $\ln[S(t)/S(0)]$, (b): horizontal dispersion trajectories

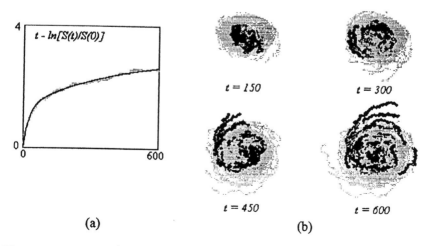

Figure 2: Dispersion behaviour of contaminants in the large eddy system with disturbances in the uniform flow ; (a): t vs $\ln[S(t)/S(0)]$, (b): horizontal dispersion trajectories

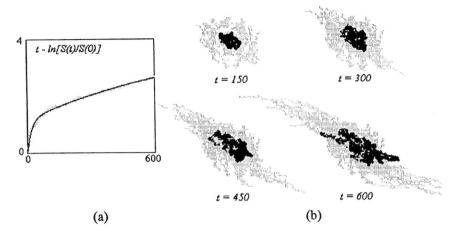

(a)   (b)

Figure 3: Dispersion behaviour of contaminants in the small eddy system with disturbances in the uniform shear flow ; (a): t vs $\ln[S(t)/S(0)]$, (b): horizontal dispersion trajectories

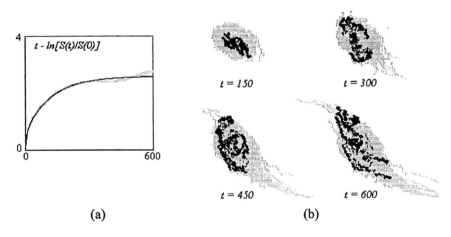

(a)   (b)

Figure 4: Dispersion behaviour of contaminants in the large eddy system with disturbances in the uniform shear flow ; (a): t vs $\ln[S(t)/S(0)]$, (b): horizontal dispersion trajectories

A method of optimising polar-orbiting weather satellite tracking

A.J. Jawad, R.J.H. Brush

Department of Applied Physics, Electronic and Manufacturing Engineering, University of Dundee, Dundee DD1 4HN, Scotland, UK

Abstract

An adapted algorithm is described to combine the conventional tracking method currently in use by the University of Dundee's satellite receiving station (programmed tracking) with received satellite data to improve satellite tracking accuracy. The new method is based upon the well-known simplex method which is adapted for use in conjunction with the satellite's orbital information to achieve this improved performance.

The method was implemented and results of the tests performed are described. The results are discussed to explain the advantages and limitations of the new method. Finally conclusions are drawn and recommendations for future enhancement of the adapted method are made.

1 Introduction

The University of Dundee's Natural Environment Research Council-funded satellite receiving station receives data from a number of polar-orbiting weather satellites like the National Oceanic and Atmospheric Administration (NOAA) series of satellites. Programmed tracking has been used successfully to track NOAA satellites. However, high tracking accuracy is required to track the more modern polar-orbiting satellites, and a new algorithm has to be developed.

Programmed tracking uses orbital information to calculate a look-up table which contains a pair of azimuth and elevation angles for each second of the pass is calculated and stored in a *pass.dat* file. The antenna is then driven to these positions at the correct time to ensure successful tracking.

The actual tracking is performed under the direction of the tracking antenna controller whose main function is to control the closed-loop antenna servo-system. This servo-system consists of a set of incremental position encoders to measure the relative position of the antenna and relay it back to the tracking controller which processes the data to produce the absolute antenna position which is then compared with the calculated (desired) antenna position. The difference between the desired and actual antenna position is the error term which is used in correcting the antenna position. This is done using a standard Proportional, Integral and Derivative (PID) control algorithm. The speed term produced by the PID algorithm is converted to a bi-polar voltage and fed to the motor controller unit and then on to the motor which drives the antenna. The tracking controller and servo-system are fully described by Jawad [1].

2 Optimisation of Satellite Tracking

A disadvantage of the programmed tracking technique is that the satellite path is calculated in advance and is followed without taking into consideration the actual position of the satellite at any given time. There will always be a slight difference between the calculated and the actual satellite position. With the more modern satellites which transmit data at high frequencies - i.e. narrow antenna beamwidth - this difference may become significant enough for the satellite signal to be lost. A loss of signal - even for a short time - may be critical in terms of the usefulness of the pass. So, an optimisation technique is needed to enable the antenna to successfully track high frequency satellites.

A simple way of determining the actual satellite position is to measure the signal strength of the signal obtained from the satellite; a greater the signal strength means a more accurate antenna pointing. A signal strength meter (S-meter) is used to measure the strength of the incoming satellite signal.

A method which uses regular geometric searching pattern known as a "regular simplex" is used to find the optimum value among a number of mutually equidistant points. The method assumes that the time interval between successive signal strength measurements is very small i.e. the satellite position is not going to change greatly between any two successive measurements, and hence the best signal value represents the best pointing accuracy. The simplex method was originally devised by Spendley et. al. [2]. A description of the method is given by Van der Schraft [3], and is summarised here.

3 The Simplex Method

The simplex method can be applied to optimise functions with many variables (dimensions). To describe the principles of this method, a function $F(x_1, x_2)$ is chosen. This choice is appropriate because it represents the dimensions (azimuth and elevation) to be optimised by this method. In the first stage of the iterative process, it is necessary to take three different points so that an equilateral triangle RBA is formed as shown in Figure 1.

Starting Simplex

The starting simplex is determined as follows: if $\underline{a}$ is the size of the edge of the simplex for n dimensions then the remaining points of the simplex are given by:

$$p = \frac{a}{n\sqrt{2}}\left(n - 1 + \sqrt{n+1}\right) \tag{1}$$

$$q = \frac{a}{n\sqrt{2}}\left(-1+\sqrt{n+1}\right) \qquad (2)$$

So in the two dimensional case p = (0.9657 a) and q = (0.2587 a). The starting triangle will then be R(0,0), B(0.9657 a, 0.2587 a) and A(0.2587 a, 0.9657 a).

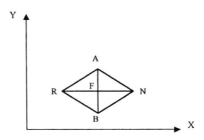

Figure 1: Two Dimensional Simplex

Generating A New Simplex

With reference to Figure 1, triangle RBA is the starting triangle and point N is the new vertex found in the first search cycle. Point N is found as a result of evaluating the function (by measuring the signal strength) at each of the first three points and then selecting the best and next best values, and rejecting the one where the function had its worst value. The new point N is determined by the line that starts at the recently rejected point and the mid point between the location where the function had the best and the second best value. This way a mirror image of the worst point is formed and a new simplex is created. When the evaluation of this last point is completed, the rejection of the worst value of the function is repeated in every new simplex.

R is rejected point and N is the reflected or mirror image of it. A line is drawn from R perpendicular to AB cutting it at point F. The line AF is projected its own length to obtain the new vertex N. The point F is, therefore, the midpoint or centre of gravity of the lines AB and RN. The co-ordinates of the new point N can be found rather easily as follows;

Assuming that A= (Ax,Ay), B=(Bx,By), R(Rx,Ry), and N=(Nx, Ny)

Since F is the mid point of AB then: $F(x,y) = 1/2((Ax + Bx), (Ay + By))$

But F is also the midpoint of RN then: $F(x,y) = 1/2 ((Rx + Nx), (Ry + Ny))$

Hence: $1/2((Ax + Bx), (Ay + By)) = 1/2 ((Rx + Nx), (Ry + Ny))$

So to find the co-ordinate of N in the X-axis $Nx = (Ax + Bx) - Rx$

Similarly, the co-ordinate of N in the Y-axis $Ny = (Ay + By) - Ry$

The procedure is a simple and iterative, but the rejection of the worst value of the function in every iteration can leave the system stranded between two points. In which case a closed cycle of the operation is entered: in triangle RBA if the $F(x,y)$ has its worst value at R, the method forces the rejection of point R and reflects it about the line AB, to give point N in the new triangle ABN. If in this new triangle, $F(x,y)$ has its worst value at point N, the application of the method would cause point N to be rejected and replaced by the original point R - effectively oscillating between points R and N.

The last situation makes it necessary to devise rules to ensure convergence:

1. To avoid oscillation, no return can be made to the point which has just been left. As a result of the rejection of the worst value of the triangle gives the previous worst value, the second worst should be rejected. When the value of the function has arrived at the optimum area, no further improvements in the value of the function is possible.

2. When oscillation occurs, optimisation is achieved by reducing the size of the simplex. So when the best value remains unchanged for more than M iterations the size is reduced. M in this case was found to be 4.

4 Adapting The Simplex Method

To apply this method on its own, an equilateral triangle will have to be formed and the signal strength is measured at each vertex before deciding where to move next. Clearly, the step size is a most important parameter; if the step size is too large then the time taken to drive the antenna will be too great to preserve the assumption that the time between successive measurements is very small. There is also a risk that too large a step may lead the system out of convergence and the signal may be lost. However, if the step is too small, then the method will be slowed down considerably. The step size is dependent upon factors like the beamwidth of the antenna in use, and the resolution of the shaft encoders.

In applying the simplex method to the tracking antenna controller a number of changes were made to adapt the method to the practicalities of the control system. One of these change was that the step size has been increased. This was done to obtain an appreciable difference in signal strength at the various vertices. What actually happened when applying the derived step size values is that the signal level at all three points was very similar. Increasing the step size to around $1.25°$ did help in obtaining variable signal strength levels without taking the process out of the beamwidth of the antenna i.e. without losing signals and without invalidating the time independence assumption.

The second change was to evaluate the starting simplex from the initial horizon position of the satellite i.e. instead of starting the first vertex at the origin (0,0) it was decided to start the simplex with the first co-ordinates obtained from the *pass.dat* file then add an offset to get the remaining vertices.

The most important change was to alter the whole simplex method strategy. Since the tracked satellite is continuously moving along a calculated path, it does not make sense to fix the triangle and drive the antenna to all three vertices while the satellite antenna is moving along - away from that particular triangle. It was decided to continuously change the triangle vertices to take the time of the pass into account. That is to say that the problem becomes tracking a moving target (the satellite) with the system always in search of the best signal strength around the path of the satellite. The whole method is changed to a series of starting simplexes as a function of the time within the pass period.

To explain this further, suppose that the starting point (taken from the *pass.dat*) of the simplex triangle in azimuth and elevation is $(300.00°, 170.50°)$. Adding a step size of 1.25° to both azimuth and elevation produces the second vertex of the triangle which has a co-ordinates of (301.25°, 171.75°). Now if a whole second has passed and, therefore, the satellite (according to the *pass.dat*) is now at co-ordinates (299.50°, 170.20°), then the third vertex co-ordinates are calculated with respect to the new location of the satellite rather than the previous location i.e. (299.50° - 1.25°, 170.20° + 1.25°). Since the vertices are all calculated relative to a predetermined origin point, a true measure of the comparative signal strength levels at the various points can still be used to determine an accurate location of the satellite. For although the simplexes are continuously moving about the path of the satellite, it is possible to compare the signal strength obtained at the second vertex of the first simplex with the signal strength obtained at the second vertex of the second simplex. This is because the offset for each vertex is constant with respect to the origin of the simplex.

This idea is applied to all vertices within the simplex including the origin i.e. as the satellite moves, so will the triangle. Ideally, the antenna should be driven to all three vertices within one second to gain an accurate comparison of the signal strength levels. However, realistically this situation is almost impossible since the mechanical parts of the system are unable to react at this rate.

5 Testing The Adapted Simplex Method

Applying the adapted simplex method as described produced signal strength values, the best of which was at the origin point of the triangle (taken straight from the *pass.dat*) This proved that the values obtained from the *pass.dat* file were accurate, but was not enough to prove the efficiency of the new method.

To prove that the new adapted method is useful in increasing the signal level, the adapted simplex method was switched on and off with the PID control algorithm for every 100 seconds of the pass. So, the first 100 seconds was tracked with the adapted simplex method then the program switched to satellite tracking using the conventional PID algorithm with a deliberate offset (see next paragraph) before switching back to the new method for the third 100 seconds. Results have shown that there is a clear and distinct improvement in signal strength when the new algorithm was used.

Another test routine was devised; rather than using the actual values obtained from the *pass.dat* file, it was decided to deliberately add an offset (an error) to these *pass.dat* angles. By adding (or subtracting) an offset to the *pass.dat* values, these values would not represent the positions where the satellite is to be found. An error of $2°$ was introduced, hence, by feeding the system erroneous data, it has been shown that the adapted simplex method works to maximise signal strength with little regard to the induced error in the initial data fed to it. This way results have shown that locations where maximum signal strength was obtained had varied from one simplex to another.

Results of testing the new algorithm are shown. A number of satellite passes were successfully tracked, and two graphs (Figures 2 and 3) from tracking a $26.5°$ maximum elevation satellite pass are included; Signal Strength vs. Time of the Satellite Pass and Mean Pointing Error vs. Time of the Satellite Pass.

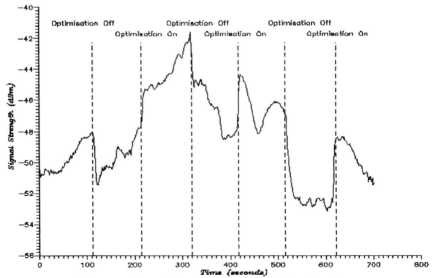

Figure 2: Signal Strength vs. Time of Satellite Pass

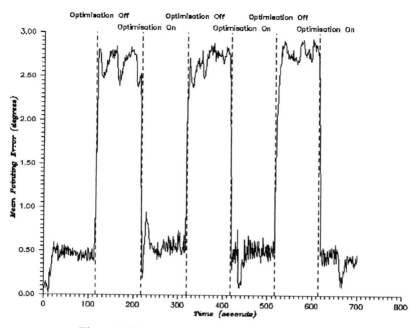

Figure 3: Mean Pointing Error vs. Time of Pass

Discussion of Results

The new method was successful in improving the satellite tracking accuracy - as shown in Figures 2 and 3. Generally, the improvement was around 5-8% on the results obtained by programmed tracking on its own using the same available equipment. This could be improved even further if the main limitation of the new method is dealt with.

In Figure 2, there was a slight signal deterioration compared with the signal obtained using programmed tracking. The satellite appears to "accelerate" when it reaches it maximum elevation, with higher elevations corresponding a greater acceleration. At zenith ($90°$) or near-zenith satellite passes the new method appears to suffer as the position of the satellite changes far too quickly for effective optimisation. However, there are several solutions to this problem, like deliberately overshooting the satellite on high-elevation passes, or changing the *pass.dat* file to account for this problem. Results of tracking high-elevation passes are presented by Jawad [1], while a discussion of the problem of tracking high elevation passes and its solution are presented by Crawford [4]. It is recommended that these limitations are tackled in future work to improve this algorithm even further.

6 Conclusions

A new adapted method to optimise polar-orbiting weather satellites is described. The method is based upon an algorithm in which an equilateral triangle is formed and the value of the function at each vertex is assessed, then the worst value is rejected and a mirror image of it is used to form the new triangle.

This method was adapted. A number of changes were made like increasing the step size, using suitable *pass.dat* angles as the starting vertex in the simplex triangle, and changing the problem to a problem of tracking a moving target.

The adapted simplex algorithm was tested and the results proved that it represented an improvement upon the conventional PID algorithm. Limitations of the new method and possible solutions were revealed.

7 References

1. Jawad, A.J., An Optimised Antenna Controller For Satellite Tracking, PhD Thesis, University of Dundee, 1995.
2. Spendley, W., Hext, G.R. & Himsworth, F.R., Sequential Application of Simplex Designs in Optimization and Evolutionary Operation, Vol. 4, 1962.
3. Van der Schraft, H., A Hill Climbing Method to Optimize Satellite Tracking, PhD Thesis, University of Dundee, 1980.
4. Crawford, P.S., Meteorological Satellite Ground Station Technology, PhD Thesis, University of Dundee, 1995.

SECTION 4:
CHEMISTRY, PHYSICS AND BIOLOGY

Multicomponent NAPL pool dissolution in saturated porous media

C.V. Chrysikopoulos, K.Y. Lee

Department of Civil Engineering, University of California, Irvine CA 92717, USA

ABSTRACT

In this paper, we study the multicomponent transport of contaminants originating from the dissolution of NAPL pools in saturated porous media. A semi–analytical procedure is introduced. The method assumes that the contaminant source input can be represented by a series of short pulses. For the duration of each pulse the equilibrium aqueous solubility of every pool constituent is considered invariant. The mole fraction, nonaqueous phase activity coefficient, and consequently, the aqueous solubility of every pool constituent are updated before each pulse period. The procedure is illustrated for a hypothetical site contaminated by three NAPL pools consisting of 1,1,2–trichloroethane (TCA) and trichloroethylene (TCE) mixtures.

INTRODUCTION

There are several studies available in the literature focusing on the migration of NAPLs and dissolution of residual segments (*Abriola and Pinder* [1]; *Imhoff et al.* [7]; *Mayer and Miller* [10]; *Powers et al.* [13], to mention a few), as well as pool dissolution (*Johnson and Pankow* [8]; *Chrysikopoulos et al.* [5]; *Pearce et al.* [12]; *Seagren et al.* [14]; *Voudrias and Yeh* [16]; *Chrysikopoulos* [4]; *Lee and Chrysikopoulos* [9]). In spite of the fact that the majority of groundwater contamination sites involve multicomponent NAPLs, only single–component systems have been studied extensively both theoretically and experimentally. In particular, the literature on multicomponent NAPL pool dissolution is rather limited.

For multicomponent NAPL systems, the dissolution rate is controlled in part by the equilibrium aqueous solubility of each individual component (*Borden and Piwoni* [3]). The aqueous solubility is described by a relationship which equates the activities in the nonaqueous (organic) and aqueous phases and is a function of the time dependent mole fraction of each component. As the mole fraction of each component changes constantly with a rate proportional to its mass transfer coefficient, the activity coefficient of each component is also changing accordingly (*Fredenslund et al.* [6]). It should be noted that the activity coefficient is a dimensionless correction term indicating the nonideality of the solution (*Stumm and Morgan* [15]). The higher the activity coefficient the greater the degree of nonideality of a solution. *Banerjee* [2] demonstrated experimentally that for a NAPL mixture containing components of similar structure, the organic phase activity coefficients can be approximated at ideal state equal to one, and the equilibrium aqueous solubility is a function of mole fraction only.

For multicomponent NAPL mixtures with components of dissimilar structure, the organic phase activity coefficient can be estimated using the software UNIFAC (UNI–Functional group Activity Coefficients, *Fredenslund et al.* [6]). The program UNIFAC was initially developed for chemical engineering applications associated with activity coefficient estimations in organic mixtures where limited or no experimental data are available.

MATHEMATICAL MODEL

Consider a multicomponent NAPL pool denser than water in a three–dimensional homogeneous porous medium. The transient transport of each dissolving component under steady–state uniform flow conditions is governed by

$$R_p \frac{\partial C_p(t,x,y,z)}{\partial t} = D_{x_p} \frac{\partial^2 C_p(t,x,y,z)}{\partial x^2} + D_{y_p} \frac{\partial^2 C_p(t,x,y,z)}{\partial y^2} + D_{z_p} \frac{\partial^2 C_p(t,x,y,z)}{\partial z^2}$$
$$- U_x \frac{\partial C_p(t,x,y,z)}{\partial x} - \lambda_p R_p C_p(t,x,y,z), \tag{1}$$

where $C(t,x,y,z)$ is the liquid phase solute concentration; x, y, z are the spatial coordinates in the longitudinal, lateral and vertical directions, respectively; t is time; R is the dimensionless retardation factor; D_x, D_y, D_z are the longitudinal, lateral and vertical hydrodynamic dispersion coefficients, respectively; U_x is the average interstitial fluid velocity; λ is a first–order decay constant; and subscript p is the component number indicator. Assuming that the thickness of the pool is insignificant relative to the thickness of the aquifer, the dissolution process for each component is described by the following mass transfer relationship, applicable at the NAPL–water interface (*Chrysikopoulos et al.* [5])

$$-\mathcal{D}_{e_p} \frac{\partial C_p(t,x,y,0)}{\partial z} = k_p(t,x,y) \Big[C_p^w(t) - C_p(t,x,y,\infty) \Big], \tag{2}$$

where $\mathcal{D}_{e_p} = D_p / \tau^*$ is the effective molecular diffusion coefficient of component p; $\mathcal{D}$ is the molecular diffusion coefficient; τ^* is the tortuosity coefficient ($\tau^* \geq 1$); $k(t,x,y)$ is the local mass transfer coefficient dependent on time and location at the NAPL–water interface; $C^w(t)$ is the equilibrium aqueous solubility, which is time dependent because the mole fraction of each component is changing as the NAPL dissolves into the aqueous phase. Hereafter, the local mass transfer coefficient is replaced by an average (or overall) mass transfer coefficient ($k_p^* = k_p(t,x,y)$).

For a NAPL pool of elliptic shape as shown in Figure 1, the appropriate initial and boundary conditions are (*Chrysikopoulos* [4])

$$C_p(0,x,y,z) = 0, \tag{3}$$

$$C_p(t,\pm\infty,y,z) = 0, \tag{4}$$

$$C_p(t,x,\pm\infty,z) = 0, \tag{5}$$

$$\mathcal{D}_{e_p} \frac{\partial C_p(t,x,y,0)}{\partial z} = \begin{cases} -k_p^* C_p^w(t) & (x-\ell_{x_o})^2/a^2 + (y-\ell_{y_o})^2/b^2 \leq 1, \\ 0 & (x-\ell_{x_o})^2/a^2 + (y-\ell_{y_o})^2/b^2 > 1, \end{cases} \tag{6}$$

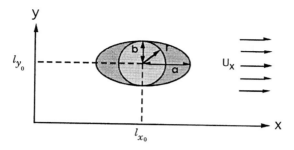

Figure 1. Plan view of a denser than water elliptic pool with origin at $x=\ell_{x_0}$, $y=\ell_{y_0}$, having major semi–axis a and minor semi–axis b. For the special case where $a=b=r$ the elliptic pool becomes a circular pool with radius r.

$$C_p(t, x, y, \infty) = 0, \tag{7}$$

where ℓ_{x_0} and ℓ_{y_0} indicate the x and y Cartesian coordinates of the center of the elliptic pool, respectively; and a, b are the major and minor semi–axes of the elliptic pool, respectively. The circular pool geometry is a special case of an elliptic pool (see Figure 1). Therefore, the appropriate source boundary condition for a circular pool is obtained by setting $a = b = r$ in (6). The solubility of a single component NAPL in terms of mole fractions can be described by (*Banerjee* [2])

$$X_s^w = \frac{X_s^o \, \gamma_s^o}{\gamma_s^w}, \tag{8}$$

where X is the mole fraction of the slightly soluble species in the appropriate phase; γ is the activity coefficient; superscripts o and w represent nonaqueous and aqueous phase, respectively; and subscript s indicates pure single component NAPL. The effect of water molecules dissolved into the nonaqueous phase is customarily neglected, and consequently the mole fraction of the contaminant in the nonaqueous phase is approximately equal to one. By assuming that the nonaqueous phase of a single component NAPL is at an ideal state with activity coefficient equal to one, the preceding equation can be reduced to

$$X_s^w = \frac{1}{\gamma_s^w}. \tag{9}$$

The solubility of each component in a multicomponent system in terms of mole fractions is given by

$$X_p^w = \frac{X_p^o \, \gamma_p^o(X_p^o)}{\gamma_p^w(X_p^o)}. \tag{10}$$

It should be noted that the activity coefficients are dependent on the mole fraction of component p in the nonaqueous phase. By dividing the preceding multicomponent solubility equation (10) with the pure single component reference solubility equation (9), yields a mole fraction ratio of the dissolving component p with respect to the mole fraction solubility of the pure component p. This dimensionless ratio is also equal to the concentration ratio of the equilibrium aqueous solubility of component p to its pure component solubility

$$\frac{X_p^w}{X_{s_p}^w} = \frac{C_p^w}{C_{s_p}} = \frac{X_p^o \, \gamma_p^o(X_p^o) \, \gamma_{s_p}^w}{\gamma_p^w(X_p^o)}, \tag{11}$$

where C_{s_p} is the pure single component saturation concentration of component p.

In view of (9), it is evident that the aqueous–phase activity coefficient is inversely proportional to the mole fraction of the NAPL. However, the majority of dissolved organic molecules present in groundwater have low solubility and they are not expected to react chemically with water; consequently, they are separated by large distances without hindering one another. Therefore, it can be assumed that interactions among NAPL molecules in the aqueous phase are small, or $\gamma_{s_p}^w \simeq \gamma_p^w$, and (11) can be reduced to

$$C_p^w = C_{s_p} \, X_p^\circ \, \gamma_p^\circ (X_p^\circ). \tag{12}$$

The experimental studies conducted by *Banerjee* [2] suggest that the nonaqueous phase activity coefficients of ideal state liquid mixtures are unaffected by the presence of cosolutes ($\gamma_p^\circ \simeq 1$). Therefore, for this special case, (12) simply reduces to Raoult's law

$$C_p^w = C_{s_p} \, X_p^\circ. \tag{13}$$

Equations (12) and (13) represent two different approaches of equilibrium aqueous solubility estimation. The nonaqueous phase activity coefficient of each component in equation (12) can be calculated using the group–contribution numerical code UNIFAC (*Fredenslund et al.* [6]).

A semi–analytical solution to the problem defined by (1) and (3)–(7) can be obtained by representing the source input as summation of consecutive pulse–type boundary conditions, assuming that the NAPL pool composition, and consequently C^w, of each component remain constant over a "small" time interval, Δt. For every pulse–type input the following analytical solution derived by *Chrysikopoulos* [4] is applicable

$$C_p(t,x,y,z) = \begin{cases} C_p^w(t)\Phi_p(t,x,y,z), & 0 < t \le \Delta t, \\[2mm] C_p^w(t)\big[\Phi_p(t,x,y,z) - \Phi_p(t - \Delta t, x, y, z)\big], & t > \Delta t, \end{cases} \tag{14}$$

where

$$\Phi_p(t,x,y,z) = \frac{k_p^\star}{2\pi \mathcal{D}_{e_p}} \int_0^t \int_{\mu_1}^{\mu_2} \left(\frac{D_{z_p}}{R_p \tau}\right)^{1/2} \exp\left[-\lambda_p \tau - \frac{R_p z^2}{4 D_{z_p} \tau}\right] \exp\left[-\mu^2\right]$$

$$\cdot \Big(\mathrm{erf}\,[n_2] - \mathrm{erf}\,[n_1]\Big) d\mu \, d\tau, \tag{15}$$

where

$$\mu_1 = (y - \ell_{y_o} + b)\left(\frac{R_p}{4 D_{y_p} \tau}\right)^{1/2}, \tag{16}$$

$$\mu_2 = (y - \ell_{y_o} - b)\left(\frac{R_p}{4 D_{y_p} \tau}\right)^{1/2}, \tag{17}$$

$$n_1 = \left[x - \frac{U_x \tau}{R_p} - \ell_{x_o} + \left\{\left[1 - \frac{(v - \ell_{y_o})^2}{b^2}\right]a^2\right\}^{1/2}\right]\left(\frac{R_p}{4 D_{x_p} \tau}\right)^{1/2}, \tag{18}$$

$$n_2 = \left[x - \frac{U_x \tau}{R_p} - \ell_{x_o} - \left\{\left[1 - \frac{(v - \ell_{y_o})^2}{b^2}\right]a^2\right\}^{1/2}\right]\left(\frac{R_p}{4 D_{x_p} \tau}\right)^{1/2}, \tag{19}$$

$$u = x - \frac{U_x \tau}{R_p} - n\left(\frac{4 D_{x_p} \tau}{R_p}\right)^{1/2}, \tag{20}$$

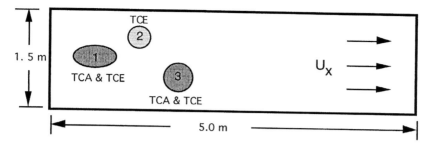

Figure 2. Schematic illustration of the hypothetical aquifer and NAPL pools.

$$v = y - \mu \left(\frac{4D_{y_p}\tau}{R_p} \right)^{1/2},$$ (21)

where τ is the dummy integration variable. The mole fraction, $X(t)$, of component p at time t is defined as

$$X_p(t) = \frac{\mathcal{M}_p(t)}{\sum_{p=1}^{\mathcal{P}} \mathcal{M}_p(t)} \qquad (p = 1, 2, \ldots, \mathcal{P}),$$ (22)

where $\mathcal{P}$ is the total number of components in the nonaqueous phase; $\mathcal{M}_p(t)$ is the number of moles of component p present in the nonaqueous phase at time t, and is defined by

$$\mathcal{M}_p(t) = \mathcal{M}'_p - \sum_{m=1}^{m_f} \frac{k_p^* \, C_p^w(m\Delta t) \, A \, \Delta t \, X_p(m\Delta t)}{(\text{Mol wt})_p} \qquad (t \geq \Delta t),$$ (23)

where $\mathcal{M}'_p$ is the initial number of moles of component p in the nonaqueous phase; $A = \pi ab$ is the surface area of the elliptic pool; m is a summation index; m_f is an integer indicating the total number of time steps (or pulses), and is defined as: $m_f = I(t/\Delta t)$ $(m_f = 1, 2, 3 \ldots)$, where $I( )$ is an integer mode arithmetic operator truncating off any fractional part of the numeric argument.

The semi–analytical solution to the multicomponent dissolution problem is obtained by summing the aqueous phase concentrations resulting from all pulse–type inputs of equal duration Δt. For m_f pulses the desired expression is

$$C_p(t,x,y,z) = \sum_{m=1}^{m_f} C_p^w(m\Delta t) \Big[\Phi_p\big(t - (m-1)\Delta t, x, y, z\big) - \Phi_p\big(t - m\Delta t, x, y, z\big) \Big]$$
$$+ C_p^w(m_f\Delta t) \, \Phi_p\big(t - m_f\Delta t, x, y, z\big).$$ (24)

MODEL SIMULATIONS

Consider three NAPL pools located at the bottom of a saturated, homogeneous sandy aquifer with unidirectional interstitial flow along the x–coordinate, as illustrated in Figure 2. Pools 1 and 3 consist of a well–mixed equimolar mixture of TCA ($\mathcal{M}' = 0.5$ mol) and

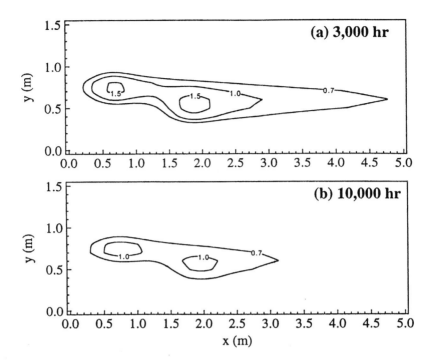

Figure 3. Cross–sections of aqueous phase TCA concentrations in the x, y plane at (a) 3,000, and (b) 10,000 hours.

TCE ($\mathcal{M}' = 0.5$ mol), and Pool 2 consists only of TCE. Pools 2 and 3 are of circular shape, whereas Pool 1 is elliptic.

Figures 3 and 4 present x, y plane cross–sections of the plume resulted from the combined dissolution of the NAPL pools shown in Figure 2, for TCA and TCE, respectively. The parameter values used for the simulations are: $a = 0.25$ m, $b = 0.18$ m, $C_s = 4.5$ g/l, $\mathcal{D}_e = D_y = D_z = 2.33 \times 10^{-6}$ m^2/h, $D_x = 1.86 \times 10^{-3}$ m^2/h, $k^* = 1.9 \times 10^{-4}$ m/h, $\ell_{x_o} = 1.6$ m, $\ell_{y_o} = 0.75$ m, $R = 1.1$, $\Delta t = 200$ h, $U_x = 0.031$ m/h, $\lambda = 0.0$h^{-1}, $y = 0.75$ m, and $z = 0.00$ m. Two different times since the initiation of the dissolution process are considered. Figure 3 indicates a decrease in TCA dissolved concentration levels with increasing time. Figure 4 illustrates a steady increase in TCE plume size and concentration levels with increasing time. These results suggest that the equilibrium solubility of TCA decreases and that of TCE increases with increasing time. The single component aqueous saturation concentration (solubility) for TCA is 4.5 g/l and for TCE 1.1 g/l (*Miller et al.*, [11]). Therefore, TCA is dissolving faster at early time leading to a lower TCA mole fractions in Pools 1 and 3, and consequently, to higher TCA nonaqueous phase activity coefficients. In view of (12), it is evident that the TCA equilibrium aqueous solubility decreases when the relative decrease in X is greater than the increase in γ. Similarly, TCE is dissolving slower at early time leading to higher TCE mole fractions, lower nonaqueous phase activity coefficients, and slowly increasing equilibrium aqueous solubilities.

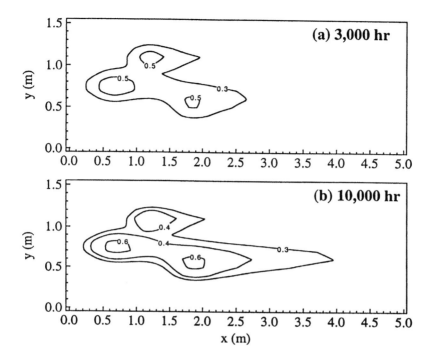

Figure 4. Cross-sections of aqueous phase TCE concentrations in the x, y plane at (a) 3,000, and (b) 10,000 hours.

SUMMARY

A model for contaminant transport originating from the dissolution of multicomponent NAPL pools in homogeneous, saturated three–dimensional subsurface formations is developed. The model assumes that the dissolution process is mass transfer limited, and each dissolved component may undergo first–order decay and may sorb onto the solid matrix under local equilibrium conditions. A semi–analytical solution is obtained by summing the aqueous phase concentrations resulting from a series of short dissolution pulses. The aqueous solubility of every pool constituent is estimated before the initiation of a pulse, and is considered constant for the duration of the pulse. The numerical code UNIFAC is employed for the evaluation of the time dependent nonaqueous phase activity coefficients.

Acknowledgements.   This work was sponsored by the U.S. Environmental Protection Agency, under award R–823579–01–0. However, the manuscript has not been subjected to the Agency's peer and administrative review and therefore does not necessarily reflect the views of the Agency and no official endorsement should be inferred.

NOTATION

a, b	major and minor semi–axes of elliptic pool, respectively, L.
A	surface area of a pool, L^2.
C	liquid phase solute concentration (solute mass/liquid volume), M/L^3.
C_s	single component aqueous saturation concentration (solubility), M/L^3.
C^w	equilibrium aqueous solubility, M/L^3.
$\mathcal{D}$	molecular diffusion coefficient, L^2/t.
$\mathcal{D}_e$	effective molecular diffusion coefficient, equal to $\mathcal{D}/\tau^*$, L^2/t.
D_x	longitudinal hydrodynamic dispersion coefficient, L^2/t.
D_y	lateral hydrodynamic dispersion coefficient, L^2/t.
D_z	hydrodynamic dispersion coefficient in the vertical direction, L^2/t.
$\mathrm{erf}[x]$	error function, equal to $(2/\pi^{1/2}) \int_0^x e^{-z^2} dz$.
$\mathrm{I}(\ )$	integer mode arithmetic operator.
k	local mass transfer coefficient, L/t.
k^*	average mass transfer coefficient, L/t.
ℓ_{x_o}, ℓ_{y_o}	x and y Cartesian coordinates, respectively, of the center of an elliptic/circular pool, L.
m	summation index.
m_f	total number of time steps.
$\mathcal{M}$	number of moles remaining in a pool, mol.
$\mathcal{M}'$	initial number of moles, mol.
n_1, n_2	defined in (18) and (19), respectively.
p	component number indicator.
$\mathcal{P}$	total number of components.
r	radius of circular pool, L.
R	dimensionless retardation factor.
t	time, t.
u	defined in (20).
U_x	average interstitial velocity, L/t.
v	defined in (21).
x, y, z	spatial coordinates, L.
X	dimensionless mole fraction.
γ	dimensionless activity coefficient.
Δt	time interval of a single pulse.
λ	decay coefficient, t^{-1}.
μ_1, μ_2	defined in (16) and (17), respectively.
τ	dummy integration variable.
τ^*	tortuosity factor (≥ 1).
Φ	defined in (15).

REFERENCES

[1] Abriola, L. M., and G. F. Pinder, A multiphase approach to the modeling of porous media contamination by organic compounds, 1, Equation development, *Water Resour. Res.*, *21*(1), 11–18, 1985.

[2] Banerjee, S., Solubility of organic mixtures in water, *Environ. Sci. Technol.*, *18*(8), 587–591, 1984.

[3] Borden, R. C., and M. D. Piwoni, Hydrocarbon dissolution and transport: A comparison of equilibrium and kinetic models, *J. Contam. Hydrol.*, *10*, 309–323, 1992.

[4] Chrysikopoulos, C. V., Three–dimensional analytical models of contaminant transport from nonaqueous phase liquid pool dissolution in saturated subsurface formations, *Water Resour. Res.*, *31*(4), 1137–1145, 1995.

[5] Chrysikopoulos, C. V., E. A. Voudrias, and M. M. Fyrillas, Modeling of contaminant transport resulting from dissolution of nonaqueous phase liquid pools in saturated porous media, *Transp. Porous Media*, *16*(2), 125–145, 1994.

[6] Fredenslund, A., J. Gmehling, and P. Rasmussen, *Vapor–liquid Equilibria Using* UNIFAC, 380 pp., Elsevier, New York, 1977.

[7] Imhoff, P. T., P. R. Jaffé, and G. F. Pinder, An experimental study of complete dissolution of a nonaqueous phase liquid in saturated porous media, *Water Resour. Res.*, *30*(2), 307–320, 1993.

[8] Johnson, R. L., and J. F. Pankow, Dissolution of dense chlorinated solvents into groundwater, 2, Source functions for pools of solvent, *Environ. Sci. Tech.*, *26*(5), 896–901, 1992.

[9] Lee, K. Y., and C. V. Chrysikopoulos, Numerical modeling of three–dimensional contaminant migration from dissolution of multicomponent NAPL pools in saturated porous media, *Environ. Geology*, *26*, 157–165, 1995.

[10] Mayer, A. S., and C. T. Miller, An experimental investigation of pore–scale distributions of nonaqueous phase liquids at residual saturation, *Transp. Porous Media*, *10*, 57–80, 1993.

[11] Miller, C. T., M. M. Poirier–McNeill, and A. S. Mayer, Dissolution of trapped nonaqueous phase liquids: Mass transfer characteristics. *Water Resour. Res.*, *26*(11), 2783–2796, 1990.

[12] Pearce, A. E., E. A. Voudrias, and M. P. Whelan, Dissolution of TCE and TCA pools in saturated subsurface systems, *J. Environ. Eng. ASCE*, *120*(5), 1191–1206, 1994.

[13] Powers, S. E., L. M. Abriola, and W. J. Weber, Jr., An experimental investigation of nonaqueous phase liquid dissolution in saturated subsurface systems: Transient mass transfer rates, *Water Resour. Res.*, *30*(2), 321–332, 1994.

[14] Seagren, E. A., B. E. Rittmann, and A. J. Valocchi, Quantitative evaluation of the enhancement of NAPL–pool dissolution by flushing and biodegradation, *Environ, Sci. Technol.*, *28*, 833–839, 1994.

[15] Stumm, W, and J. J. Morgan, *Aquatic Chemistry*, 2nd ed., 780 pp., Wiley, New York, 1981.

[16] Voudrias, E. A. and M. F. Yeh, Dissolution of a toluene pool under constant and variable hydraulic gradients with implications for aquifer remediation, *Groundwater*, *32*(2), 305–311, 1994.

Effect of drop-phase oxidation on raindrop acidification due to washout of sulfur dioxide

S.Shiba, Y. Hirata
*Department of Chemical Engineering, Osaka University,
Toyonaka, Osaka 560, Japan*

Abstract

For the purpose of investigating the effect of the drop-phase oxidation on acid rain formation by washout of gaseous pollutants, a mathematical model based on physico-chemical consideration has been constructed and the dynamics of the acidification has been simulated numerically with use of the model. The model deals with the below-cloud scavenging of the pollutant gases, considering such factors as: (1) the drop-phase chemistry; and (2) the mass transfer at the interface between the drop and the atmosphere. The results of the model simulation have disclosed that: (1) in the early stage of acidification bisulfate ion HSO_3^- is the major species to acidify the drop; and (2) with lapse of time SO_4^{2-} produced by the drop-phase oxidation increases gradually and eventually dominates the raindrop acidity. It is suggested that such a very high acidity as exceeds the acidity equilibrated with the ambient $SO_2(g)$, which is often detected in drizzle and fog, should be reasonably attributed to the increase of SO_4^{2-} in the drops.

1 Introduction

The quality of the rainwater in urbanized area gets worse and worse in these days. This is because urbanization is growing so fast and the atmospheric pollutants emitted from the urban area are increasing explosively. Accordingly the pollutants are increased to deteriorate the water quality. Among such atmospheric pollutants acidic species as sulfur dioxide and nitrogen oxide are important, because they form acid rain. Acid rain has been given considerable attention due to the great impact to the various environments, that is, it has not only human health effects but also a number of other effects, including material corrosion and ecosystem interference such as acidification of lakes and streams and interference with forest productivity.

It is well known that acid rain is caused mainly by such air pollutants as sulfur dioxide and nitrogen oxide. From the global viewpoint, however, the amount of the sulfur dioxide emitted to the air is much greater than that of nitrogen oxide. Then in this study the acid rain due to sulfur dioxide is treated exclusively. Sulfur dioxide is absorbed by cloud or raindrops to dissociate into anions and cations in two steps and acidify the drops. However, since the dissociation of $SO_2(g)$ in the drop is limited to the equilibrium concentration to the ambient $SO_2(g)$, in order to attain a higher acidity than the equilibrium value, it is necessary that $SO_2(g)$ dissolved in drops is oxidized to sulfate before the deposition on the ground. The highly soluble hydrogen peroxide $H_2O_2(g)$, which results from HO_2 by the photolysis of carbonyl compounds (e.g., HCHO) in dry atmosphere (Mason[1]), is a common chemical species and can work for $SO_2(g)$ oxidation in rain drops. The model developed here, which treats the concurrent absorption of $SO_2(g)$ and $H_2O_2(g)$ and the liquid phase chemical reaction, can well simulate the dynamics of acidification of raindrops in this situation, incorporating the pollutant mass transfer rate between the air and the raindrops.

2 Chemical transformation of acidifying gases in rain drop

$SO_2(g)$ is chemicaly transformed by liquid phase reactions, i.e., dissociation and oxidation, after incorporation into rain drops. The dissociation process are typically quite fast and the oxidation process could be considerably slower. If $SO_2(g)$ is absorbed by raindrops, after the two steps of dissociation, the resulting concentration in drops reach equilibrium with a given air phase concentration. For absorption of $SO_2(g)$ the sequence of process can be described as follows (Liss and Slinn[2]):

$$SO_2(g) + H_2O \rightleftharpoons SO_2(aq) \qquad (\mathcal{H}_1) \qquad (1)$$

$$SO_2(aq) \rightleftharpoons HSO_3^- + H^+ \qquad (K_1) \qquad (2)$$

$$HSO_3^- \rightleftharpoons SO_3^{2-} + H^+ \qquad (K_2) \qquad (3)$$

where $\mathcal{H}_1$ = distribution coefficient (= 30.32); and K_1 and K_2 = first and second dissociation constants, respectively (= 1.74×10^{-2}M and 6.24×10^{-8}M). Consequently sulfur dioxide exists in raindrops as physically disolved sulfur dioxide, $SO_2(aq)$, and in dissociated form as bisulfite ion, HSO_3^-, and sulfite ion, SO_3^{2-}. These values of coefficient and constants are measured at 25 C. When $H_2O_2(g)$ is absorbed, the resulting concentration in drops also reach equilibrium as follows:

$$H_2O_2(g) + H_2O \rightleftharpoons H_2O_2(aq) \qquad (\mathcal{H}_5) \qquad (4)$$

$$H_2O_2(aq) \rightleftharpoons HO_2^- + H^+ + H_2O \qquad (K_5) \qquad (5)$$

where $\mathcal{H}_5$ = distribution coefficient (= 1.73×10^6); and K_5 = dissociation constant (= 1.84×10^{-12}M). On the time scale of the diffusion process, the equilibrium can be treated as being instantaneously established.

If $SO_2(g)$ and $H_2O_2(g)$ are absorbed concurrently, HSO_3^- is oxidized by $H_2O_2(aq)$ to become sulfate ion, SO_4^{2-}, as follows (Maahs[3]):

$$HSO_3^- + H_2O_2(aq) \rightarrow SO_4^{2-} + 2H_2O + H^+ \qquad (k^* = k[H^+]) \qquad (6)$$

where k^* = reaction rate; and k = reaction rate constant (= $5.2 \times 10^7 s^{-1}M^{-2}$). Because hydrogen peroxide is often present in low concentrations compared with those of $SO_2(g)$, the conversion rate of $SO_2(g)$ to SO_4^{2-} may be limited by the concentration of $H_2O_2(g)$. Rain water acidity, which is represented by the concentration of H^+ or by pH, reflects a balance between the concentration of acidic anions and those of basic cations. SO_4^{2-} is the acidic anion as well as HSO_3^- and SO_3^{2-} and is supposed to contribute to rain water acidity.

3 Working equations of acidification model

The washout processes of pollutant gases are controlled by such various physical and chemical factors as the drop diameter, fallling velocity, inner circulation flow (Levich[4]), gas solubility, liquid phase reactions, diffusion in drop and concentrations of chemical species. In this study, however, the complex interacting wahout system is represented by chemical equailibrium relations, oxidation reaction, electro-neutrality condition, and mass balance equations to build a tractable model, providing that rigid spherical raindrops fall through the air of uniform gas concentration with the terminal velocity.

The notations used to represent concentrations of chemical species in drops and in the air are:

$$(C_1, C_2, C_3, C_4, C_5, C_6, C_7) = ([SO_2(aq)], [HSO_3^-], [SO_3^{2-}], [SO_4^{2-}],$$
$$[H_2O_2(aq)], [H^+], [HO_2^-]) \qquad (7)$$

$$(C_{1G}, C_{5G}) = ([SO_2(g)], [H_2O_2(g)]) \qquad (8)$$

From eqns (1)-(5) three equilibrium relations held in raindrops can be taken as

$$K_1 \cdot C_1 = C_2 \cdot C_6 \qquad (9)$$

$$K_2 \cdot C_2 = C_3 \cdot C_6 \qquad (10)$$

$$K_5 \cdot C_5 = C_7 \cdot C_6 \qquad (11)$$

The charge neutrality condition for anions and cations in drops is

$$C_6 - C_2 - 2C_3 - 2C_4 - C_7 - K_W/C_6 = \alpha \qquad (12)$$

where K_W = dissociation constant for water (= $10^{-14}M^2$); and α = constant estimated by initial concentrations of ions in drops. Although the concentration of chemical species in raindrops can be estimated by the diffusion model described by partial differential equations (Shiba[5]; Shiba and Hirata[6]), it is considerably time-consuming task to compute the time variation of the concentations in raindrops especially for small drops whose travelling time is long. Therefore, to make the computation time shorter in this study the model is constructed by simple ordinary differential equations as shown below.

With use of the macroscopic mass balance of C_i in a raindrop the governing equation for chemical species can be taken as follows:

$$\frac{d(V \cdot C_i)}{dt} = V \cdot R_i + S \cdot k_{Gi} \left(C_{iG} - \frac{C_i}{\mathcal{H}_i} \right) \tag{13}$$

$$C_i = C_{i0} \quad \text{at} \quad t = 0 \tag{14}$$

where t = time; V = volume of drop $(= \pi D^3 /6)$; S = surface area of drop $(= \pi D^2)$; k_{Gi} = gas phase mass transfer coefficient; $\mathcal{H}_i$ = distribution coefficient; D = drop diameter; and R_i = reaction terms given as follows:

$$R_1 = -k_{1+}C_1 + k_{1-}C_2 C_6 \tag{15}$$

$$R_2 = k_{1+}C_1 - k_{1-}C_2 C_6 - k_{2+}C_2 + k_{2-}C_3 C_6 - kC_6 C_2 C_5 \tag{16}$$

$$R_3 = k_{2+}C_2 - k_{2-}C_3 C_6 \tag{17}$$

$$R_4 = kC_6 C_2 C_5 \tag{18}$$

$$R_5 = -k_{5+}C_5 + k_{5-}C_7 C_6 - kC_6 C_2 C_5 \tag{19}$$

$$R_6 = k_{1+}C_1 - k_{1-}C_2 C_6 + k_{2+}C_2 - k_{2-}C_3 C_6$$
$$+ k_{5+}C_5 - k_{5-}C_7 C_6 + kC_6 C_2 C_5 \tag{20}$$

$$R_7 = k_{5+}C_5 - k_{5-}C_7 C_6 \tag{21}$$

As in most acid rain C_6 ranges from 10^{-6}M (pH 6) to 10^{-3}M (pH 3), it can be assumed that C_2 is much greater than both of C_1 and C_3 and that C_5 is much higher than C_7. Making use of these assumptions the governing equations can be reduced to simple dimensionless forms as

$$\frac{d\hat{C}_i}{d\hat{t}} = \hat{R}_i + \hat{T}_i \quad (i = 2, 4 \text{ and } 5) \tag{22}$$

$$\hat{C}_i = \hat{C}_{i0} \quad \text{at} \quad \hat{t} = 0 \quad (i = 2, 4 \text{ and } 5) \tag{23}$$

where $\hat{R}_i$ = reaction terms and $\hat{T}_i$ = mass transfer terms. $\hat{R}_i$ and $\hat{T}_i$ are given by

$$(\hat{R}_2, \ \hat{R}_4, \ \hat{R}_5) = (-\hat{k}\hat{C}_6\hat{C}_2\hat{C}_5, \ \hat{k}\hat{C}_6\hat{C}_2\hat{C}_5, \ -\hat{k}\hat{C}_6\hat{C}_2\hat{C}_5) \tag{24}$$

$$(\hat{T}_2, \ \hat{T}_4, \ \hat{T}_5) = \left[3\text{Bi}_1 \left(\mathcal{H}_1\hat{C}_{1G} - \frac{\hat{C}_{6i}}{\hat{K}_1}\hat{C}_2 \right), \ 0, \ 3\hat{D}_5\text{Bi}_5 \left(\mathcal{H}_5\hat{C}_{5G} - \hat{C}_5 \right) \right] \tag{25}$$

where $\hat{C}_{6i}$ = value of $\hat{C}_6$ at the drop surface.

The dimensionless variables and constants used in the above equations can be written as

$$\hat{C}_i = \frac{C_i}{\mathcal{H}_i C_{1G}} \tag{26}$$

$$\hat{t} = \frac{4t\mathcal{D}_1}{D^2} \tag{27}$$

$$\hat{\mathcal{D}}_i = \frac{\mathcal{D}_i}{\mathcal{D}_1} \tag{28}$$

$$\mathrm{Bi}_i = \frac{k_{Gi}D}{2\mathcal{H}_i\mathcal{D}_i} \tag{29}$$

$$\hat{k} = k\frac{D^2(\mathcal{H}_i C_{1G})^2}{4\mathcal{D}_1} \tag{30}$$

$$\hat{\mathcal{H}}_i\hat{C}_{iG} = \frac{\mathcal{H}_i C_{iG}}{\mathcal{H}_1 C_{1G}} \tag{31}$$

$$\hat{K}_i = \frac{K_i}{\mathcal{H}_1 C_{1G}} \tag{32}$$

where Bi_i is the Biot Number for species i and it controls the mass transfer rate of gas between raindrops and the air. Biot Number can be estimated by

$$\mathrm{Bi}_i = \mathrm{Sh}_{iG}\frac{\mathcal{D}_{iG}}{\mathcal{D}_1}\frac{1}{2\mathcal{H}_i} \tag{33}$$

$$\mathrm{Sh}_{iG} = 2 + 0.6\mathrm{Re}_G^{1/2} \cdot \mathrm{Sc}_{iG}^{1/3} \tag{34}$$

$$\mathrm{Re}_G = \frac{uD}{\nu_G} \tag{35}$$

$$\mathrm{Sc}_{iG} = \frac{\nu_G}{\mathcal{D}_{iG}} \tag{36}$$

where Sh_{iG} = Sherwood Number; Re_G = Reynolds Number; Sc_{iG} = Schmidt Number; ν_G = kinematic viscosity of gas; and u = falling velocity of raindrop. u in cm/s is given by

$$u(D) = 958\left\{1 - \exp\left[-(D/0.177)^{1.147}\right]\right\} \tag{37}$$

From the above relations Bi_1 for $SO_2(g)$ and Bi_5 for $H_2O_2(g)$ are formulated as functions of Re_G as

$$\mathrm{Bi}_1 = 217.4 + 72.97\mathrm{Re}_G^{1/2} \tag{38}$$

$$\mathrm{Bi}_5 = 3.557 \times 10^{-3} + 1.060 \times 10^{-3}\mathrm{Re}_G^{1/2} \tag{39}$$

Figure 1 shows the relationships between Biot Number and drop diameter. Bi_1 is about 10^4 or 10^5 times as much as Bi_5. This is the reason why the mass transfer rate between the air and drops of $SO_2(g)$ is much greater than that of $H_2O_2(g)$.

Once $\hat{C}_2$, $\hat{C}_4$ and $\hat{C}_5$ are known, the remainders $\hat{C}_1$, $\hat{C}_3$, $\hat{C}_6$ and $\hat{C}_7$ are easily calculated from the charge neutrality condituion and the equilibrium relations. They are given as follows:

$$\hat{C}_6 = \frac{1}{2}\left(\hat{C}_2 + 2\hat{C}_4 + \hat{\alpha} + \sqrt{(\hat{C}_2 + 2\hat{C}_4 + \hat{\alpha})^2 + 4(2\hat{K}_2\hat{C}_2 + \hat{K}_5\hat{C}_5 + \hat{K}_w)}\right) \tag{40}$$

$$(\hat{C}_1, \ \hat{C}_3, \ \hat{C}_7) = (\hat{C}_6\hat{C}_2/\hat{K}_1, \ \hat{K}_2\hat{C}_2/\hat{C}_6, \ \hat{K}_5\hat{C}_5/\hat{C}_6) \tag{41}$$

The computational algorithm of the concentrations in the raindrop at every time step is summarized as: 1) With the appropriate time increment and old known values of C_i's integrate the governing eqn (22) to obtain new $\hat{C}_2$, $\hat{C}_4$ and $\hat{C}_5$; 2) Substituting thus obtained new $\hat{C}_2$, $\hat{C}_4$ and $\hat{C}_5$ into eqn (40), compute $\hat{C}_6$; and 3) Using these new $\hat{C}_2$, $\hat{C}_4$, $\hat{C}_5$ and $\hat{C}_6$, calculate $\hat{C}_1$, $\hat{C}_3$ and $\hat{C}_7$ by eqn (41).

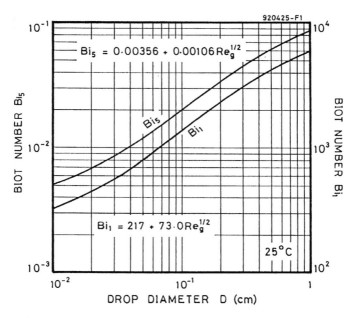

Figure 1: Variation of Biot Number with D (Bi_1: $SO_2(g)$; and Bi_5: $H_2O_2(g)$).

4 Acidification of raindrop

The acidification process of raindrops has been simulated for various drop diameters, the cloud bottom height being fixed at 500 m. The initial concentrations in the raindrops are: $C_2 = 10^{-7}M$; $C_4 = 10^{-10}M$; $C_5 = 10^{-6}M$; and $C_6 = 2.5 \times 10^{-6}M$ (pH = 5.6). The relationships between drop acidity in pH and drop diameter D in cm are depicted in Figures 2 and 3.

In Figure 2 the curves are shown parametrically in $H_2O_2(g)$ concentration and $SO_2(g)$ concentration is fixed at 50 ppb. As is supposed, the smaller the drop size is, the lower the pH value becomes, that is, the higher the acidity becomes. Also the higher $H_2O_2(g)$ concenration is, the smaller pH value becomes. However, in the range of small size ($D < 0.2$ cm) the manner of the curves of pH variations with drop size is greatly different from curve to curve depending on their $H_2O_2(g)$ concentrations. For large size ($D > 0.3$ cm) all of the curves take almost same manner of the variation, that is, pH is nearly independent on $H_2O_2(g)$ concentration. This means that the sensitivity of pH to $H_2O_2(g)$ concentration is much higher in small size region ($D < 0.2$ cm) than in large one ($D > 0.3$ cm). The sensitivity of $[H^+]$ to D is much greater than that impressed visually from the pH figure (Figure 2), because pH value is the logarithm of inverse of $[H^+]$ to base 10.

The curve for 0.1 ppb of $H_2O_2(g)$ has a plateau of pH value of about 4.45. The plateau spreads out to a width from 0.05 cm to 0.15 cm. The value of H^+ (i.e., C_6) for this plateau may correspond to the quasi-equilibrium concentration equilibrated approximately to the ambient $SO_2(g)$, because $H_2O_2(g)$ concentration is low to neglect the oxidation effect on H^+ production in this size region. About other curves there cannot be seen such plateau like this and their pH values drop dramatically with decrease in size. The high acidity observed often in drizzle ($D \leq 0.05$ cm) supports these sudden falls of curves in pH.

The higher $H_2O_2(g)$ concentration is, the more remarkable the decrease in pH becomes. This is because the production rate of SO_4^{2-} (i.e., C_4) by the oxidation is intensified with increase in both the concentrations of $H_2O_2(aq)$ (i.e., C_5) and H^+ (i.e., C_6). Since from eqn (40) C_6 is increased wih growth of C_4 and since from eqn (6) the production rate of C_4 is catalysed by increase in C_6, C_6 in raindrops is auto-multiplied during the fall in the air.

The curves in Figure 3 are shown parametrically in $SO_2(g)$ concentration but $H_2O_2(g)$ is fixed at 0.1 ppb. As is supposed from Figure 2 all these curves have plateaus and the acidities attain the quasi-equilibrium state, since $H_2O_2(g)$ concentration is low enough to neglect the effect of the oxidation in this size region. pH values, however, decrease abruptly in small size drop ($D < 0.05$ cm) to show clearly the oxidation effect. The sensitivity of pH to the variation of parameter [i.e., $SO_2(g)$ concentration] seems to be not so clear. It, however, may be seen that the sensitivity is rather higher in large size ($D > 0.1$ cm) than in small size ($D < 0.05$ cm). The variation of the sensitivity is not so drastic and in contrast with Figure 2 the sensitivity varies rather in reverse manner to Figure 2 with respect to the drop size.

In Figure 4 variations of C_2, C_4, C_6 and $2C_4/(C_2 + 2C_3)$ with drop size are shown. It is apparent that both SO_4^{2-} concentration (i.e., C_4) and H^+ concentration (i.e., C_6) increase suddenly with decrease in drop size especially in drizzle size ($D \leq 0.05$ cm). It should be noted that the increase in H^+ concentration enhace the production of SO_4^{2-} and vice versa. Without oxidation the major contributor to the acidification of raindrops is HSO_3^- (i.e., C_2) produced by dissociation reaction. This is true also in the early stage of acidification in which the effect of oxidation is small. The value of $2C_4/(C_2 + 2C_3)$ shows which of S(VI) (i.e., SO_4^{2-}) and S(IV) (i.e., HSO_3^- and SO_3^{2-}) is the major contributor to drop acidification. At $D = 0.1$ cm the ratio is about 0.12 ($\ll 1.0$) and the contribution of HSO_3^- (i.e., C_2) is overwhelmingly great ($C_2 \gg C_3$), but at $D = 0.02$ cm the ratio becomes about 4.4 and the contribution of SO_4^{2-} (i.e., C_4) grows to be very important.

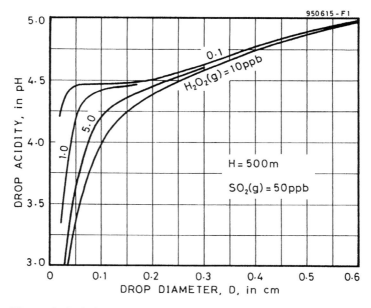

Figure 2: Relations between raindrop acidty pH and diameter D.

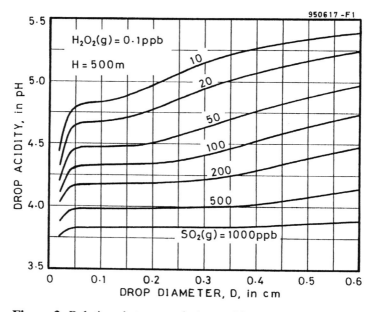

Figure 3: Relations between raindrop acidty pH and diameter D.

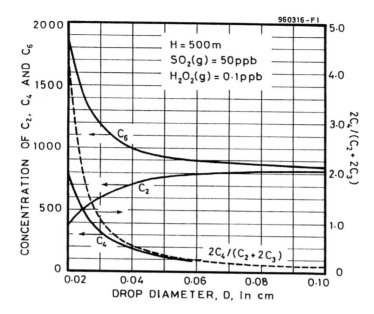

Figure 4: Variation of C_2, C_4, C_6 and $2C_4/(C_2 + 2C_3)$ with raindrop diameter D.

5 Conclusions

With use of the mathematical model developed in this study the following conclusions may be drawn from the simulations:

1. Comparing the Biot Number of $SO_2(g)$ with that of $H_2O_2(g)$ (Figure 1), it has been proved that the nonsteady mass transfer rate of $SO_2(g)$ is much greater than the rate of $H_2O_2(g)$, though $H_2O_2(g)$ is said to be more soluble than $SO_2(g)$ in equilibrium state.

2. Then the conventional prediction of acidity based on the equilibrium concentration is supposed to be inadequate to nonsteady problems and then inadequate to the problems which is dependent on travelling time t. The same is true to the problems which is dependent on drop size D, because the independent variable D can be transformed to travelling time t via falling velocity $u(D)$ with use of the relationship between t and D as: $t =$ (cloud bottom hight) $/u(D)$.

3. The effect of $H_2O_2(g)$ concentration on raindrop pH, in other words the sensitivity of pH to $H_2O_2(g)$ concentration, is greater in small size drops than in large ones (Figure 2). On the other hand the effect of $SO_2(g)$ concentration on raindrop pH is greater rather in large size drops than in small size drops, although this is not so clear(Figure 3).

4. In small size drops especially in drizzle size the contribution of SO_4^{2-} (i.e., oxidation of HSO_3^-) to the acid rain formation becomes more and more important with lapse of time (Figure 4). As well as small drops, drops falling long distance also produce significant amount of SO_4^{2-} to contribute to the acidification, since the residence time (i.e., reaction time) in the air becomes long enough to produce SO_4^{2-}.

5. Uunder the existence of the oxidant like $H_2O_2(g)$, washout of 50 ppb $SO_2(g)$ in about 500 m fall by 0.1 cm drops can raise the drop acidity up to pH 4.5 from pH 5.6 of natural cloud water. This means that the acid rain of pH 4.5 experienced in Japan may be easily formed even if there is no long-distance transport of acidic pollutants from outside of Japan.

References

[1] Mason, B. J. *Acid Rain*, Oxford University Press, Oxford, 1992, pp.14-15.

[2] Liss, P. S. and Slinn, W. G. N. *Air-Sea Exchange of Gases and Particles*, NATO ASI Series, Series C, Mathematical and Physical Series No.108, D. Reidel Publishing Co., Dordrecht, 1983, p.206.

[3] Maahs, H. G. Kinetics and Mechanism of the Oxidation of S(IV) by Ozone in Aqueous Solution with Particular Reference to SO_2 Conversion in Nonurban Tropospheric Clouds, *J. Geophysical Research*, 1983, Vol.88(C15), pp.10721-10732.

[4] Levich, V. G. *Physico-Chemical Hydrodynamics*, Prentice-Hall Inc., Englewood Cliffs, N. J., 1962, pp.395-402.

[5] Shiba, S. Acid Rain Formation by $SO_2 - H_2O_2$ Absorption with Sequential Drop-Phase Chemical Reactions, *Computer Techniques in Environmental Studies IV*, Zannetti, P. (Ed.), Computational Mechanics Publications, Southampton, 1992, pp.45-60.

[6] Shiba, S. and Hirata, Y. Effect of Ambient Ammonia on Acid Rain Formation Caused by Sulphur Dioxide, *Computer Techniques in Environmental Studies V*, Vol.1; Pollution Modeling, Zannetti, P. (Ed.), Computational Mechanics Publications, Southampton, 1994, pp.57-66.

A dynamic model for the chemistry of Manganese at River Eume

P. Bello,[a] J.A. Souto,[a] J.J. Casares,[a] T. Lucas[b]
[a]Department of Chemical Engineering, University of Santiago de Compostela, 15706 Santiago de Compostela, Spain
[b]C.T. As Pontes, Endesa, 15320 As Pontes, Spain

Abstract

A dynamic model of the behaviour of Manganese species (Mn^{2+} and MnO_x) along a river has been developed and tested with real data. The model follows the same approach to the problem as other well established models like MASAS (Ulrich et al. [1]). The one-box model adopted here reproduces the kinetics of the various processes involved in the sedimentation and dissolving of Mn species: oxidation and reduction, interface transport and sedimentation, combined with flow terms.

The model has been applied to the simulation of the Mn chemical behaviour along the River Eume, from As Pontes Power Plant (main source of Mn) to a dam, 20 km below. Historical data over two years were used for validation purposes and, though no flow modeling was considered (as it corresponds to a one-box model), the results of the dynamic model agree to the Mn^{2+} measurements at the dam outlet; as a conclusion, the dynamic model developed can be used to simulate the future Mn concentration at the river for different conditions.

1 Introduction

The load of heavy metals in the water of a dam can change the biological equilibrium into it, and downstream. However, the evolution of these chemical species in dams is very slow, with a seasonal cycle, so it's difficult to find tendencies from the measures of heavy metals concentrations along several years. Water quality models in equilibrium (Allison et al. [2]) provides stationary solutions, so time-dependent processes are not considered. A dynamic model that includes internal and external processes affecting these species can help to understand the behavior of pollutants in the dam, and after calibration and validation, provides a good tool for the estimation of concentration tendencies.

Some models, such as WASP (Ambrose and Martin [3]) provide a complete tool for dynamic water quality modeling of typical water pollutants (chemical, biological); but, because of its complexity, some pollutants as Mn, with non-linear chemistry, are not included. On the other hand, box models (MASAS, Ulrich et al. [1]) are a good solution when flow and pollutant dispersion are not significant processes for the time scale considered, and the chemistry of the pollutant is the main force.

Manganese reactions in natural waters can produce dissolved (Mn^{2+}) and solid species (Mn oxides), in various conditions, so Mn deposition is extremely affected by the chemistry. In this work, a one-box model for Mn behaviour in natural waters is developed and applied. The model includes the most significant redox kinetics mechanism of Mn in water, as the main dynamic process that affect Mn deposition in the dam, so the influence of other species (H^+, O_2, organic compounds) in the Mn deposition can be explained.

2 The model

In natural waters, Manganese can be present in dissolved or solid form, following a redox mechanism affected for several conditions. Dissolved Mn is presented as Mn^{2+}, and solid Mn is mainly presented as oxides of Mn (MnO_x), with different oxidation states; some Mn complexes could appear as solid matter, but usually they are not significant. So, a dynamic model must consider two phases for Mn: dissolved and solid, and a mass balance should be applied to both phases.

For dissolved phase, mass balance includes the following processes,

- Mn^{2+} flow inlet, as the amount at the tail (entrance) of the dam.

- Mn^{2+} flow outlet, as the amount at the head (exit) of the dam.

- Chemical reaction rate in dissolution, as oxidation rate of Mn^{2+}.

- Transfer rate from dissolved to solid phase.

For the solid phase, a similar mass balance includes,

- MnO_x flow inlet, as the amount of MnO_x in the solids at the tail (entrance) of the dam.

- MnO_x flow outlet, as the amount of MnO_x in the solids at the head (exit) of the dam.

- Chemical reaction rate in dissolution, as reduction rate of MnO_x.

- Transfer rate from solid to dissolved phase.

- Deposition rate of solids with MnO_x.

These mass balances can be summarized in two equations,

$$\frac{d[Mn^{2+}]}{dt} = E_{Mn^{2+}} - S_{Mn^{2+}} - R_{Mn^{2+}} - J_{trans}$$
$$\frac{d[MnO_x]}{dt} = E_{MnO_x} - S_{MnO_x} - D_{MnO_x} - R_{MnO_x} + J_{trans}$$

$$(1)$$

with all terms expressed in moles/L·day. Submodels for each process involved were formulated in order to obtain a global model for the dynamic of Manganese; but, the main effort was centered in chemical kinetics of Mn. The interphase transfer model applies a modified expression from the Wehrli et al. [4] equation,

$$J_{trans} = -k[s.s.]\frac{d[Mn^{2+}]}{dz}$$

where the value for k is 824 L·dm/mol·day. In the expression considered, the suspended solids concentration [s.s.] is used instead of the solids porosity, as a measure of the active surface of suspended solids. For Mn^{2+} gradient, several concentration profiles were measured in the summer of 1995, and the value obtained for the gradient was similar in each profile, around 2.6 (mg/L)/m.

For deposition rate of suspended solids a constant value of 0.7 m/day was assumed. These two parameters depend on the flow, but the chemical behaviour of Mn at the river was considered more important.

3 Chemistry of Mn

The chemistry of Mn in natural waters is determined by different competitive redox reactions; these processes are thermodynamically favourable, but kinetics is an important problem. Manganese in natural waters usually reacts in several days, and its reaction rates depend on several factors, oxygen solubility, pH, organic compounds, microorganisms, that affect oxidation and/or reduction of Mn species in different ways.

3.1 Oxidation of Mn²⁺

The result of Mn^{2+} oxidation in water is a set of Manganese oxides (denoted as MnO_x), where x varies from 1.3 to 2.0 (Eary and Schramke [5]). Oxidation is faster in alkaline solutions, and reaction rate increases as MnO_x are produced, so an autocatalytic effect is to be considered. Stumm and Morgan postulated a general mechanism (Hsiung and Tisue [6]) in three steps for the autocatalytic oxidation of Mn^{2+},

Homogeneous oxidation:

$$Mn^{2+}(aq) + O_2(aq) \xrightarrow{\text{slow}} MnO_2(s)$$

Adsorption:

$$Mn^{2+}(aq) + MnO_2(s) \xrightarrow{\text{fast}} Mn(II).MnO_2(s)$$

Heterogeneous oxidation:

$$Mn(II).MnO_2(s) + O_2(aq) \xrightarrow{\text{slow}} 2MnO_2(s)$$

The reaction rate for this process can be expressed as,

$$-\frac{d[Mn^{2+}]}{dt} = k_0[Mn^{2+}] + k[Mn^{2+}][MnO_x]$$

where k_0 and k are rate constants for the oxygenation reaction, and equal to

$$k_0 = 10^{9.76}[OH^-]^2 P_{O_2}$$
$$k = 10^{6.97}[OH^-]^2 P_{O_2}$$

Other mechanisms including Fe species (Eary & Schramke [5]) and microorganisms (Wehrli et al. [4]) have been studied, but in the model analyzed, the Stumm & Morgan mechanism was adopted as the most relevant.

3.2 Reduction of oxidized Mn

Organic compounds are the most important reduction agents of oxides and hydroxides of Mn in natural waters. Stone and Morgan [7] studied these processes for different organic compounds (alcohols, hydroquinones, aromatics and organic acids) and they derived a expression for MnO_x reduction rate, at pH=7.2,

$$\frac{d[Mn^{2+}]}{dt} = k_x[Red]([MnO_x]_0 - [Mn^{2+}])$$

where $[Red]$ is the organic compounds concentration and $[MnO_x]_0$ is the initial MnO_x concentration. For hydroquinones, Stone and Morgan [8] found other pH-dependent expression, as follows,

$$\frac{d[Mn^{2+}]}{dt} = k_1[H^+]^{0.46}[QH_2]^{1.0}([MnO_x]_0 - [Mn^{2+}])$$

where $[QH_2]$ is the hydroquinones concentration.

pH dependent reduction rate expression is presented in this work, that applies to organic reductors,

$$\frac{d[Mn^{2+}]}{dt} = k_2[H^+]^x[Red]^y([MnO_x]_0 - [Mn^{2+}])$$

where x and y are the reaction orders, and $k_2 = 1 \cdot 10^{-8}$ moles/L· day. In this equation, the difference $[MnO_x]_0 - [Mn^{2+}]$ is equal to MnO_x concentration at any time.

4 Results

The mathematical expressions that correspond to the different processes involved in Manganese dynamics are included in the general overall mass balance as written in equations (1), so a set of two differential equations is obtained,

$$\frac{d[Mn^{2+}]}{dt} = \frac{I_{Mn^{2+}}}{V} - \frac{Q_{sal}}{V}[Mn^{2+}] - k_0[Mn^{2+}] - k_1[Mn^{2+}][MnO_x] +$$

$$+ k_2[H^+]^x[Red]^y[MnO_x] - \left(-k[s.s.]_{sal}\frac{d[Mn^{2+}]}{dz}\right)$$

$$\frac{d[MnO_x]}{dt} = \frac{I_{MnO_x}}{V} - \frac{Q_{sal}}{V}[MnO_x] - D_{MnO_x} + k_0[Mn^{2+}] + k_1[Mn^{2+}][MnO_x] -$$

$$- k_2[H^+]^x[Red]^y[MnO_x] + \left(-k[s.s.]_{sal}\frac{d[Mn^{2+}]}{dz}\right)$$

In one-box models, the water system is assumed to behave as a continuous stirred reactor, so the concentration of the different species in the volume is equal to the outlet concentration. But, chemical processes of Mn in the dam occur mainly near the sediments, because of the higher concentration there, so the Mn species concentrations should be as similar as possible to the actual concentration in the dam. From several measurements of Mn^{2+} and suspended solids at different heights in the dam, two linear relationships between outlet concentrations and the average concentration where found,

$$[Mn^{2+}]_{reaction} = K_1[Mn^{2+}]$$

$$[MnO_x]_{reaction} = K_2[MnO_x]$$

where $K_1 = 1.13$ and $K_2 = 1.19$. These expressions were included in the reaction rate terms only.

For solving the ODE system proposed, a Fortran 77 program was developed, using the IMSL math library; the IVPRK subroutine, that includes a Runge-Kutta method, was applied. The maximum relative error considered was $5 \cdot 10^{-3}$. In this conditions, the program only needs around 2-3 CPU seconds on a Fujitsu VP2400/10 for a 2-years simulation.

This one-box model was applied for the simulation of Mn behavior at River Eume, between the discharge of water with Mn and the head of the dam 15 km below. The discharge comes from a water treatment plant that cleans waters of an open-pit coal mine. The Mn concentration in the discharge is around 0.8 mg/L, and the water flow can reach above 20 Hm^3/month. The river flow is around seven times that value, so Mn is dispersed rapidly.

The sediments in the dam are specially active for Manganese dynamics, because the sediments with Mn are mainly amorphous species (goetites), so Manganese can be resuspended easily. When the chemical conditions are favourable, this Manganese (mainly MnO_x) can be reduced to Mn^{2+}, which then is dissolved in the surrounding water.

As the main result of the simulation performed, daily values of Mn^{2+} and MnO_x at the dam outlet was obtained from June 1992 to November 1994. At figure 1, estimated Mn^{2+} concentration is compared to monthly averages of Mn^{2+} measurements. The model can get similar tendencies, with some quantitative differences, specially in months with important flow variations at the river.

Because Mn concentration in suspended solids is unknown, estimated MnO_x concentration is compared to monthly averages of suspended solids [s.s.] concentration at figure 2. Estimated tendencies are similar again, but in this case it seems that the model is slower than the real processes; this problem can be overcome modifying the assumed value for the constant sedimentation rate for the suspended solids.

Figure 3 shows estimated Mn^{2+} concentration and pH measurements. The model reproduces the natural tendency of Mn to reduction (Mn^{2+} increase) at low pH. Figure 4 shows estimated vs. measured Mn^{2+} concentration (monthly averages). Most of the points are close to the right value, as only five points (months) present important deviations along 2 years. The flow variations is the main factor for these differences, because the one-box model is not a flow model.

5 Conclusions

A one-box model for Manganese dynamic in natural waters has been developed and .applied to a river-dam system at River Eume. The model is mainly focused in the redox kinetics of Mn in water, in order to study the influence of different factors, like pH, on this behaviour. A two ODE system has been solved using a Fortran 77 program and standard IMSL routines.

The model shows reasonable agreement with experimental measurements, except for strong varying flow situations. Results for a more than two years simulation are compared to measurements, and the estimated Mn^{2+} concentration at the dam outlet is correct except in five months. The kinetics redox mechanisms applied reproduce the influence of chemical factors such as pH. So, for the long-term evaluation of Mn dynamics, the model could be used to study the influence of changing chemical conditions (pH, dissolved O_2, organic compounds) in the dam, and therefore it is possible to forecast the Mn^{2+} concentration for the future evolution of the river.

A multi-box model and a flow model are suggested for solving some differences between measured and estimated concentrations.

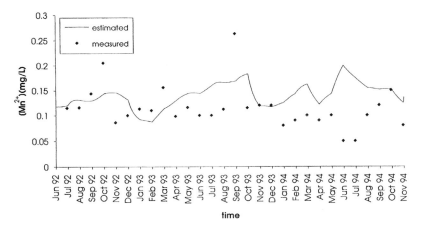

Figure 1: Measured (monthly) and estimated (daily) Mn^{2+} concentration, July 1992-November 1994, for the dam outlet at River Eume.

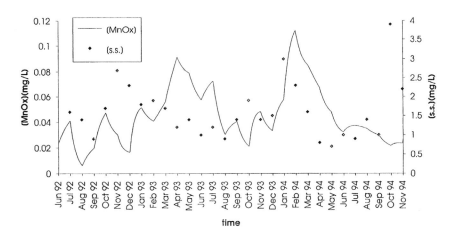

Figure 2: Measured (monthly) suspended solids concentration and estimated (daily) MnO_x concentration, July 1992-November 1994, for the dam outlet at River Eume.

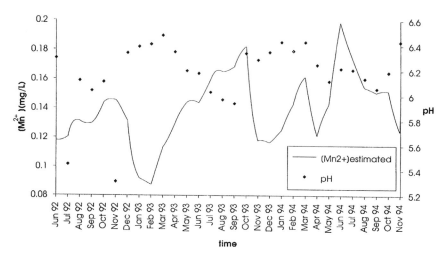

Figure 3: Measured (monthly) pH and estimated (daily) Mn^{2+} concentration, July 1992-November 1994, for the dam outlet at River Eume.

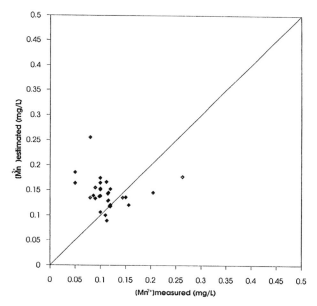

Figure 4: Estimated vs. measured Mn^{2+} concentrations (monthly averages).

Acknowledgements

The computational time assigned and technical support of the Centro de Supercomputación de Galicia are gratefully acknowledged. This work has been financially supported by the Empresa Nacional de Electricidad, S.A..

References

1. Ulrich, M., R.P. Schwarzenbach & D.M. Imboden. MASAS - Modelling of Anthropogenic Substances in Aquatic Systems on Personal Computers - Application to Lakes. *Environmental Software*, 1991, **6**, 34-38.

2. Allison, J.D., D.S. Brown & K.J. Novo-Gradac. *A Geochemical Assessment Model for Environmental Systems*. U.S. Environmental Protection Agency, 1991.

3. Ambrose, R. B., & J.L. Martin. *The Water Quality Analysis Simulation Program, WASP*. Center for Exposure Assessment Modeling (CEAM), U.S. Environmental Protection Agency, 1993.

4. Wehrli, B., G. Friedl & A. Manceau. Reaction Rates and Products of Manganese Oxidation at the Sediment-Water Interface. *Aquatic Chemistry: Interfacial an Interspecies Processes*. Advances in Chemistry Series, American Chemical Society, 1995, **244**, 111-134.

5. Eary, L.E. & J.A. Schramke. Rates of Inorganic Oxidation Reactions Involving Dissolved Oxygen. *Chemical Modeling of Aqueous Systems II*, A.C.S. Symposium Series, American Chemical Society, 1990, **416**, 379-393.

6. Hiung, T.-M. & T. Tisue. Manganese Dynamics in Lake Richard B. Russell. *Environmental Chemistry of Lakes and Reservoirs*. Advances in Chemistry Series, American Chemical Society, 1994, **237**, 499-524.

7. Stone, A.T. & J.J. Morgan. Reduction and Dissolution of Manganese(III) and Manganese(IV) Oxides by Organics: 2. Survey of the Reactivity of Organics. *Environmental Science and Technology*, 1984, **18**, 450-456.

8. Stone, A.T. & J.J. Morgan. Reduction and Dissolution of Manganese(III) and Manganese(IV) Oxides by Organics: 1. Reaction with Hydroquinone. *Environmental Science and Technology*, 1984, **18**, 617-624.

Multivariate cluster analysis of trace elements and mineralogical components from some rural soils

M.M. Jordán Vidal,[a] J. Mateu Mahiques,[b] A. Boix Sanfeliu[a]

[a]Universitat Jaume I, Departament de Ciències Experimentals, Campus Borriol, 12071 Castellón, Spain

[b]Universitat Jaume I, Departament de Matemàtiques, Campus de Penyeta Roja, 12071 Castellón, Spain

Abstract

During the 1994-95 two-year period, a research project has been carried out to characterise the mineralogical and chemical composition of some rural soils located close to five clay outcrops that supply the Castellon (spanish province) ceramic industry. A total of 50 soil samples were collected in accordance with statistical criteria and their chemical elemental composition were determined by ICP-AES. The mineralogical composition of each sample was discovered using X-ray diffraction. Semi-quantitative mineralogical analysis was done following the Chung[3] methodology. A hierarchical cluster analysis using Ward method has been carried out to detect statistical significative groups (clusters) and a discriminant analysis of proximities to obtain probabilities. Univariate statistical descriptions have been computed for all variables.

Statistical procedures have been implemented using the software package SPSS for Windows release 5.0.1. This is regarded as a useful software for environmental studies.

1 Introduction

In the province of Castellón, the expansion of agriculture toward the very sloping hillsides, deforestation, shepherding and fundamentally the forest fires are the causes that have produced the loss of soil, in some instances irreversible, and the road to desertization. This situation gets worse with ceramic clays extraction in the study areas, since the impact on soil can be qualified as negative and serious due to ground occupation in making hollows, building access tracks and rural roads and accumulating material stocks on the ground. Furthermore, these soils can be contaminated by solids in suspension contribution, organic material, fats and lubricants derived from the machinery used in the extractive processes.

All over this, during the biennium 1994-95 the Crystallography and Mineralogy area of the Experimental Sciences Department of the Jaume I University accomplished an investigation to characterize the mineralogical and chemical composition, from majority elements to traces, of rural soils next to five clays outcrops that supply the ceramic industry of the province of Castellón. The studied area extends to the townships of Forcall (series FM), Argelita (series LA), Zucaina (series ZC), La Jana (series LA) and Cervera (series CE).

2 Experimental

A total of 50 samples were selected, 10 for each deposit. Each clasified and numbered sample and of an approximate quantity of 5 Kg has been selected for its study in the laboratory.

2.1 Mineralogical Analysis

The procedure employed in the mineralogical analysis is the breaking-up method described by Hathaway[7]. The customary methods in the study of the clays (Carrol [2]) have been practiced, on the treated samples and without treatment making orientated aggregate preparations (normal, heated to 550 °C during 2 hours and treatment with ethylenglicol during two hours previously to the difractometric record).

For the orientated aggregates obtainment, we disperse the clay fraction of each one of the samples in distiled water (Ballbe [1]). To apply this technique the record of the X-ray graphs at ambient temperature (18-22 °C) was accomplished using the powder difractometer Siemens D-5000 with Bragg-Bretano geometry, graphite (monocromador) and twinkle detector. Also, a graph for the identification of the mineralogical components of the total fraction was accomplished. The spectra were taken with of Cu, K (1,54052) radiation to 40 KW and 20 mA. A normal graph from 4 to 70 of 2 using Ni filter , 0,050 steps, constant of time of 3 seconds and incidence splits combination(1 mm), of divergence (1 mm) and detector of 0,2 mm with continuous record modality was accomplished.

The identification of the present mineral phases in each one of the samples has been accomplished through the utilization of the computer program EVE of the Socabim company of the IJA and the JCPDS index cards. Through the mentioned program the elimination of corresponding difractometer background, the diffraction peaks selection and phases search was accomplished, afterwards completed with the JCPDS diffraction index cards consultation.

A diffraction detected phases quantification has been accomplished. This is not in absolute terms quantification, but it is comparative between the different samples, reflecting the percentage of each substance respect to the total identified substances (Chung [3]).

2.2 Chemical Analysis

The instrumental analytical technique used to know the studied samples chemistry composition has been the inductive coupling plasma emission spectroscopy (ICP-AES). The inductive coupling plasma spectrometer (ICP) used corresponded to the Jobin-Yvon company, model JY-38 VHR.

The samples were put on dissolution with hot concentrated HCl, diluting afterwards with water until an appropriate volume. The used standards were containing the assault reactives of the sample to eliminate the counterfoil effect. In each one of the resulting solutions of the described protocol the following elements were determined: Al, C, Mg, K, N, Ti, Fe, Mn, Cu, Zn, Co, Ni, V, Sr, P, B, and S.

2.3 Statistical Methodology

Cluster analysis

Cluster analysis is concerned with the discovering group structure amongst the cases of a n by p matrix. Two general references are Gordon[4] and Hartigan[6]. Almost all methods are based on a measure of the similarity or dissimilarity between cases. A *dissimilarity coefficient* d is symmetric (d(A,B)=d(B,A)), non-negative and d(A,A) is zero. A *similarity coefficient* has the scale reversed. Dissimilarities may be a *metric*:

$$d(A,C) \leq d(A,B) + d(B,C)$$

or an *ultrametric*:

$$d(A,B) \leq max(d(A,C),d(B,C))$$

but need not be either.

There are several families of similarity and dissimilarity measures to produce clusters. Similar objects should appear in the same cluster and dissimilar objects in different clusters. A well known interobject measure of similarity is the correlation coefficient between a pair of objects measured on several variables. *Correlational measures* represent similarity by the correspondence of patterns accross the characteristics (X variables) and are not influenced by differences in scales between objects. But correlational measures are rarely used because emphasis in most applications of cluster analysis is on the magnitudes of the objects, not the patterns of values. *Distance measures* of similarity, which represent similarity as the proximity of observations to one another across the variables in the cluster variate, are the similarity measure most often used. These measures, like Euclidean, Gamma, Pearson..., are significantly affected by differences in scale so before computing them, it is needed to standardize the data across the measured attributes. The most commonly used is the *Euclidean distance*.

On the other hand, various linkage methods can be used to compute the distance of one object or cluster from another and to determine whether the two should be merged in a given step. The *Centroid Linkage* method uses the average value of all objects in a cluster as the reference point for distances to other objects or clusters. The *Ward's method* (Ward [8]) resembles Centroid Linkage but adjusts for covariances.

Cluster algorithms can be classified into two general categories: *hierarchical* and *nonhierarchical*. We'll just discuss the former. Hierarchical procedures involve the construction of a hierarchy of a treelike structure. There are basically two types of hierarchical clustering procedures: *agglomerative* and *divisive*. In the agglomerative methods, each object or observation starts out as its own cluster. In subsequent steps, the two closest clusters are combined into a new aggregate cluster. Eventually, all individuals are grouped into one large cluster. This process is usually shown in a tree graph or dendrogram. In divisive methods, we begin with one large cluster containing all observations. In succeeding steps, the observations that are most dissimilar are split off and made into smaller clusters.

Multiple discriminant analysis

Discriminant analysis is the appropriate statistical technique when the dependent variable is categorical and the independent variables are metric. In many cases, the dependent variable consists of two groups or classifications. In other instances, more than two groups are involved, such as a three-group classification. Discriminant analysis is capable of handling either two groups or multiple groups. When two classifications are involved, the technique is referred to as *two-group discriminant analysis* (Hair et al [5]). When three or more classifications are identified, the technique is referred to as *multiple discriminant analysis* (MDA).

Discriminant analysis involves deriving a *variate*, the linear combination of the two (or more) independent variables that will discriminate best between a priori defined groups. Discrimination is achieved by setting the variate's weights for each variable to maximize the between-group variance relative to the within-group variance. The linear combination for a discriminant analysis, also known as the *discriminant function*, is derived from an equation that takes the following form:

$$Z = W_1 X_1 + W_2 X_2 + \ldots + W_n X_n$$

where

Z=Discriminant score
W_i=Discriminant weight for variable i
X_i=Independent variable i.

Suppose that g is the number of groups, and for each case we know the group (assumed correctly). We can then use the group information to help reveal the structure. Let W denote the within-group covariance matrix, that is the covariance matrix of the variables centred on the group mean, and B denote the between-groups covariance matrix , that is of the predictions by the group means. Let M be the gxp matrix of group means, and G be the nxg matrix of group indicator variables (so g_{ij}=1 if and only if case i is assigned to group j). Then the predictions are GM. Let $\bar{x}$ be the means of the variables over the whole sample. Then the sample covariance matrices are

$$W = \frac{(X - GM)^T (X - GM)}{n - g} \; , \; B = \frac{(GM - 1\bar{x})^T (GM - 1\bar{x})}{g - 1}$$

If there are more than two groups in the dependent variable, discriminant analysis will calculate more than one discriminant function. As a matter of fact, it will calculate g-1 functions. Each discriminant function will calculate a discriminant score . In the case of a three-group dependent variable, each object will have a score for discriminant functions one and two, allowing the objects to be plotted in two dimensions, with each dimension representing a discriminant function.

2.4 Computer Comands

The SPSS software package provides many useful statistical techniques to look, explore, analyze and present environmental data. Some of them are used in this paper and the SPSS comands to run these methods are provided here.

a) Cluster analysis

PROXIMITIES
variable names /MATRIX OUT ('C:\WINDOWS\TEMP\spssclus.tmp')
/VIEW=CASE
/MEASURE=SEUCLID /PRINT NONE /STANDARDIZE=NONE.
CLUSTER /MATRIX IN ('C:\WINDOWS\TEMP\spssclus.tmp') /METHOD WARD
/PRINT SCHEDULE /PLOT DENDROGRAM.
ERASE FILE='C:\WINDOWS\TEMP\spssclus.tmp'.

b) Discriminant analysis

DISCRIMINANT /GROUPS=grouping variable /VARIABLES=variable names
/ANALYSIS ALL /STATISTICS=MEAN STDDEV UNIVF COEFF
/CLASSIFY=NONMISSING POOLED.

c) Analysis of variance and T-tests

ANOVA VARIABLES=variable names BY grouping variable /MAXORDERS
2 /METHOD UNIQUE /FORMAT LABELS.
T-TEST   GROUPS=grouping   variable   /MISSING=ANALYSIS
/VARIABLES=minority
/CRITERIA=CIN(.95).

3 Results and Discussion

The X Ray Difractograms analysis of the clay fraction shows the predominantly illitic character of all the analyzed samples. Chlorite and the Kaolinite are also present. Quartz is also detected as integrating the clay fraction.

The semiquantitative analysis of the total fraction has stated a great similarity in the LA and ZC series mineralogical composition. Both possess a strong illitic character (40-54 %), a content in quartz around 30 % and a total percentage of Clorita + Caolinita around 10 %. The medium content in Hematites is found about 5 %. The potasic feldspars prevail as compared to the plagioclases though in no case surpass 7 %. The samples belonging to the CE series possess a high content in clay's minerals. The Quartz does not exceed 21% in any case The Illita and Clorita+Caolinita percentages are located around 45 % and 30 % respectively. In most of the samples it is not detected the Hematites, Dolomite or Plagioclases presence. The Calcite content is similar to the LA and ZC series, no surpassing 2 %. These samples can be classified as illitic -caolinitic clays with under content in sand fraction. The series JT is characterized by a percentual similarity concerning Illite and Caolinite+Clorite located around 30 %. The content in Quartz is variable, oscillating between 27 % in JT1 sample and 44 % in JT8 and JT9 samples. This series is the one which possesses a greater content in carbonates as much in the form of Calcite as Dolomite reaching a 14 % in the JT7 sample for the Calcite and a 3 % for the Dolomite. On the contrary in JT8, JT9 and JT10 samples only have indicia. The FM series is quite similar to LA and ZC series, whose results have been discussed previously. However, it emphasizes the presence of potasis feldspars as important part of the sand fraction.

About the mineralogy of the sandy fraction, at the sight of the results, we observe that this fraction is constituted mainly by Quartz followed quantitatively by the Illite, that in some samples and in the fraction fine sand, arrives to surpass even the Quartz percentage in the same fraction.

Between the analyzed trace elements emphasize by its relative importance and "abundance" (> 100 ppm) the following elements: P, B, S and Mn. Consequently, the trace element analysis results has been mainly focused in the discussion of the presence of these 4 elements in each one of the considerate series. The phosphorus analysis (P) states that its content varies sensibly of some samples to other, no existing a clear concentration of this element in given series. Something similar occurs when it is analyzed the Barium (B), whose content varies from 145 ppm analyzed in the CE5 sample and 506 ppm of LA4. It is not observed a concentration of this element in a given series either. The S is an essential element in the nourishment of the plants, and results indispensable for the cysteine and certain vitamins synthesis. When it is analyzed the S, is observed the concentration of this element in the CE series, arriving to surpass 5000 ppm in the CE2 sample and being its concentration above 3000 ppm in CE5, CE6 and CE7. In the rest of analyzed series the concentrations are found far below 300 ppm. Thus, the samples of the LA series only exceed 100 ppm in the LA2 sample, while the samples of the ZC and FM series are found below 200 ppm and the corresponding samples to the JT series do not reach 200 ppm. The manganese analysis (Mn) permitted to observe a meaningful increase of this element in samples of the CE (643 ppm in CE7) and FM (573 ppm in FM3) series, as well as an important decrease of this element when samples of the ZC series are analyzed. The LA and JT series possess variable concentrations according to the analyzed sample.

A statistical cluster analysis of data provided by elemental chemical analysis (ICP-AES) as well as by quantitative mineralogical analysis (Chung [4]) has been accomplished. Since the analyzed data suggest the existence of more than a form of soil, from the geological age point of view, it has been turned to the cluster or grouping analysis and to the discriminant analysis for its clasification. Previously, univariate descriptions of the variables have been calculated.

In first place, it was accomplished a cluster analysis of the samples, through the Ward's hierarchical agglomerative method based on the euclidean distance. Two types of dendrograms have been accomplished: the first corresponds to the grouping of each one of the samples according to its chemistry composition (Fig.1) and the second corresponds to the grouping of these according to its mineralogical composition (Fig.2).

The samples classification according to its chemistry drove to establish three well differentiated groups. As checking, a discriminant analysis was accomplished. Group I consisted of ZC series and LA series. The group II gathered samples of the FM and JT series, being the FM series samples closer to the group I than the ones corresponding to the JT series. It is convinient to indicate that the groups I and II could be included in only one group since, as it is shown in the dendrogram, the differences are minimal. In fact, after a T-test, there are only meaningful differences between both groups for the Mn, Zn, Co and Th variable. Finally, to indicate that the group III is constituted by the samples of the CE series. These samples have some typical compositional characteristics by its high content in S and alkalinoterreous elements. The averages and standard deviations for each group and only for trace elements come given in table 1. Differences between groups I and II with respect to III, are confirmed with a new T-test for independent samples given a statistical significance value of p=0.01.

The classification or grouping of the samples attending to its mineralogical composition permitted to establish three groups, see the dendrogram of the Fig.2, confirmed by the applied discriminant analysis. This grouping coincided with the accomplished one attending to the chemistry composition of the samples solely in the establishment of the group I (ZC series + LA series). The group II gathered solely the samples of the JT series with the exception of the FM1 sample that was included in this group. The group III gathered the samples of the CE and FM series.

To confirm the existence of meaningful differences between the three formed groups, an analysis of the variance (ANOVA) for each one of the variables under study was made. In all the variables, except in F (Feldspars) in the one which is reached a statistical significance level of p=0.56, there are statistical significant differences between the three groups formed by the cluster analysis with very close statistical significances to the p=0.0. The univariates descriptions for each group and variable come given in table 2.

Table 1. Univariate descriptions for groups of chemical trace elements.

Element	Group I mean	Group I std.dev	Group II mean	Group II std.dev	Group III mean	Group III std.dev
Mn	116.87	72.16	310.38	173.48	419.71	159.41
Cu	4.66	4.26	28.89	51.61	27.17	19.78
Zn	24.69	11.3	50.7	32.27	59.78	23.7
Co	1.16	1.86	7.96	3.83	8.35	3.51
Ni	22.08	7.1	31.9	20.34	28.83	11.6
V	76.47	26.9	69.75	40.97	72.16	24.36
Sr	75.42	21.75	125.64	98.8	274.96	233.71
P	234.71	82.4	301.41	91.67	497.91	153.87
Ba	355.57	98.11	315.43	132.88	367.02	148.31
S	84.41	69.62	138.52	81.76	4175.61	925.71

Table 2. Univariate descriptions for groups of mineralogic elements.

Element	Group I mean	Group I std.dev	Group II mean	Group II std.dev	Group III mean	Group III std.dev
Q	32.05	4.17	18.89	3.21	32.8	8.52
Cc	1.45	1.39	0.77	0.64	4.5	4.32
D	0.65	0.74	0.0	0.0	0.9	0.99
M + I	47.1	4.97	43.83	2.95	26.7	6.81
C + K	9.1	4.22	29.22	3.45	28.1	6.24
Hm	4.95	3.06	2.44	2.45	2.4	1.77
F	4.7	2.73	4.55	2.09	3.7	2.45

References

1. Ballbé, E. y Martinez, S. (1985). «Métodos de diferenciación de caolinita y cloritas». Acta Geológica Hipánica. V.20, n° 3-4, pp. 245-255.
2. Carroll, D. and Starkey, H. (1971). «Reactivity of clay minerals with acide and alkalis». Clays and Clay Minerals. pp. 321-33.
3. Chung, F.H. (1974). «Quantitative interpretation of X-ray diffraction patterns of mixtures I. Matrix-flushing method for quantitative multicomponent analysis». J. Appl. Cryst., 7, 519-525.
4. Gordon, A.D. (1981). Classification: Methods for the Exploratory Analysis of Multivariate Data. London: Chapman and Hall.
5. Hair, Joseph F., Anderson, Rolph E., Tatham, Ronald L. and Black, William C. (1995). Multivariate Data Analysis with Readings.Fourth Edition 745 pp. Prentice Hall, New Jersey.
6. Hartigan, J.A. (1975). Clustering Algorithms. New York: John Wiley and Sons.
7. Hathaway, J.C. (1956). «Procedure for clay mineral analysis used in the Sedimentary Petrology laboratory of the U.S. Geological Survey». Clay Min. Bull, 3, pp 8-13.
8. Ward, J.H. (1963). Hierarchical Grouping to Optimize an Objective Function. Journal of the American Statistical Association, 58, 236-244.

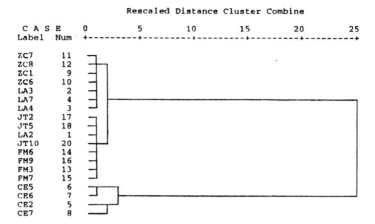

Figure 1: Dendrogram using Ward method for chemical elements.

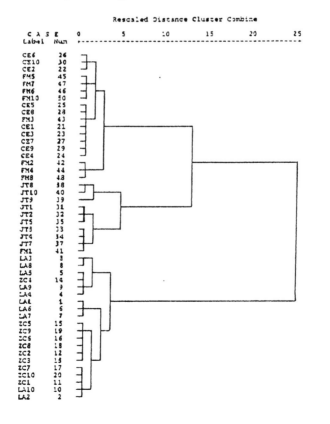

Figure 2: Dendrogram using Ward method for mineralogic elements.

A comparison of 2-D and 3-D pre- and post-processing techniques in groundwater flow modeling

S.A. Sorby, A.S. Mayer, C.G. Tallman, J.G. Johnson
Department of Civil and Environmental Engineering, Michigan Technological University, Houghton, Michigan, USA

Abstract

In the past, the use of 3-D groundwater models has been hindered by the lack of adequate pre- and post-processors for entering data and interpreting results. In order to overcome these difficulties, an interface between I-DEAS solid modeling software and MODFLOW has been developed. The pre-processing capabilities of the software interface allow the user to build a geologic model from field data, discretize the model into a finite-difference grid and define boundary conditions and material properties on the grid. Simulation is accomplished by the execution of MODFLOW within the framework. The results of the simulation can then be visualized with 3-D color images.

This paper describes the development and application of the software interface with full 3-D capabilities. The software interface is demonstrated with an example problem. The same problem is analyzed by "traditional" 2-D methods using popular 2-D pre- and post-processors. A comparison is made between these two types of analysis.

1 Introduction

Groundwater modeling is an increasingly important area in environmental engineering, in response to the deterioration of groundwater quality. For most problems encountered in groundwater modeling, numerical methods have become widely recognized and accepted means of solution. One of the disadvantages of groundwater modeling methods is the vast amount of data required to solve and interpret problems. One of the most widely accepted finite-difference codes for groundwater modeling is MODFLOW (McDonald and Harbaugh, 1988), which was developed by the USGS. While this model is a powerful analysis tool, use of this and other public domain codes is limited by a lack of an adequate, graphical pre- and post-processor. This deficiency makes the groundwater modeling task cumbersome, and results are often difficult to interpret.

The lack of an adequate pre- and post-processor for three-dimensional groundwater flow analysis has serious implications, because several simulations often are required for an effective characterization of a groundwater flow problem. Multiple simulations are needed because it is often the case that the input data is uncertain and/or incomplete. Thus, there is often a need to perform a series of simulations to test the sensitivity of the results as a function of the possible range of data values. An additional limiting factor is the ability to visualize the output results, that is, the patterns of groundwater flow as functions of space and time. A typical model simulation results in the output of hundreds to tens of thousands of numerical values. The objective of this project, is to develop a software interface between a groundwater modeling code (MODFLOW) and a fully interactive, graphical pre- and post-processor in order to enhance the groundwater modeling visualization capabilities and increase the productivity of current groundwater modeling efforts.

The most widely-used groundwater flow model in the U.S. is MODFLOW (McDonald and Harbaugh, 1988). The MODFLOW model is based on a three-dimensional, finite difference solution to the groundwater flow equation. The model can be used to simulate flow in unconfined and confined aquifers and provides for flows associated with external stresses such as wells, areal recharge, evapotranspiration, drains and streams. The model source code is written in FORTRAN. The MODFLOW model was originally written in 1984, but has been updated as progress has been made in the field of groundwater flow modeling.

Application of the MODFLOW model involves the superposition of a finite difference grid over the groundwater aquifer of interest. The aquifer is thus subdivided into a series of rectangular or cubic cells, defined by nodes at the center of the cell or at each corner. The unknown of interest, *i.e.*, groundwater pressure head, is solved for at each nodal location. For groundwater flow solutions, the following parameters must be defined at each node or cell: hydraulic conductivities for each coordinate dimension, aquifer thickness, aquifer storage coefficients and withdrawal or injection rates. In addition, boundary and initial conditions must be defined as needed.

I-DEAS (Integrated Design Engineering Analysis Software) was developed by the Structural Dynamics Research Corporation (SDRC, 1992) in the early 1980s as a geometric modeler and mesh generator for finite-element analysis. The geometric database manager in I-DEAS enables a user to create a file translator for use with virtually any other external finite element or finite difference program. Data files for external solvers can be written from I-DEAS and converted to their required format for execution. If the source code is available, the external solver can also be modified to output a file which contains the results of the numerical analysis in standard I-DEAS format. In this way, the user can take full advantage of the meshing and post-processing capabilities of I-DEAS while still using the solver which is most suited to his or her needs.

2 Framework Development

In this project, a framework has been developed to link the MODFLOW code with I-DEAS software. The framework has been developed in the following steps. First, programs were written using the I-DEAS programming language for automation of I-DEAS pre-processing capabilities including the creation of a geologic model from raw site data, application of a three-dimensional finite-difference grid to this model and designation of all boundary and initial conditions, external stresses, and material properties. Second, additional programs were developed to write data from I-DEAS into the format needed for input to MODFLOW. These programs are written in the C programming language and they can be executed directly from within I-DEAS. Third, the source code for MODFLOW was modified so that the results of the groundwater analysis can be output to I-DEAS readable files in addition to the standard MODFLOW output files. Finally, a program was created to run within I-DEAS to read in the results of the numerical analysis and automate some of the I-DEAS post-processing commands.

The pre-processor allows the user to graphically build a three-dimensional geologic model and designate relevant hydrogeologic properties. The pre-processor was constructed using I-DEAS programming language and is menu- and query-driven. The programs written in this manner are invisible to the user. I-DEAS programming files are written as a combination of I-DEAS menu picks, menu and input commands (where users are queried for input data) and statements which are similar to typical FORTRAN statements. A portion of an I-DEAS program file is shown in Figure 1.

```
K : #pmod:
K : #menu "Parameter to Modify:" choice 0 4 ,
K : "A-All" "L-Stress Period Length" ,
K : "N-Number of Time Steps" "TSM-Time Step Multiplier"
K : #if (Z_INP_STAT eq 0) then goto sp
K : #if (Z_INP_STAT eq 1) then goto imod
K : #if (Z_INP_STAT ne 3) then goto pmod
K : #if (choice eq 1) then #modfl=1; #index=num; #goto len
C : ** get a new value for the parameter
K : #nv:
K : #input "Enter new value:" newv
K : #if (Z_INP_STAT eq 0) then goto imod
K : #if (Z_INP_STAT eq 1) then goto pmod
K : #if (Z_INP_STAT ne 3) then goto nv
K : #if (choice eq 2) then #perlen=newv; #goto writ
K : #if (choice eq 3) then #nstp=newv; #goto writ
K : #tsmult=newv
C : ** write new data
K : #writ:
K : /MO LAB 1
K : DON
K : XF perlen
K : YF nstp
K : ZF tsmult
```

Figure 1: Example I-DEAS Program File

In this program file, lines beginning with **C :** are comments, and lines beginning with **K :** are command lines. Command lines which do not start with a # are strings of I-DEAS menu picks. **Menu** commands insert a menu on the screen which is identical to standard I-DEAS menus from which the user can graphically pick options. **Input** commands query users at the I-DEAS prompt line and assign a variable name to the user input. I-DEAS program files are written so that there is a main program and several sub-routines.

The data file format required by the groundwater model was analyzed and a file translator program was created to be run from within I-DEAS. The file translator draws upon the grid, material property, and other data generated from pre-processing and writes out the data files suitable for direct input to MODFLOW. This activity is also invisible to the user. The user can then execute MODFLOW from within the framework, without exiting I-DEAS.

A link between the output of the groundwater modeling code and I-DEAS was developed. The link involves the addition of subroutines to the source code of the model such that a universal file is generated as a part of its output. Universal files are ASCII files which contain geometric information in a format which is acceptable to I-DEAS. In this way, the results of the groundwater modeling can be read directly into I-DEAS for post-processing. All of the post-processing capabilities are standard I-DEAS options and can be automated with the use of program files if desired.

3   Example Problem

In the following section, the MODFLOW/I-DEAS (M/I) framework is applied to an example problem. The example problem is also analyzed with a frequently-utilized two-dimensional (2-D) pre-processor and a 2-D post-processor, to provide a comparison with the M/I framework. The example problem is taken from the MODFLOW manual (McDonald and Harbaugh, 1988). This problem is composed of three layers, with each layer comprising a 75,000 ft. by 75,000 square divided into a 15-row by 15-column grid. Each grid block, or finite-difference cell, forms a square that is 5,000 ft. per side. The flow input is derived from recharge, while the flow outputs consist of drains, extraction wells, and a constant head boundary. The simulation was conducted at steady state. Further details can be found in McDonald and Harbaugh (1988).

The first step in the M/I framework is to construct the geologic model. The geologic model is constructed by entering coordinates corresponding to contacts between different geological materials. Well logs, which are the most likely source for these coordinates, are displayed when the horizontal and vertical coordinates are entered numerically. Figure 2 shows a set of hypothetical well logs corresponding to the example problem and Figure 3 shows the layers corresponding to the well log data. A finite difference grid is overlain onto the geological model. The grid is a subdivision of the problem domain into rows, columns, or layers, resulting in collection of grid blocks. Figure 4 shows a 3-D view of the finite-difference grid for the example problem. The dimensions of the grid blocks are set by numerical inputs; the grid can be further refined by graphical selection of additional rows, columns or layers.

With the 2-D preprocessor, no provision is made for constructing a geologic model directly from well logs. Instead, the number of layers and their thicknesses must be predetermined and entered numerically in the preprocessor. The finite-difference grid is constructed by entering the number of rows, number of columns, and the horizontal dimensions of the grid. Figure 5a and 5b shows two of the model layers created by the 2-D preprocessor. Each layer is displayed one at a time in an areal view only.

The next step in pre-processing involves assigning the physical properties of the aquifer system. In both the M/I framework and the 2-D pre-processor, the properties may be defined cell-by-cell with numerical inputs or by graphically grouping blocks with identical material properties together and entering the group properties. Boundary and initial conditions are applied in a similar manner. External stresses, (e.g. pumped wells or drains) are entered by graphical

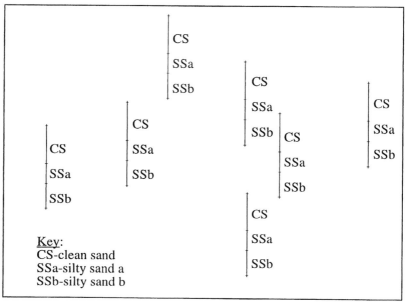

Figure 2: Display of well logs in the M/I framework

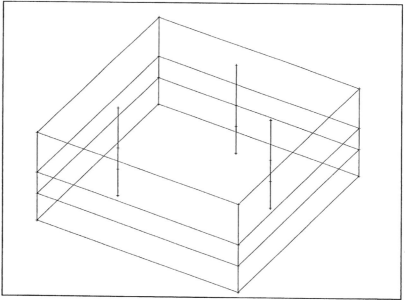

Figure 3: Model layers constructed from well logs in the M/I framework

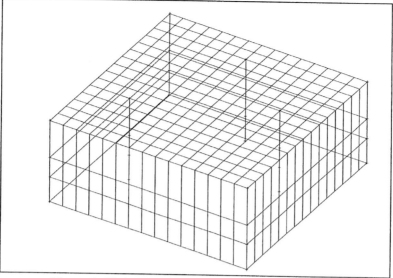

Figure 4: Finite difference grid devised in M/I framework.

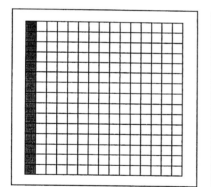

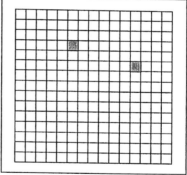

Figure 5a: Finite-difference grid for layer 1 in 2-D pre-processor. Shaded squares indicate constant-head boundary conditions.

Figure 5b: Finite-difference grid for layer 2 in 2-D pre-processor. Shaded squares indicate well locations.

selection of the stress location and numerical entry of the stress magnitude with both the M/I framework and the 2-D pre-processor. Figure 6 shows the locations of external stress as displayed by the M/I framework.

In the M/I framework, assignment of physical properties, boundary and initial conditions, and external stresses is accomplished using graphical techniques

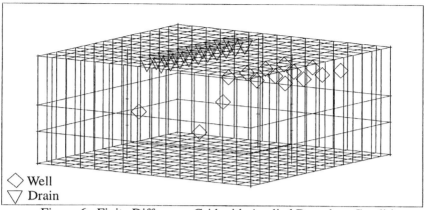

◇ Well
▽ Drain

Figure 6. Finite Difference Grid with Applied Boundary Conditions

combined with pull-down menus and simple queries. Portions of the graphical display can be magnified or cut away so that small-scale definitions of the grid, material properties, stresses or any other modeling property can be made. Colors, gray-scales, or pattern variations are used to distinguish different properties by individual blocks or groups of blocks; symbols are used to designate items such as wells. With the 2-D pre-processor, only entire, individual layers can be displayed, one at a time. Numerical inputs of physical properties is accomplished in a separate screen where the grid cannot be displayed.

The last task in pre-processing involves numerical entry of the parameters that control the execution of MODFLOW, such as designating a steady-state or transient simulation or designating various output options. When the pre-processing is completed, a data translation program is executed to create input data sets for MODFLOW. The program creates files that follow the specific format required for input to MODFLOW. However, the M/I framework and 2-D pre-processor follow similar procedures for this task. The M/I framework allows for execution of MODFLOW from within the main menu of the framework, while MODFLOW must executed outside of the 2-D pre-processor.

A numerical output file consisting of groundwater heads (potentials) and drawdowns (potential losses) is produced by the execution of MODFLOW. This file is post-processed within the M/I framework so that results can be viewed as color-coded, three-dimensional images. These images can be cut away and magnified to view areas of interest. Figure 7 displays the distribution of groundwater heads near a pumped well. The images can be viewed as a shaded image, in which the results are smoothed into a continuum, or as a solid contour image, with each color or shade representing a range of groundwater heads. The user also is able to show only those elements which have a groundwater head greater than or less than a specified value. The post-processing results from different simulation times can be stored and used in an animated sequence, if desired.

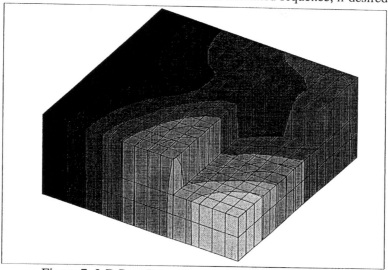

Figure 7. 3-D Post-Processing-Contours of Constant Head

A separate software package must be utilized to post-process and view the results if the 2-D pre-processor is utilized and the MODFLOW output must be manipulated in order to be accessed by most 2-D post-processing packages. A typical post-processor displays the heads as contour maps or surfaces that are functions of two dimensions. Individual two-dimensional images must be displayed one at a time. Figure 8 shows the distribution of groundwater heads for each layer produced by a typical 2-D post-processing software package.

4 Conclusions

In this project, a framework was developed to link MODFLOW with I-DEAS, a graphical pre- and post-processor. The M/I pre-processor allows the user to graphically build the geologic model and grid from field data. Boundary and initial conditions can be input and the data exported to an external finite difference code for processing. Post-processing of the model output can then be achieved within the framework. Application of the M/I framework to an example problem is compared with an application of conventional 2-D pre- and post-processors. The application demonstrates the significant advantages of the M/I framework: (1) a geologic model can be built from field data, (2) the material properties, external stresses, and boundary conditions can be assigned

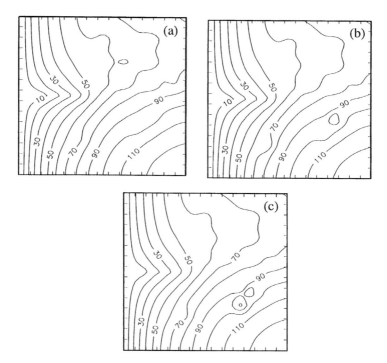

Figure 8: Contours of equal groundwater heads for (a) layer 1, (b) layer 2, (c) layer 3 as displayed by the 2-D post-processor.

to the model grid while viewing the grid from any perspective and with portions of the gird cutaway, and (3) the model outputs can be displayed with multiple 3-D views and a variety of coloring and contouring schemes.

5 Acknowledgment

This work has been supported by the Michigan Research Excellence Fund.

6 References

Structural Dynamics Research Corporation, "I-DEAS User's Guide," Structural Dynamics Research Corporation, Milford, Ohio, 1992

McDonald, M. G. and A. W. Harbaugh, "Techniques of Water-Resources Investigations of the United States Geological Survey: Chapter A1, A Modular Three-Dimensional Finite-Difference Ground-Water Flow Model," U.S. Government Printing Office, Washington, D.C., 1988.

A software tool for water resources sharing planning problems

E. Avogadro, A. Boccalatte, R. Minciardi

DIST - Department of Communication Computer and System Sciences, University of Genoa, Via Opera Pia 13, 16145 Genoa, Italy

Abstract

The paper describes the general architecture of a decision support system, designed for the solution of water resources sharing planning problems, taking into account both quantity and quality issues.

The considered system was inspired by a real case study belonging to a general class of problems. The considered planning problem refers to the water resources sharing in a basin having a general topology, with one or more reservoirs, where several users (for irrigation, for hydropower generation, for drinking water supply) compete for the water resources.

The water quality aspects are taken into account by simulating the concentrations of pollutants along the main river trunk, with the objective of checking if pre-specified standards are fulfilled.

The decision support system is composed by modules which interact among themselves and with the decision makers who are involved in the planning process. Each module of the decision support system is described in detail in the paper, and attention is focused on the information flow among the various blocks.

1 Introduction

Water resources management problems have received considerable attention in the last three decades, both as regards the quantitative aspect of the management problem, and the necessity of ensuring an adequate quality level of the water. From the quantitative viewpoint, the main approaches reported in the literature (see, for instance, [1], [2], [3]) regard the deterministic as well the stochastic modelling of optimization problems related to water distribution. More recent literature regarding water resources management, pays a particular attention has been paid to those aspects regarding the application of advanced mathematical programming techniques [4], the multi-objective formalization of water resources distribution [5, 6], or the development of optimal control policies, possibly in an uncertain setting [7].

On the counterpart, in the literature about water quality models (see, for instance [8, 9, 10]), a major emphasis has been put on modelling and analysis rather than on decisional issues. This is mainly due to the really complex nature of the physical system under concern, in which several submodels interact at different time and space scales.

In the literature surveyed the quantity and quality aspects of water resources management have been treated generally as separate problems, even though there are some examples [11, 12] where a unified treatment of these two aspects has been attempted. This separation may be attributed to two types of reasons. First, the models to be used for the two different issues have quite a different structure. Second, the uncertainties affecting the two kinds of models are of different nature: for water quantity management models, the uncertainties generally refer to the random nature of some quantities, such as water stream flows, natural inflows, etc., whereas for water quality models, the uncertainties mostly regard the dynamical structure of the system model.

Actually, the necessity of an integration between the quantitative and the qualitative issues of water resources management cannot be overemphasized, since for instance, in most cases it is the exploitation of the water resources that induces the presence of risky or poor conditions of the water quality.

The objective of the present work is that of presenting the general design of a decision support system for the analysis and solution of water resources management problems of a class which is relatively frequent in northern Italy, and particularly in basins which are close to the Alpine range. Clearly, a modular architecture of the decision support system allows a ready replacement of a submodel with another one, possibly structurally different, which may be considered as more realistic for a new case study.

The decision support system described in this paper is oriented towards the solution of (off-line) planning problems regarding the sharing of the water resources among various possible competitive users. No on-line information is assumed to be available for this purpose, and the objective is that of determining reference values around which an on-line decisional policy should work. In the present formulation, the planning problem is addressed in a completely deterministic version.

2 The necessity of a Decision Support System

The solution of water resources sharing problems requires the synthesis and analysis of operating policies for managing water quantity and quality issues. To this end, the use of a computerized system is essential, in order to solve the optimization problems, to simulate the system behaviour, and then to allow the final users, i.e. the decision makers, to efficiently interact with the decision support system, also with a visual display of the performances of the policies. Moreover, the computer support system should assist the decision makers in finding the best compromise among the various competitive water users.

Generally speaking, DSS are computerized systems that assist the decision makers in dealing with ill-structured problems. Such systems facilitate the development and evaluation of alternative courses of action for the decision maker, who is allowed to use an interactive language to combine data from databases with potential models and explore the resulting solutions. Traditional DSSs include a set of tools that supports the storage, manipulation and access of data, and the process of fitting these data into formal models; they include also a set of methods and algorithms which are used to solve decision models. Modern DSSs use advances in computing and information technologies to organize and automate the process of alternative evaluation and selection into a flexible fully integrated, interactive, user-friendly computing environment.

Decision makers are frequently overwhelmed by the vast amounts of information which they must consider. Often, they are forced to make partially informed decisions which ignore critical issues because of the complexity of the situation being analyzed, and thus they are unable to identify actual optimal outcomes.

DSSs can play a crucial role in the decision making process by allowing the decision maker to navigate large amounts of information quickly and to explore interrelationships between factors which may influence the decision.

3 The Architecture of the proposed Decision Support System

For the particular class of problems under concern, the conceptual architecture in Figure 1 is proposed. Let us proceed starting from the 'external' level, i. e., that concerning the interaction with the user, and then moving towards the 'inner' modules, which are not directly accessible to the user.

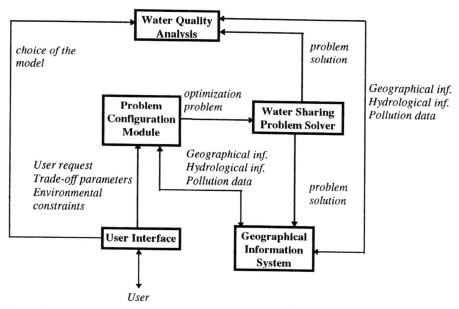

Figure 1 - The conceptual structure of the Decision Support System

The User Interface has the function of providing a user-friendly access to all the various modules of the system. The interaction with the user interaction takes place via a menu-driven command system.

As it has already been remarked, the considered class of problems is characterized by the necessity of taking into account several, generally conflicting, optimization objectives. In this connection, it has to be noted that most of the effective multi-objective analysis techniques require some form of interaction between the decision maker(s) and the mathematical procedure for the problem solution. This interaction may consist in an iterative presentation of solutions to the decision maker, who is then requested to express his/her evaluation in relation to each of such solutions. Each solution is modified in light of the concerns of the decision maker(s), and the process is repeated until a solution is obtained, which can be considered satisfactory.

In the evaluation of the solutions successively provided by the decision support system, it may be necessary to correlate the information used by the mathematical decision procedure (which has generated the solutions) with some other pieces of information, which cannot be used from the above procedure.

In this connection, the use of a Geographical Information System (GIS) turns out to be essential. Generally speaking, a GIS has the main function of storing, manipulating, and displaying geographical data. Besides, the GIS has the essential function of storing and making accessible all information needed to formalize the decisional problem, apart from the information coming directly from the system users. Essentially, such information refers to geographical aspects of the basin under concern, as well as to the hydrology of the basin (river flows, etc.) and to the location, type and size of the pollutant sources along the river.

All the above information is made available to the problem configuration module which has the function of formalizing mathematically the water sharing optimization problem; such a formalization requires also the use of the information provided directly from the decision maker. The latter information is relevant, for instance, to the requests of the water resources expressed by the various users, and to the weight coefficients (trade-offs parameters) that the decision maker assigns to such requests (taking into account the relative importance of the water users). Such information may also refer to environmental constraints coming from laws and regulations.

The problem configuration module formalizes an optimization problem which is then solved by the water sharing problem solver. This is simply a mathematical programming problem (more specifically, a linear programming one, as it will be shown later on), which has to be solved via the use of a suitable software package. Actually, the reason why it is convenient to think of the problem configuration and the water sharing problem solver as two separate modules, is just that of separating the problem formalization from the problem solution, which can be accomplished by means of one among a set of linear programming codes which are commercially available (for instance the LINDO or CPLEX tools).

The solution of the water sharing problem is provided to the GIS and to the water quality analysis modules. The latter includes one or more water quality models, which can be used to evaluate the environmental impact on the basin of the decisions taken about the water resources sharing. The choice about the models to be used is made by the decision maker, through the user interface; the evaluation of the water quality in the river (or, more generally, in the basin) takes place mainly by simulation and needs the use of information provided from the GIS module and from the water sharing problem solver module.

The final evaluation, from the environmental and from the "economical" point of view, of the solution obtained is made by the decision maker, by using the interface module. To this end, the graphical facilities of the GIS module are essential. There is also the possibility that such a module integrates the information embedded in the obtained solution with information which has not been used in the problem formalization (for example, information regarding the land use or the location of settlements of particular relevance).

The user interface has to be provided with all features necessary to allow an easy navigation in the overall system. In the following, three of the modules represented in Figure 1, namely the problem configuration module, the water quality analysis module, and the GIS, will receive particular attention.

4 The Problem Configuration Module

As already pointed out, the function of this module is that of assembly all information necessary to the formalization of the water resources sharing problem. Such information is then provided to the problem solver module, which is simply a mathematical programming tool.

Given the generality of the structure of the proposed DSS, the problem configuration module can be applied to formalize problems referring to basins having different topologies and complexity, e.g., as regards the number of water streams, of reservoirs, of water users. In the following, for the sake of clarity we explicitly refer to a specific case study, presently under development at our Department, and referring to a pre-Alpine basin in the north of Italy, in the Piemonte region. Such a case study can be considered representative of water resources planning problems which are common in northern Italy.

The considered basin consists of the main trunk of a river flows from a reservoir, which has also a second separate outflow intended for drinkable water supply and hydropower generation. The river is characterized by a single space variable y, ranging from 0 to L. The flow of the river is considered piecewise constant, over the river length. That is, a given set of sections of N points y_i, i=1,...,N ($y_0=0$ and $y_N=L$), are fixed a priori, each one corresponding to variations of the river flow. Such variations are due to: either a) the confluence of other minor natural water streams or of sewerage water coming from industrial plants or civil settlements $y_i \in \Lambda$ or b) the water withdrawal for irrigation, industrial or other uses $y_i \in \Phi$.

The confluence sections may be furtherly partitioned into four different subsets, namely:

-Λ_a including the sections y_i which correspond to the confluence of water streams whose quantity and quality are not controllable;

- Λ_b including the sections y_i which correspond to the confluence of water streams whose quantity is controllable and quality is not;

- Λ_c including the sections y_i which correspond to the confluence of water streams whose quality is controllable and quantity is not;

- Λ_d including the sections y_i which correspond to the confluence of water streams whose quantity and quality are both controllable;

As regards the withdrawal points, they also will be partitioned in two sets:

- Φ_a including the sections y_i for which the water withdrawn is not returned to the river (irrigation uses);

- Φ_b including the sections y_i for which the water withdrawn is (totally or partially) returned to the river (industrial uses); the returned water is in general characterized by pollutant concentrations different from those of the water withdrawn.

The time variable is discretized by considering a set of time intervals (whose length is, for example, equal to one month) (t_j, t_{j+1}), $j=0,...,$ T-1, being $t_0=0$ and t_T the last instant in which the system is observed. The length of such time intervals may be fixed by the system user. Then, $\overline{Q}_i(t_j)$ represents the river flow, in the time interval (t_j, t_{j+1}), in the reach (y_i, y_{i+1}). $\overline{B}_i(t_j)$ is the variation of flow (positive for $y_i \in \Lambda$, nonpositive for $y_i \in \Phi$) in the time interval (t_j, t_{j+1}), at the section y_i. $\overline{D}_i(t_j)$ is the water flow withdrawn and $\overline{E}_i(t_j)$ is the water flow returned at sections $y_i \in \Phi_b$ in the time interval (t_j, t_{j+1}). It is assumed that $\overline{E}_i(t_j)=\eta_i\overline{D}_i(t_j)$ being $\eta_i \leq 1$ a fixed nonnegative parameter. Obviously, $\overline{B}_i(t_j)=\overline{E}_i(t_j)-\overline{D}_i(t_j)$, for $y_i \in \Phi_b$.

Finally each variable $\overline{B}_i(t_j)$, for $y_i \in \Lambda_b \cup \Lambda_d$ may be decomposed as $\overline{B}_i(t_j)=\overline{B}'_i(t_j)-\overline{B}''_i(t_j)$, where $\overline{B}'_i(t_j)$ represents the value of the inflow in absence of water withdrawals, and $\overline{B}''_i(t_j)$ is the value of the overall water withdrawals before the confluence into the main river trunk.

As previously mentioned, there are two water outflows from the reservoir, namely $\overline{Q}_0(t_j)$, which is the river initial flow, and $\overline{Z}(t_j)$, which is the flow of the water which is withdrawn from the reservoir and not returned to the river; finally, $\overline{S}(t_j)$ is the volumetric content of the reservoir in the time interval (t_j, t_{j+1}).

Any formalization of water resources sharing planning problem must have the objective of optimizing the compromise among the various possible uses of water. Such a problem which is essentially a multi-objective one. In the designed structure for the module under concern, the original problem is reduced to a single objective formalization through the so called "goal programming" method [13] based on the proper specification of the aspiration levels and the asymmetric weights for the various water resources users.

More specifically, the above formalization leads to the following mathematical programming problem:

$$\min_{\xi} \sum_{j=0}^{T-1} \left\{ \begin{array}{l} \tau_j \max\left[Z_j^* - Z(t_j),0\right] + \sum_{i:y_i \in \Phi_a} \left\{ \beta_{i,j} \max\left[B_{i,j}^* - \overline{B}_i(t_j),0\right]\right\} + \\ + \sum_{i:y_i \in \Phi_b} \left\{ \gamma_{i,j} \max\left[D_{i,j}^* - \overline{D}_i(t_j),0\right]\right\} + \sum_{i:y_i \in \Lambda_b \cup \Lambda_d} \left\{ \delta_{i,j} \max\left[(B_{i,j}'')^* - \overline{B}_i''(t_j),0\right]\right\} \end{array} \right\}$$

(1)

s.t.

$$S(t_{j+1}) = S(t_j) - \overline{Q}_0(t_j)(t_{j+1} - t_j) - \overline{Z}(t_j)(t_{j+1} - t_j) + P(t_j) \qquad (2)$$

$$\overline{Q}_i(t_j) = \overline{Q}_{i-1}(t_j) + \overline{B}(t_j) \qquad (3)$$

$$\overline{Q}_i(t_j) \geq Q_i^{min}(t_j) \qquad\qquad i=1,...,N, j=0,1,...,T-1 \qquad (4)$$

$$\overline{B}_i''(t_j) \le v_i \overline{B}_i'(t_j) \qquad i: y_i \in \Lambda_b \cup \Lambda_d, j=0,1,\dots,T\text{-}1 \qquad (5)$$

$$\overline{B}_i(t_j) \le \chi_i \overline{Q}_{i-1}(t_j) \qquad i: y_i \in \Phi_a, j=0,1,\dots,T\text{-}1 \qquad (6)$$

$$\overline{D}_i(t_j) \le \pi_i \overline{Q}_{i-1}(t_j) \qquad i: y_i \in \Phi_b, j=0,1,\dots,T\text{-}1 \qquad (7)$$

$$\overline{B}_i(t_j) = (1-\eta_i)\overline{D}_i(t_j) \qquad i: y_i \in \Phi_b, j=0,1,\dots,T\text{-}1 \qquad (8)$$

$$S(t_j) \ge S^{min}(t_j) \qquad j=0,1,\dots,T\text{-}1 \qquad (9)$$

$$S(t_T) = S(t_0) \qquad j=0,1,\dots,T\text{-}1 \qquad (10)$$

$$\underline{\xi} \ge 0 \qquad (11)$$

where

$S(t_0)$ may be considered known; $P(t_j)$ represents the amount of water entering the reservoir during the time interval (t_j, t_{j+1}), $j=0,1,\dots,T\text{-}1$;

Z^*_i, $B^*_{i,j}$, $D^*_{i,j}$, $(B''_{i,j})^*$ are the aspiration level respectively of the average flow delivered for drinkable water and/or hydropower generation uses, for withdrawals in sections $y_i \in \Phi_a$, for withdrawals in sections $y_i \in \Phi_b$, and for the overall water flow withdrawn for each $y_i \in \Lambda_b \cup \Lambda_d$;

$Q_i^{min}(t_j)$ and $S^{min}(t_j)$ are respectively the "vitality minimum" for the river flow and the minimum allowed reservoir content;

the quantities v_i, χ_i, π_i, η_i are coefficients fixed and known;

the quantities τ_j, $\beta_{i,j}$, $\gamma_{i,j}$, $\delta_{i,j}$ are constants which have the function of taking into account the importance of the various water users as well as their determination in specifying their aspiration levels;

$\underline{\xi}$ is the decisional vector defined as

$$\underline{\xi} = \text{col} \begin{bmatrix} Z(t_j), j = 0,\dots,T-1; \\ \overline{Q}_i(t_j), i = 0,\dots,N-1; \\ -\overline{B}_i(t_j), i: y_i \in \Phi_a, j = 0,\dots,T-1 \\ \overline{D}_i(t_j), i: y_i \in \Phi_b, j = 0,\dots,T-1 \\ \overline{B}_i''(t_j), i: y_i \in \Lambda_b \cup \Lambda_d, j = 0,\dots,T-1 \\ S(t_j), j = 0,\dots,T-1 \end{bmatrix} \qquad (12)$$

At this point, it is sufficient to note that, by using standard mathematical programming devices, it is possible to convert the cost to be minimized into a linear one, thus making the overall optimization problem become a linear programming one.

5 Water Quality Analysis

As regards the water quality models, a variety of choices are possible, different for the mathematical complexity of the involved differential equations as well as for the chemical compounds taken into account [8, 9, 10]. As the proposed system is intended to be applied to solve planning problems, the models we are interested in are those describing the system in a time-stationary setting. As regards the chemical compounds, attention is presently moving from the traditional Biochemical Oxygen Demand (BOD) Dixolved Oxygen (DO) analysis towards more complex models taking into account also toxic metals and mutagenic compounds. In any case, the modular architecture mentioned in the introduction will allow the choice of the water quality sub-model most suitable for the particular case study under concern.

A series of computer packages have been made available for easy configuration and calibration of water quality sub-models; among them, one can cite the QUAL2EU [14], the MIKE 11 [15], the WODA [16] packages.

In the proposed decision support system the US. EPA QUAL2E tool has been selected for the simulation of water quality along the river. QUAL2E is a comprehensive and versatile stream water quality model . It can simulate up to 15 water quality constituents in any combination desired by the user. The model is applicable to dendritic streams that are well mixed. It assumes that the major transport mechanism, advection and dispersion, are significant only along the main direction of flow (longitudinal axis of the stream). It allows for multiple waste discharges, withdrawals, tributary flows, and incremental inflow and outflow. QUAL2EU can operate either as a steady-state or as a dynamic model, making it a very helpful water quality planning tool. In steady-state analysis, QUAL2 performs the system simulation over time periods during which both the stream flow in river basins and input waste loads may be considered constant. In this case, it can be used to study the impact of waste loads (i.e., of their magnitude, quality and location) on instream water quality.

6 Geographical Information Systems

Geographical Information Systems (GIS) are computerized (software and hardware) systems aiming at providing a number of tools to code, store and retrieve data about aspects of the earth's surface. GIS have the ability to display and graphically summarize both the input data for the analytical models and the results of application of management models using those data. Generally, GIS are composed by several hardware components (scanner, plotter, Personal Computer, etc.) and software tools to manage data provided by external devices.

Because the data can be accessed, transformed, and manipulated interactively in GIS, they can serve as a test bed for studying environmental processes, for analyzing the results of trends, or for forecasting the possible results of planning decisions. Using GIS in such a manner, it is possible for planners and decision makers to explore a range of possible scenarios and to obtain an idea of the consequences of a course of action.

The ability to display the results graphically improves the man-machine interaction which is generally accepted as being an integral part of multi-objective water resources analysis. GIS should not be considered as a means of providing final answers to complex water resources planning issues, but they should be seen as an important tool of Decision Support Systems by which information on the basin issues is transferred to the decision-maker for his/her considerations. Within this framework, GIS or any computer aided system should not be considered as a means of obtaining the answers, but more properly as a means for identifying objectives or goals, constraints, etc. of problems which are not well defined. In [17, 18, 19] examples may be found of the application of GIS to water resources management problems.

A primary role of a GIS is to facilitate the whole process by upgrading data input, improving data accessibility, allowing a better interpretation of results. A comprehensive analysis of the water resources in a particular river basin will require the consideration of all aspects regarding the water resources.

An important tool for the decision maker is the map of the region of interest. In particular, there are two types of map that can be treated by GIS: thematic maps and Digital Elevation Models.

A thematic map can be defined as a set of points, lines, and areas that are defined both by their location in space with reference to a coordinate system and by their non-spatial attributes. The map legend is the key linking the non-spatial attributes to the spatial entities. Non-spatial attributes may be indicated visually by colors, symbols or shading, the meaning of which is defined by the legend. The non-spatial attributes of a region could represent the different uses of the land, different types of land/water users, etc., associated to different colors, or shaded regions, on the map.

On the counterpart, unlike land-use, the landform is usually perceived as a continually varying surface that cannot be modelled appropriately only by the thematic maps. Any digital representation of the continuous variation of relief over space is known as a Digital Elevation Model, whose most important uses are the storage of elevation data for digital topographic maps and three dimensional display of landforms for design and planning the location of dams, waste water treatment plants, etc.. Besides it can also serve as background for displaying thematic information or for combining relief data with thematic data such as soils, land-use or vegetation. A DEM can also provide data for image simulation models of landscapes.

The integration of the GIS with the decisional architecture described in this paper is useful to provide the possibility of evaluating the impact of the outcomes of the planning procedure over the territorial area under concern, taking into account issues which cannot be modelled in the quantitative decisional procedure. To this end, the most reasonable choice seems that of using the ARC/INFO version for PC.

7 Conclusions

This paper reports the general guidelines for the development of a computerized Decision Support System designed to assist in decisions regarding water resources management in basins having a certain degree of complexity. The novelty of the proposed system is in the attempt to combine quantity and quality issues, and to integrate advanced software tools for the solution of mathematical programming problems with established codes for the simulation of water quality in rivers.

The whole system is presently under development; in the same time, the application to a specific case study is carried out. The objective is to obtain a fully integrated system which can be operational for a large class of water resources management problems.

References

1. Loucks D. P., Stedinger J. R. & Haith D. A. *Water Resources Systems Planning and Analysis*, Prentice-Hall, Inc., New Jersey, 1981.
2. Kottegoda N.T. *Stochastic Water Resources Technology*, The MacMillan Press LTD, London, 1980.
3. Haimes Y.Y. *Hierarchical Analyses of Water Resources Systems*, McGraw Hill Inc., USA, 1977.
4. Sylla C. A subgradient-based optimization for reservoirs system management, *European Journal of Operational Research*, 1994, **76**,28-48.
5. Goicochea A., Stakhiv E. Z. & Li F. Experimental evaluation of Multiple Criteria Decision Models for Application to Water Resources Planning, *Water Resources Bulletin*, 1992, **28**, 89-102.
6. Hipel K.W. Multiple Objective Decision Making in Water Resources, *Water Resources Bulletin*, 1992, **28**, 3-12.
7. Yakowitz S. Dynamic Programming Applications in Water Resources, *Water Resources Research*, 1982, **18**, 673-696.
8. Rinaldi S., Soncini-Sessa R., Stehfest H. & Tamura H. *Modeling and Control of River Quality*, Mc Graw Hill, Great Britain, 1979.
9. Beck M. B. *Systems Analysis in Water Quality Management*, Pergamon Press, Canada, 1987.
10. Biswas A.K. *Models for Water Quality Management*, Mc Graw Hill, United States of America, 1981.
11. Abadie J. & M'Silti H. La Programmation Mathematique Multicritere et la Gestion des Ressources en Eau, *Recherche operationelle/Operations Research*, 1988, **22**, No. 4, 363-385.
12. Jacovkis P.M., Gradowczyk H., Freisztav A. M. & Tabak G. A Linear Programming Approach to Water-Resources Optimization, *Methods and Models of Operations Research*, 1989, **33**, 341-362.
13. Chankong V. & Haimes Y.Y. *Multiobjective Decision Making*, Elsevier Science Publishing Co., New York, 1983
14. Brown L. C. & Barnwell JR. *The Enhanced Stream Water Quality Models, QUAL2E and QUAL2E-UNCAS: Documentation and User's Manual* - U.S. EPA Environmental Research Laboratory, Athens, GA, EPA/600/3-87/007, 1987.
15. *MIKE 11, User Manual and Reference Manual* - WS Atkins Consultants Ltd Wood.
16. Kraszewski A. & Soncini-Sessa R. *WODA-A Computer Package for the Identification and Simulation of BOD-DO River Quality Model*, Cooperativa Libraria Universitaria del Politecnico, Milano, 1984 (in Italian).
17. J.D. Milton Geographical Information Systems: an alternative for manipulating multi-attribute spatial data in land-use planning, *Proceedings of the Workshop on the Impact of RiverBank Erosion and Flood Hazards in Bangladesh*, Dacca, Bangladesh, 1985.
18. W.M.Grayman Tutorial-geographic and spatial data management and modelling, Computer Applications in Water Resources, *Proceedings of ASCE*, Buffalo, 1985.
19. R.H. Berich A micro-computer GIS for water resources planning, Computer Applications in Water Resources, *Proceedings of ASCE*, Buffalo, 1985.

Multiphase flow simulation overcoming current restrictions of single-phase groundwater models

G. Zangl, M. Stundner

LISTEN AND TALK Environmental Services, A-2500 Baden, Austria

Abstract

The application of numerical computer models, who can handle multiphase flow, represents a reasonable alternative to the use of single phase flow models. An introduction is given to the theory of multiphase flow and afterwards some practical applications are shortly described.

Introduction

Numerical computer models are widely used in groundwater and environmental engineering. In most cases, this numerical models are designed to calculate flow of only one fluid phase. Several limitations are the result of this simplification. Risk analysis of leaks in a landfill barrier or the planning of a site-remediation, where the contaminant is present in an own fluid phase, cannot be determined with this models.

Multiphase flow models are developed in the petroleum industry to calculate flow of oil, gas and water in hydrocarbon reservoirs. With the development of faster computers, reservoir simulation became state of the art for all reservoir engineers and a valuable tool for reservoir production predictions. Multiphase models are based on the same principal equations and can be adapted to the needs of groundwater and environmental concerns.

To understand the concept of multiphase flow, a short derivation of the governing equations for a finite difference model are written in the next chapter, followed by practical applications.

1 Conservation of mass

Equations used to model isothermal multiphase flow in porous media are obtained by combining conservation of mass with D'Arcy's Law. These equations along with appropriate constraints, constitutive relations, and boundary and initial conditions can be solved by approximate numerical techniques. Consider a system with 3 fluid phases and one dimensional linear flow (see Figure 1). $x_{i \pm 1/2}$ are the block boundaries, x_i is the node of the block. The mass balance for the block volume of block i is :

$$\Delta t \left(q_{p_{i-\frac{1}{2}}} - q_{p_{i+\frac{1}{2}}} - q_{p_i}^w \right) = M_{pi}^{n+1} - M_{pi}^n \tag{1}$$

where Δt is a time increment (time step) and q_p's are mass flow terms. The indices $i \pm 1/2$ stand for mass flow across block boundaries (flow terms) and index i stands for the block itself. The term q^w represents a source or sink, depending on + or - .

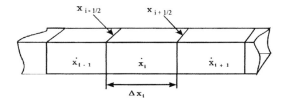

Figure 1: One dimensional block model

M_i^{n+1} represents the mass of phase p in the control volume i at the end of the time step and with the superscript n at the beginning of this time increment. The term q_p can be rewritten in

$$q_{p_{i \pm \frac{1}{2}}} = \left(A \rho_p u_p \right)_{i \pm \frac{1}{2}} \tag{2}$$

where A is the cross-sectional area, ρ_p is the density of phase p and u_p is the velocity of phase p. In this equation it is assumed that flow across boundaries is only by convection. Now we relate these velocities to pressure gradients through D'Arcy's Law:

$$u_p = -\frac{k\,k_{rp}}{\mu_p} \left(\frac{\partial p_p}{\partial x} - \gamma_p \frac{\partial z}{\partial x} \right),$$

$$\gamma_p = \rho_p g \tag{3}$$

k is the absolute permeability of the porous medium, k_{rp} is the relative permeability to phase p, μ is the viscosity of phase p, p_p is the pressure, g the acceleration due to gravity and z the vertical depth. The right hand side of equation (1) is called accumulation term and can be rewritten as

$$M_{p_i} = (V\Phi\rho_p S_p)_i \tag{4}$$

where V is the bulk block volume, ϕ is the reservoir porosity and S_p is the saturation of phase p in block i.

The main difference to single phase models appears in the calculation of phase saturations in the discretisized blocks instead of hydraulic heads. The sum of all phases in one block must be equal to the total pore volume of this block. As the saturations are expressed in terms of fraction of the pore volume, the sum of all saturations has to be equal to one.

$$S_{water} + S_{gas} + S_{oil} = 1 \tag{5}$$

If we combine equation (1) with equation (3) and (4) we get the form:

$$-\sum_p \left[\nabla(\rho_p u_p) - (\rho_p q_p)\right] = \sum_p \frac{\partial}{\partial t}\left(\Phi\, \rho_p S_p\right) \tag{6}$$

This equation can be rewritten with differential operators to

$$\Delta\sum_p \left[\Delta\left(T_p \Delta p_p - T_p \gamma_p \Delta z\right)_i - q_{p_i}^w\right] = \sum_p \Delta\left(V\Phi\rho_p S_p\right)_i \tag{7}$$

where

$$T_{p_{i\pm\frac{1}{2}}} \equiv \left(\frac{A\rho_p k k_{rp}}{\mu_p \Delta x}\right)_{i\pm\frac{1}{2}} \quad \text{and} \quad \Delta x_{i\pm\frac{1}{2}} \equiv \Delta x_i \tag{8}$$

The compressibility of the fluids is implemented in the form of the formation volume factor B_p. This factor is defined as the ratio of the volume of phase p at some specified pressure conditions to the volume of that phase at standard (stock tank) conditions.

$$B_p = \frac{V_p}{V_{pST}} \tag{9}$$

where the subscript ST means at standard (stock tank) conditions.

This was the derivation of one dimensional flow, which can easily be rearranged and developed to three dimensions. Equation (7) can be solved with the standard numerical methods. Most three phase reservoir simulators solve two saturations and the pressure simultaneously. This method is called FIME (=Fully implicit method).

2 Saturation dependent variables

2.1 Capillary pressure concept

The presence of two phases with different densities and viscosities leads to different phase potentials, which often cannot be ignored. Therefore we have to introduce the capillary pressure concept. The data for the capillary pressure are read from a function, either measured in laboratory tests, or by use of Leverett's J-function to scale Pc data:

$$J = \frac{P_c}{\sigma}\sqrt{\frac{k}{\Phi}} \tag{10}$$

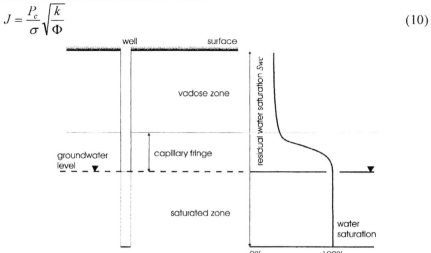

Figure 2: Capillary pressure curve

The Pc curve is usually drawn with an infinite slope at $Sw = Swc$, for a numerical simulation model, the value of Swc has to be a finite number.

2.2 Relative permeability concept

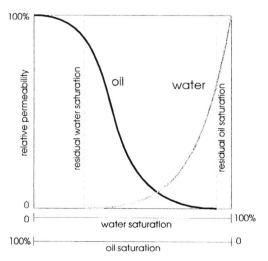

Figure 3: Two phase relative permeability curves

If a porous medium is filled with two phases, one phase restrains the flow of the second phase. Obviously, the permeability of the flowing phase decreases with increasing amount of the other phase in pore space. The concept of relative permeabilities contributes to this effect. Mostly, the two phases don't have the same physical properties. The relation of the different viscosities divided by the relative permeabilities is called the mobility ratio. This factor is responsible for the way, how these fluids flow together in a porous medium.

3 Productivity Index

Modeling of an oil reservoir by numerical methods implies the use of grid blocks, whose horizontal dimensions are much larger than the diameter of a well. As a result, the pressure calculated for a block containing a well, p_o, is greatly different from the flowing pressure of the well p_{wf}. Peaceman stated, that the calculated pressure for a well block should be the areal average pressure in the portion of the reservoir represented by the block. D'Arcys law, written in radial coordinates, leads to the relation

$$q_p^w = PI(p_o - p_{wf}) / \mu_p \tag{11}$$

and the definition of the productivity index PI

$$PI = \frac{2\pi k h}{\ln \frac{\sqrt{\Delta x \Delta y / \pi}}{r_w} - \frac{1}{2} + s} \tag{12}$$

where h is the saturated zone thickness, r_w is the well radius and s the skin effect of the well. Δx and Δy are the block length and block width. With the appropriate selection of the parameters used for PI, the well rate and the flowing pressure are adjusted, so that it is possible to calculate well tests with numerical models, without a fine discretisation which uses the well radius as internal boundary.

4 Local Grid Refinement

This option allows to divide blocks and enables a fine discretisation where it is needed. Under adherence of some rules, it is possible to create up to 64 new refined blocks from one fundamental block.

This method helps in modeling heterogenities, boundary conditions and, as the grid refinement can be used on already existing models, allows to implement new underground buildings (like slurry walls) or other areas of interest, easily.

In this way, a fine discretisation in areas, where it is not needed, is avoided. If the reservoir is vertically strongly stratified, instead of introducing one additional layer for a clay lens, the existing blocks can be divided in this area (Figure 4).

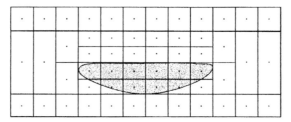

Figure 4: Modeling geological heterogenities with vertical grid refinement

Because of this block division, the equation matrix becomes irregular, which cannot be solved with the well known direct and iterative solution methods. The solver, which is used to calculate the saturation and pressure equations was especially programmed for this kind of simulation grids.

5 Applications

The first example will show the advantage of using the well productivity index to correlate the calculated block pressure and the flowing well pressure at a given rate. The second example shows the water saturation propagation underneath a leak in a land fill sealing. In the third example a DNAPL fluid migration through the unsaturated and the saturated zone is calculated and compared to actual measured values.

5.1 Well Test Simulation

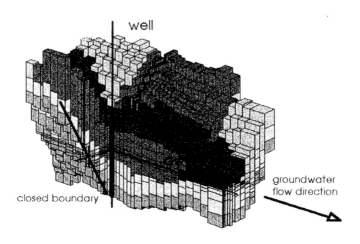

Figure 5: part of the 3D block model to simulate a pumpig test

An alpine valley with very heterogeneous geological formations was used as landfill without a barrier layer. A pumping test should give information about permeabilities and boundary conditions for further hydrological investigations.

The width of the valley is approximately 80 meters, so that the noflow boundaries play an important role in the pressure response of the well. Because of the complex vertical sequence of different geological formations, only a three dimensional numerical model could lead to an useful interpretation of the hydrological situation.

Due to a lack of long time period groundwater level measurements, this pumping test was used to calibrate the groundwater simulation model in this area. With the implementation of the well equation (12) the drawdown of the pumping period could be successfully simulated not only in the pumping well, but also in all measured neighboring monitoring wells (Figure 6).

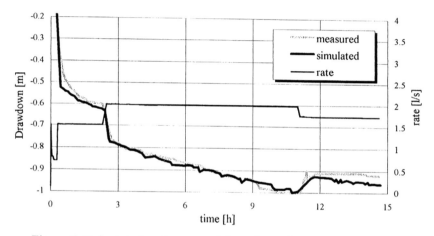

Figure 6: Calculated and measured drawdown during a pumping test

The pressure drawdown was calculated and measured in Pascal and afterwards converted to an equivalent water height in meters. In the beginning of the pumping test, the simulated well pressures differ from the measured pressures because of the variations of the pumping rate and wellbore effects. After one hour of pumping, the calculated well pressure corresponds with the measured data, even if the pumping rate is changed. At the end of the test the maximum deviation of the calculation is 4 centimeters.

5.2 Planning of Landfill Liners

The design of landfill liners has to be adjusted with the site geology. Therefore guidelines for engineers help to come to a decision, which liner system should be used, but in each case careful calculations can optimize the liner composition and safety.

If a leakage occurs in a landfill liner, waste-water migrates through the unsaturated zone into the groundwater. A multiphase flow model is able to calculate the path of the contaminated water in the unsaturated zone, which makes it possible to estimate the risk of a leakage and to design the monitoring system.

Figure 7 shows a cross section through a planned landfill. There is no groundwater flowing under the waste disposal facility. Water is flowing vertical through a leak in the liner system into the natural soil. It is deviated to a horizontal flow direction when the top of a natural geological barrier is reached. The water saturation of this soil is calculated with a multiphase flow model and visualized in Figure 7.

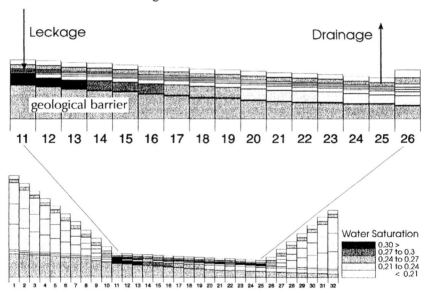

Figure 7: Cross-section trough a landfill liner system

5.3 Simulation of a DNAPL Spill

F. Schwille started in 1975 to examine the migration of NAPLs (Non Aqueous Phase Liquids) in a porous medium in laboratory tests. In 1984 he published lab tests of dense NAPLs on their way through the unsaturated zone, the capillary fringe and the saturated zone.

We could successfully simulate one of these laboratory experiments with a three phase simulation model (Figure 8). We took a lab test to show the ability of a multiphase flow model, because all uncertainties of the data aquisition vanish. Permeability and porosity are homogenous distributed and the properties of the infiltrated fluid are well known.

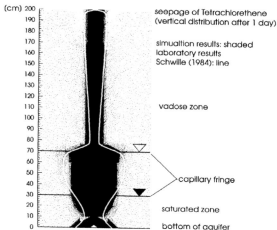

Figure 8: Calculated and measured distribution of tetrachlorethene in a laboratory test

Conclusions

The theoretical background of multiphase flow models is well known from the petroleum industry, where multiphase simulation studies are state of the art. With some small changes, these models can be adapted to the needs of groundwater modeling.

These numerical models are a valuable tool, not only for site-remediations, where NAPL fluid flow occurs, but also in hydrological investigations and in the planning of all kinds of operations related to water migration through the unsaturated zone.

Nomenclature

Latin symbols

A	[m²]	area of the cross-section
B_p	[-]	formation volume factor
g	[m/s²]	gravity constant
J	[-]	Leverett - function
k	[m²]	permeability
k_{rp}	[-]	relative permeability of phase p
M_p	[kg]	mass of phase p
P_c	[Pa]	capillary pressure
p_p	[Pa]	phase pressure

PI		productivity index
r_w	[m]	well radius
q_p	[kg/s]	flux term between blocks
q_p^w	[kg/s]	source or sink
s	[-]	skin effect
S_p	[-]	saturation of a phase p
T_p		phase transmissibility
u_p	[m/s]	phase velocity
V	[m³]	blockvolume

V_p	[m³]	volume of phase p in block i
V_{pST}	[m³]	volume of phase p at standard conditions
x		direction of coordinate

Φ	[-]	porosity
μ_p	[Pa s]	viscosity of phase p
ρ_p	[kg/m³]	density of phase p
σ		surface tension

Greek Symbols

Δt	[s]	time step length
Δx	[m]	block length
Δy	[m]	block width
Δz	[m]	block heigth

Indices

i		block number in x-direction
p		mobile phase
n		number of time step
w		well index

References

1. Aziz, K., Settari, A. *Petroleum Reservoir Simulation*, Applied Science Publishers, 1979
2. Collins, R.E. *Flow of Fluids through Porous Materials*, The Petroleum Publishing Company, Tulsa, OK, 1976
3. Crichlow, H.B. *Modern Reservoir Engineering - A Simulation Approach*, Prentice-Hall, Englewood Cliffs, NJ, 1977
4. D'Arcy, H. *Fontaines Publiques de la Ville de Dijon*, Victor Dalmont, Paris, 1856
5. Heinemann, Z. E., Brand, C., Munka, M., Chen, Y.M. *Modeling Reservoir Geometry with Irregular Grids*, SPE paper no. 18412 , 1989
6. Heinemann, Z. E. *"Advances in Gridding Techniques"*, prepared for Fifth International Forum on Reservoir Simulation, Muscat, Oman , 1994
7. Kinzelbach, W. *Numerische Methoden zur Modellierung des Transports von Schadstoffen im Grundwasser*, R. Oldenburg Verlag, Paul Parey, Hamburg, 1987
8. Peaceman, D.W. *Interpretation of well-block pressures in numerical reservoir simulation*, Society of Petroleum Engineers Journal 18, pp183-194, 1978
9. Schwille, F. *Leichtflüchtige Chlorwasserstoffe in porösen und klüftigen Medien-Modellversuche*, Besondere Mitteilungen zum Deutschen Gewässerkundlichen Jahrbuch Nr. 46, Bundesanstalt für Gewässerkunde, Koblenz, 1984
10. Stundner, M., Zangl, G. *Mathematische Simulationsmodelle helfen Mineralölschäden zu beschreiben und deren Sanierung zu optimieren*, In Erarbeitung der Methodik zur verursacherbezogenen Zuordnung von Öl-kontaminationen Teil 1: Machbarkeitserhebung, Forschungszentrum Seibersdorf, 1994

SECTION 5:
SURFACE AND GROUNDWATER HYDROLOGY

GIS in water resources: stormwater master plan

T.V. Hromadka II

Senior Managing Editor, Failure Analysis Associates, Newport Beach, California, and

Professor of Mathematics and Environmental Studies, Department of Mathematics, California State University, Fullerton, CA 92634-9480, USA

Abstract

A Graphics Data Base Management System is developed for use with computerized Master Plans of Drainage and Storm Water Plans. The Storm Water Plans are prepared according to particular governing agency specifications involving multiple hydrologic and hydraulic modeling options, integrated into a single software package. Data bases are prepared for graphical representation of streets, land uses, hydrologic soil groups, rainfall, master plan system elements, topographic, and other data, as well as computational results developed from the Master Plan of Drainage computer model. Two applications are available; an integrated package enabling editing and upgrading of the Master Plan, and another package designed to publish and distribute the Graphics and the Master Plan of Drainage data bases to the public in an access-only data base retrieval environment. The opportunities provided by such public information programs are significant, in that the entire Master Plan becomes available to the public in an easy-to-read and easy-to-use environment. The public therefore becomes an integral and important member of the Master Plan team, exchanging input and easing the way for acceptance of the project.

INTRODUCTION

The advent of graphical display software and linkage to hydrology / hydraulics modeling is gaining in popularity (e.g., Bergman and Richtig (1990); Djokic and Maidment (1991)). The several problems typically encountered in storm water master plans now have computer packages available to provide solutions (e.g., Hromadka, 1983; Hromadka et al, 1985 and 1987). Interface of hydrological solutions to hydraulics models has also been integrated into computer software packages, providing a single software source to solve coupled hydrology/hydraulics problems (Hromadka, 1988).

In this paper, an integrated hydrology / hydraulics / planning / deficiency-analysis Master Plan of Drainage computer model, which simultaneously performs several master planning and engineering analysis tasks, is linked to a graphics data base management program developed for the purpose of easy display of data and graphics without distributing expensive proprietary software. The computer Master Plan of Drainage and Storm Water Plan modeling approach evaluates each link of the Master Plan of Drainage for deficiencies with respect to several defined street flow criteria, and determines mitigation measures of parallel and replacement systems. Because different hydraulic systems have different flow characteristics, hydrology estimates are recomputed as the master plan is developed. Although small areas (less than about one square mile) are modeled by a rational method technique, the computer model integrates the small area hydrology techniques with the unit hydrograph technique for tributary catchment areas greater than one square mile.

The resulting Master Plan of Drainage and Storm Water Plan is represented by a simple to use (and inexpensive to distribute) Graphics Data Base Management System (or GDBMS) which allows for rapid communication of master plan data and estimates in graphical form. Two applications are developed:

Application 1: Graphical representation, storage, and editing via an AutoCAD environment, wherein hydrologic, planning, topographic, and geographic data are accessible for processing in AutoCAD, and thence transferable to the Master Plan of Drainage computer model, with access to a data base retrieval system; and

Application 2: Graphical representation of data, and access to a data base retrieval system, which is noneditable, and which can be published and distributed to the public.

In the following, each major element of the above described GDBMS will be discussed. An application of GDBMS to an example Master Plan of Drainage and Storm Water Plan (hereafter, simply Master Plan of Drainage) will be used to demonstrate graphical display opportunities.

COMPUTERIZED MASTER PLAN OF DRAINAGE AND GRAPHICS DATA BASE MANAGEMENT SYSTEM

The total Master Plan of Drainage software package and data base system contains numerous elements and components that span several technical fields including data base management, geographic information systems (or GIS), hydrologic/hydraulic computer modeling, graphical data base management, flood control engineering and planning, among others. In the following, a brief survey of the key elements of the total software package is provided.

Coupled Hydrologic Modeling Technique
Most flood control agencies at the city, county or state level require specific procedures for the calculation of flood flow quantities. Often the procedure may involve the use of two or more estimates, depending on conditions such as watershed size. In Southern California, several county flood control districts require use of two flood flow estimation techniques depending on catchment area. The rational method for areas smaller than about one square mile. The design storm unit hydrograph method for areas larger than about one square mile. The transition between techniques has been coupled into an integrated computerized Master Plan of Drainage model, for the first time, enabling the development of an integrated hydrologic computer model with one pass of the analysis, rather than two separate studies. As a result of coupling hydrologic techniques in a single computer model, a single system is available for use in preparing Master Plans of Drainage and upgrading the master plan, thereby greatly reducing the complexity, review and cost involved.

The Master Plan of Drainage software contains internal editing and computational elements that involve 152 hydraulic and hydrologic submodels and global modeling commands. The software enables analysis of an integrated open channel or closed conduit flood control system on a study-wide basis (Hromadka et al, 1987, 1993).

Graphical Data Base
Primary hydrologic parameters used in the Master Plan of Drainage computer model are land use, hydrologic soil group, rainfall, and hydrologic subarea topographic data such as area, length of water course, and elevation. In general, a study is discretized into subareas that are 10 to 20 acres in size. These subareas require definition with respect to each of the parameters listed above. Additionally, maps are needed in order to communicate these data. By obtaining in digital form or actually digitizing the land use maps, hydrologic soil group maps, rainfall maps, and subarea maps, not only is a digital/graphical representation available for display, but the data can then be processed

by a "polygon processor" (described below) in order to partition the subareas into the intersection of all of the graphical layers (e.g., Hromadka et al, 1993). Further enhancement possibilities exist, such as automated elevation data extraction from topographic maps (Jenson and Domingue, 1988) that significantly simplify complex storm water model data base construction. Geographic location is provided by use of street layout layers, right-of-way maps and freeway maps. The graphics data base is used to prepare hard-copy maps for reports, as well as graphical layers for display on the computer monitor.

Polygon Processor

The use of geographic information systems (GIS) has become widespread in many facets of engineering and planning, among other fields. A key element of a GIS is the ability to intersect graphical layers so that several forms of information are resolved into "cells" wherein all parameters are constant.

In the Master Plan of Drainage, each subarea requires definition of land use, hydrologic soil group, rainfall subarea size, and the proportions of each within the subarea. The polygon processor performs this important task, and then develops a data base for use in the Master Plan of Drainage computer model. The subarea data are stored in tabular format, on a subarea basis, indexed according to subarea number (Hromadka et al, 1993; Smith, 1993). Thus, the retrieval of a specific subarea number will access these several data, automatically developed by the polygon processor (for specific issues, see Smith and Brilly, 1992).

Master Plan of Drainage Data Base

The Master Plan of Drainage may be represented, in a data base form, as a collection of nodes (specific points along the catchment flood control system), and subareas (10 to 20 acres in size). All information computed by the Master Plan of Drainage, such as deficiency system mitigation needs, flow quantities, hydraulic properties, streetflow characteristics, flood control system characteristics, hydrologic parameters, and costs, among others, are stored in agency-designed tabled form in a data base indexed according to node number, link number, and subarea number. Data entered directly into the data base such as flood control system history, age, and so forth are also stored. Once the data base is assembled, it may be linked to the graphical data base which displays the digital graphical layers constructed for the polygon processing (i.e., multiple use of a data base form), while allowing easy access to the Master Plan of Drainage data base.

Modeling analysis programs access the data base, in turn, supplying computational results.

HydroGraphics Management System

The graphical data base and retrieval software and the Master Plan of Drainage hydrologic/hydraulic computer software are coupled to form the Graphics Data Base Management System. Each of the above software packages are developed specifically for this application, and do not require use of other software packages.

Two applications are available. Application 1 is the actual Graphics Data Base Management System which includes all the features of Application 2, but also includes the ability to upgrade the Master Plan of Drainage due to changes in system requirements, land use, hydrologic parameters, among other factors. Because the agency can perform the upgrade, the master plan can be kept current, enabling up-to-date drainage fee assessment to be developed.

The second application, or Application 2, enables publication of the Master Plan data base for distribution to the public. Using "slides" (i.e., screen images stored in the graphics data base), the entire study can be resolved into graphics slides of about one-half square mile in size, showing hydrologic master plan nodes, subareas, links, streets, land use, hydrologic soil group, among other designed data. Each slide is indexed to successively larger maps so that by selecting an appropriate grid from the monitor, one is able to navigate through the city to a select point. Additionally, each slide is cross-referenced to a Master Plan of Drainage data base map that stores all the data associated to the slide appearing on the monitor. Figure 1 shows a slide of the entire data base "Index" map which appears on the monitor. Data base operating commands are displayed at the top of the monitor screen, enabling the user to access the next level of the graphics slides. In this application, the map depicts the entire 45-square mile watershed catchment area relevant to the urbanizing City of Yucaipa, in the State of California. A grid system for selecting the next screen level (shown in more detail) is also shown in Figure 3. To access a particular grid slide, say D2, one first clicks onto the top "Grid Number" option.

This simple application provides significant communication opportunities for the agency to communicate with both the public and the technical sectors. The engineering and planning communities can access the data base for other technical needs, and also inspect the Master Plan of Drainage without reviewing the usual report documents (which typically run several volumes). The public can inspect the Master Plan, and access information that would otherwise be unavailable. The cost of Application 2 is simply the cost of a 3.5-inch disc, or a micro-floppy. Figures 2 and 3 show successive steps through an application made for the City of Yucaipa, California by Boyle Engineering Corporation.

Computer System Requirements

The necessary computer hardware needs (e.g., IBM 486 system or equivalent) for a study of 100 square miles involves the data base of some 5,000 subareas, 6,000 nodes, 5,000 links, and about 250,000 pieces of information. The total graphics and Master Plan data base requirements is in the 40 to 50 megabyte range.

Hard Copy Products and Mapping

Hard copy maps and reports are readily prepared using the several constructed data bases. Consequently, once the graphical data bases are assembled, drafting time is significantly reduced by using hard copy printouts. Similarly, report technical appendices can be prepared using the Master Plan of Drainage data base.

Application to Sewer, Water, Environmental Systems

Extension of the Graphics Data Base Management System to use in sewer, water, and environmental systems is straightforward and has been accomplished in several applications in San Bernardino County, California. Both Applications 1 and 2 follow the procedural steps described for Master Plans of Drainage. A key element to use with other systems is the availability of an integrated processing model -- such as readily available integrated sewer and water system models. In the application described herein, the development of an integrated Master Plan of Drainage computer model was a crucial step in the evolution of the Graphics Data Base Management System approach for Master Plans of Drainage.

CONCLUSIONS

A graphics data base management system for computerized Master Plans of Drainage has been developed. Two applications are available which enables the agency to upgrade the Master Plan in the future, and to publish the Master Plan in computer graphics form for distribution to the public. Because of the ease of communication opportunities afforded by this approach, the utility in Agency public information programs may be significant.

REFERENCES

1. Bergman, H. and Richtig, G., 1990, Decision support model for improving storm drainage management in suburban catchments, *Proc. of the Fifth International Conference on Urban Storm Drainage, Osaka, Japan,* 1429-1434.

2. Djokic, D. and Maidment, D.R., 1991, Terrain Analysis for urban stormwater modelling, *Hydrol. Proc.,* 5, 115-124.

3. Hromadka II, T.V., 1983, Computer Methods in Urban Hydrology, Rational Methods and Unit Hydrograph Methods, Lighthouse Publications.

4. Hromadka II, T.V., Durbin, T.J., and DeVries, J.J., 1985, Computer Methods in Water Resources, Lighthouse Publications.

5. Hromadka II. T.V., McCuen, R.H., and Yen, C.C., 1987, Computational Hydrology in Flood Control Design and Planning, Lighthouse Publica-tions.

6. Hromadka II, T.V., 1988, Computational Hydraulics of Irregular Channels, Lighthouse Pub-lications, 270 pgs.

7. Hromadka II, T.V., McCuen, R.H., Durbin, T.J., and DeVries, J.J., 1993, Computer Methods in Water Resources and Environmental Engineering, Lighthouse Publica-tions, 450 pgs.

8. Jenson, S.K. and Domingue, J.O., 1988, Extracting topographic structure from digital elevation data for geographic information system analysis, *Photo. Eng. and Remote Sensing,* 54 (11), 1593-1600.

9. Smith, M.B., 1993, A GIS-based distributed parameter hydrologic model for urban areas, *Hydrol., Proc.,* 7, 45-61.

10. Smith, M.B., and Brilly, M., 1992, Automated grid element ordering for GIS-based overflow, *Photo. Eng. and Remote Sensing,* 58 (5), 579-585.

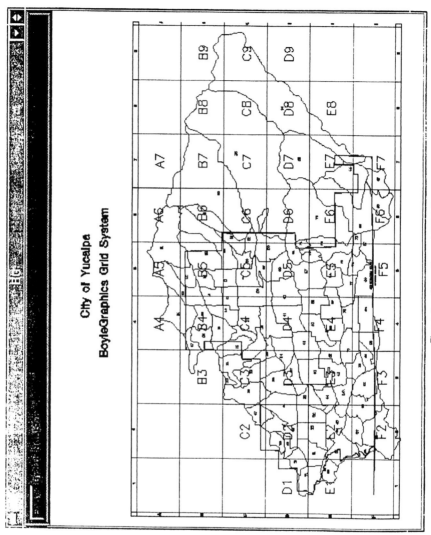

Figure 1.   Cover Page of GDBMS

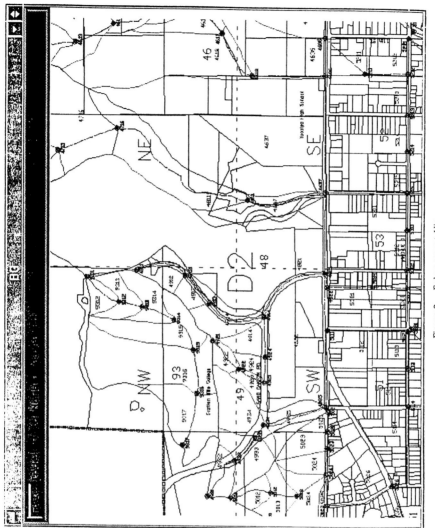

Figure 2.   Enlargement View

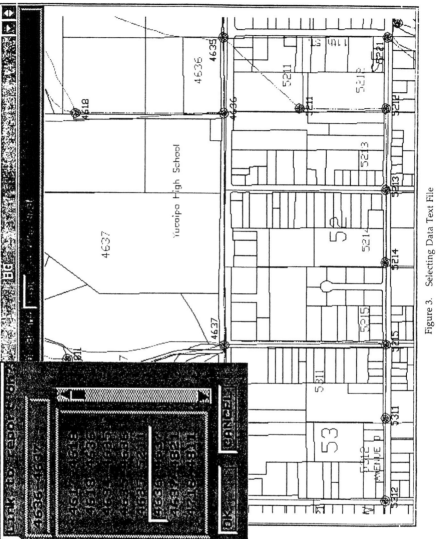

Figure 3.   Selecting Data Text File

Time-domain tuned rainfall runoff models optimised using genetic algorithms

J.G. Ndiritu, T.M. Daniell
Postgraduate Student and Senior Lecturer, Department of Civil and Environmental Engineering, The University of Adelaide, Adelaide SA 5005, Australia

Abstract

The genetic algorithm method has been investigated for the calibration of a rainfall runoff model as part of research utilising the time-domain in rainfall runoff modelling. The genetic algorithm method has proved robust and the need to limit the number of model parameters to minimise difficulties in model calibration has been found unnecessary. A routine that shifts and varies the size of the search space as optimisation proceeds has been included. This feature which is intended to make the optimisation more robust and fine tune the search has been found to improve the calibration significantly. The calibration of an 8 year long 6 and 7 day time increment flow series of a 27 km^2 Australian catchment gave a correlation coefficient of 0.93 and a volume ratio of 0.89. In a theoretical case using a runoff series generated by the model as the 'actual' data, the calibration achieved a near perfect fit.

1 Introduction

Optimisation procedures are grouped into global and local search methods or alternatively into either direct and derivative methods. Global optimisation procedures have the significant advantage of starting the search from several points in comparison to local ones in which the search starts from a single point. Methods consisting of repetitions of local procedures starting from different positions in the search space are effectively global search methods. An example is the multisimplex method [1], [2]. Direct methods use the objective function values while derivative methods use response surface derivatives of the first and/or higher order. As a consequence derivative methods encounter more difficulties than direct ones in situations where the response surface is rough. Johnston and Pilgrim [3], Duan *et al* [2] and Bates [1], note the main challenges in conceptual rainfall runoff model calibration as the presence of regions of attraction in the search space, minor local optima, roughness of the response surface, long curved ridges along the response surface, parameter interdependence and the indifference of the objective function to variations in the parameters.

The genetic algorithm has been used in many optimisation problems as Goldberg [4] indicates. Wang [5] reports the application of the genetic algorithm in the calibration of a 7 parameter rainfall runoff model. The genetic algorithm method has the two advantages of being global and direct and is computationally not complicated. It was therefore chosen for model calibration in a study investigating the utilisation of the time domain in rainfall runoff modelling. The genetic algorithm is based on the principle of survival of the fittest and is thus a practical application of the theory of evolution. Following is a qualitative explanation of how the method works.

A population of individuals is generated from the search space. An individual is a coded string consisting of coded substrings. The substrings represent the variables to be optimised. In rainfall runoff model calibration, the substrings represent model parameters. In pipe network optimisation, some of the substrings could be representing pipe diameters [6]. The performance of each of the individuals is then evaluated. A new population is created in a manner that gives the higher performance individuals of the current population a better chance of being used in generating the new individuals. This process is repeated many times over with the effect of preserving and combining the higher performance traits within the individuals and removing the poor performance traits from the population. This results in optimisation. A few random changes or mutations are introduced to increase the chances of creating some high performance traits not yet present in the population.

In the following example a Fortran program was developed to implement the optimisation. The details of this implementation are described elsewhere [7,8].

2 Calibration of Rainfall Runoff Model

2.1 Model Description

The model has been developed to investigate the utilisation of the time-domain in rainfall runoff modelling and uses simple empirical expressions to represent the catchment process. The model is described in equations 1 to 8. In the equations, ra_x, pet_x and run_x are the rainfall, potential evapotranspiration and computed runoff in period x respectively, $p1$ to $p14$ are model parameters and x is the 6 or 7 day long period ($1 \leq x \leq xm$). Parameter $p1$ determines to which weighted net rainfall in the past the current runoff depends. The catchment wetness $p7.sl_x^{p8}$ is divided into a maximum of three states where sl_x is the sum of the weighted net rainfalls upto period $x-1$. Parameters $p9$ and $p12$ are optimised to specify these states.

$$sl_x = \sum_{j=x-p1}^{x-1} \left(\frac{(nra_j)^{p2}}{(x-j+1)^{p3}} \right)$$

(1)

$$sl_x \geq 0$$

where

$$p4 \leq \left(nra_j = ra_j - p6.pet_j\right) \leq p5$$

(2)

If

$$p7.(sl_x)^{p8} \geq p9$$

(3)

then

$$run_x = p7.(sl_x)^{p8} + p10.p7.(nrai_x)^{p11}$$

(4)

where

$$nrai_x = ra_x - p6.pet_x$$

(5)

If

$$p12 \leq p7.(sl_x)^{p8} < p9$$

(6)

then

$$run_x = p7.(sl_x)^{p8} + \left(\frac{p10}{p13}\right).p7.(nrai_x)^{p11}$$

(7)

else

$$run_x = p14.p7.(sl_x)^{p8}$$

(8)

2.2 Objective Function

An objective function (OBF) of the form of equation 9 was used for optimisation

$$\text{Minimise} \quad \frac{Const.}{(xm-p1+1)} \left(\sum_{x=p1}^{xm} \frac{(run_x - arun_x)^{in}}{(run_x + arun_x)^{1/in}} \right)$$

(9)

where $arun_x$ is the actual runoff, Const. is a constant to scale the OBF into a reasonable range for computation purposes and in is an index determining the relative importance given to low and high flows. In all the calibrations, an in value of 1.2 was used.

2.3 Historical data case

Daily rainfall, potential evapotranspiration and runoff data from Scott Creek catchment for the period 1970 to 1977 was used. Scott Creek is located in the Onkaparinga basin in South Australia and exhibits distinct wet and dry seasons. For the period of the study, the average annual rainfall was 965mm, the potential evapotranspiration 1085mm and the runoff 155mm.

The data was summed into 6 day and a few 7 day durations so that each month contained five such durations. In leap years, the months were assumed to alternately consist of 30 days (five 6 day periods) and 31 days (four 6 day and one 7 day period). In other years, the only difference was the 12th month which consisted of 30 and not 31 days.

The following four calibrations were carried out using the initial parameter ranges specified by $PAL_{i,0}$ and $PAU_{i,0}$ in Table 1, where $PAL_{i,t}$ is the lower limit of the i th parameter in evaluation t and $PAU_{i,t}$ is the upper limit of the i th parameter in evaluation t, with t in this case $=0$.

- case 1 with no parameter range modification; $\beta_1=0$, $\beta_2=0$, $\beta_3=0.5$; where β_1 is the search space shift parameter; β_2 and β_3 are the search space reduction parameters
- case 2 with parameter range reduction and centralising only; $\beta_1=0$, $\beta_2=0.2$, $\beta_3=0.35$;
  - case 3 with parameter range shift only; $\beta_1=0.1$, $\beta_2=0$, $\beta_3=0.5$ and
  - case 4 with both parameter range reduction and shift; $\beta_1=0.1$, $\beta_2=0.2$, $\beta_3=0.35$.

Table 1. Initial parameter ranges and least objective function values obtained in calibrations

Parameter	p1(p1*)	p2	p3	p4	p5	p6	p7
PAL(i,0)	13(1.000)	0.500	0.500	-50. 000	50. 000	0.500	0.001
PAU(i,0)	43(4.000)	1.000	1.500	0.000	100.000	2.000	0.200
Case 1	24(2.197)	0.935	1.200	-3.068	99.728	0.554	0.058
Case 2	20(1.717)	0.954	0.808	-1.083	74.368	0.465	0.039
Case 3	23(2.008)	0.776	1.083	-0.090	95.610	0.438	0.070
Case 4	19(1.635)	0.498	0.940	1.023	133.052	0.429	0.019
Ideal Data	19(1.672)	0.553	0.952	0.985	112.465	0.424	0.027

p1=Integer value of $(10 \times p1*)+3$

Parameter	p8	p9	p10	p11	p12	p13	p14
PAL(i,0)	0.400	1.000	4.000	1.000	0.200	2.000	0.600
PAU(i,0)	1.000	5.000	10.000	2.000	2.000	10.000	1.000
Case 1	0.998	2.007	7.207	1.635	0.550	4.002	0.604
Case 2	1.578	1.820	9.774	1.729	-0.461	8.201	0.869
Case 3	1.946	1.812	5.360	1.843	-0.842	7.981	0.957
Case 4	3.300	1.895	20.359	1.936	-0.779	7.225	1.005
Ideal Data	3.021	1.879	14.500	1.935	-1.129	7.233	0.860

The correlation coefficients between the calibrated and the actual sequences and the volume ratios (volume of calibration series/volume of actual series) for the four cases are presented in Table 2 while Figure 1 compares the variation of the *OBF* value with the number of evaluations. Figure 2 gives a graphical presentation of the actual series and the calibration result of case 4. The measures of range reduction $rr_{i,t}$ and the range shift $rs_{i,t}$ were evaluated using equations 10 and 11.

$$rr_{i,t} = \frac{\left(PAU_{i,t} - PAL_{i,t}\right)}{\left(PAU_{i,0} - PAL_{i,0}\right)} \tag{10}$$

$$rs_{i,t} = \frac{\left(PAU_{i,t} + PAL_{i,t}\right) - \left(PAU_{i,0} + PAL_{i,0}\right)}{2\left(PAU_{i,0} - PAL_{i,0}\right)} \tag{11}$$

The variation of the two measures during optimisations are presented in Figures 4 and 5 for selected parameters.

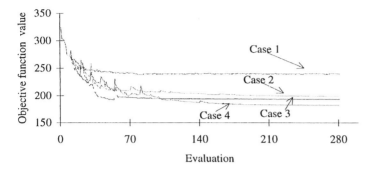

Figure 1 Variation of Objective Function Value with number of evaluations

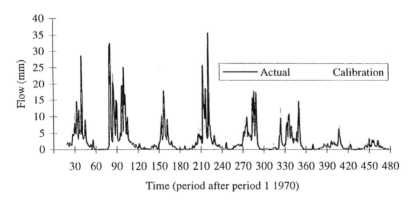

Figure 2 A comparison of the calibration and actual runoff sequence of Case 4

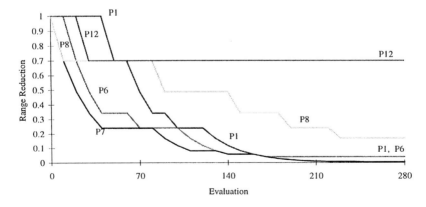

Figure 3   Range Reduction in Case 2 for selected parameters

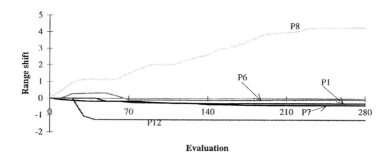

Figure 4 Range Shift for selected parameters for Case 2

Table 2. Correlation coefficients and volume ratios obtained in calibrations

Calibration	Case 1	Case 2	Case 3	Case 4
Correlation coefficient	0.902	0.912	0.922	0.927
Volume ratio	0.909	0.909	0.903	0.893

2.4 Ideal Data case

The runoff series obtained from the calibration of case 4 was used as the 'actual' series. The optimisation was then aimed at obtaining the global optimum which is the parameter set for case 4 in Table 2. The variation of the *OBF* with the number of evaluations is shown in Figure 5 and the parameter set giving the least *OBF* value given in Table 2.

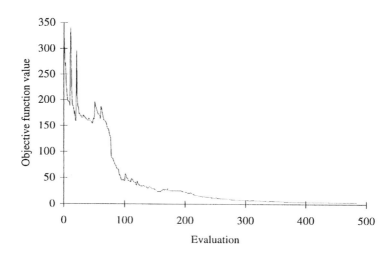

Figure 5 Variation of Objective Function Value with number of Evaluations

3 Discussion

The results of Table 2 and Figure 2 give evidence of a successful calibration. This is attributed to both the model conceptualisation and the optimisation procedure. Figure 1 indicates that the two operations of parameter range reduction and range shift were useful since case 4 which applied both gave the least *OBF* values. Individually, the range shift achieved a better result than the range reduction. It was also observed that the initial reduction of *OBF* (up to evaluation 50) was greatest in case 3 which involved range shift only. It may therefore be advisable to delay the range reduction operation until improvements in *OBF* have reduced considerably. There is the danger of reducing the range excessively without an adequate search especially if parameter β_1, β_2 and *tc* are not selected carefully. It is observed from Figure 4 that the range shifts for case 3 involving range shift only were lower than those of case 4 which included both operations. It is therefore unlikely that there was premature range reduction in the calibrations. The standard deviations of the scaled parameter values obtained in the last 100 evaluations (evaluation 181 to 280) were used as a measure of the extent of convergence. The scaling was done by dividing the parameter values with their mean values. Figure 6 presents the standard deviations of the scaled parameters for cases 3 and 4.

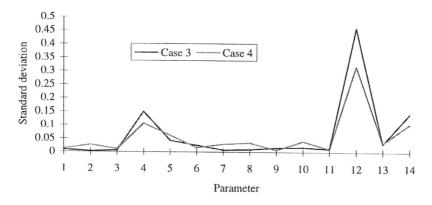

Figure 6 Standard Deviations of Scaled Parameter Values for Evaluations 181 to 280

Parameters *p4*, *p12* and *p14* exhibited notably higher standard deviations. The non-convergence of *p12* and *p14* is supported by the fact that the process described by equation 8 did not take place as all the catchment wetness values ($p7.sl_x^{p8}$ in equation 6) were greater than the *p12* values. It is thus probable that the following two approaches could be applied to modify the model without an impairment of performance;

• use a simpler model that excludes the the processes described by the three parameters thereby reducing the required effort in optimisation; and

• use the three parameters to model other catchment processes not presently included which could have a chance of promoting better performance. For instance, two parameters could be used to model the degree of evapotranspiration loss. The present model uses only $p6$.

It is however noted that the importance of model components varies among catchments and also depends on the relative importance assigned to high and low flows. It may therefore not be approprite to dismiss a model component without a comprehensive study.

By evaluation run 484 in the ideal data case, the global optimum value had been obtained for parameter $p1$ and the values for parameters $p3$, $p4$, $p6$, $p9$, $p11$, $p13$ and $p14$ were quite close to the global optimum values. The fit between the calibrated and the actual runoff series was almost perfect with a correlation coefficient and a volume ratio of 1.000 and 1.002 respectively. The genetic algorithm was therefore highly successful in capturing the information contained in the ideal runoff sequence. Significant improvements of the historical data calibration are therefore more likely to be obtained from modifications to the model structure rather than attempts to improve the optimisation procedure. Investigations aimed at improving the efficiency of optimisation to save computation space and time would however be appropriate.

4 Conclusions

A description of the genetic algorithm method and its successful application to the calibration of a rainfall runoff model have been presented. Two operations, one that shifts the parameter search space to increase robustness and the other that reduces and centralises the search space to fine-tune the calibration were developed and tested. Both operations were found to give significant improvements. The genetic algorithm proved to be robust and it seemed that the model structure was the factor that set the limit of performance. It was also explained how the results of the genetic algorithm optimisation could be used to guide improvements to the structure of the rainfall runoff model.

As these conclusions are based on results of only one catchment, a more extensive study involving other catchments of varying hydrological characteristics is being undertaken. The selection of the optimisation parameters and the development of methods for modifying the parameter search space was relatively intuitive although some trial runs were made.

References

1. Bates, B. C., Calibration of the SFB model using a Simulated Annealing Approach, *International Hydrology and Water Resources Symposium, Adelaide,* 1-6, 1994.

2. Duan, Q., Sorooshian, S., and Gupta, V., Effective and Efficient Global Optimization for Conceptual Rainfall-Runoff Models, *Water Resour. Res., 28(4),* 1015-1031, 1992.

3. Johnston, P. R., and Pilgrim, D. H., Parameter Optimization for Watershed Models, *Water Resour. Res., 12(3),* 477-486, 1976.

4. Goldberg, D. E., *Genetic Algorithms in Search, Optimization and Machine Learning,* Addison Wesley, 1989.

5. Wang, Q. J., The Genetic Algorithm and Its Application to Calibrating Conceptual Rainfall-Runoff Models, *Water Resr. Res., 27(9),* 2467-2471, 1991.

6. Murphy, L. J., Simpson, A. R., and Dandy, G. C., *Pipe Network Optimization Using an Improved Genetic Algorithm,* Dept. of Civil and Env. Eng., University of Adelaide, Research Report No. R 109, 1993.

7. Ndiritu, J.G. and Daniell, T.M. Rainfall Runoff Model Optimised Using Genetic Algorithms. 23rd Hydrology and Water Resources Symposium, Hobart, Australia, 21-24 May 1996.

8. Ndiritu, J.G. and Daniell, T.M. Rainfall Runoff Model Calibration by the Genetic Algorithm Method. submitted to the *ASCE Journal of Computing in Civil Engineering,* 1996.

Appendix. Notation

β_1	Search space shift parameter
β_2, β_3	Search space reduction parameters
Const	Constant for factoring objective function value to reasonable range
in	Index of relative importance allocated to low and high flows
np	Number of parameters
$nrai_x$	Net rainfall in period x
nra_x	Modified net rainfall in period x
$OBF_{j,t}$	The objective function value of individual j in evaluation t
p1,p2,..p14	Rainfall runoff model parameters
$PAL_{i,t}$	Lower limit of the i th parameter in evaluation t
$PAU_{i,t}$	Upper limit of the i th parameter in evaluation t
pet_x	Potential evapotranspiration in period x
ra_x	Rainfall in period x of rainfall time series
$rr_{i,t}$	A measure of search space reduction for parameter i in evaluation t
$rs_{i,t}$	A measure of search space shift for parameter i in evaluation t
run_x	Calibrated runoff in period x
$arun_x$	Actual runoff in period x
sl_x	Sum of weighted modified net rainfalls in period x
tm	Maximum allowed number of evaluations
xm	Length of hydrological time series

A parametric hydrodynamic model of a complex estuary

J.B. Hinwood,[a,b] E.J. McLean[a,b]

[a]Mechanical Engineering, Monash University, Clayton, Victoria, Australia

[b]Geography, Deakin University-Rusden, Clayton, Victoria, Australia

1.   Introduction

Parametric modelling of an estuary aims to produce cost effective and robust computer models for use by environmental managers. These requirements lead to a different approach to the developments of the last few decades which have emphasised detailed representation of the physics of the estuary.

A scheme of classification of numerical models for the prediction of flow in tidal waters, including estuaries, was given by Hinwood and Wallis [11]. They classified the models into an array according to the number of spatial dimensions, the level of physics represented, and whether the model was eulerian or lagrangian. The spatial dimensions may be three, two (in plan or elevation), one (length) or zero (a single well-mixed box) dimensional. The levels of physics may be as follows:

(i)   Hydrodynamic - uses momentum equations and hence can find velocities in new configurations. The energy equation may be used in one dimensional hydrodynamic models.

(ii)   Kinematic - uses mass conservation equations and empirical data on velocity profiles and water levels to solve for velocities.

(iii) Transport - uses the convection-diffusion equation and data on velocities to solve for pollutant transport.

Hinwood and Wallis [12] found models of most types, and more have been developed since then. All of the models use finite difference or finite element schemes dividing the estuary into large numbers of elements, all but a couple solving the primitive equations.

A hydrodynamic model must be used for prediction of flow velocities and longitudinal salinity distributions in applications such as substantial upstream diversion of flow or channel modifications or where data are sparse. On the other hand, where the tidal and river flows are known with sufficient accuracy, a kinematic model may be used. In both cases a transport model is then used to compute the pollutant distributions.

The averaging involved in one or zero dimensional models makes them too crude for many applications but frequently there is no need for a three-dimensional model and a two-dimensional (in elevation) model may be used. In such models the computation of flows over many tide cycles requires extensive programming and data preparation and appreciable computer time and power. These cost and time penalties reduce the utility of these models for resource management, planning of data collection, and other interactive tasks where the highest precision is not essential.

To overcome these disadvantages, a parametric hydrodynamic model has been developed and is described in this paper. The application of this model to the Snowy River estuary is discussed.

2. The Parametric Hydrodynamic Model

2.1 Modelling Concept
It is presumed that the flow in the estuary is forced by the river flow and the presence of salt water at the seaward limit. Tidal forcing may also be present. Boundary conditions depend on the flow model adopted for each reach, as discussed below. Data required are the upstream river flows, tidal range and period, channel dimensions, bed and interfacial resistance coefficients, and eddy diffusion coefficients.

2.2 The General Model
The model uses published analytical solutions of specific flow situations to describe the hydrodynamics. These solutions include homogeneous flow, the salt wedge, and some partially mixed flows. While the presently available solutions have been used, the model is modular and improved solutions may be substituted as they become available. For each identified reach of the estuary, an appropriate flow is selected. The solutions are applied to each reach, to build up a set of linked flows describing the whole estuary. The model has been developed for a branched, but not looped channel system, with one point of discharge to the sea.

The analytical solutions have been obtained only for estuarine reaches with very simple geometry, most being restricted to uniform rectangular channels. In order to use the solutions in a real estuary, each reach is divided into a few segments over which the restrictions are adequately met. The solutions are used in a manner analogous to finite element equations over each of the segments.

The flows are assumed to be quasi-steady, with the salt or pollutant concentrations changing with time via the unsteady convection-diffusion equation. This restricts the model to slowly changing forcing, but permits the simulation of changes from an initial salinity distribution following changes in the forcing flows or tides, including response to a flood hydrograph.

The different analytical solutions used are described in the following sections and conditions applied at the junctions are then outlined.

2.3 Kinematic Model For Velocities
For all flow regimes, the kinematic model utilises the continuity equation in the form

$$\frac{\partial Q}{\partial x} + A_s \frac{\partial y}{\partial t} = 0 \qquad (1)$$

where Q is the volumetric discharge, y is the water depth, $A_s(y)$ is the water surface area of the segment, x is the longitudinal coordinate and t is time.

Where a sectionally averaged velocity is required it is given by

$$V = Q/A \tag{2}$$

where A is the cross-sectional area of the segment. Non-uniform velocity calculations depend on the flow regime in the reach, as described below.

2.4 Hydrodynamic And Transport Models

Models for the different flow regimes considered are given in the following subsections.

2.4.1 Unstratified Flow

In reaches of an estuary which are well mixed vertically and laterally, the well established equations of homogeneous channel flow may be used:

(i) In uniform steady flow in a wide channel, the depth, y_0, and the average velocity, V_0, are given by

$$y_o = \left(\frac{q^2 f}{8gS_o}\right)^{1/3} \qquad V_o = \frac{q}{y_o} \tag{3}$$

where q is the discharge per unit breadth, f is the bed friction factor, g is the acceleration of gravity and S_o is the bed slope. If necessary, the velocity distribution over the depth may be described by a seventh-power law or another boundary layer expression.

Depth-averaged pollutant concentrations, C, are then found from the depth-integrated convection-diffusion equation

$$\frac{\partial C}{\partial t} + V \frac{\partial C}{\partial x} - K \frac{\partial^2 C}{\partial x^2} = 0 \tag{4}$$

where the longitudinal dispersion coefficient, $K = 6V\sqrt{f/8}$. The solution of equation(4) for steady flow is

$$C = C_o \left(1 - e^{x/a}\right) \tag{5}$$

where $a = Ky/V$. Solutions may be superimposed to accommodate multiple sources. Alternatively, equation (4) may be solved by finite differences using velocities and depths from the hydrodynamic solution. Finite difference solution for each segment is required if V is not the same in all segments, which is the usual case.

(ii) In uniform steady flow in a general prismatic channel, expressions equivalent to equation (3) are available for a range of channel shapes (Chow, [2]). Fischer et al [4] give expressions for K for channels of arbitrary cross-section and for estuaries in which tidal motion affects the coefficient in the tidally-averaged form of equation (4). Equations (4) and (5) are unaltered in form.

(iii) For a uniform rectangular cross section with linear friction, tidally forced at one or both ends, Ippen and Harleman [14] provide equations which replace equation (3).

(iv) For a tidally forced basin connected to the sea by a short channel, Glenne et al [5] provide an equation to replace equation (1).

For all other cases data on water levels must be provided, then equation (1) is used to solve for V as in section 2.2.

2.4.2   Salt Wedge There are several solutions for the form and dimensions of the salt wedge. The model described in section 3 used the equations of Keulegan [15] for the salt wedge length and that of Farmer and Morgan [3] for the interface profile. While these equations are widely used, it is believed that the following equations are slightly more accurate (Hinwood, [9]).

The depth of the salt wedge at the sea, d_o, may be found in terms of the upstream densimetric Froude number, $F_o = V/\sqrt{gD_o\,\Delta\rho/\rho}$:

$$d_o = D_o\left(1 - m\,F_o^{\,2/3}\right) \tag{6}$$

where $\Delta\rho/\rho$ is the relative density difference between the layers, and $m = 1$ for uniform flow in the upper layer and no flow in the lower layer, or $m = 0.85$ suggested for boundary layer flow in the upper layer and minor entrainment.

The shape of the interface may be obtained from the following implicit equation (Officer, [17]):

$$\frac{x}{D_o} = \frac{F_o^{\,2}\,n^2}{3k}\left(1 - \frac{16}{9}n + \frac{19}{18}n^2 - \frac{8}{105}n^3 + \dots\right) \tag{7}$$

where x is the longitudinal coordinate measured seawards from the toe of the wedge, $n = d/D_0$ is the relative depth of the salt wedge and k is an interfacial friction factor. At the sea, $d = d_0$ and $x = Lw$, the length of the salt wedge.

The velocity distribution in each layer has been assumed uniform in the example of section 3, but could be calculated from boundary layer equations.

Pollutant or salt transport in each layer may be found from equation (5) or by finite difference solution of equation (4). The mean motion of the lower layer due to entrainment has been neglected in deriving equations (6) and (7) but could be estimated using the numerical results of figure 7 of Arita and Jirka [1].

2.4.3   Partially Mixed Regimes Since the relative importance of buoyancy flux and tidally-induced mixing vary along the estuary a single analytical model is not usually adequate. Rattray and Hansen [6,7,18] recognised this. They obtained a sequence of three analytical solutions, each restricted by its own assumptions of geometry and diffusion coefficients for a stratified estuary in which mixing was significant. For application to a particular estuarine regime, they have simplified the two equations and have chosen convenient forms of the momentum and mass diffusion coefficients, as follows:
(i)   In Rattray and Hansen [18] they developed similarity solutions for the seaward reach of an estuary of uniform breadth and variable depth, valid for low river inflows.
(ii) In Hansen [6] they developed solutions for the central reach of an estuary in which there is a nearly constant salinity difference from top to bottom along the reach.
(iii) In Hansen and Rattray [7] they generalised the method of (i) and obtained solutions for the central and for the inner (upstream) reaches, subject to particular functional forms for the variations of the breadth, river flow and the diffusion coefficients.

2.4.4   Surface Plumes Studies of surface plume flow over stationary underlayers are being developed for use in the model. Extensions of the present model will utilise slightly modified forms of the solutions of Stolzenbach and

Harleman [19]; more recent reviews return to this work while complaining about its assumptions.

2.5   Selection Of Regime For Each Reach

While the most comprehensive scheme of classification of estuarine flow regimes is that of Hansen and Rattray [8], it is impractical to apply in modelling since it requires advance knowledge of the downstream stratification and the flow of the entrained salt water. The scheme of Ippen and Harleman [14] is readily applied and is used here as the principal criterion. It uses the parameter G/J, which may be interpreted as the ratio of the tidal energy available for mixing to the energy required to mix the fresh water over the depth. Based on an inspection of their figure 1, the following criteria have been adopted:

$$G/J < 30 \qquad\qquad \text{sharply stratified}$$
$$30 < G/J < 100 \qquad\qquad \text{partially mixed}$$
$$100 < G/J \qquad\qquad \text{well mixed.} \qquad (8)$$

Once a sharply stratified flow is indicated it is tested to confirm the presence of a salt wedge by checking the densimetric Froude number at the upstream end of the reach and the (approximate) length of the salt wedge:

$$F_o < 1 \quad \text{and} \qquad\qquad Lw > 1 \qquad (9)$$

If a wedge is not indicated, a surface plume is assumed and similar tests to equations (9) are made. If a surface plume is not indicated a two-layer exchange flow is assumed.

2.6   Treatment of Junctions between Reaches

At junctions of reaches, mass and energy conservation rules are applied. The water levels of each reach at a junction are made the same. For low speed flows or where the speed in each reach is the same, this is equivalent to conservation of energy. This assumption does not generally restrict the applicability of the model.

The sums of the fresh water and salt water inflows to the junction are made equal to zero. If more than one reach at a junction is stratified, the salinities of the bottom waters are made the same in each reach, and equal to the maximum at the junction. A corresponding condition is applied to the surface waters.

2.7   Computational Procedure

To establish a model, a file of segment dimensions and coefficients within each reach is created. A file of reaches made up of these segments is then defined and the expected flow regime is entered for each reach. The selection of both reaches and segments is critical to the efficient operation of the model. Selection is based on "natural" divisions considering both hydrodynamics and geomorphology (McLean and Hinwood, [16]). The ordering of the file of reaches defines the sequence of computation, which is up the main stream from the sea in the cases studied to date.

Computations are started by entering the forcing river flows and sea level data to calculate the sectional-average velocities. The first pass of the main computations solves for the salinity distributions using the expected regimes. For this step the junction relations are defined; for the example in section 3 this was done for each junction individually but the process has been made part of the model software. At the end of this step, the regimes are reassigned if necessary, and the computations repeated. Time is advanced and the sequence repeated as required.

The kinematic and transport models use first order forward time differences and upwind spatial differences. A test based on Courant number and another on rate of change of variables are applied to check that conditions for accurate solution are met and, if not, execution is terminated.

2.8   Attributes Of The Model
The attributes of the model have been discussed in Hinwood and McLean [10], but in summary are as follows:

•   The model is able to obtain salinity values within each estuarine segment with sufficient accuracy for assessment of water use and ecological impact.
•   The output is at temporal and physical scales compatible with data generated by other disciplines and can be readily integrated into multidisciplinary studies.
•   The model may be run interactively on a basic personal computer.
•   The initial segmentation and setting of limits of validity are critical and require expert input.
•   After its establishment the model may be run by managers and others not expert in modelling or hydrodynamics.
•   The model is robust and includes traps to prevent use outside its range of validity.

3.   Application To The Snowy River Estuary

3.1 Model Schematisation
The general model discussed above was set up for the Snowy River Estuary, selecting reaches and flow regimes from field data. The model was developed to provide an interactive tool to enable planners to gain maximum information from the data collected in a field study of the Snowy River estuary (Hinwood *et al*, [13]). The estuary, shown in figure 1, is complex with two tidal lakes, tidal lagoons and three stream inflows. Under low to moderate flow conditions there is a salt wedge in the upper Snowy River, while the Brodribb River and Corringle Creek are well mixed except near slack water. There is a weak surface plume in the lower Snowy river but this was approximated as a well mixed reach in the model.

Field data on bathymetry and sediments were used to identify significant geomorphological boundaries in the estuary. These were then assessed for their hydrodynamic significance by reference to longitudinal profiling of salinity and temperature under varying fluvial flows into the estuary. It was observed that, noticeable changes in estuarine configuration usually coincide with changes in the processes operating in adjacent reaches. For example, the transition between the flood-tidal delta (segment q) and the mud basin (segment o) and the separation between the latter and the shallower upstream tidal channel are obvious segment boundaries where the magnitude of model flows will reflect real processes within the estuary. These locations were selected as reach boundaries and, in the case of major morphological and hydrodynamic change, location in the model segmentation for change to a different analytical solution.

While some hydrodynamic changes were coincidental with geomorphic change, others show considerable locational variation and are not reflected by dramatic dimensional or sedimentological change. For example, the upstream limit of salt incursion in the salt wedge (segments e to a) must be treated as a model variable with segment boundaries selected, in this case, on purely dimensional grounds. While, at the micro-level, circulation patterns in individual segments may be complex, e.g. the pattern of influx of tidal water into and across Lake Corringle, other factors, such as local wind stirring of the very shallow lake, allow it to be treated as a single segment with homogeneous properties in a model of this scale.

Selection of estuarine segments reflecting real and observable estuarine characteristics permits sensible assessment of model results, especially with reference to the suitability of regimes selected for each reach (see section 2.4). Appropriate identification of process boundaries, both sharp and transitional, in the estuary is essential for models of this type.

3.2 Model Calibration

The model was calibrated by adjusting the friction factor in the salt wedge reaches and dispersion coefficients until agreement was obtained between predicted and measured salinities. The final values of the coefficients did not differ appreciably from the values given in section 2 and the references cited there and, as both high and low water salinities were matched, there are grounds for cpnfidence in the model. Figure 2 shows the match of depth-mean salinities at high water for a verification run, made following calibration based on data from a different field trip. Again the agreement is good. The bottom figure shows the measured and simulated density differences at high tide; note that in this version of the model only the upper Snowy River Estuary is modelled as a salt wedge reach.

The model runs in less than 0.2 secs/tide cycle on an IBM 486DX, making it attractive to explore the consequences of a wide range of management options.

4.   Conclusions

Most estuaries display a range of salinity regimes simultaneously at different locations. The extent of salt intrusion into such an estuary may be modelled relatively simply and very economically by the use of a parametric model.

Equations are given for several estuarine flow regimes including well-mixed flow and the salt wedge. The modelling method has been applied to the estuary of the Snowy River to predict depth-mean salinities and the upstream salt wedge profile and length.

5.   References

1.   Arita, M. and Jirka, G.H. (1987) Two-layer model of saline wedge II: Prediction of mean properties, J. Hydraul. Eng., 113:1249-63.
2.   Chow, V.T. (1959) Open-Channel Hydraulics, McGraw-Hill.
3.   Farmer, H.G. and Morgan, G.W. (1953) The salt wedge, Proc. 3rd Conf. Coast. Eng., ASCE, pp 54-64.
4.   Fischer, H.B., List, E.J., Koh, R.C.Y., Imberger, J. and Brooks, N.H. (1979) Mixing in Inland and Coastal Waters, Academic Press.
5.   Glenne, B., Goodwin, C.R. and Glanzman, C.F. (1971) Tidal choking, J. Hydraul. Res., 9:321-333.
6.   Hansen , D.V. (1964) Salt balance and circulation in partially mixed estuaries, Proc Conf. on Estuaries, Jekyll Island, 45-51.

7. Hansen, D.V. and Rattray, M., Jr. (1965) Gravitational circulation in straits and estuaries, J. Marine Res., 23 104-122
8. Hansen, D.V. and Rattray, M., Jr. (1966) New dimensions in estuary classification, Limnology and Oceanography, Vol. 11, No. 3
9. Hinwood, J.B. (1995) Short salt wedges and the limit of no salt wedge, Proceedings of Hydrodynamics of Estuaries and Coastal Seas, Margaret River, 1992. Amer. Geophys. Union
10. Hinwood, J.B. and McLean, E.J. (1992) Development of a parametric estuarine model for Gippsland resource planning, Gippsland Basin Symposium, 300-327.
11. Hinwood, J.B. and Wallis, I.G. (1975a) Classification of models of tidal waters, Proc. Amer. Soc. Civil Eng., J. Hydraulics Division, 101:1315-1331.
12. Hinwood, J.B. and Wallis, I.G. (1975b) Review of models of tidal waters, Proc. Amer. Soc. Civil Eng., J. Hydraulics Division, 101:1405-1421.
13. Hinwood, J.B., Watson,J.E., McLean,E.J. and Pollock,T.J. (1989) Snowy River Estuary Study, report to Dept. Water Resources, Victoria, Australia
14. Ippen, A.T. and Harleman, D.R.F. (1966) Estuary and Coastline Hydrodynamics, McGraw-Hill.
15. Keulegan, G.H. (1966) The mechanism of an arrested saline wedge, pp 546-574 of Ippen and Harleman (1966).
16. McLean, E.J. and Hinwood, J.B. (1993) Model segmentation in estuaries using geomorphological, sedimentological and hydrodynamic evidence, Proc.11th Australasian Conf. on Coastal & Ocean Engineering, Inst. Eng. Aust, 625-630.
17. Officer, C.B. (1976) Physical Oceanography of Estuaries, Wiley.
18. Rattray, M. and Hansen, D.V. (1962) A similarity solution for circulation in an estuary, J. Mar. Res. 20:121-133.
19. Stolzenbach, K.D. and Harleman, D.R.F. (1973) Three-dimensional heated surface jets, Water Resources Research, 9:129-137.

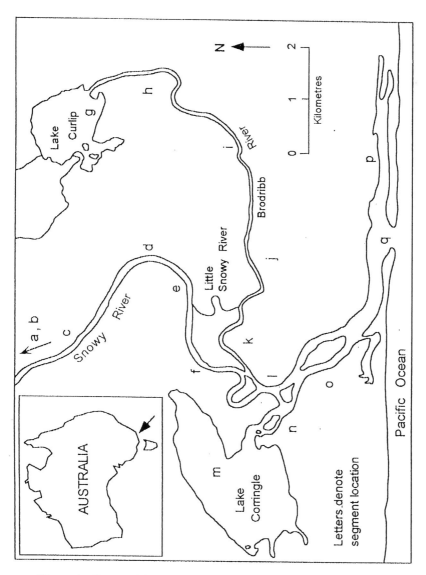

Figure 1   Snowy River Estuary showing computational segments

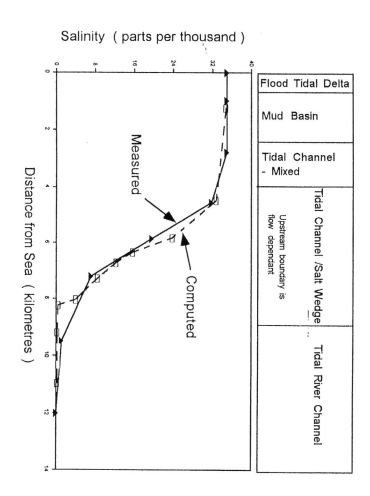

Figure 2   Longitudinal profile of depth-mean salinity for independent verification data set

FEFLOW: a finite element code for simulating groundwater flow, heat transfer and solute transport

L. Koskinen, M. Laitinen, J. Löfman, K. Meling, F. Mészáros

Technical Research Center of Finland, VTT Energy, P.O.Box 1604, 02044 VTT, Finland

Abstract

The FEFLOW code is a finite element program package developed at VTT Energy to model flow, solute transport and heat transfer in coupled and non-coupled, steady-state and transient situations, as well as in deterministic and stochastic modes.

The code offers a novel finite element technique to model groundwater phenomena in fractured crystalline rock. Linear and bi-quadratic one-, two- and three-dimensional finite elements can be used for describing engineered and natural bedrock structures. One of the solute transport models implemented in the package is capable of taking into account matrix diffusion as well. Highly convective cases are handled with different kinds of upwind schemes. The system of linear algebraic equations emerging from the standard Galerkin approximation can be solved with a direct frontal solver, as well as with an array of iterative solvers partly from the NSPCG package. The nonlinear algebraic equations resulting from coupled cases are solved with the Picard iterative approach with options for relaxation. The discretization of time is based on a simple finite difference scheme. For each result quantity to be determined, the code offers a wide selection of nodal boundary conditions including prescribed values, sources, sinks and/or fluxes. These may be constant or a function of time. Hydraulic properties of the bedrock features may also be constant or vary with depth. Besides the finite element analysis code the FEFLOW package comprises several programs to compute derived quantities (like flow paths and flow rates) and to facilitate generic modelling tasks.

The code has been tested in a series of test cases, and verified in the international HYDROCOIN project. Main application areas of the FEFLOW package have been site investigations and safety analyses undertaken by the Finnish power company Teollisuuden Voima/Posiva Oy operating two nuclear power plants. It has also been employed to simulate various hydraulic disturbances and solute transport phenomena in the Äspö Hard Rock Laboratory, Sweden.

1 Introduction

Teollisuuden Voima/Posiva Oy carries out investigations and safety analyses concerning the fractured, crystalline bedrock for selected sites in Finland for the final disposal of spent nuclear fuel. Modelling the subsurface flow and other groundwater-related phenomena is a part of these investigations. In response to the specific needs of the modelling studies, the FEFLOW package was developed at VTT Energy.

Traditional approaches for describing the fractured rock media are commonly based on fracture network models or discretization of the domain for finite difference or finite element analyses. In the latter the large, deterministic fracture zones are usually implemented by introducing anisotropy for the cells/elements coinciding with the surface of the fracture zones.

Along with the common functionalities of a standard finite element analysis package, FEFLOW offers a novel approach in modelling the fractured rock media. In the finite element meshes various types of elements are used intermixed. Three-dimensional (3D) elements describe the rock blocks. Embedded in the 3D elements and defined by a subset of their nodes, 2D elements along the surfaces of identified fracture zones describe bedrock structures of primarily 2D nature. A set of advanced routines have been developed to facilitate the required element mesh creation process and consequently the addition or removal of embedded two-dimensional meshes can be performed flexibly. One-dimensional elements are used for modelling engineered structures like shafts, tunnels or repositories. Thus the geometry of the fracture zones, often decisive for the local groundwater regime, can be modelled with reasonable accuracy (Figure 1).

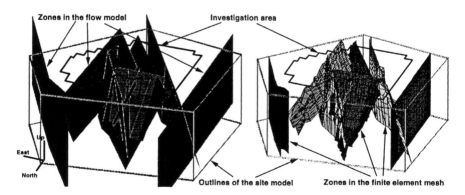

Figure 1. Modelling fracture zones with two-dimensional finite elements at the Kivetty site, Finland. Only a subset of all identified fracture zones is displayed.

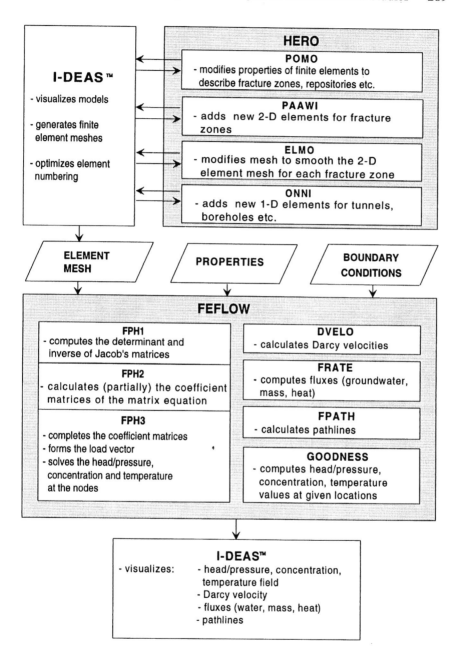

Figure 2. Components and data flow in the FEFLOW package.

The most important inputs to the code are the geometry and the hydraulic properties of the bedrock structures, which are obtained from geological investigations, and the initial as well as the boundary conditions describing the prevailing hydraulic and other conditions affecting flow, transport and heat transfer.

Primary result quantities that are computed with the code are the hydraulic head/pressure, concentration and temperature fields, whilst derived result quantities are Darcy velocity, flow rates through specified surfaces and flow paths. The I-DEAS commercial finite element pre- and postprocessor and an array of in-house developed modelling tools facilitate the analysis of results (Figure 2).

The above three result quantities can also be coupled in any combination. Thus eg density-affected flow can be studied, when the density of the groundwater is assumed to vary according to the concentration of some dissolved matter in it, or due to a non-ignorable heat source. In the case of coupled problems nonlinearity of the algebraic equations is handled with the Picard iterative method with options for underrelaxation.

More information on the FEFLOW package and its applications are available at links on the WWW home page of VTT Energy:

http://www.vtt.fi/ene/enehome.htm

2   Model equations

The mathematical models of the phenomena taking place in subsurface flow that are implemented in the FEFLOW package take the form of partial differential equations. These are solved numerically with conventional Galerkin approximation, standard upwind method, streamline upwind Petrov-Galerkin (SUPG) method, or discontinuity capturing SUPG (Brooks & Hughes, 1992). The system of algebraic equations emerging from the above techniques are solved by a direct frontal method, or with an iterative solver (conjugate gradient method for the symmetric and a Lánczos method for non-symmetric problems from the NSPCG package (Oppe et al., 1988)). All problems are typically solved with the equivalent continuum (EC) approach, but for certain transport problems the dual-porosity (DP) concept has also been implemented.

With regard to the global system matrix associated to the model, large symmetric problems (flow modelling) as well as strongly non-symmetric cases (material transport in the presence of hydraulic disturbances causing convection) can be handled with FEFLOW efficiently. The package has been streamlined and made available on various computer hardware architectures to process a large number of cases for sensitivity analyses, model calibrations and transient analyses.

2.1 Flow equations

In case the density of the groundwater is constant, the flow equation is solved for the hydraulic head h [m] (Bear, 1979):

$$\nabla(K\nabla h) - Q = S_s \frac{\partial h}{\partial t} \tag{1}$$

where: K is the hydraulic conductivity tensor [m/s],
Q is the flow rate per unit volume as sources and sinks [1/s],
S_s is the specific storage [1/m] and
t is time [s].

Simulations taking into account the varying density of water are based on the flow equation describing the pressure p [Pa]:

$$\nabla\left(\frac{\rho k}{\mu}\nabla(p + \rho g z)\right) - \rho Q = \phi\frac{\partial \rho}{\partial t} \tag{2}$$

where: ρ is the density of water [kg/m^3],
k is the permeability tensor [m^2],
μ is the viscosity of the water [kg/m/s],
g is the gravitational acceleration (9.81 m/s^2),
z is the elevation relative to the reference plane [m] and
ϕ is the total porosity [-].

2.2 Transport equations

Two distinctly different concepts have so far been implemented in the FEFLOW package for solute transport. The EC model (Huyakorn & Pinder, 1983) describes the concentration c [kg/m^3] as:

$$\nabla(D\nabla c) - \nabla(qc) + Q_{in}c_{in} - Q_{out}c = \phi_f\frac{\partial c}{\partial t} \tag{3}$$

where: c is the concentration [kg/m^3],
D is the dispersion tensor (incl. convection and diffusion [m^2/s],

q is the Darcy velocity [m/s],
Q_{in} is the the term for sources [1/s],
c_{in} is the concentration in the inflowing water [kg/m^3],
Q_{out} is the term for sinks [1/s] and
ϕ_f is the flow porosity [–].

In DP model (Huyakorn et al. 1983) the computational domain is decomposed into fractures with flowing water (Equation 4) and matrix blocks with stagnant water (Equation 5), in which the concentration c is described as:

$$\nabla(D\nabla c) - \nabla(qc) + Q_{in}c_{in} - Q_{out}c - (1 - \phi_f)\Gamma = \phi_f\frac{\partial c}{\partial t} \tag{4}$$

and

$$\frac{\partial}{\partial z'}(D'_e\frac{\partial c'}{\partial z'}) = \phi'\frac{\partial c'}{\partial t} \tag{5}$$

where: Γ is mass transfer term

$$\Gamma = -\frac{1}{a}(D'_e\frac{\partial c'}{\partial z'}\mid_{z=0}) \tag{6}$$

a is the half thickness of the matrix block [m],
c' is the concentration in the matrix blocks [m^3/s],
D'_e is the effective diffusion coefficient [m^2/s] and
ϕ' is the the porosity in the matrix blocks [–].

2.3 Heat transfer

FEFLOW offers the concept suggested by Huyakorn & Pinder (1983) for modelling the distribution of temperature T [K]:

$$\nabla(\lambda\nabla T) - \nabla(\rho_w c_w qT) + \rho_{in}Q_{in}c_{w,in}T_{in} - \rho_w Q_{out}c_w T - H =$$
$$= \frac{\partial}{\partial t}((\rho c)_{rw}T) \tag{7}$$

in which

$$(\rho c)_{rw} = (1 - \phi)\rho_r c_r + \phi \rho_w c_w \qquad (8)$$

where: λ is the heat conductivity tensor [W/m/K],
ρ_w is the density of water [kg/m^3],
ρ_r is the density of rock [kg/m^3],
ρ_{in} is the density of the inflowing water [kg/m^3],
c_w is the specific heat of the water [J/kg/K],
c_r is the specific heat of the rock [J/kg/K],
T_{in} is the temperature of the inflowing water [K] and
H is the heat source [W/m^3].

2.4 Coupling the flow, transport and heat transfer

In coupled analyses certain parameters of an equation depend on the result quantity of the other equation, and the outcome is obtained by iteration. These parameters are:

- the density:

$$\rho = \rho_0 + a_c c + a_T(T - T_0) \qquad (9)$$

- the viscosity:

$$\mu = \mu(T) \qquad (10)$$

- the Darcy velocity:

$$q = -\frac{k}{\mu}(p + \rho g z) \qquad (11)$$

- the dispersion tensor:

$$D_{ij} = \varepsilon_T |q| \delta_{ij} + (\varepsilon_L - \varepsilon_T)\frac{q_i q_j}{|q|} + D_e, \qquad i, j = 1, 2, 3 \qquad (12)$$

- the heat conductivity tensor:

$$\lambda = \lambda(q) \qquad (13)$$

where: ρ_0 is the density of the fresh water [kg/m^3],
a_c is the coefficient of density dependence of solute concentration [–],
a_T is the thermal expansion coefficient [kg/m^3/K],
ε_T is the transversal dispersion length [m],
ε_L is the longitudinal dispersion length [m],
δ_{ij} is the Kronecker delta function [–], and
D_e is the effective diffusion coefficient [m^2/s].

3   Applications

The FEFLOW package has been employed in numerous site characterization studies (eg Taivassalo & Mészáros, 1994), safety analyses (eg Vieno et al., 1992) and in various projects of the Äspö Hard Rock Laboratory in Sweden (eg Taivassalo et al., 1994, Löfman & Taivassalo 1995). These sites typically included 20-40 fracture zones with complex geometries and properties affecting flow and solute transport.

3.1 TVO/Posiva site investigations

In a former phase of site characterization works that was completed in 1992, three sites had to be selected from five (Kivetty, Olkiluoto, Romuvaara, Syyry and Veitsivaara) for further studies. Groundwater flow modelling was part of these investigations, and was expected to produce the following measures:

- hydraulic headfield over the investigation area,

- average groundwater flux (calculated at 100, 500 and 900 m),

- Darcy velocity and

- groundwater flow routes.

On the basis of geological investigations finite element models were built for all the sites on both regional and site scales. Field measurements of the hydraulic properties, as well as of the hydraulic responses to pumping tests facilitated the calibration of these models. Sensitivity analyses were used to establish the importance of uncertainties associated to the input data for the models. Finally, the uncertainties of the model output were discussed, also indicating a plausible range for the quantities of interest.

Kivetty, Olkiluoto and Romuvaara are the present candidate locations for a possible future spent fuel repository and more detailed site investigations have continued at these sites since 1993. The FEFLOW code has been further developed, and is currently employed in simulations addressing the saline subsurface waters at the Olkiluoto site.

3.2 Äspö HRL projects

The Äspö Hard Rock Laboratory was initiated by SKB (Sweden), and is now a joint international effort to study subsurface phenomena and deep rock behaviour.

The FEFLOW package has so far been employed in the following modelling works:

- a six-borehole pumping test (the withdrawal holes were pumped consecutively)

- simulating the hydraulic impact of the excavation of the access tunnel and shafts to the underground laboratory

- simulating the development of the freshwater lens under the island of Äspö caused by the land uplift that Scandinavia currently experiences, and

- a multiple-hole tracer retention understanding experiment.

4   Concluding remarks

The idea of modelling large, deterministic fracture zones with sets of 2D elements proved successful in numerous analyses, and has become an established modelling technology at VTT Energy. Comparison with other approaches (Gustafson & Ström, 1995) shows that complex problems can be addressed with the FEFLOW package efficiently both in terms of the quality of results and the modelling effort required for them. As well as the basic concept of the package, the following four, more general issues of methodology are highlighted.

Frequent communication with field investigators who produce the data used as model input has always proved particularly fruitful in the conceptualization phase.

Quality assurance calls for computer-readable input data for building models of any complexity. Batch (ie non-interactive) processing of as many phases of the modelling work from model development through simulations to postprocessing as possible may also contribute to reliable and reproducible results.

Large, nonetheless numerically simpler problems, like modelling hydraulic disturbances of the groundwater regime especially of transient nature can be handled effectively with the iterative solvers developed for symmetric, sparse systems. However, modelling more complicated phenomena, like convection-dominated material transport, may require a direct solver to obtain reliable results.

In order to understand the model output, post-processing the results from several aspects with tools ranging from simple summary tables to sophisticated visualization software is indispensable.

References

1. Bear, J. *Hydraulics of Groundwater,* McGraw-Hill, Israel, 1979.

2. Brooks, A. N. & Hughes, T. J. R. Streamline Upwind/Petrov-Galerkin Formulations for Convection Dominated Flows with Particular Emphasis on the Incompressible Navier-Stokes Equations, *Computer Methods in Applied Mechanics and Engineering,* **32,** 199–259.

3. Gustafson, G. & Ström, A. *The Äspö Task Force on Modelling of Groundwater Flow and Transport of Solutes. Evaluation Report on Task No 1, the LPT2 Large Scale Field Experiments,* SKB International Cooperation Report 95-05, Stockholm.

4. Huyakorn, P.S., Lester, B.H. & Mercer J.W. An Efficient Finite Element Technique for Modelling Transport in Fractured Porous Media, 1. Single Species Transport, *Water Resources Research,* 1983, vol. 19. **3,** 841–854.

5. Huyakorn, P.S. & Pinder G.F. *Computational Methods in Subsurface Flow,* Academic Press Inc., Orlando, 1983.

6. Löfman, J. & Taivassalo, V. *Simulations of pressure and salinity fields at Äspö,* SKB International Cooperation Report 95-01, Stockholm.

7. Oppe, T.C., Wayne D.J. & Kincaid, D.R. *NSPCG Users's Guide, Version 1.0,* Center for Numerical Analysis, University of Texas, 1988.

8. Taivassalo, V., Koskinen, L., Laitinen, M., Löfman, J. & Mészáros, F. *Modelling the LPT2 Pumping and Tracer Test at Äspö,* SKB International Cooperation Report 94-12, Stockholm.

9. Taivassalo, V. & Mészáros, F. *Simulation of the Groundwater Flow of the Kivetty Area,* Nuclear Waste Commission of Finnish Power Companies, Helsinki 1994.

10. Vieno, T., Hautojärvi, A., Koskinen, L. & Nordman, H. *TVO-92 Safety Analysis of Spent Fuel Disposal,* Nuclear Waste Commission of Finnish Power Companies, Helsinki 1992.

Landslide risk: 2D modelling of hydraulic consequences

N. Gendreau, P. Farissier

Cemagref, Division Hydrologie-Hydraulique, 3bis, Quai Chauveau, CP 220, 69336 Lyon Cedex 09, France

1. Presentation

A landslide may happen in the French Alps. A very huge mass of materials may fall (100 hm^3 maximum) and would fill a part of the valley, where a river flows. A natural dam may be created. A reservoir may appear upstream (backwater phenomenon). The existing of a tunnel may mitigate its volume, or even remove it. Over a certain water depth in the reservoir and constraints, the natural dam may fail (breaking up phenomenon) and the water may then flow downstream in the river valley. A strong rainfall event may be one of the possible cause of the landslide and a flood may probably be concomitant with it. Moreover, several successive material falls may happen and generate a water wave. It would have impacts upstream to a village and a housing estate.

A global analysis, involving socio-economic, historical, geographic, urban, hydrologic, geological aspects gives answers to the directly concerned inhabitants but also to the whole valley. We present here the topographic and hydraulic modelling, in order to dimension the tunnel and to evaluate the impacts of the successive material falls.

The main constraints for the tunnel building, except its cost and its building technique, concern its working during the flood, with a possible pressure flow. Moreover, for security reasons, it has to be dimensioned to minimise the water depth and the surface of the residual reservoir.

Concerning the successive material falls, we have an hydraulic shock. There is a discontinuity of the water depth. The hydraulic modelling is done in 2 dimensions in order to take into account the wave propagation and the impacts on the village and the estate housing upstream.

2. Methodology

To modelize hydraulic phenomena, it is necessary to have a complete data processing from topographical data to hydraulic results mapping. Specific tools are needed to take into account hydraulic constraints. In a 2D hydraulic modelling, these aspects are all the more important.

2.1. The mesh

A numerical topographical modelling has to be built from topographical data, taking into account hydraulic specificity.
The hydraulic elements, topologically perpendicular to the flow, are the cross-sections, on which are identified specific points (river axis, minor/major boundary...) to define trajectory functions. These lasts are cubic splines.

$$D_i = a(P_{i+1} - P_i)$$
With P : cross-section object
        D : derived vector
        a : parameter

A steady quadrangular mesh is defined by linear interpolation between cross-sections and on cross-sections. The PF3D software, developed by the Cemagref Institute, allows the building of the mesh. The PF3D software is used as a pre-processor for hydraulic calculations and as a post-processor for results mapping.

2.2. Hydraulic modelling

The calculation software RUBAR20, developed by the Cemagref Institute, allows a hydraulic simulation based on the bidimensionnal Saint-Venant shallow water equations, especially when the hydraulic characteristics temporal variations are considerable (floods for example). The software uses the finite volume method applied to quadrilaterals and triangles. The used numerical scheme is the Van Leer scheme with second order in time and space resolution. The equations are solved with an explicit method. It allows to have no downstream boundary conditions, and to have torrential flows.

Equation of Mass Conservation :

$$\frac{\partial h}{\partial t} + \frac{\partial U}{\partial x} + \frac{\partial V}{\partial y} = 0$$

Equations of Momentum Conservation

$$\frac{\partial U}{\partial t} + \frac{\partial\left(\frac{U^2}{h} + g\frac{h^2}{2}\right)}{\partial x} + \frac{\partial\left(\frac{UV}{h}\right)}{\partial y} = -gh\frac{\partial Z}{\partial x} + D\left(\frac{\partial}{\partial x}\left(h\frac{\partial\left(\frac{U}{h}\right)}{\partial x}\right) + \frac{\partial}{\partial y}\left(h\frac{\partial\left(\frac{U}{h}\right)}{\partial y}\right)\right) - gU\frac{\sqrt{(U^2 + V^2)}}{C^2 h^2} + \tau_{w1}$$

$$\frac{\partial V}{\partial t} + \frac{\partial\left(\frac{UV}{h}\right)}{\partial x} + \frac{\partial\left(\frac{V^2}{h} + g\frac{h^2}{2}\right)}{\partial y} = -gh\frac{\partial Z}{\partial y} + D\left(\frac{\partial}{\partial x}\left(h\frac{\partial\left(\frac{V}{h}\right)}{\partial x}\right) + \frac{\partial}{\partial y}\left(h\frac{\partial\left(\frac{V}{h}\right)}{\partial y}\right)\right) - gV\frac{\sqrt{(U^2 + V^2)}}{C^2 h^2} + \tau_{w2}$$

With h : Water depth

U : x axis flow rate, equal to the product of velocity and water depth

V : y axis flow rate, equal to the product of velocity and water depth

g : Acceleration of gravity

Z : Bottom height

D : Viscosity coefficient

C : Chézy coefficient ($C = K.h^{1/6}$)

τ_{w1} and τ_{w2} : Wind's constraints

To solve these equations, boundaries conditions (inlet hydrograph for example), initial conditions and Strickler coefficients have to be defined. Moreover, specific hydraulic structures have to be described usually with a Discharge rate - water depth relationship (tunnel for example).

3. Application to hydraulic consequences of a landslide

3.1. Building of the mesh

River valley topographical data come from a photogrammetric analysis on which we superimpose hypotheses concerning the landslide (7 and 25 hm³). The mountain stability is weakened, and subsequently, a material fall of several millions of square meters (1 and 2 hm³) happens.

3.2. Hydraulic hypotheses

It is difficult to be exhaustive for the study of the phenomena because the hypotheses field is considerable. Then, the field of the possible events is limited using two hypotheses :

- The natural dam created by the landslide will remain in place
- The tunnel, with a circular diameter, will be built before the landslide and will work in an optimal way.

The tunnel has to be dimensioned to leave the smallest residual reservoir as possible, with realistic diameters. The material fall will propagate a wave upstream and downstream.

Figure 1 : 3D Mesh

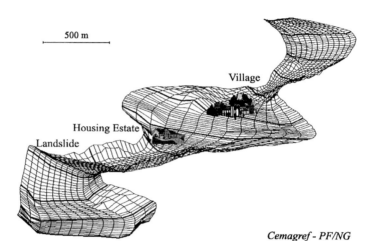

500 m

Village

Housing Estate

Landslide

Cemagref - PF/NG

3.2.1. Boundary and initial conditions hypotheses

The upstream hydraulic condition is a 200 years return period flood, hypothesis for the calculation of the ideal dimensions of every hydraulic structure for the city downstream (150 000 inhabitants).

The explicit numerical scheme allows to have no downstream boundary condition.

The other boundary conditions, corresponding to the mountainous slopes, have a reflection hypothesis for calculation.

3.2.2. The tunnel hydraulic hypotheses

The planned tunnel is dug in the left side mountainous slope. This tunnel has two functions during and after the landslide :

- To maintain the river flow in the valley
- To maintain road communications through the natural dam

Its hydraulic working after the landslide, concomitant with the 200 year return period flood, is supposed to happen like following :

① At first, there is a free surface flow in the tunnel

② Subsequently, the tunnel is pressure

The relationship between the water depth at the inlet mesh and the corresponding discharge through the tunnel results from a steady flow calculation of the Manning-Strickler equation for the free surface flow (Equation 1) and from a steady flow calculation of the Bernoulli equation using the Colebrook equation for the pressure flow (Equation 2).

$$Q = \sqrt{\frac{2gS^2(z-z_s-D)}{1+\xi+\dfrac{\lambda L}{D}}} \qquad (1)$$

$$Q = K.\sqrt{I}.S.R_h^{2/3} \qquad (2)$$

With R_h : Hydraulic Radius
        g : Acceleration of gravity
        z : Free surface water height at the tunnel inlet
        z_s : Free surface water height at the tunnel outlet
        D : Tunnel diameter
        ξ : Tunnel inlet local discharge loss coefficient

λ : Colebrook's friction coefficient ($\dfrac{1}{\sqrt{\lambda}} = -2.log\left(\dfrac{\left(21/K\right)^6}{3.71D}\right)$)

L : Tunnel length
S : wetted surface
I : Tunnel average slope
K : Tunnel Strickler coefficient

Figure 2 : the flow in the tunnel

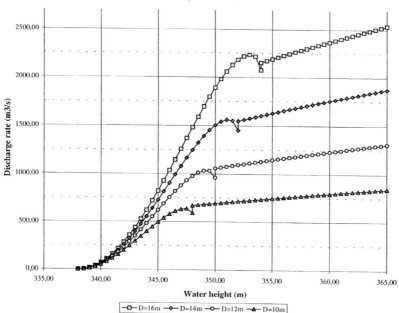

3.2.3. Material falls modelling hypotheses

It is impossible to modelize directly the material fall itself in the residual reservoir. Then, we suppose that there is a complete transfer of the potential energy. The bottom of the river valley suddenly loses its shape with the material fall and the water surface too, in the same way. From this instant, there is a wave propagating upstream and downstream.

3.3. Analysis

3.3.1. Tunnel's diameter calculation

The tunnel working depends very much on the upstream water height. As the tunnel is circular, the free surface water discharge increases and then decreases with the water height at the end (as soon as the water depth is over 0.95D). At last, when the tunnel is in charge, the discharge varies less with water height. The diameter is an important element for the tunnel capacity.

The tunnel's diameter is chosen to have a maximum water height around 346 m, 347 m upstream the natural dam.
Three different hydrological scenarii are used, function of different ways of upstream reservoirs management. These reservoirs are used for hydroelectric energy

The water height in the residual reservoir is the same whatever the chosen hydrological scenarii, if we calculate the ideal dimensions of the tunnel, that are 10 m, 13 m or 16 m. The residual reservoir stretches out from the natural dam to the housing estate. There is no consequence to the upstream village.

3.3.2. Wave propagation due to a material fall

The maximum water height in the residual reservoir (347.5 m) is reached after 29 h. From this instant, the topography is suddenly changed but the water depth is kept. The material fall volume is then taken into account and the simulation continues. A 2D calculation allows us a good description of the phenomenon.

The simulated material falls have a 30 to 40 m maximum height. The water depths on the material falls decrease, whereas at first they increase for a short while all around and then stabilise to their initial values.

Figure 3 : Outer layer results

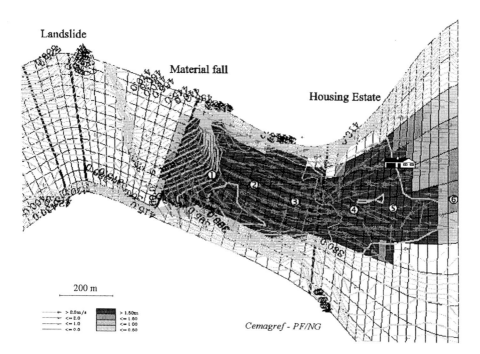

The water level recorders nearest to natural dam (1, 2 and 3), as long as the river valley is narrow, are unstable. They show a wave propagation in all the directions with secondary oscillations due to the reflection to the mountainous slopes. Moreover, the wave water depths are large, since they space out from 2 m to 10 m.

On the other hand, as long as the river valley is wider, the wave propagation appears with fewer oscillations, absorbed very quickly (water level recorder 5 and 6). The water depths are only around 0.5 to 4 m. The wave propagates well on residual reservoir but as soon as the initial water depth is shallow it stops rapidly.

The material falls have no impact to the village. Only the housing estate is concerned by the wave, but in a relatively weak way. The material fall in the residual reservoir produces an increase of the water depth in the housing estate between 30 cm and 85 cm, depending to the simulation.

The phenomenon is short. It lasts only a few minutes : between 5 mn and 10 mn. The wave velocity is high : around 10 m/s.
For the different simulations, there is no overflow over the natural dam.

Figure 4 : Material fall impacts

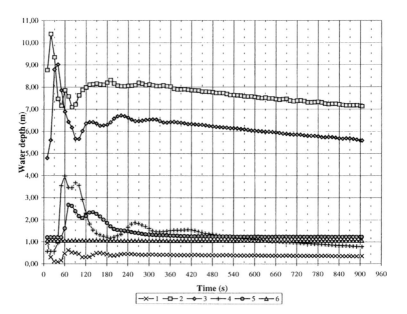

3.4. Conclusion

The different retained hypotheses were always going in the way of security. There are realistic but rather pessimistic :

- The landslide happens during a 200 years return period flood
- The tunnel is circular : there is no optimisation of the relationship depth-discharge function of the shape
- The material falls happen when the residual reservoir reaches its maximal water height.

The calculated tunnel dimensions show that with reasonable diameters, it is possible to have a residual reservoir with a shallow water depth. The retained diameters are from 10 m to 16 m depending of hydrological scenarii, with in any case a free surface flow.

For all the different studied scenarii, the numerical results show that the upstream village is not reached by the wave. Only the housing estate (around forty houses) is concerned.

4. Conclusion

Due to specific topographical conditions, the hydraulic modelling has to be done in 2 dimensions. It requires tools that can take into account this aspect for the making of the mesh, the solving of the bidimensionnal hydraulic equations and the results mapping.

And it is more complex when there is, like this application, when there are torrential flows and considerable hydraulic characteristics temporal variations

The modelling in 2 dimensions allows us to have a good description of the phenomena, especially for the wave generated by the material fall in the residual reservoir. The non-steady global modelling allows us to take into account the different components of this complex system, with an inlet hydrograph, the tunnel working and the shock of the material fall.

5. References

FARISSIER P., *Etude d'un modèle cartographique adapté à la simulation des écoulements en rivières*, Université Claude Bernard de Lyon 1, Thèse de doctorat, 1993

FARISSIER P.,GIVONE P., *Modèle de cartographie automatique des zones inondées*, 100 p, 1992

PAQUIER A., *Modélisation et simulation de la propagation de l'onde de rupture de barrage*, Université Jean Monnet de Saint Etienne, Thèse de doctorat, 1995

VILA, J.P., *Modélisation mathématique et simulation numérique d'écoulements à surface libre*, La houille blanche 6/7, pp. 485-489, 1984

SECTION 6:
METEOROLOGY AND
CLIMATOLOGY

Ozone modelling over a large city by using a mesoscale Eulerian model: Madrid case study

R. San José,[a] J. Cortés,[a] J. Moreno,[a] J.F. Prieto,[a] R.M. González[b]

[a]Group of Environmental Software and Modelling, Computer Science School, Technical University of Madrid, Campus de Montegancedo, Boadilla del Monte, 28660 Madrid, Spain
[b]Department of Meteorology, Faculty of Physics, Complutense University of Madrid, Ciudad Universitaria, 28040 Madrid, Spain

Abstract

An Air Quality Model has been applied over the mesoscale urban area of Madrid (Spain). The Model is composed on different modules: a mesoscale meteorological module REMEST which solves the Navier-Stokes equation system over the Madrid domain. This model is running under the non-hydrostatic option and is the main code of the system. The model solves the prognostic equations by using an Eulerian approach and integrated an eulerian transport equation which is solved simultaneously. The meteorological module provides the wind speed, direction, temperature and humidity on a three dimensional domain at very short time steps (a few seconds). The model uses a 14 different landuse types which are obtained by using the REMO module which uses the information provided by the LANDUSE satellite image over the area with 30 m spatial resolution. The model transport the emitted pollutants into the domain and deposit them over the domain. The emission module EMIMA takes into account the point, line and area emissions over the area. Special importance is given to the biogenic emissions which are obtained by using the satellite landuse classification for caducous, perenneal and mixed terrain. The emission module considers the EPA and CORINAIR emission factors and is 250x250 m spatial resolution and 1 hour temporal resolution. The deposition module DEPO uses the resistance approach for obtaining the deposition velocity for different species and the photochemical module CHEMA uses the CBM-IV extended mechanism which is solved numerically by using the SMVGEAR method.

The model is applied over the June, 5-9, 1995 period and the results are compared to the meteorological and pollution data obtained by the Madrid Metropolitan Network and the rural stations surrounding the city over the 80x100 km domain from the Meteorological Institute and the Madrid Autonomous Community Network. Results compare satisfactory with the observed data, however, further investigations are necessary particularly on the deposition mechanisms and the numerical algorithms. Much more computer power is also needed particularly under the new parallel platforms.

1 Introduction

The formation of ozone in the ambient atmosphere is separated in time and space from emission sources of precursors (volatile organic compounds, VOC, and NO_x). Photochemical air pollution is an environmental problem that is both pervasive and difficult to control. The necessity to have reliable three dimensional mathematical model which allow to predict the ozone three dimensional fields for 24-120 hours is particularly important in the future. Policy makers can find it an essential tool for taking decisions. Public information will be benefited for these mathematical tools.

An important element of any approach directed at attempting to improve the situation is a reliable means for predicting the air quality impacts of alternative emission control measures. While many different methods have been developed, the most comprehensive and technically defensible approach has been to use mathematical models that describe, in detail, the physical and chemical processes responsible for the chemical transformation, transport and fate of pollutants in the atmosphere. In the last two decades rapid progress has been made in this direction. Three dimensional mathematical models have been developed that simulate these phenomena to calculate the evolution of ozone and notable among these are: the Urban Airshed Model (Reynolds et al., 1979 [1], Morris and Myers, 1990 [2], System Applications, Inc., 1990 [3]), CIT (MacRae et al., 1982 [4, 5]) and the Regional Acid Deposition Model (Chang et al., 1987 [6]). The California Air Resources Board Airshed model (CALGRID) (Yamartino et al., 1992 [7]) was developed to upgrade and modernize the Urban Airshed Model (UAM) by implementing state-of-the-science improvements in the model. In the European side, the EZM EUMAC Zooming Model (Moussiopoulos, 1994 [8]) is the most important representative of the prognostic mesoscale models. This model is the base of the ANA model which is the model of this contribution. Previously, the E3DUSM (San José et al., 1994 [9]) and NUFOMO (San José et al., 1995 [10]) models are references for the ANA model. These models are mesoscale atmospheric prognostic models which are applied over Madrid Area. The former models and applications were made for one day, August, 15, 1991. This contribution show the results for a longer period and with a much more consolidated system.

ANA stands for Atmospheric mesoscale Numerical pollution model for regional and urban Areas. Figure 1 shows a scheme of the different modules of the ANA system. This model is composed on several different codes such as: the CHemical Model for Atmospheric processes (CHEMA), the DEPOsition model (DEPO), the REmote sensing MOdel (REMO), the EMIssion model for MAdrid Area (EMIMA) and the REgional MESoscale Transport model (REMEST). All of this modules are in fact independent models which can be applied for specific purposes. In this contribution the set of all this modules which is called ANA is applied for one ozone episode over the Madrid Area during the 5-9, June, 1995 period. June, 5 is Monday and June, 9 is Friday.

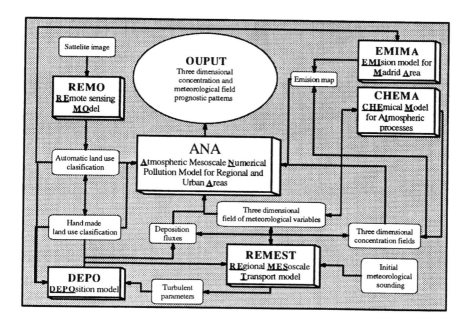

Figure 1.- ANA Model for Air Quality Diagnostic and Prognostic studies.

The mesoscale meteorological model is based on the MEMO model (Flassak, 1990 [11]) which is a non-hydrostatic three dimensional meteorological prognostic mesoscale model characterized by the completeness, consistency, robustness and flexibility. The model has a strict preservation of conservativity properties and it uses algorithms which allow estimating the numerical error associated. The Navier-Stokes partial differential equations which describe the prognostic and diagnostic features for the three wind speed components and for the temperature and humidity are solved into the MEMO model. The transport equation is based on the Eulerian approach that solves numerically the advection-diffusion equation on a staggered grid. The form of the equation we adopted is

$$\underbrace{\frac{\partial c_s}{\partial t}}_{I} + \underbrace{\frac{\partial (u_i c_s)}{\partial x_i}}_{A} = \underbrace{\frac{\partial}{\partial x_i} \left[K_{c_{ij}} \frac{\partial c_s}{\partial x_j} \right]}_{D} + \underbrace{S_{c_s}}_{E} + \underbrace{P_{c_s}(c)}_{CH_1} + \underbrace{L_{c_s}(c)}_{CH_2} \tag{1}$$

where c_s denotes the gas phase concentrations of pollutants, u_i are the wind velocity components and K_c is the eddy-diffusivity for scalars (K-theory is used here). The I term is the inertia or storage, A corresponds to the advection, D to diffusion, E to emission (for point and area sources inside the domain), and $CH1$ and $CH2$ stands for production and loss terms, respectively, regarding gas-phase chemistry.

The 3D eulerian model uses the same computational domain (σ-coordinates, expanding vertical grid, etc.) as the mesoscale meteorological model, but the time step used was $\Delta t = 30$ s. The temporal discretization adopted makes use of the 2nd Order Adams-Bashforth scheme. The vertical diffusion is implemented with the Crank-Nicholson method. For the A term a modification of the original 1D TVD (total variation diminishing) method for the three dimensional case was introduced by Harten, 1986 [12]. This method achieves a great reduction in the undesirable numerical diffusion but we should point out that this spurious diffusion is not completely removed.

The emission model is a high resolution emission inventory (spatial and temporal) and this information is provided by EMIMA. EMIMA is an emission model of atmospheric pollutants in a domain centered in the Madrid metropolitan area following the CORINE methodology. The pollutants which have been taken into account in the current version of the model (2.0) are: SO_2, NO_x and anthropogenic and biogenic VOC's (isoprene and monoterpenes). The model domain comprises an area of 80 km x 100 km with 5,108,144[11] inhabitants and more than 2,000,000 vehicles. The number of towns is 210. The city of Madrid (capital of the Nation) with more than 3,000,000 people is the main source of anthropogenic emissions and its influence is key for understanding the atmospheric environment of the area. The time resolution is one hour and the sources classification is: line, point and area. The output used in this contribution is 2 km x 2 km for use in the numerical photochemical air quality system (ANA). It is also possible to obtain the emission per grid cell per pollutant and per hour. The Area sources consider emissions from small industries, tertiary, domestic consumption, biogenic emissions and urban/suburban traffic. The line emissions consider the emissions from the national roads in the Area (six and one important secondary road in the South) and the two Road Rings in Madrid City (M-30 and M-40).

Six types of vehicles and five types of fuel are considered. Finally, the point emissions consider the emissions of large industries with more than 100 Tm SO_2/year.

The emission model EMIMA considers 14 types of land use which are also used in the mesoscale meteorological model MEMO and the resistance deposition model DEPO included into the ANA system. The land use types are: caduceus forest, perennial forest, mixed forest, olive, garden, bush, vineyard, fruit, pasture ground, rice, dry land, inland water, urban land and suburban land. We have computed the emissions from the road traffic by the following the expression:

$$E = n.r.c.f \qquad (2)$$

where E is the emitted pollutant in g, n is the number of travelling vehicles of each type, each hour, r is the mean distance travelled by vehicle in km, c is the mean consumption by vehicle in l/km and f is the emission factor for each pollutant given in g/l. The value of n is based on yearly averaged data and then distributed equally on each day. This number is corrected with a time dependent factor using information from the Traffic Department of Madrid City. We assume that the traffic during weekends and holidays is 0.65 % on respect of the traffic during working days. The following expression is used for calculating the number of vehicles per cell during working days:

$$n = people \ in \ the \ cell \ x \ 0.3 \ x \ x \ fcnv(hour) \ x \ NumVehi(type, landuse) \qquad (3)$$

where NumVehi(type,landuse) is a function of the type of vehicle and the landuse (urban or suburban where the cell is classified). The number of vehicles per person is assumed to be 0.3. Isoprene emissions depend strongly on the light and on the temperature. The effects of these two parameters are critical on the parameterization of the emission factors for isoprene and monoterpene. Increasing temperature from 25 C to 30 C can cause a 70% increase in isoprene emissions and doubling the available photosynthetically active radiation (PAR) can increase isoprene emissions by 100%. The isoprene emission factor is calculated by the following expression:

$$E\left(T, PAR\right) = \frac{E_f E_T\left(T, PAR\right)}{E_T\left(30C, 400\mu Em^{-2}s^{-1}\right)} \qquad (4)$$

where,

$$log_{10}E_T\left(T, PAR\right) = \frac{a}{1 + e^{-b(T-c)} + d} \qquad (5)$$

and $E_T(30, 400) = 20, 27$. E_f is a specified emission rate factor at 30 C and 400 $\mu Em^{-2}s^{-1}$ and it depends on the corresponding vegetation class. $E_T(T, PAR)$ is the isoprene emission rate obtained from Tingey et al.'s curves [12] for the specified conditions. The α-pinene and the sum of other monoterpenes, the emission rate does not depend on the light (they are emitted during day and night). It is considered to depend only on the temperature. The land use classification is obtained using the information provided by the LANDSAT-5 satellite. The methodology to obtain this information is based on Ormeño et al. [13] This methodology has been compared with hand-made maps and by sensitivity analysis by using the integrated air quality system ANA for land use sensitivity [14] with satisfactory results. The advantage of using remote sensing information is high because of the extraordinary man power involved in hand-made land use classification.

The deposition flux is based on the resistance approach. In this contribution a simple aerodynamic approach is used. The chemical processes are represented in this model by the CHEMA code. This code is CBM-IV based mechanism [15] which is composed by 31 active species, 68 kinetic reactions and 11 photoprocesses. This chemical mechanism [16] is based on the species approach. We have included the isoprene reactions which increase the total number of active species up to 34 and the number of reactions up to 72. the system of chemical reactions is solved by using the SMVGEAR [16] method. This is a highly accurate method where the diffusion is strongly reduced. The chemical mechanism is solved every 1800 seconds by using small and variable timesteps. The chemical system is solved for the 80 x 100 km^2 domain centered in Madrid with 2000 m spatial resolution. The vertical system has 25 different layers. Finest layers are those close to the surface. A total of 50000 cells are solved during 1800 seconds and advected and diffused by the meteorological module. Ozone appear as a secondary product of the system. The total anthropogenic VOC emissions are splitted (in volume units) following the proposal suggested by the Mechanism Comparison Group (EUROTRAC project) (personal communication, F. Kirchner, Oct'95, Fraunhofer Institut, IFU).

2 Madrid Case Study

Th domain is quite rough because of the Guadarrama mountains located in the north and northwest part of the area. This will create a general mountain and anti-mountain winds which are also combined by the Guadarrama and Jarama rivers which will create the valley and anti-valley winds. Figure 2 show the wind patterns for 0h00, 8h00, 16h00 and 24h00 for the second day of the simulation (June, 6th, 1995). These patterns show the mountain and valley flows according to the meteorological situation for the June, 5th-9th, 1995. The meteorological conditions are characterized by a local low pressure over the iberian Peninsula with anticyclonic synoptic conditions for the north and northwest part of Europe. Pressure values are around 1012 mb for all the week and no significant rain was reported for all the week. Under these conditions, the five days simulation is expected to simulate the meteorological conditions quite well because the wind flows are driven by local topography and diurnal heating cycle.

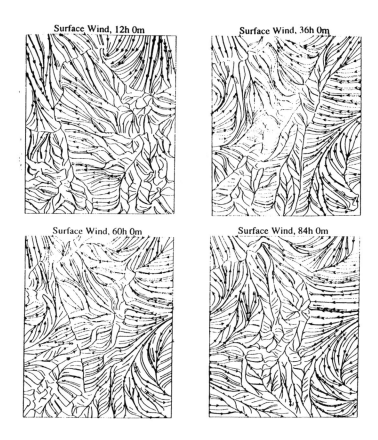

Figure 2.- Simulated wind patterns for the four days case study.

The meteorological simulation was initialized by using the wind, temperature and humidity vertical soundings at Madrid International Airport located on the northeast area of the domain. The initial values given for the different species were those obtained by the Madrid Municipality pollution network for the day before (June, 4th, 1995). Two vertical soundings were used (0h00, 12h00, June, 5th, 1995) for running all the 120 hours simulation. The simulation was performed on the Fujitsu vector machine VP-2400 of the Supercomputer Center of Galicia (Spain). The model requires 110 MB RAM and 4 CPU-hours per simulated day.

Figure 3 show the surface ozone patterns for June 6th, 1995 at 0h00, 8h00, 16h00 and 24h00. The 16h00 pattern show the high ozone values outside of the urban area because of the absence of NO_x to consume it. The white spot is located in the Madrid urban metropolitan area. Unfortunately, ozone information in rural areas is not existing yet. Some preliminary data taken by mobile stations have confirmed this simulated data. Figure 4 show the simulated and observed ozone data. The stations are located in the Metropolitan Area. The SREMP station is a station maintained by the authors of this contribution and during June, 8th, 1995 the station was not running because of a failure in the power supply. The Madrid Metropolitan network is calibrated every 20 days and our station was calibrated after the study and the results were into the 3-5 % range.

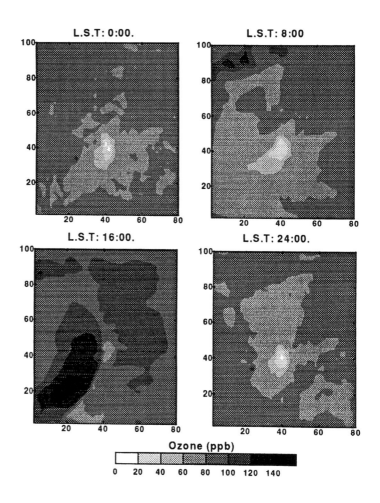

Figure 3.- Simulated surface ozone patterns for June, 6th, 1995.

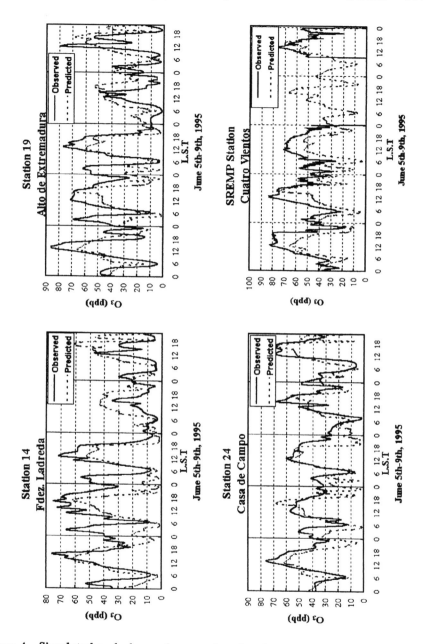

Figure 4.- Simulated and observed ozone data for the June, 4-9th, 1995 period for different stations. SREMP station is a station supported by our group.

3 Conclusions

We have presented the comparison between ozone observations in Madrid Metropolitan Area and ozone predictions by using the model ANA which is a complex Air Quality Prognostic and Diagnostic Model. The results are very hopefully however much more observations are required to compare the results and much more simulations are needed. Furthermore, the computer power requirements continue to be a critical demand for these type of studies. The use of new massive parallel computers will become a necessity in the future. The model has shown the capability to predict the ozone patterns for different locations in a mesoscale area.

Acknowledgements

The authors wish to thank the Supercomputer Center of Galicia (Spain) for the use of the Fujitsu VP-2400. The Madrid Metropolitan Council, the Madrid Environmental Agency and the Spanish Meteorological Institute for the meteorological and pollution data. Prof.-Dr. N. Moussiopoulos from the Aristotle University (Greece) and Dr. Jacobson and Dr. Turco from Los Angeles University (USA) for the MEMO and SMVGEAR base codes. The SREMP ozone data has been obtained under the EU-EV5V-CT93-316 contract by the European Union.

References

1. Reynolds S.D., Tesche T.W. and Reid L.E. "An introduction to the SAI airshed model and its usage", SAI report EF78-53R4-EF79-31, 1979.

2. Morris R.E. and Myers T.C. "User guide for the Urban Airshed model". Vol I, User's Manual for UAM (CBM-IV), EPA-450/4-90-007 A, 1990.

3. Systems Applications, Inc. "User's Guide for the urban Airshed Model", SYSAPP-90/018. Systems Applications, Inc., San Rafael, CA 94903, 15 may 1990.

4. McRae G.J., Goodin W.R. and Seinfeld J.H. "Numerical solutions of the atmospheric diffusion equation for chemically reacting flows", J. Comput. Phys. 45, 1-42, 1982.

5. McRae G.J., Goodin W.R. and Seinfeld J.H. "Development of a second generation mathematical model for urban air pollution I- model formulation", Atmospheric Environment 16, 679-696, 1982.

6. Chang J.S., Brost R.A., Isaksen I.S.A., Madronich S., Middleton P., Stockwell W.R. and Walcek C.J. "A three-dimensional Eulerian acid deposition model: physical concepts and formulation", J. Geophys. Res. 92, 14681-14700, 1987.

7. Yamartino R.J., Scire J.S., Carmichael G.R. and Chang Y.S. "The CALGRID mesoscale photochemical grid model- I. Model formulation", Atmopsheric Environment, 26A(8), 1493-1512, 1992.

8. Moussiopoulos N. "The EUMAC Zooming Model (EZM): An introduction", in *The EUMAC Zooming Model: Model Structure and applications*, Ed. N. Moussiopoulos EUROTRAC, International Scientific Secretariat, Garmisch-Partenkirchen, March, 1994, 7-21, 1994.

9. San José R., Rodríguez L., Moreno J., Sanz M. and Delgado M. "Photochemical air pollution model for the Madrid area", *Computer Techniques in Environmental Studies, Vol I: Pollution Modelling* Ed. P. Zannetti, 79-88, 1994.

10. San José R., Marcelo L.M., Moreno B. and Ramírez-Montesinos A, "Ozone modelling over a large city by using a mesoscale eulerian meteorological and transport model: Madrid case study", *21st NATO/CCMA International Technical Meeting on Air Pollution Modelling and its Applications*, Ed. American Meteorological Society, 132-139, Preprints, 1995.

11. Flassak Th. "Ein nicht-hydrostatisches mesoskaliges Modell zur Beschreibung der Dynamik der planetaren Grenzschicht", VDI Verlag, Dusseldorf, pp. 203. 1990.

12. Harten A. "On a large time-step high resolution scheme". Mathematics of Computation, 46, No. 174, pp. 379-399, 1986.

13. Ormeño S., Madrona H., Castillo M., Hernández J., Delgado M. and San José R. "Methodology for landuse type classification of multiespectral images", *Computer Techniques in Environmental Studies, Vol II: Environmental Systems*, Ed. P. Zannetti, 295-302, 1994.

14. San José R., Sanz M.A., Moreno B., Ramírez-Montesinos A., Hernández J. and Rodríguez L. "Anthropogenic and biogenic emission model for mesoscale urban areas by using LANDSAT satellite data: Madrid case study", Proceedings EUROPTO Series, *Air Pollution and Visibility Measurements, Vol: 2506* Ed. SPIE-The International Society for Optical Engineering, 286-297, 1995.

15. Gery M.W., Whitten G.Z., Killus J.P. and Dodge M.C. "A photochemical kinetics mechanism for urban and regional scale computer modelling", Journal of Geophysical Research, 94, D10, 12925-12956, 1989.

16. Jacobson M.Z. and Turco R.P. "SMVGEAR: A sparse-matrix, vectorized gear code for atmospheric models", Atmospheric Environment, 28, 2, 273-284, 1994.

A simple algorithm for the diurnal evolution of the atmospheric boundary layer height

P. Martano, A. Romanelli
CNR ISIAtA, Via Monteroni, 73100 Lecce, Italy

Abstract

A simple algorithm is described, that estimates the diurnal evolution of the height of the atmospheric boundary layer using a reduced data set of surface layer parameters, as obtained for example from standard surface automatic stations.

The routine is based on the combination of the Batchvarova-Gryning model (1991) for daytime and on both Tennekes-Niewstadt (1981) and Yamada (1979) models, in a slightly modified form, for night-time.

Hourly values of temperature, friction velocity and potential temperature scale (or sensible heat flux) in the surface layer need only to be supplied as input data.

The calculated surface inversion profile at sunrise gives an estimate of the temperature gradient on the top of the growing daytime mixed layer, so that only a rough evaluation of the lapse rate in the free atmosphere remains eventually to be given, where necessary in the final part of the growth.

The routine has shown satisfactory performances when compared with sodar measurements.

1 Introduction

The tremendous improvements of the computational capabilities of portable computers in the last years led to a change of perspectives in meteorological numerical modelling.

Complex meteorological codes, once able to run over large main frames only, are now available for desk workstations, when not even for personal notebooks, at least in the field of regional or mesoscale atmospheric modelling.

This brings the three-dimensional modelling of the troposphere towards a large portion of even non-specialised potential users, such as in hazards forecasting, agricultural applications, environmental management and pollution modelling.

In many applications involving human activities (as air pollution modelling or crops management)   a reasonable description of the structure of the atmospheric boundary layer (ABL) plays an essential role.

A great effort has been made in the last decades to achieve a description the vertical structure of the ABL trough a reduced set of scale parameters, and to connect this simplified parameterisation to the results of routine surface measurements (see for example: Stull[1], 1990). The result has been a simple one dimensional description of the turbulent ABL based on a set of height, temperature and velocity scales that can be directly related to the turbulent fluxes at the earth surface. These fluxes can be measured directly or, in turn, calculated on the basis of routine surface meteorological measurements.

A number of proposed routines already exist to obtain turbulent surface fluxes and typical scale parameters of the ABL from data of windspeed, temperature, radiation and humidity (see for example Van Ulden-Hostlag[2], 1985).

One dimensional models for the vertical structure of the ABL can then be inserted as part of full 3-D mesoscale tropospheric modelling, or used separately when full 3-D modelling can be unsuitable or time consuming.

This can happen for example when climatological modelling or screening analysis are required, where long term time series of one-point data for a limited number of (surface) quantities can make the application of full scale 3-D models quite awkward for both time consuming and data limitations problems.

Here it is presented a simple routine for the calculation of the time-depending ABL height in the whole diurnal cycle that is able to use one-point surface data only, in the framework of simple one-dimensional models.

The aim was to implement a simple fast-running code to be used in all situations in which only one point data series are available and vertical sounding is not possible (for example as input routine in climatological dispersion models for an improved dynamical estimate of the ABL height, instead of the usually employed equilibrium similarity formulas).

The routine attempts to join together three already existent models for respectively Convective Boundary Layer (CBL) growth (Batchvarova-Gryning, 1991[3]), Stable Boundary Layer (SBL) decay (Tennekes-Niewstadt, 1981[4]), and Surface Inversion Height (SIH) development (Yamada, 1979[5]): the interplay among the cyclic development of these structures allows to eliminate almost totally the need of data from vertical sounding, that are required when the models are used separately.

2 Short description of the basic models

2.1 The CBL model

The Batchvarova-Gryning[3] (BG) model is based on the well known one-dimensional approach for the growth of an inversion topped CBL due to the previous works by Carson[6] (1973), Tennekes[7] (1973) and Zilitinkevich[8] (1974).

In the Carson-Tennekes model the time evolution of the three basic unknowns describing the CBL growth model, say h (CBL height), Δ (top-layer temperature gap), and $(d\theta/dt)_{ml}$ (averaged mixed layer heating) is controlled by the potential temperature flux divergence throughout the CBL $(<w\theta>_s - <w\theta>_h) / h$, where indexes s and h indicate surface and top layer quantities, respectively.

A reasonable parameterisation of the top-layer temperature flux $<w\theta>_h$ as a function of the surface flux $<w\theta>_s$, the friction velocity u*, and the CBL height h, allows the BG model to get a single equation for the evolution of h:

$$[\frac{h^2}{(1+2a)h-2bkL}+\frac{cu_*^2T}{\gamma g[(1+a)h-bkL]}]\frac{dh}{dt}=\frac{<w\theta>_s}{\gamma} \tag{1}$$

where $L = - u_*^3 T/(kg <w\theta>_s)$ (Obukhov length), T is a representative CBL temperature g is the gravity acceleration and k the Von Karman constant.

The parameters a, b, c are experimentally determined constants (a=0.2, b=2.5, c=8 in BG), and the temperature lapse rate at the CBL top γ is the only non-surface quantity that remains to be given as model input.

2.2 The SIH model

Yamada's 1979[5] model for the surface inversion height h_i is based upon the equation for the surface cooling (θ is the potential temperature), as due to sensible $(<w\theta>)$ and radiative (R) heat fluxes:

$$d\theta(z)/dt = - d<w\theta>/dz - dR/dz \tag{2}$$

where dR(z)/dz is simply parameterised as a power-law profile proportional to the surface cooling:

$$- dR/dz = C (d\theta_s /dt) (1- z/h_i)^{r'} \qquad 0 < C < 1 \tag{3}$$

Inserting eqn. (3) in eqn. (2), writing down the potential temperature profile again in the form of a power law (indexes s and hi indicate values at surface and inversion height respectively) :

$$\theta(z) = \theta_{hi} - (\theta_{hi} - \theta_s)(1-z/h_i)^{n'} = \theta_{hi} - \Delta\theta_i (1-z/h_i)^{n'} \tag{4}$$

after a straightforward integration of eqn. (2) between $z=0$ and $z=h_i$ and the main assumption of neglecting the cooling at the inversion top (say $|d\theta_{hi}/dt|$ $<< |d\theta_s/dt|$), the following equation is obtained:

$$dh_i / dt = [h_i (d\theta_s /dt) [1-C(n'+1)/(r'+1)] - (n'+1)<w\theta>_s] / \Delta\theta_i \tag{5}$$

Yamada's choices for the parameters are n'=3, r'=1, C=1, means the choice of an equilibrium profile for the temperature (Sorbjan[9], 1986), and a surface cooling totally due to radiative effects. The choice of r'=1 appears to be due essentially to simplicity and consistency reasons (Martano-Romanelli[10]).

2.3 The SBL model

The Tennekes-Nieuwstadt[7] (TN) model for the evolution of the turbulent SBL gives rise to an exponential decay for the height h towards an equilibrium height h_e, with a time constant inversely proportional to the surface cooling $d\theta_s /dt$.

The model can be slightly modified, adding a term to the temperature equation (eqn. (9c) in TN) to make it consistent with eqn.(2), obtaining the following system of starting equations (corresponding to eqns. 9a,b,c in TN):

$$du/dt = fv - d<uw>/dz \tag{6a}$$

$$dv/dt = -f(u-G) + d<vw>/dz \tag{6b}$$

$$d\theta/dt = -d<w\theta>/dz - dR/dz \tag{6c}$$

where u and v are the windspeed components parallel and normal to the geostrophic windspeed G respectively, <uw> and <vw> the components of the Reynolds stress and f the Coriolis parameter.

Describing the vertical profiles of turbulent stress, temperature flux, windspeed and potential temperature as power laws scaled on h, and following the same steps as in TN, it can be shown that the following set of equations can be obtained (Martano-Romanelli[10]):

$$dh/dt = (h_e - h) / \tau \tag{7a}$$

$$\tau = 2(B/A)[\beta D/(2B)-1] [\Delta\theta / (d\theta_s /dt)] \tag{7b}$$

$$h_e = - Ri_f P / [A (g/T) (d\theta_s /dt)] \tag{7c}$$

where $\Delta\theta$ is the top-bottom SBL temperature difference, T the surface air temperature, $\beta=Ri_f/Ri_b$ (Ri_f is the flux Richardson number, Ri_b the SBL bulk

Richardson number), and P is the vertically averaged turbulent energy production in the SBL :

$$P = \frac{1}{h}\int_0^h v f G dz \tag{8}$$

The numerical coefficients A,B,D,β are known functions of the power exponents of the scaled vertical profiles.

It results that the presence of a (power-law) radiative term R and the actual values of the power law exponents only determine the values of the coefficients of the expressions for the equilibrium height h_e and relaxation time scale τ in the eqns. (7), but the general form of an exponential decay towards an equilibrium height h_e (like in TN model) is maintained, joint with the functional forms of h_e and τ (eqns.(18),(19),(20) in TN).

3 The proposed algorithm

3.1 Basic equations

In stable conditions the routine calculates the evolution of the surface inversion through the expressions:

$$h_i(t_i) = [h_i(t_{i-1}) - F/E] \exp[E(t_i - t_{i-1})] + F/E \tag{9a}$$

$$E = (d\theta_s/dt)[1 - C(n'+1)/(r'+1)]/\Delta\theta_i \tag{9b}$$

$$F = (n'+1)<w\theta>_s/\Delta\theta_i \tag{9c}$$

that is the analytical solution of eqn.(5) to be used between two subsequent time steps (when the surface conditions are assumed to be constants).

As remarked by Nieuwstadt[11] h_i is increasing in time with surface cooling for C=1 (and n'>r', Martano-Romanelli[10]). The temperature flux is supposed to be known from the surface data, as well as $d\theta_s/dt$ is directly calculated by the temperature (hourly) data trend.

To estimate $\Delta\theta_i$ it is supposed that the boundary layer passes trough a neutral condition at the time at which the surface temperature flux reverse its sign. At this time the potential temperature should be approximately constant through the whole layer, so that the surface temperature at the time t_{ss} when $<w\theta>_s$ becomes negative (that will be called 'sunset', while the time t_{sr} when $<w\theta>_s$ becomes positive will be called 'sunrise') represents also the top layer potential temperature.

Assuming again (Yamada[5]) that the subsequent cooling does not affect the top layer temperature, $\Delta\theta_i$ can be given as $\Delta\theta_i(t) = T(t_{ss}) - T(t)$.

The values of the parameters n', and C are chosen as in Yamada's model, while r' has been adjusted by comparison between the routine output and experimental data from sodar measurements (see section 4).

The SBL height is calculated in a similar way by the expression:

$$h(t_i) = [h(t_{i-1}) - h_e] \exp[-(t_i-t_{i-1})/\tau] + h_e \tag{10a}$$

that again is the analytical solution of eqn.(8a) between two subsequent time steps.

The expressions for τ and h_e are chosen as:

$$\tau = -(1/a_1) [\Delta\theta / (d\theta_s /dt)] \tag{10b}$$

$$h_e = (u_*/f) / [a_2(h_e/L)+a_3] \tag{10c}$$

Eqn.(10b) is formally identical to eqn.(7b) but the parameter a_1 has been determined experimentally (see section 4). This preserves the general characteristic of the exponential decay for the SBL height as well as the functional form of the time constant τ (that have a quite general validity, as noted at in the previous section), but allows a proper calibration of the model with respect to the environmental conditions. This solution has been preferred also because the use of two height scales (h and h_i) to characterise the SQL structure shows some mathematical inconsistencies in the framework of a similarity-based theory, as discussed elsewhere (Martano-Romanelli[10]), that makes somehow dubious the use of calculated coefficients A,B,D,β of eqn.(7b).

The expression for h_e, eqn. (10c), has been proposed by Derbyshire[12] as an extension of the well-known Zilitinkevich-Nieuwstadt formula for the SBL equilibrium height in strongly stable conditions (L→0) : $h_e = a_2^{-1/2} (u_*L/f)^{1/2}$. It shows the expected asymptotic behaviour in a close to neutral boundary layer (L→∞): $h_e = a_3^{-1} (u_*/f)$. As for a_1 , a_2 and a_3 have been experimentally determined, with results in reasonable agreement with the expected asymptotic values available in the literature (see section 4).

Martano and Romanelli[10] also discussed the equivalence between eqn. (10c) and eqn.(7c), that would be cumbersome to use over true experimental data ($d\theta_s/dt$ is not necessarily negative in situations in which <wθ> is negative and the evaluation of P is difficult).

$d\theta_s/dt$ was estimated as described above and $\Delta\theta$ is given by the expressions:

$$\Delta\theta = \Delta\theta_i (1-[1-(h/h_i)]^3) \quad \text{for } h<h_i$$

$$\Delta\theta = \Delta\theta_i \quad \text{for } h>h_i$$

that is, consistently with the choice of a cubic temperature profile scaled by h_i in the SIH model.

Last, the CBL height is calculated numerically solving equation (1), in which L and u* are considered known as (hourly) surface data.
The potential lapse rate at the layer top γ, to which h is quite sensitive, is estimated in the following way:

$$\gamma = (d\theta/dz)_{z=h} = (\theta_* /kh) (1+5h/L)(1-h/h_i)^2 \qquad \text{for } h<h_i \ (t_{sr})$$

$$\gamma = \gamma_{\text{free atmosphere}} \qquad \text{for } h>h_i \ (t_{sr})$$

where $\theta_* = - <w\theta>_s / u_*$ (potential temperature scale).
The former expression comes from a cubic temperature profile with the proper surface sublayer scaling added (Sorbjan, 1986[9]). The latter means that an estimate of γ in the free atmosphere (out of the boundary layer) must be given as external input.
The choice of h_i at 'sunrise' as discriminating height means that the lapse rate is calculated within the nocturnal inversion before it has been completely 'burnt out' by the growing CBL, so that the choice of $\gamma_{\text{free atmosphere}}$ generally affects only the last (and slowest) part of the CBL growth.   .

3.2 Routine characteristics.

The routine calculates the boundary layer height increment Δh in a chosen time interval $\Delta t=t_i-t_{i-1}$ changing the surface input data as well as they are available (typically hourly).
As already shown, eqn.(1) is applied in convective ($<w\theta>_s > 0$) conditions and eqns.(10) in stable ($<w\theta>_s < 0$) conditions, always starting from the last found height $h(t_{i-1})$, so that the routine can work with arbitrarily long time series of input data.
For $<w\theta>_s < 0$ eqns. (11) are also applied in the same way to calculate the SIH, that is always initialised with $h_i (t_{ss}) = 0$.
The required input values are time series of the friction velocity u_*, the surface air temperature T, the potential temperature scale θ_* (or, alternatively, the surface temperature flux $<w\theta>_s$) and an estimate of the free atmosphere lapse rate $\gamma_{\text{free atmosphere}}$, which, being quite insensitive to the boundary layer changing conditions, does not need to be given as frequently as the surface data (in practice the routine appears to work reasonably well with only one constant estimate of $\gamma_{\text{free atmosphere}}$, see next section).
An initialising value of h_0 must of course be given to start up the routine. A possibility to reduce the arbitrariness of the choice is to start the calculation at the end of the night, setting $h_0 = h_e(u_*,\theta_*,T)$ from eqn (10c).

4. Comparison with experimental data.

Figure 1 shows the results of applying the described routine to a data set obtained from surface measurements of windspeed, temperature and radiation. Experimental hourly time series were collected in July 1995 by an automatic micrometeorological station of the 'Osservatorio di Fisica e Chimica della Terra e dell'Ambiente', University of Lecce, Italy. At same time and location a phased array doppler sodar was used to obtain the turbulent height scale of the atmospheric boundary layer.

As quoted in the introduction, it is well-known (Van Ulden-Hostlag, 1985[2]) that surface windspeed temperature and radiation data allow to obtain estimates of the parameters u_* and θ_* of the surface layer. Here the routine FLXLN3 from the Beljiaars-Hostlag[13] code has been used.

The coefficients a_1, a_2, a_3, were chosen to best fit the experimental values of the turbulent height (with the physically required constraint to be all positive): the obtained values are $a_1=20$, $a_2=1$, $a_3=5$. The last two values are in agreement with the suggestions from the literature (see for example Zilitinkevich[14], 1989, where from $a_2 \sim 1$-2 and $a_3 \sim 2$-5 can be inferred).

The obtained value for a_1 implies a decay time τ typically between few minutes and one hour.

The value for the free atmosphere lapse rate was estimated by some previously obtained soundings for several days in the same month, where it appears to be almost constant in time above the boundary layer with a value $\gamma_{\text{free atmosphere}} = 0.007$ °C/m. This constant value was then selected, that appeared also to best fit the routine performance on the CBL height data.

Lacking experimental data about the SIH, the CBL growth, that depends on the SIH value at 'sunrise', has also been used to get a best value for the radiative profile exponent r'. The value r'=1.5 has then been chosen (r'=1 in Yamada's paper[3]).

From fig.1 the modelled height appears to follow the general trend of the experimental data in a reasonable way, especially in the framework of a long term 'climatological' use.

It appears that some shortcomings come from the use of the expression for h_e, that is underestimated in low wind conditions, but overestimated in strong wind conditions. No chance to find a better agreement with different values of a_2 and a_3 if the physical condition a_2, $a_3 > 0$ must hold (that is A different expression related to the expected asymptotic behaviour of h_e).should be proposed to get a better agreement, but also doubts about the comparison between measurements and model results cannot be excluded (for example: sensitivity of the sodar measurements to turbulence intermittence in strongly stable conditions).

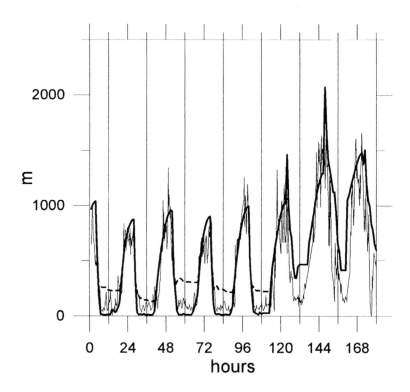

Figure 1 - Comparison between sodar measurements and model output (bold). The dashed line is an example of the computed inversion height. Time starts at 12.00 LST of 13 July 1995.

References

1. Stull R.B., 1988: *An introduction to boundary layer meteorology.*- Kluwer Academic Publishers.

2. Van Ulden A.P., Hostlag A.A.M., 1985: Estimation of atmospheric boundary layer parameters for diffusion applications. J. of Climate and Appl. Met. 24, pp.1196-1207.

3. Batchvarova E., Gryning S., 1991: Applied model the growth of the daytime mixed layer. - Boundary Layer Meteorol., 56, pp.261-274

4. Tennekes H., Nieuwstadt F.T.M., 1981: A rate equation for the nocturnal boundary layer height. - J. of the Atmos. Sci., 38, pp. 1418-1428.

5. Yamada T., 1979: Prediction of the nocturnal surface inversion height.- J. of Appl. Met., 17, pp.526-531.

6. Carson D.J., 1973: The development of a dry inversion-capped convectively unstable boundary layer. - Q. J. R. Met. Soc. 99, pp.450-467

7. Tennekes H., 1973: A model for the dynamics of the inversion above a convective boundary layer.- J. of the Atmos. Sci, 30, pp.558-567

8. Zilitinkevich S.S., 1974: Comments on 'A model for the dynamics of the inversion above a convective boundary layer'.- J. of the Atmos. Sci, 32, pp.991-992

9. Sorbjan Z., 1986: On similarity in the atmospheric boundary layer. - Boundary Layer Meteorol., 34, pp. 377-397.

10. Martano P.& Romanelli A.: 1995: A composed full-day routine for the calculation of the time-dependent height of the atmospheric boundary layer from surface layer parameters.- Submitted to Boundary Layer Meteorology.

11. Nieuwstadt F.T.M., 1980: A rate equation for the inversion height in a nocturnal boundary layer. - J. of Applied Meteorology, 19, pp. 1445-1447.

12. Derbyshire S.H., 1990: Nieuwstadt's stable boundary layer revisited. - Q.J.R. Met. Soc., 116, pp.127-158

13. Beljiaars A.C.M., Hostlag A.A.M., 1990: A software library for the calculation of surface fluxes over land and sea.- Environmental Software, vol.5, n.2, pp.60-68.

14. Zilitinkevich S.S., 1989: Velocity profiles, the resistence law and the dissipation rate of mean flow kinetic energy in a neutrally and stably stratified boundary layer.- Boundary Layer Meteorol., 46, pp.367-387.

The interaction of the aerodynamic roughness length with the atmospheric boundary-layer

S.D. Wright,[a] L. Elliott,[a] D.B. Ingham,[a] M.J.C. Hewson[b]
[a]Department of Applied Mathematical Studies, University of Leeds, Leeds, UK
[b]British Textile Technology Group, Shirley House, Wilmslow Road, Manchester, UK

Abstract

The direction of transportation of airborne pollutants is directly affected by their ability to penetrate vertically upwards into the atmosphere. Many processes, such as the density stratification within the atmospheric boundary-layer, and the buoyancy associated with the emitted pollutant, will assist or inhibit the vertical motion of the pollutant. In this paper the inducement of vertical fluid flows within the atmospheric boundary-layer, by a change in surface roughness, is considered. These fluid flows may assist the upwards vertical motion of the pollutants and hence, through the action of the Coriolis force, affect the direction of transport of the pollutant. It is shown that a small change in surface roughness induces vertical fluid flow in all regions of the atmospheric boundary-layer.

1 Introduction

The structure of the earth's boundary-layer is similar to the boundary-layer that can be generated in a wind tunnel, being highly turbulent and hence diffusive, with a distinctive inner and outer region. Within the inner region the flow is mainly influenced by the physical characteristics of the terrain whereas in the outer region, unlike any laboratory generated boundary-layer, the earth's rotation is important. This is of crucial importance when considering the dispersion of airborne pollutants since the Coriolis force causes the mean wind direction to change with height. This suggests that the ability of a pollutant to penetrate vertically upwards into the outer region of the atmospheric boundary-layer will have significant consequences on the direction in which it is transported.

If an accurate prediction of the distance and the direction in which a pollutant is transported is to be achieved then the flow within the atmospheric boundary-layer needs to be accurately modelled. The underlying terrain will directly affect the transport of the pollutant by the interaction of the boundary-layer with the large scale topology but also, in an indirect way, through the aerodynamic roughness which influences the variation of the mean wind speed direction with height. In the outer region the Coriolis/pressure balance is important and hence a mechanism to account for this must be included within the model.

To model the fluid flow over terrain where there is a step change in roughness, the continuity equation and the Reynolds averaged momentum equations have been solved by the method of Patankar [4] and Van Doormal et al. [9]. A first-order closure scheme, namely the mixing length formulation, a one and a half closure scheme, the k-l formulation, and a second-order closure, the k - ε formulation, have been used to model the eddy viscosity.

2   Governing Equations

Using the following non-dimesionalisations

$$x = L\tilde{x}, \quad y = L\tilde{y}, \quad z = H\tilde{z}, \quad \rho = M\tilde{\rho}/L^3, \quad P = M f G/L^2 \tilde{P}$$
$$u = G\tilde{u}, \qquad v = G\tilde{v}, \qquad w = W\tilde{w}, \qquad t = \frac{L}{G}\tilde{t}, \qquad v = v_c \tilde{v} \tag{1}$$

where G is the magnitude of the geostrophic wind, L and H suitable horizontal and vertical length scales, W a vertical velocity scale, v_c a characteristic viscosity and f the Coriolis parameter then if the position within the flow domain is defined as $\underline{x}$ and the fluid velocity as $\underline{u}$, the continuity equation becomes

$$\underline{\tilde{\nabla}}.\underline{\tilde{u}} = 0 \tag{2}$$

where W is defined as

$$W = \frac{GH}{L} \tag{3}$$

Using the shallow convection Boussinesq approximation of Spiegal and Veronis [7] the density can be written in the form

$$\tilde{\rho} = \tilde{\rho}_o(\tilde{z}) + Pa\, \tilde{\rho}_1(\tilde{x},\tilde{y},\tilde{z}) \tag{4}$$

where $\tilde{\rho}_o$ is the mean density, $\tilde{\rho}_1$ an induced perturbation and Pa an undetermined parameter. Similarly to the density, it can be deduced that the pressure can be written as

$$\tilde{P} = \int_{z'=0}^{z'=\tilde{z}} \tilde{\rho}_o(z')g'dz' + \tilde{P}_1(\tilde{x},\tilde{y}) + Ro\, \tilde{P}_2(\tilde{x},\tilde{y},\tilde{z}) \tag{5}$$

where the first term on the right hand side of expression (5) is due to the hydrostatic pressure, the second the geostrophic pressure and the third a perturbation that is zero far upstream and downstream of the roughness change. Substituting expression (5) into the momentum equations and Reynolds averaging, yields

$$
\begin{aligned}
\frac{\partial \tilde{u}}{\partial \tilde{t}} + \tilde{u}_j \frac{\partial \tilde{u}}{\partial \tilde{x}_j} &= -\frac{1}{\tilde{\rho}_o}\frac{\partial \tilde{P}_2}{\partial \tilde{x}} + \frac{\left(\tilde{v} - \tilde{v}_g\right)}{Ro} - \frac{Q\tilde{w}\cot(\varphi)}{Ro} + \\
&\quad \frac{1}{Re}\left\{ \frac{\partial}{\partial \tilde{x}}\left(\tilde{v}\left(\frac{\partial \tilde{u}}{\partial \tilde{x}}+\frac{\partial \tilde{u}}{\partial \tilde{x}}\right)\right) + \frac{\partial}{\partial \tilde{y}}\left(\tilde{v}\left(\frac{\partial \tilde{u}}{\partial \tilde{y}}+\frac{\partial \tilde{v}}{\partial \tilde{x}}\right)\right) + \frac{1}{Q^2}\frac{\partial}{\partial \tilde{z}}\left(\tilde{v}\left(\frac{\partial \tilde{u}}{\partial \tilde{z}}+Q^2\frac{\partial \tilde{w}}{\partial \tilde{x}}\right)\right) \right\}
\end{aligned}
\tag{6}
$$

$$\frac{\partial \tilde{v}}{\partial t} + \tilde{u}_j \frac{\partial \tilde{v}}{\partial \tilde{x}_j} = -\frac{1}{\tilde{\rho}_o} \frac{\partial \tilde{P}_2}{\partial \tilde{y}} - \frac{(\tilde{u} - \tilde{u}_g)}{Ro}$$

$$\frac{1}{Re}\left\{ \frac{\partial}{\partial \tilde{x}}\left(\tilde{v}\left(\frac{\partial \tilde{v}}{\partial \tilde{x}} + \frac{\partial \tilde{u}}{\partial \tilde{y}} \right) \right) + \frac{\partial}{\partial \tilde{y}}\left(\tilde{v}\left(\frac{\partial \tilde{v}}{\partial \tilde{y}} + \frac{\partial \tilde{v}}{\partial \tilde{y}} \right) \right) + \frac{1}{Q^2}\frac{\partial}{\partial \tilde{z}}\left(\qquad Q^2\frac{\partial \tilde{w}}{\partial \tilde{y}} \right) \right\}$$

(7)

$$Q^2\left(\frac{\partial \tilde{w}}{\partial t} + \tilde{u}_j \frac{\partial \tilde{w}}{\partial \tilde{x}_j} \right) = -\frac{1}{\tilde{\rho}_o}\frac{\partial \tilde{P}_2}{\partial \tilde{z}} - \frac{\tilde{\rho}_1}{\tilde{\rho}_o}g' + \frac{Q}{Ro}\tilde{u}\cot(\varphi) +$$

$$\frac{Q^2}{Re}\left\{ \frac{\partial}{\partial \tilde{x}}\left(\tilde{v}\left(\frac{\partial \tilde{w}}{\partial \tilde{x}} + \frac{1}{Q^2}\frac{\partial \tilde{u}}{\partial \tilde{z}} \right) \right) + \frac{\partial}{\partial \tilde{y}}\left(\tilde{v}\left(\frac{\partial \tilde{w}}{\partial \tilde{y}} + \frac{1}{Q^2}\frac{\partial \tilde{v}}{\partial \tilde{z}} \right) \right) + \frac{1}{Q^2}\frac{\partial}{\partial \tilde{z}}\left(\tilde{v}\left(\frac{\partial \tilde{w}}{\partial \tilde{z}} + \frac{\partial \tilde{w}}{\partial \tilde{z}} \right) \right) \right\}$$

(8)

where $\left(\tilde{u}_g, \tilde{v}_g \right)$ are the components of the geostrophic wind and φ a given latitude on the earth. The non- dimensional parameters are

$$Q = \frac{H}{L}, \; Ro = \frac{G}{fL}, \; Re = \frac{GL}{v_c}, \; v_c = \frac{H^2 G}{L} \; and \; g' = \frac{gH}{fGL}$$

(9)

Three turbulence closure models have been employed, namely the mixing length, the k-l and the k-ε models. The form of the mixing length used in the mixing length and k-l models is

$$\frac{1}{\ell} = \frac{1}{\kappa z} + \frac{1}{\ell_o}$$

(10)

which takes into account that the mixing length behaves linearly near the ground but reaches a maximum size, ℓ_o, in the Ekman layer. In expression (10) κ is the von-Karman constant and equal to 0.4. To incorporate this behaviour into the k-ε model the modification of Apsley [1] is employed.

Unlike Blom et al. [2], Rao et al. [4], Shir [6] and Taylor [8], no boundary-layer approximations have been used, as these are not valid in the vicinity of the roughness change and it is this region that is of most interest. Also a pressure perturbation to the global geostrophic and hydrostatic pressure balance has been included in the form of $\tilde{P}_2(\tilde{x},\tilde{y},\tilde{z})$ in expression (5). This takes into account the local pressure fluctuations that are induced by the underlying topology and the turbulent structure of the atmospheric boundary-layer.

The governing equations (6), (7) and (8) have to be solved subject to the appropriate boundary-conditions. At the upper boundary of the solution domain, i.e. $\tilde{z} = 1$, the horizontal components are set equal to the magnitude of the geostrophic wind components

$$\left(\tilde{u}, \tilde{v}\right)\Big|_{\tilde{z}=1} = (1,0) \tag{11}$$

and a zero gradient boundary condition enforced on the vertical velocity, i.e.

$$\partial\tilde{w}\Big/\partial\tilde{z}\Big|_{\tilde{z}=1} = 0 \tag{12}$$

Physically this means that the vertical flow is constant with height and can be deduced because the streamlines within the upper reaches of the atmospheric boundary-layer are parallel. At the lower boundary of the solution domain the logarithmic law of the wall is used to calculate the wall shear stress, i.e.

$$\tilde{v}\,\partial\tilde{u}\Big/\partial\tilde{z}\Big|_{\tilde{z}=0} = \tilde{v}\tilde{u}_1\Big/\tilde{z}_1\,\ln\!\left(H\tilde{z}_1\Big/z_{roughness}\right) \tag{13}$$

where the subscript '1' represents the evaluation of the quantities at the first mesh point within the solution dom   As there is no normal fluid flow through the lower boundary of the solution domain

$$\tilde{w} = 0\Big|_{\tilde{z}=0} \quad \text{for} \quad -\infty < \tilde{x}, \tilde{y} < \infty \tag{14}$$

For flow over uniform terrain the velocity components in the $\tilde{x}$ and $\tilde{y}$ directions will be unchanged and no vertical flows will be induced in the fluid flow field. Hence by setting $\partial/\partial\tilde{x} = \partial/\partial\tilde{y} = 0$, a one-dimensional solution to the governing differential equations (6), (7) and (8) was obtained. This was then used as the inflow boundary condition for the two-dimensional situation. Far downstream of the roughness change the fluid flow should behave as if it was flowing over uniform terrain with the modified surface roughness value and hence zero normal gradient boundary conditions are applied at the outflow.

3   Results

The governing differential equations (6), (7) and (8), along with either the mixing length, the k-l or the modified k - ε models were solved numerically using the SIMPLEC method of Patankar [4]. In order to illustrate the results obtained we set the parameters as follows

$$z_o = 0.02\text{m} , \ G = 20\text{ms}^{-1} , f = 1.176 \times 10^{-4}\text{s}^{-1} , \ L = 85000\text{m}$$
$$H = 1900\text{m}, \quad m = z_1/z_o = 5, \quad \ell_o = 30\text{m} \tag{15}$$

and used a Cartesian mesh locally refined around the area of the roughness change.

Taylor [8] noted that the wind speed adapted to the new surface roughness very rapidly near the surface and then spread up from the surface with increasing values of $\tilde{x}$ suggesting that an internal boundary-layer is forming. The work of Blom et al. [2] further validated that an internal boundary-layer forms after the roughness change and was an extension of the work of Panofsky and Townsend [3], who were unable to obtain an adapted internal boundary-layer after the roughness change. The magnitude of the horizontal wind speed as a function of height is shown in figure 1 for the mixing length, the k-l and the modified k-ε models.

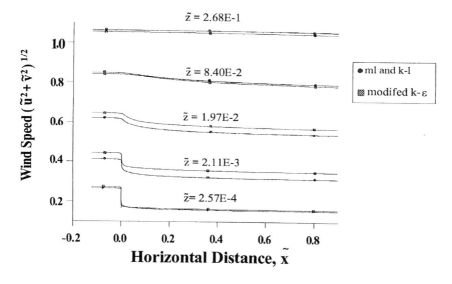

Figure 1. The variation of the magnitude of the horizontal wind speed with height, $\tilde{z}$, for the mixing length, the k-l and the modified k-ε models for the parameters
Ro = 2, m = 5, Re = 2001.

It is observed that the variation in the magnitude of the horizontal wind speed as a function of height, when calculated from the mixing length and the k-l models, are in excellent agreement whereas the results obtained from the modified k - ε model are in reasonable agreement with those of the mixing length and k-l models. The formation of an internal boundary-layer can be seen to be occurring in figure 1 with the boundary-layer adapting to the new surface roughness quickly near the surface. Within the surface-layer the predicted wind speed, calculated form the modified k - ε model, is slightly higher at any given height and horizontal distance than is predicted from the k-l and mixing length, but in the Ekman layer, it is slightly lower.

Figures 2 and 3 show the variation in the vertical fluid velocity, $\tilde{w}$, at the first mesh point above $\tilde{z} = 0$ and at a height $\tilde{z} = 1.77 \times 10^{-2}$ respectively.

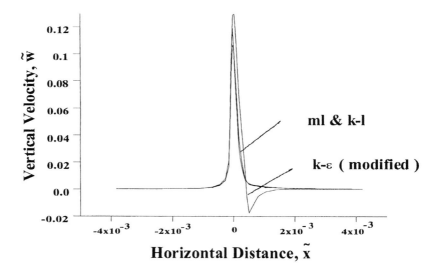

Figure 2. The variation of the vertical fluid velocity, $\tilde{w}$, with horizontal distance, $\tilde{x}$, at the
first mesh point above $\tilde{z}$ = 0 for the parameters Ro = 2, m = 5 and Re = 2001.

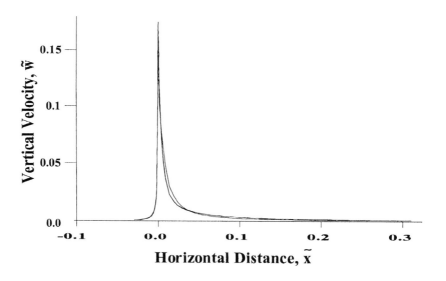

Figure 3. The variation of the vertical fluid velocity, $\tilde{w}$, with horizontal distance, $\tilde{x}$, at a
height, $\tilde{z}$ = 1.77 x10^{-2} for the parameters Ro = 2, m = 5 and Re = 2001.

Figure 2 shows that only a small amount of fluid is displaced vertically near the ground.
This is confined to a very small region just before the change in surface roughness,

where the fluid is forced upwards. A negative vertical velocity can be seen from the results obtained using the modified k-ε model just after the roughness change and this corresponds to the presence of a separation region after the roughness change.

Figure 3 shows that a far more significant amount of fluid is being forced upwards in the region above the surface-layer of the boundary-layer. The difference in the mass flow rates between the two undisturbed states, i.e. the one-dimensional solution for a roughness z_a and for a roughness z_b, indicates that fluid flows across the upper boundary of the solution domain.

Figure 4 shows the flow of fluid across the upper boundary of the solution domain with the effects of the roughness change felt before it has occurred with fluid flowing out of the boundary-layer in the region $\tilde{x} < 0$.

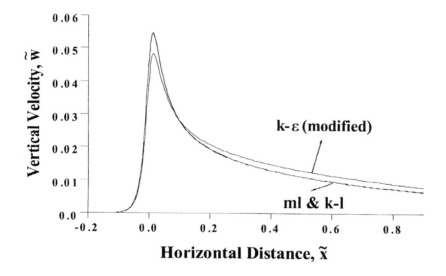

Figure 4. The variation of vertical fluid velocity, $\tilde{w}$, at the upper boundary of the solution domain, $\tilde{z} = 1$, as a function of the horizontal distance, $\tilde{x}$, for the parameters Ro = 2, m = 5, Re = 2001.

The maximum vertical component of the fluid velocity, which passes across the upper boundary of the solution domain, occurs in the region immediately after the roughness change but doesn't decay quickly, indeed there is still a significant flow of fluid across the upper boundary of the solution domain at $\tilde{x} = 1$.

4   Conclusions

Using specialised techniques for solving the full Navier-Stokes equations, the effects of a change in surface roughness have been investigated. It was found that an internal boundary-layer formed after the roughness change with small amounts of fluid, in the vicinity of the roughness change, being displaced vertically upwards near the ground. In the surface-layer a larger amount a fluid was displaced vertically upwards, whereas at the top of the solution domain, i.e. $\tilde{z} = 1$, fluid was still passing across this boundary at large distances downstream of the roughness change. These induced vertical velocities may affect the direction of the transportation of any airborne pollutant by the action of the Coriolis force.

5   References

1. Apsley D.D. PhD Thesis, 1995, Department of Mechanical Engineering, University of Surrey.
2. Blom J. and Wartena L. The influence of changes in surface roughness on the development of the turbulent boundary-layer in the lowest layer of the atmosphere, *Journal of the Atmospheric Sciences*, 1969, 26, 255 - 265.
3. Panofsky H.A. and Townsend A.A. Change of terrain roughness and the wind profile. *Quarterly Journal of the Royal Meteorological Society*, 1964, 90, 147 - 155.
4. Patankar S. V. *Numerical heat transfer and fluid flow*, Hempshire, Washington D.C., 1980.
5. Rao K. S, Wyngaad J.C. Coté O. R. The structure of the two-dimensional internal boundary - layer over a sudden change of surface roughness. *Journal of the Atmospheric Sciences*, 1974, 31, 738 - 746.
6. Shir C.C. A Numerical computation of air flow over a sudden change in surface roughness, *Journal of the Atmospheric Sciences*, 1972, 29, 304 - 310.
7. Spiegal E.A. and Veronis G. On the Boussinesq approximation for compressible fluids, 1960, *Journal of Astrophysics*, 1960, 131, 442 - 447.
8. Taylor P.A. On wind and shear stress above a change in surface roughness. *Quarterly Journal of the Royal Meteorological Society*, 1969, 95, 77 - 91.
9. Van Doormal J. P. and Raithby G. D. Enhancement of the SIMPLE method for predicting incompressible flow, *Numerical Heat Transfer*, 1984, 7, 147 - 163.

An automatic pluviogram reducer

R. Leonardi, C. Rafanelli, T. Montefinale
CNR Instituto di Fisica dell'Atmosfera, 31 P. le Luigi Sturzo, 00144 Rome, Italy

Abstract

The meteorological data recording (rain, temperature, pressure and so on) using mechanical recorders occurs drawing the trend, trace, on a graduated paper by a pen-nib linked to instrument transducer. Manual techniques or hand-digitiser readings to reduce those traces are used. The vast amount of work to reduce a lot of pluviograms with high precision becomes prohibitive.

This paper shows a software, written in 'C' and running under Extended DOS, for fully automatic reading of scanner digitised pluviograms. So that it is possible to recover the large papery archives existing in many organisations and institutions.

The pluviogram digitised in a graphic file is submitted to a series of elaborations concerning image-enhancement, geometric transformation, scaling and filtering of the background, to clear the graduated scale and other spurious marks isolating the rain trace. Finally the rain amount versus time is recorded on file.

Key words : rain-gauge, pluviogram, image processing, software

1 Introduction

Nearly all countries make some effort to collect and publish the measurements of rainfall, collected with specified instruments named rain-gauges or pluviometer. Their number and the density varie greatly from country to country and even within the same country. The most dense networks have been established, as may be expected, in areas where there is great economic significance in local variation of rainfall and generally seem to be related to the density of population. In the more developed and densely populated countries as Italy, there have been fairly good networks for 70 to 100 years. Obviously after a few years the manuscript records become voluminous and their management becomes a problem.

The goal of the work is to give a new digital method to read pluviographs for long historical series management.

In this paper only the tipping bucket rain-gauge paper recorder data are considered. This instrument write a mark on the chart a continuous trace in time stepped every 0.2 mm, height of the fallen rain volume, fig. 1,3,4. To know the intensity of rainfall or rate of precipitation, usually expressed in millimetre per hour, an operator reads the trace on these charts and makes up a report. This method adds random errors valued at about a 7% due to the training and tiredness of operators.

Also for a net of a few instruments this is a very long and tedious work so that considering the general purpose of data, the daily amount of rain only is generally reported. This for many scientific uses is a hard limitation that this digital method contributes to overcome, in fact the pluviographic charts are read by the software with a few minutes step and data recorded on a daily file. So that users can examine every time interval, multiple of unit interval.

The paper describes a software to produce rain data archives starting from digitised pluviogram image even if of short quality. To standardise and speed the process the image, obtained putting the hard-copy of a pluviographic chart in a scanner device, the values of brightness, contrast and resolution are always fixed.

Scanned image is recorded and is part of the archive for possible further analyses. The procedure treats the image full automatically and measures the rain intensity without adding significant errors besides those due to the pluviometre itself.

2 Software technique

This software elaborates a digital image to clear as much as possible the background signal to evidence the rain trace. The rain amount is measured versus time and the results are recorded in a file.

To produce the original digitised image, the pluviogram paper, approximately 7*20 cm, is obtained by a monochromatic scanner at 150 dpi, with 256 grey scale tones. The resolution and image greys depth are chosen as a trade off between the image quality and a relatively small file dimension. With this setting a typical pluviogram is represented by a 600*1800*1 byte image, producing a GIF file about 200-500 Kbyte.

The flow chart of the software here proposed has 7 mainly item :

1 - reading and showing image
2 - image enhancement and region selection
3 - image de-warping
4 - grid removing

5 - thresholding
6 - rain signal extraction
7 - final result and file recording

2.1 Reading and shoving image

At the beginning the pluviogram is shown on the screen, fig. 1. The resolution is settled at 800∗600 pixels so that the image is shown in 1:1 pixel to pixel scale. At right hand there is a *tote area* showing parameters of the figure and other informations related to the elaboration progress. The operator can scroll the image and read the co-ordinates and value of each pixel pointing it by mouse.

2.2 Image enhancement and region selection

The overall quality of available original pictures to be scanned is often poor. In fact reproductions from microfilm or photo-copy are usually available instead of original pluviograms. So the pictures may be too light or too dark,

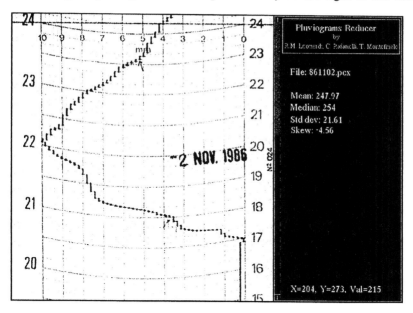

Figure 1 : An example of screen hard copy. The *tote area* on the right shows the image parameters.

much or less contrasted and also zoned because not uniformly illuminated. Moreover somebody may have wrote, by a pencil, fig. 1, fibre pen, fig. 4a, or stamp, fig. 3a, some notes or marks. Marks and not uniform areas affect the histogram of the frequencies of the grey tones. This is important feature because the analysis of this histogram makes possible to separate the pixels of the rain

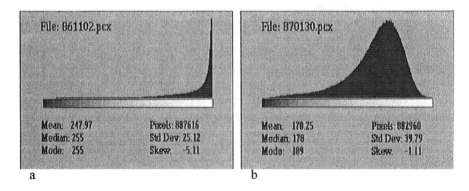

Figure 2: Two histograms of grey tones of scanned pluviograms. Picture a) is linked to fig. 3a. Histogram b) is linked to fig. 4a. The horizontal scales are 0 to 255 from left to right. In vertical axes the frequencies.

signal, in the dark region of histogram, from the background, on lighter values, Di Zenzo [5].

As an example the fig. 2 shows the histograms of the picture 3 and 4. The fig. 2a is linked to the light image of figure 3a with little of dark tones and many white values. On the contrary the fig. 2b shows grey tones shifted on the darker level.

As first step a routine scans the image to search for wide dark zones and remove those setting the values at a local mean of a larger area. Than the equalisation of the background level is obtained subtracting the local mode to each pixel and adding a fixed value of 180 that is the mode of a good and well scanned original, the effects are evident in figures 3b and 4b, respectively compared with 3a and 4a. The dimensions of area to computing the local mode is a compromise between processing stability and computing time.

A modified average filter, with 3∙3 mask to reduce the noise without blurring contours, complete the operation of preliminary treatment, Cappellini [4].

The enhanced image is shown and the operator selects the region to be analyse choosing by the mouse the four extreme of a quadrangular shaped area, in witch fall the scales of rain and time. This is the first of the only two operator's requested actions.

2.3 Image de-warping

The scanned image can be not correctly squared for the optical aberrations and the grid scale of pluviogram is represented in cylindrical co-ordinates. So that is convenient to square and re-scale into Cartesian co-ordinates the image (de-warping), Russ [3]. Moreover to avoid some problems linked to the algorithm used for grid removing, discussed below, the image area is normalised in size multiple of power 2, for a typical 24 hours pluviogram to 512∙2048 pixels. The

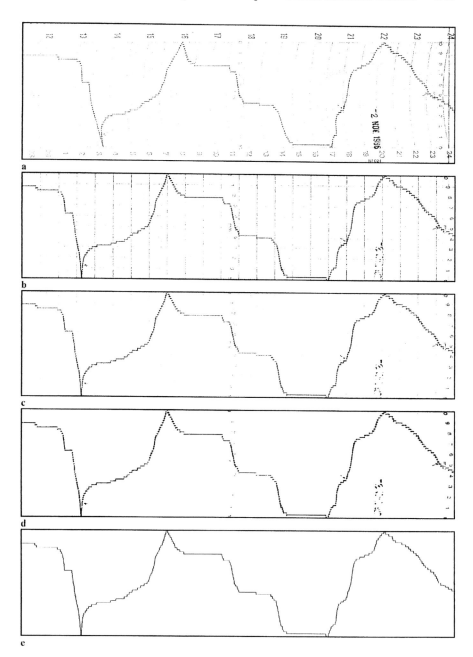

Figure 3 : Example of pluviogram whit light background. From top to bottom ; a) original scanned pluviogram; b) after image enhancement; c) result of grid removing; d) effect of applying the optimum threshold ; e) resulting rain trace.

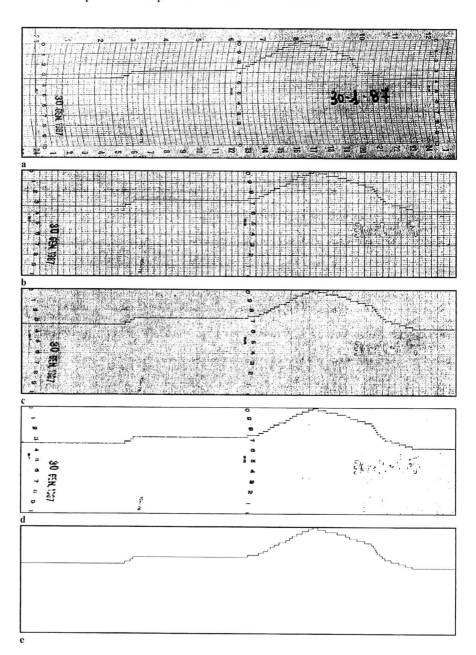

Figure 4 : Example of pluviogram whit dark and not uniform background. From top to bottom ; a) original scanned pluviogram; b) after image enhancement; c) result of grid removing; d) effect of applying the optimum threshold ;e) resulting rain trace.

figures 3b and 4b show the pluviograms after enhancement of § 2.2 and de-warping algorithms application.

2.4 Grid removing

The background grid now consists of periodic vertical and horizontal lines and the rain trace is a non periodic signal. It is possible clear away the grid lines applying a bi-dimensional Fast Fourier Transform (FFT) technique, Press et al.[2]. So that it is possible to build a filter to clear the only periodic marks. Moreover, in the same step, a 2nd order Butterworth low pass filter, Russ [3], is applied to reduce noise and ringing effects, Melli [1]. The fig. 3c and 4c show the results of filtering.

2.5 Thresholding

The operation of thresholding transforms a multi-grey level image in a dual colour map. The all grey pixels whit value lower than threshold are *objects*, the higher are *background*, Russ [3]. The questions are how to define the optimum level of threshold and if there is a single threshold good all over the image.

From samples examined in this paper it is possible to note that also after the image pre-processing it is not possible clearly identify the levels of the *objects*, because the images are still noisy and the background level still diversified over the image itself.

The optimal threshold value can not be fixed for all images nor unique in the same image, Kittler et. al. [6]. It is necessary to compute it from the image itself, considering the statistical properties of grey levels for various zones in each image. In this software, the thresholds of 4 areas in the image corners are chosen and a bi-linear interpolation solve the problem. In figures 3d and 4d are show the effects of the thresholding in the two sample pluviograms.

2.6 Rain signal extraction

At this point of analysis, the image is constituted by the rain signal and by spurious marks due to residuals of original grid, pencil and pen notes. It needs other filters to remove the isolated pixels, filling the little holes of traces and separate the true rain trace from other *objects*.

A process of skeleton, Russ [3], is used to reduce the *objects* to single pixel width traces. Following, the splitting of branching is made. Then techniques of pattern recognition are utilised to clear away the irregular shapes.

The true trace pursuing begin at one of extreme of image, finding the first pixel of a trace and following it until the last pixel. Here the nearest start points of other traces are searched applying a strategy that move forward or laterally, considering the relative distances, the previous and next directions and the reversing of motion at end-scale of the traces. All the *objects* that for own morphological characteristics are not possible to be considered rain trace are set *background*. Finally all remaining *objects* are jointed to produce the continuous rain trace, see fig. 3e and 4e.

2.7 Final result and file recording

The digital rain signal is showed on the screen, over-imposed on to the original image to evaluate the goodness of result. So the operator can choose to accept the pattern and an ASCII file is recorded with the name of station, the date and cumulative rain versus time every 5 minutes. This time interval is chosen because the amount of rain is indicated by the ordinate change, fig 1,3,4, between two consecutive pixels along time axis. Taking into consideration the digitalisation of image, five minutes correspond at about 7 pixels of abscissa. The size of ASCII file produced is typically of 3 Kbyte per day.

In this release, if the operator rejects the pattern the file will not be created. In the tests carried out this case has never occurred. However, in the future, it will be implemented a manual method to permit to locate correctly bad parts of trace.

3 Concluding remarks

It is to be noticed that today also other kinds of pluviometers are operated, those store observations directly on solid state memory or transmit them in real time. But the procedure of this paper is yet useful as for the tipping bucket nets as for management of the historical series existing, some of them is a merely storage of pluviographic charts because not completely read.

The software is tested with many pluviograms obtained as A4 size hardcopy from microfilms of Italian Air Forces archive, having good performances in precision and time. The actual release, running on a 486 DX2-66 PC under Extended DOS, allows a fully processing of a scanned 24 hour pluviogram in about 4 minutes.

The future evolution of the software aims at the improvement of the techniques of pattern recognition, at the enhancement of thresholding and at portability on other platforms.

The algorithms here implemented are written in ANSI "C" and with very little modifications, may be applied to other kind of recorders as thermometer or barometer and so on.

Acknowledgements

The authors are grateful to personnel of C.N.M.C.A. of Italian Air Force for the pluviograms supply and for good suggestions on the analysis. The authors are grate also to PFT2 that supported the research.

Bibliography

1 Melli, P. , *L'elaborazione digitale delle immagini*, Ed. Franco Angeli, Milano Italy, 1991.

2 Press W.H., Teukolsky S.A., Vetterling W.T. & Flannery B.P., *Numerical recipes in C*, Cambridge Univ. Press, Cambridge UK, 1992.

3 Russ J. C., *The image processing handbook*, CRC Press, London UK, 1992.

4 Cappellini V., *Elaborazione numerica dlle immagini*, Ed. Boringhieri, Torino Italy, 1985.

5 Di Zenzo S., Advances in image segmentation, *Image and vision computing*, vol 1, n° 2, pag. 93-97, 1983.

6 Kittle J, Illingworth, Foglein J., & Paler K., An automatic thresholding method for waveform segmentation, Digital Signal Processing - 84, ed. V. Cappellini & A.G. Constantinides, *Proceeding of Digital Signal Processing 84*, Florence, Italy, 1984, Elsevier Sci. Pub., 1984.

Sea-breeze modelling of middle Adriatic sea coast

G. Latini, F. Polonara, G. Vitali

Department of Energetics, University of Ancona, Via Brecce Bianche, I-60100 Ancona, Italy

Abstract

Italian power plants scenario is about to change because of the possible introduction of new small-sized mixed-fuel plants to be scattered around the territory. In order to evaluate the increased pollution risk connected with these new plants, the present work analyzes the effects of geographical driving forces of a particular environment, the Adriatic-sea coast, characterised by a series of valleys facing the sea. As a first step the analysis has been carried out by searching the steady state solution for a multiple valley system under sea- and river-breeze conditions. To perform this analysis the software STAR for thermofluids modelling has been employed. Afterwards the flow field elements have been tracked to show if a pollutant can migrate from one valley to the adjacent ones.

1 Introduction

Italy is experiencing a gradual renewal of power plant scenario involving the planning of a number of little mixed-combustible power plants. Quite a large amount of italian coasts consists of adjacent valleys facing the sea and their lower section are all potential sites for those plants.

Air pollution in a coastal area is controlled in periods with significant solar radiation by land-sea breeze circulation. In the case of a valley facing the coast, sea-breeze couples to river-breeze (Zhong et al.[12]).

Dynamics of a sea-breeze circulation system has been summarized by Hsu[4] and it is still a subject of wide interest from both modelling and observation viewpoints (De Oliveira & Fitzjarrald[2], Kuwagata et al. [5], Prakash et al. [8]).

Object of interest has been also the analysis of the Coastal Internal Boundary Layer dynamics and depth (Venkatram[11], van Wijk et al. [10]) although little attention has been paid to sea-breeze spatial structure.

Stull[9] has pointed out as synoptic flow may influence the breeze convective cell, changing its circular path in a spiral-like one, allowing a pollutant to be transported along the coast-line.

The present work aims at analysing the hypothesis that a similar effect could result from the particular topography of a series of adjacent valleys, and so forth that breeze events could be related to pollution migration from one valley to the next one.

2 The Problem

The middle Adriatic sea coast is characterized by a series of adjacent valleys with the axis approximately orthogonal to the coast-line. The basic topography has been sketched through a simple model without branchings and just a gorge at its end. The idealized valley has a base slope, a sinus-like cross-section and lateral hills whose height is growing linearly from the coast-line to the gorge.

The valley-ground altitude, z_g, is described by the following expression:

$$z_g = x\left\{\frac{b}{l}+\frac{(m-b)}{2l}\left[1-\cos\left(\pi\frac{|y-0.5w|}{0.5w}\right)\cdot\alpha\right]\right\} \qquad (1)$$

where:
  x = distance from the coast,
  y = distance from valley side,
  w = valley width (20 km),
  l = valley length (40 km),
  b = valley elevation at gorge (500 m),
  m = mountains elevation near the gorge (1000 m),
  α = gorge opening parameter (1).

The first task of the present study has been to obtain a steady state solution for a free-convection problem whose driving forces are the temperature differences among sea surface and valley ground, the slope-driven buoyancy and a low intensity geostrophic wind. For this purpose the surface temperatures have been respectively set to 10°C (the sea) and 30°C (the soil) and the inversion layer (at 3000 m) has been set at 0°C. A gestrophic wind blowing toward 45° at that height was set to 3 m·s⁻¹.

In order to study the periodical structure of the system, five equal valleys have been put side by side and a cyclic boundary condition has been chosen to take into account the lateral flow re-entering between the first and the last valley side. The resulting computational domain is 100 km large (5 valley 20 km each) and 80 km long (40km on-shore + 40km off-shore) with a rectangular grid of 2 km by 2 km. The vertical node spacing has been obtained by dividing the distance from ground to the inversion cap into 30 equal intervals (so that over the sea, the z intervals are all of 100 m). The bottom of the domain is shown in figure 1.

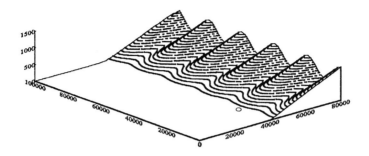

Figure 1: Topography of the bottom of computational domain.

3 The Analysis

The problem has been analysed with the aid of STAR (from Computational Dynamics Ltd[1]), a recently developed package for thermofluids analysis based on a finite volume solver written in FORTRAN and a powerful graphic user interface for pre-processing (setting up geometry, computational mesh and computation controls) and post-processing (results displaying and flow elements tracking).

3.1 The Equations

STAR is able to treat the integral conservation equations for mass (ρ is dry-air density), momentum (U is averaged value and u' is its turbulence component), enthalpy ($H = c_p \cdot \Delta T$ is averaged value and h' is its turbulence component):

$$\frac{\partial}{\partial t}\rho + \frac{\partial}{\partial x_j}\left(\rho U_j\right) = 0 \tag{2}$$

$$\frac{\partial}{\partial t}\left(\rho U_i\right) + \frac{\partial}{\partial x_j}\left(\rho U_j U_i - \tau_{ij}\right) = -\frac{\partial}{\partial x_i}p + \rho g \delta_{ij} \tag{3}$$

$$\frac{\partial}{\partial t}\left(\rho H\right) + \frac{\partial}{\partial x_j}\left(\rho U_j H - \tau_{ij}U_i - F_j\right) = \left(\frac{\partial}{\partial t} + U_j\frac{\partial}{\partial x_j}\right)p + s_H \tag{4}$$

where the Newtonian turbulent fluid constitutive relation is assumed:

$$\tau_{ij} = \mu\left[\frac{\partial}{\partial x_j}U_i + \frac{\partial}{\partial x_i}U_j - \frac{2}{3}\frac{\partial}{\partial x_k}U_k\delta_{ij}\right] - \rho\cdot\overline{u'_j u'_i} \qquad (3a)$$

and where stress term for momentum and enthalpy turbulence are:

$$-\rho\overline{u'_j u'_i} = \mu_t\left[\frac{\partial}{\partial x_j}U_i + \frac{\partial}{\partial x_i}U_j - \frac{2}{3}\frac{\partial}{\partial x_k}U_k\delta_{ij}\right] - \frac{2}{3}\rho K\delta_{ij} \qquad (3b)$$

$$F_j = C\frac{\partial}{\partial x_j}T - \rho\overline{u'_j h'} = C\frac{\partial}{\partial x_j}T + \frac{\mu_t\left[\partial H/\partial x_j\right]}{Pr_t} = (C+C_t)\frac{\partial}{\partial x_j}T \qquad (4a)$$

where C and C_t are standard and turbulent thermal conductivity coefficients. Coriolis force has been included only by the above mentioned geostrophic wind (as a top boundary layer condition).

Enthalpy equation includes wall conduction sources (wall at costant temperature T_w) with a standard thermal resistance (r_w) model:

$$s_H = \frac{(T - T_w)}{r_w} \qquad (4b)$$

For the purposes of this analysis, air moisture has been neglected, so the ideal gas law has been used ($\Re$ is dry air gas costant):

$$\frac{p}{\Re} = \rho T \qquad (5)$$

The supplied turbulence energy $\left(K = \overline{u'_i u'_i}/2\right)$ conservation equation accounting for buoyancy is:

$$\frac{\partial}{\partial t}(\rho K) + \frac{\partial}{\partial x_i}(\rho U_i K - G_i) = \mu(P_{ij} + P_{ib}) - \rho\varepsilon - \frac{2}{3}\mu_t\left(\frac{\partial}{\partial x_j}U_i + \rho K\right)\frac{\partial}{\partial x_j}U_i \qquad (6)$$

where:

$$G_j = (\mu + \mu_t)\frac{(\partial H/\partial x_j)}{Pr_t} \qquad (6a)$$

$$P_{ij} = \left(\frac{\partial}{\partial x_j} U_i + \frac{\partial}{\partial x_i} U_j - \frac{2}{3} \frac{\partial}{\partial x_k} U_k \delta_{ij} \right) \frac{\partial}{\partial x_j} U_i \tag{6b}$$

$$P_{ib} = \frac{-g \, \partial \rho / \partial x_j}{\rho \, Pr_t} \tag{6c}$$

and Pr_t is the turbulent Prandtl number.

The turbulent dynamic viscosity μ_t and turbulence dissipation ε are related to turbulence length scale L by the following relations (K-L model, by Launder & Spalding[7]):

$$\mu_t = f_\mu \mu^{1/4} \rho K^{1/2} L \tag{6d}$$

$$\varepsilon = \frac{c_\mu \mu^{3/4} K^{3/2}}{L} \tag{6e}$$

where f_μ and c_μ are empirical constants.

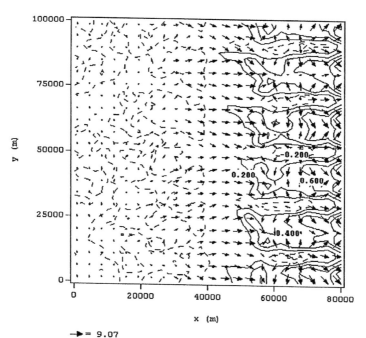

→ = 9.07

Figure 2a: Flow field at z-node 2; arrows show the x-y velocity components and contours the vertical velocity component; dashes indicate a negative value.

The equations above are discretised in STAR by the finite volume method (Hirsh[3]) and have been solved using the PISO algorithm (Issa[6]) applying the suggested relaxation and solver tolerances for buoyancy driven problems.

The system variables initialization is based on a Mass Consistent built-in procedure; since no K field initial values were available, its value was set by the system to 2.5%.

3.2 The Stationary Solution

The resulting steady state flow field is shown by 2 sections (figure 2a-b) parallel to valley ground corresponding to z-nodes 2 and 25 (that is 100 m and 2400 m on the sea), they allow to identify the main convective cells related to sea-breeze ranging from about 15 km off-shore to 15 km on-shore. The ground level (fig 2a) wind intensities grow significantly from the coast-line (x=40 km) to a maximum of about 9 m·s^{-1} near the front.

The inland direction is mantained to the thermals line (the front) where arrows show the flow divergence and continuos contour lines (some vertical velocity values are reported) show the growth of vertical wind component.

The counter sea-breeze aloft can be observed in figure 2b; the dotted lines show clearly the descent zone off-shore (near x=15 km, 5 km from coastline).The intensity of the flow field is in agreement with usually observed sea-breeze values (Hsu[4]).

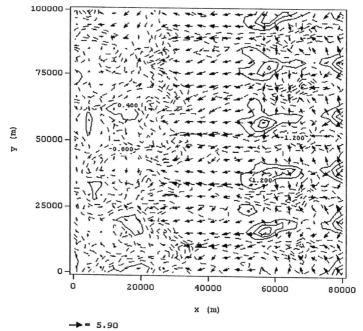

Figure 2b: Flow field at z-node 25; arrows show the x-y velocity components and contours the vertical velocity; dashes stand for a negative value.

3.3 Tracking of Flow Field

To analyse the flow field a procedure has been developed to follow the path of a weightless particle inside the discrete field, based on the integration of the equation:

$$\frac{\partial}{\partial t} x_p = U(x_p) \tag{7}$$

The method is based on a variable time-step approach evaluating the time-interval so as to move the particle of a distance smaller than the local cell dimension (each direction):

$$\Delta t = \min\left\{\frac{U_i}{\Delta x_i}; \ i = 1, 2, 3\right\} \tag{8}$$

where U_i is the i-velocity component and Δx_i the local i-width of the cell.

Three particles have been tracked, each one in a different valley; all have been released at an initial height of 100m, at centerline of the valley and 10 km inland. The path projections obtained are shown in figure 3 and 4a,b (respectively x-y, y-z, x-z). The view from above (fig 3) shows mainly the effect of the boundaries on the simulation: the path of the third particle shoulld be the cleaner from the numerical viewpoint.

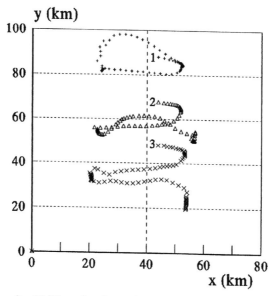

Figure 3: X-Y projection of the fluid element paths;
numbers (1 to 3) show the start points.

Both back and lateral projections (4a,b) show the characteristic circular shape of a convection cell; they also show clearly how a particle could move to an adjacent valley to a location similar to its initial one, an effect due to the fact that valley flow fields are not insulated one another.

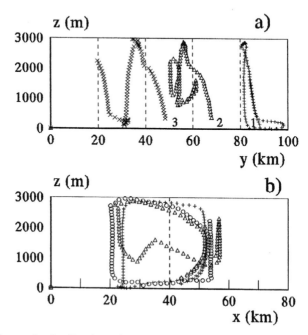

Figure 4 a,b: Back and Lateral projection of the paths;
numbers 1 to 3 show the initial particle locations.

4 Conclusions

Even if the analysis is exclusively based on flow simulation, it has shown how pollution in valleys facing coastal areas can not be considered a local problem.

Actually pollution particles can migrate from a valley to an adjacent one, hills being unable to operate any valley flow separation.

The present analysis is the first step in the study of a particular area and a test for a simulation tool, STAR, which appears to be powerful enough both to obtain qualitative diagnostic information and to plan further experimental measurements.

5 References

1. Computational Dynamics Ltd, *STAR- Ver. 2.3 Manual.* Computational Dynamics Ltd. 317 Latimer Road - London W106RA, 1995.

2. De Oliveira, A.P., Fitzjarrald, D.R., The Amazon River Breeze and the Local Boundary Layer: II linear analysis and modelling, *Boundary Layer Met.*, 67,75-96, 1994.

3. Hirsh C. *Numerical Computation of Internal and External Flows*, Vol. 1, John Wiley & Sons, 1990.

4. Hsu, S.A. *Coastal Meteorology*, Academic Press, Inc., 1988.

5. Kuwagata, T., Kondo, J. & Sumioka M., Thermal effects of the Sea Breeze on the structure of the Boumndary Layer and the Heat Budget over Land, *Boundary Layer Meteorology* 67,119-144, 1994.

6. Issa, R.I. Solution of the Implicitly discretised fluid flow equations by Operator-Splitting, *J. Comp. Phys.*, 1986, 62, 66-82.

7. Launder, B.E., & Spalding, D.B.   The Numerical Computation of Turbulence Flow, *Comp.Meth. in Appl. Mech. & Eng.*,1974, 3, 269-279.

8. Prakash, J.W.J. Ramachandran R., Nairy, K.N., Gupta, K.S., Kunhikrishna, P.K., On the structure of the sea-breeze fronts observed near the coastline of Thumba - India, *Boundary-Layer Met.* 59,111-124, 1992.

9. Stull, R.B. *An Introduction to Boundary-Layer Met.*, Kluwer Academic Publishers, 1988.

10. van Wijk, A.J.M., Beljaars, A.C.M., Holstag, A.A.M., Turkenburg, W.C., Diabatic Wind Speed profiles in coastal regions: comparison of an Internal Boundary Layer model with observations, *Boundary Layer Met.* 51,49-75, 1990.

11. Venkatram, A., An Examination of methods to estimate the height of Coastal Internal Boundary Layer, *Boundary Layer Met.* 36,149-1565, 1986.

12. Zhong, S., Leone, J.M. & Takle E.S. Interaction of sea breeze with a river breeze in an area of complex coastal heating, *Boundary-Layer Met.*, 1991, 56, 101-139.

High resolution, near field, meteorological wind analysis system for emergency response and other applications

R.M. Cionco, J.H. Byers

U.S. Army Research Laboratory, ATTN:AMSRL-BE-S, White Sands Missile Range, New Mexico 88002-5501, USA

Abstract

In most emergency response activities, it is essential to know how the local meteorological conditions and terrain can affect inadvertent and accidental releases and other operations. A better representation of the near field meteorology will ensure a more accurate and dependable on-site handling of situations of hazardous materials. This means including the presence and effects of land feature morphology (i.e. vegetation, buildings, etc) as well as terrain in your analysis. To transform meteorological models/codes into information for response and assessment during emergency activities, some seven tasks may be implemented with a high resolution, micrometeorological code in-hand. The user can then analyze areas affected as having adverse and favorable impacts at and downwind of the hazardous release site. The results of the effects analysis can be interpreted on-site for customized use and assessment information. This wind simulation and effects analysis system was used successfully during the planning stages and on-site, real-time during the conduct of the recent MADONA Field Study. The described methods can easily be adapted to a variety of other field activities where the wind is influenced by the presence and configuration of variable terrain, and building and vegetation features within the domain.

1. Introduction

A high resolution meteorological simulation and analysis system of codes implemented on a high performance lap top was tested during a recent field study in a situation similar to an emergency response-type operation. The high resolution wind model and its complimentary effects (impact) analysis method, called HRW, 1.1, was run real-time[1] using on-site, real-time meteorological measurements to initialize the wind code as both smoke and SF_6 plumes and puffs were released within a very localized area. The purpose of this field study, called MADONA[2], was to collect meteorological and aerosol concentration data required to test and evaluate our library of models.

With the simulation and analysis system installed on our lap top along with the digitized terrain elevation and land feature morhology data, we responded to the release of smoke and tracer materials by preparing model initialization input data from on-site, real-time meteorological data and then simulated the high resolution, mean quantity wind and temperature fields for the near field domain of the field study. Doing each of the simulation and analysis tasks real-time while the source created a continuous plumes and instantaneous puffs, in essence was emulating the meteorological activities and components for an emergency response scenario (obviously without the usual expediency and confusion of the real event). The same approach can be applied also to similar sets of field activities and operations when aerosol are released into the atmosphere and their downwind behavior must be quantified and analyzed for their control or the protection of a community in the near field domain.

Other types of field activities and operations that can be addressed with these methods are: the behavior of aerial-released aerosols and drift analysis of pesticide spray; the management of smoke produced during forest service slash and burn operations; the behavioral studies of the release and travel of spores and disease in agricultural and forested areas; the general release of aerosol that must be quantified, analyzed, and monitored for control, health, and safety, and similar other operations.

Note that near field in this concept refers to areas of the order of 5Km by 5Km with the source located within these boundaries. High resolution equates to a computational grid of 100m in x and y for the 5Km by 5Km area, but may also range from 40m to 400m as the domain size changes from 2Km by 2Km to 20Km by 20Km.

2. Discussion

To transform meteorological model/code output into information for response and assessment during emergency and the above mentioned activities, the following tasks may be satisfied with a high resolution micrometeorological wind flow model/code in-hand appropriate for emergency response: (1) identify digitized terrain data sets and digitized land feature morphology data sets (optional); (2) identify source of meteorological input data (fixed installation, safari, or central office); (3) simulate high resolution wind fields for emergency events; (4) analyze terrain, land feature, and thermal effects upon the wind and other meteorological variables; (5) use criteria appropriate for the "operational or hazardous" event to classify and further identify areas of concern/impact about the hazardous release; (6) visualize, interpret and depict the meteorological fields and effects analyses; (7) analyze areas of effect for adverse and favorable impacts at and downwind of the hazardous release site. The user can then interpret the results of the effects analysis on-site for customized use and assessment information.

This method can easily be adapted to a variety of emergency response situations where the wind is influenced by terrain and land feature morphology. It also can be a useful simulation tool in preparedness planning, training, and later for post-analysis. This wind simulation and effects analysis method[1] was used successfully during the planning stages and again in the conduct of the MADONA Field Study[2]. With the digitized terrain/land feature data installed in the lap top, on-site field observations were transformed real-time into code inputs for high-resolution wind field simulations.

3. Model and method

Two microscale models can be considered for this high resolution analysis of flow over complex terrain and optional within and above land use features. The first code, HRW[3], analyzes horizontal wind and temperature fields over a local area of some 5 km by 5 km. The second code, C-CSL[3], couples the same HRW surface layer flow with canopy flow for a similar domain. This meteorological coupling adds vertical structure to the simulated horizontal wind field. For the general purpose being addressed here, the HRW code is chosen to provide the simulated meteorological fields for the effects/impact analysis described herein because it addresses a high-resolution, small domain for near field analyses with minimum inputs and because it is also fast, friendly, and easy to implement and view results.

3.1 The high resolution wind model
The HRW model is a two dimensional, diagnostic, time independent model that simulates the wind flow over a grided area of 5 km by 5 km with a nominal spacial resolution of 100 meters. The code is initialized with single values of time-averaged surface layer wind speed, wind direction, temperature, and an estimate of buoyancy by applying these values at each grid point in the computational array. Simulation results are obtained by a direct variational relaxation of the wind and temperature fields in the surface layer. The solution is reached when the internal constraint forces imposed by the warped terrain surface, thermal structure and requirement for flow continuity are minimized. The procedure makes use of Gauss' Principle of Least Constraints [4] which requires these forces to be minimized in order to satisfy the equations of motion. When applied to the surface layer, this procedure also requires the use of empirical wind and temperature profiles. As mentioned earlier, the computational domain size can range from 2 km by

2 km to 20 km by 20 km with grid resolutions varying from 40 meters to 400 meters respectively. The vertical thickness of the computational layer is designed to be 1/10th the magnitude of the grid size. A grid size of 100 m therefore produces simulated fields at the 10 m level as a result of integrating through the thickness of this layer.

The initialization of the wind code requires surface layer values of wind speed, wind direction, and temperature at one location at the 10 m level in addition to an upper air temperature-pressure-height profile to estimate the surface domain's bulk buoyancy. Terrain elevation data and land use feature information in a digital format are also required. For output, the following simulated x,y fields are computed at the top of the 10 m layer: (a) u and v wind components, (b) potential temperature, (c) friction velocity, (d) wind power law exponent, and (e) the Richardson Number. The vertical wind profile also can be calculated at each grid point through the computational layer. Note that all simulated values represent a five minute average and are also valid for a period of up to one hour for this microscale surface layer domain.

An example showing a comparison of a simulated horizontal mean wind field and the concurrent observed field of 14 sites is shown in Figure 1 for a computational grid of 100 m in x and y. A streamline analysis of the simulated vector field is presented rather than the vector field. The agreement is quite reasonable, however, the reader must be advised that although the direction agreement is in-scale, the observed speed vectors are not to the same scale as the simulated field.

3.2 The effects/impact method

The effects/impact analysis method is developed to identify and quantify the degree and character of the effect (impact) of wind, terrain, and land use features upon a field event or operation. The concept of 'impact' refers to the resultant effect of the wind speed increasing or decreasing notably or the wind changing direction significantly from the initialization field somewhere in the domain because of direct interaction with the changing terrain and land use features and thermal buoyancy. The method does this by applying operational criteria to the resultant effect of the simulated field versus the initial field from the HRW code. During the MADONA field study, criteria were established specifically in regard to the release site of smoke and SF_6 aerosol and the downwind travel area of the plumes and puffs for a reasonable measurements setup of concentration amounts.

The method involves analyzing the newly simulated field in comparison to the initialized field:

EFFECT = FINAL SIMULATED FIELD - INITIAL FIELD

More specifically, the wind speed effect and the wind direction effect are determined separately. The Wind speed Effect (E_{ws}) is the difference between the values for the final simulated wind speed field (S_s) and the initial wind speed field (S_i):

$$E_{ws} = S_s - S_i \qquad (1)$$

The Wind Direction Effect (E_{wd}) is the difference between the values for the final simulated wind direction field (D_s) and the initial wind direction field (D_i):

$$E_{wd} = D_s - S_i \qquad (2)$$

A qualitative assessment of the Total Effect can be made next combining the effects of the two difference fields. To quantify these effects, a set of appropriate operational criteria is established for the emergency or similar type of activities.   By example, three levels of Effect are defined for MADONA diffusion experiments:

LIGHT if:

$$E_{ws} < 10\% \text{ of } S_i, \quad \text{and} \quad E_{wd} < 10° \text{ of } D_i \qquad (3)$$

MODERATE if:

$$E_{ws} \text{ is } 10\% \text{ to } 50\% \text{ of } S_i, \quad \text{and} \quad E_{wd} \text{ is } 10° \text{ to } 30° \text{ of } D_i \qquad (4)$$

SEVERE if:

$$E_{ws} > 50\% \text{ of } S_i, \quad \text{and} \quad E_{wd} > 30° \text{ of } D_i \qquad (5)$$

The visualization scheme is next implemented to overlay a color coded map of areas of these three impact levels onto plots of the terrain and land feature morhology and wind streamlines. It should be noted that you, the user, set the "operational" criteria and the visualization technique to fit your specific requirements be they emergency response or otherwise.

4. Results

To illustrate the impact (effects) analysis technique, several simulations made during the MADONA field study are selected to represent examples of two levels of impact. Examples of light-to-moderate and severe impact cases are presented to demonstrate where it was feasible to release and measure the aerosol plumes and puffs.

Light-to-moderate impact areas are shown in Figure 2 initialized with a mean wind of 3.0 m/s from the west at 250 degrees which nearly parallel the terrain features causing a light to moderate effect of terrain upon the meteorology. The plot for the impact analysis of this case indicates that a wide area of light wind effects (white areas without symbols) has opened up in the area of the Bowl where the aerosol generators were located. Moderate impact areas (gray and hatched areas) are observed immediately to the south of the Bowl, along the ridge and hill tops. Several small areas of severe impact (black and cross-hatched areas) are indicated at the highest elevations. The interpretation is that in the more uniform, lower elevation of the Bowl, the plumes would behave as expected and measurement systems (lidars) could be located with confidence. The tracer plume would not be accelerating or decelerating notably (<10%) to cause dilution , pooling, or accumulation. Nor would it change direction (<10°) significantly and therefore cause possible endangerment to unexpected areas. In the limited moderate and severe areas of this simulation, the plume behavior would sufficiently change travel in both speed and direction to produce areas of greater effect or adverse impact. These more notable changes could define where adverse conditions such as pooling and stagnation are and also alert the emergency team where vulnerable locations may be.

A severe impact case is shown in Figure 3. This simulation indicates severe adverse wind effects in large areas throughout the entire domain. The initial mean wind speed and wind direction for this simulation was 1.0 m/s and 180 degrees. As would be expected for these light speed conditions, the streamline plot indicates meandering winds, twisting and turning and flowing from the south. Although an attempt to release "puffs" was made, the tests were aborted because a dependable wind speed and direction could not be maintained for a reasonable time period. The field notes show this to be a no test day. The simulation verifies this to be a correct decision. The adverse impact plot indicates virtually no contiguous area large enough that would be suitable for a valid test. In the case of inadvertent and accidental releases, the entire area could be vulnerable to serious problems because of the high degree of occurrence of acceleration, deceleration, and highly variable directional changes. Once the source is located, one can then consider a more limited area of travel and, therefore, the potential behavior of the 'dispersion' of the aerosol plume.

5. Conclusion

It is shown for activities of an emergency nature as well as other applications that the near field situation can be simulated and analyzed in an efficient and highly representative way given appropriate, customized operational criteria. The field testing of HRW and its complimentary impact analysis method was successfully accomplished in real-time during a recent set of meteorological and diffusion field trials. The implementation of this simulation and analysis approach onto a high performance lap top allows you to make a quick assessment on-site and real-time of the impact of the wind and terrain/land use features effect upon your emergency response activity. As shown by the above simulations, solutions, and analyses, the wind simulation and impact technique can be a useful tool for on-site use . The system can work equally as good for support of other activities that are notably influenced by their interactions with wind and terrain effects. Activities such as the management of smoke released during forest service slash and burn operations, the aerial release and drift analysis during pesticide spray operations, studies of the transport of spores and disease over forested and agricultural areas, the general diffusion of aerosols for areas of pooling, accumulation, dilution, and higher concentration and dosage locations specifically for the near field and other similar applications.

References

1. R. M. Cionco and J. H. Byers, High-resolution wind field simulation run real-time the MADONA field study, *Proceedings of the 11th Symposium on Boundary Layers and Turbulence,* Charlotte, NC, American Meteorological Society, Boston MA, USA, 1995.

2. R. M. Cionco et al, An overview of MADONA: A multi-nation field study of high-resolution meteorology and diffusion over complex terrain, *Proceedings of the 11th Symposium on Boundary Layers and Turbulence,* Charlotte, NC, American Meteorological Society, Boston, MA, USA, 1995.

3. R. M. Cionco, Modeling Wind fields and Surface layer Profiles over Complex Terrain and Within Vegetative Canopies, *Forest Atmosphere Interactions*, ed B. A. Hutchison and B. B. Hicks, D. Reidel Publishing Co., Netherlands, 1985.

4. C. Lanczos, *The Variational Principles of Mechanics,* 2nd Edition, The University of Toronto Press, Toronto, Canada, 1962.

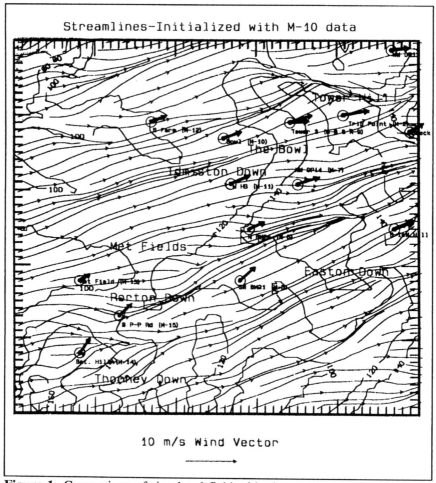

Figure 1. Comparison of simulated field with observations.

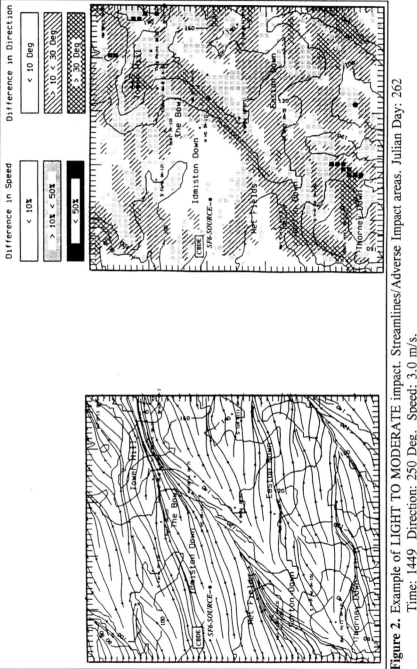

Figure 2. Example of LIGHT TO MODERATE impact. Streamlines/Adverse Impact areas. Julian Day: 262
Time: 1449   Direction: 250 Deg.   Speed: 3.0 m/s.

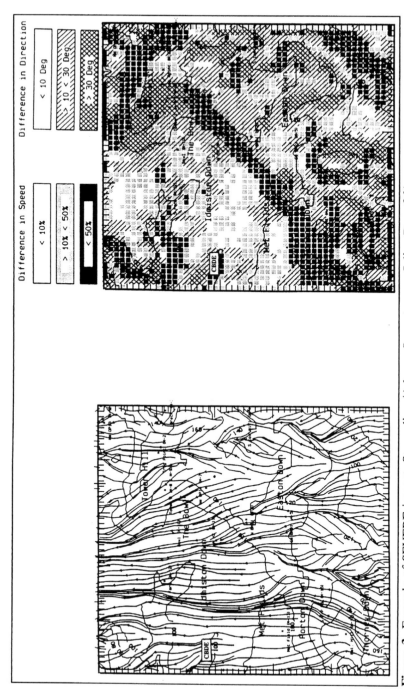

Figure 3. Example of SEVERE impact. Streamlines/Adverse Impact areas. Julian Day: 261
Time: 1415   Direction: 180 Deg.   Speed: 1.0 m/s.

SECTION 7:
FLUID DYNAMICS

Modeling environmental flows with adaptive methods

D.W. Pepper, D.B. Carrington, Z. Shi

Department of Mechanical Engineering, University of Nevada, Las Vegas, NV 89154-4027, USA

Abstract

An h-adaptive finite element model is used to calculate environmental transport problems associated with atmospheric dispersion or groundwater transport in saturated or unsaturated porous media. A coarse mesh is first created; the mesh refines and unrefines locally as the concentration front advances in time. Petrov-Galerkin weighting is used for the advection terms, along with mass lumping. The model is written in C/C++ and runs under WINDOWS on PCs; a WINDOWS based mesh generator also accompanies the model, allowing the user to quickly create an initial mesh with appropriate boundary conditions. Current research includes simulating windfields and species transport over the Nevada Test Site, as well as groundwater flows at the Savannah River Site and the proposed Yucca Mountain Repository Site.

1. Introduction

The employment of adapting, unstructured meshes permits one to accurately solve large problems with a reduced number of nodal points (degrees of freedom). This is accomplished by concentrating (refining) nodes in those regions where most activity takes place, and unrefining in regions where solutions are smooth. Efficient mesh adaptation depends upon the choice of element type, handling of interface transitions, and rapid mesh refinement and unrefinement operations during the transient solution.

Adaptive techniques have been historically used to calculate compressible flows and shock locations in the aerospace field[1]. Inclusion of these techniques in several commercial fluid engineering programs are now being advertised in the literature. Research into using adaptive methods is now being investigated in earnest; however, much yet needs to be done before the techniques can be widely used with ease.

The use of h - adaptation for environmental flows and species transport appears particularly attractive[2-4], especially when dealing with complex terrains over large spatial scales. Future research will include implementation of p-adaptive schemes, and coupled h-p adaptation[5].

2. Finite Element Method

2.1 Groundwater Flow and Species Transport

2.1.1 Saturated Porous Media

The governing equations for groundwater flow and subsurface transport of toxic material are well known[6,7]. In groundwater transport, the diffusion coefficients are important in describing subsurface movement. Groundwater flow (or head) is quite slow, and is usually modeled with simple, linear forms of the equations of motion (compared to the nonlinear equations describing atmospheric winds).

The matrix equivalent forms of the governing equations for groundwater head and species transport are[2]

$$[M]\{\dot{h}\} + [K_h]\{h\} = \{F_h\} \tag{1}$$

$$R_d[M]\{\dot{\chi}\} + [D_\chi]\{\chi\} + [A(V)]\{\chi\} = \{F_\chi\} \tag{2}$$

where h is head (m), R_d is retardation, χ is concentration (gm/m^3), V is the vector velocity field (m/day), and D is the directionally dependent dispersion tensor (m^2/day). An explicit forward-in-time Euler scheme used to advance the concentration in time. The matrix coefficients (denoted by [ ]) and column vectors ({ }) are integral relations based on the value of the shape function and its derivatives with respect to x, y, and z. The dot above the variable refers to the time derivative, and all variables are approximated as trial functions (denoted by the caret), e.g., concentration becomes

$$\hat{\chi}(x,y,z,t) = \sum N_i(x,y,z)\chi(t) \tag{3}$$

where N_i is the shape (or basis) function.

The hydrodynamic dispersion tensor can be defined as[7]

$$D = \begin{cases} \alpha_T V + (\alpha_L - \alpha_T)\dfrac{v_i v_j}{V} + \mu & i = j \\[2mm] (\alpha_L - \alpha_T)\dfrac{v_i v_j}{V} & i \neq j \end{cases} \tag{4}$$

where i,j = 1,2,3 denotes x,y,z directions, respectively, v_i and v_j are the velocity components in the respective directions, $V = [(v_1)^2+(v_2)^2+(v_3)^2]^{1/2}$, α_L is the longitudinal dispersion coefficient (m²/day), α_T is the transverse dispersion coefficient, and μ is the molecular diffusion coefficient (m²/day). Velocity components are determined from the relations

$$V=-(k\cdot\nabla h)/\theta_p \qquad (5)$$

where θ_p is effective porosity, k is the hydraulic conductivity (m/day), and h is the height of the water table (head-m) above an impermeable base.

In order to reduce numerical dispersion, a Petrov-Galerkin technique is used for the advection terms[8]. The use of this weighting function selectively eliminates the dispersive computational noise associated with steep gradient resolution, and works very effectively when coupled with h-adaptive methods[3,4]. The weighting for the advection term is altered to the form

$$W_i = N_i + \frac{\alpha h_e}{2V}(V\cdot\nabla N_i) \qquad (6)$$

where $\alpha = \coth \beta/2 - 2/\beta$, $\beta = Vh_e/2D_e$, h_e is the average element length, and D_e is an effective diffusion calculated from

$$D_e = \frac{V^T\cdot D\cdot V}{V^2} \qquad (7)$$

2.1.2 Unsaturated Porous Media

The matrix equivalent forms of the governing equations for unsaturated pressure head and species transport are[9]

$$C_\psi[M]\{\dot\psi\}+[K_\psi]\{\psi\}=\{F_\psi\} \qquad (8)$$

$$[M]\{\dot\chi\}+[D_\psi]\{\chi\}+[A(V)]\{\chi\}=\{F_\chi\} \qquad (9)$$

where ψ is pressure head (m), and $D_\psi \equiv D(\psi)$ is the directionally dependent dispersion tensor (m²/day). Due to the nonlinear nature of Eq. (8), Picard iteration is used to achieve convergence for pressure head. A Petrov-Galerkin scheme is also used to weight the advection terms, and an explicit forward-in-time Euler scheme used to advance the transient solution. Hydraulic conductivities are calculated using approximations discussed in Guymon[10] and Istok[9], i.e.,

$$K(\psi) = K_s e^{-a\psi} \qquad (10)$$

where K_s is the saturated value and a is an empirical coefficient[10].

2.2 Atmospheric Flow and Species Transport

2.2.1 Diagnostic

Prior to solving the atmospheric equations of motion, an objective analysis scheme is used to produce a *diagnostic* windfield. Data from towers are used to create an initial sparse windfield approximation. The surface windfield is constructed from measured data by interpolation to an initial coarse mesh using inverse distance-squared weighting. Once the surface level flow field has been established, the upper level wind data are interpolated to the 3-D grids. A variational formulation (elliptic equation) is then solved for the potential and the velocities "corrected" to ensure mass consistency.

The variational statement consists of a set of Lagrangian multipliers which are calculated from the interpolated windfield. The weak formulation yields the matrix equivalent equation[12]

$$[K_\lambda]\{\lambda\} = \{F_\lambda\} \tag{11}$$

where $\{F_\lambda\} \equiv \nabla \cdot V$ is the mass conservation applied over all node points, V is interpolated velocity vector (m/s), and λ is the Lagrangian multiplier. Once the values of λ are calculated, the velocities are adjusted to ensure mass consistency throughout the entire flow field, i.e.,

$$\{V\}^{n+1} = \{V\}^n + \frac{1}{2\alpha_i^2}\nabla\lambda \tag{12}$$

where α_i are Gauss precision moduli based on standard deviations associated with atmospheric stability[13]. The diagnostic flow thus acts as the initial condition for the forecasted wind field, i.e., the *prognostic* flow predictor.

2.2.2 Prognostic

The matrix equivalent forms of the governing equations for atmospheric motion and energy are written as[13,14]

$$[M]\{\dot{V}\} + [K]\{V\} + [A(V)]\{V\} + C^T\{p\} = \{F_V\} \tag{13}$$

$$[M]\{\dot{\theta}\} + [K_\theta]\{\theta\} + [A(V)]\{\theta\} = \{F_\theta\} \tag{14}$$

$$[M]\{\dot{\chi}\} + [K_\chi]\{\chi\} + [A(V)]\{\chi\} = \{F_\chi\} \tag{15}$$

where V is the velocity vector (m/s), C^T is the gradient operator, θ is potential temperature (°C), [K] refers to the stiffness matrix, and χ is species concentration (gm/m³). The matrix expressions are described in Pepper[13]. An explicit Euler scheme is also used to advance the solution in time. Mass lumping is employed, along with reduced integration (when applicable).

The modeling of atmospheric turbulence and resulting forms of closure are quite varied[15]. In order to simplify the complexities associated with turbulence modeling, a simple expression for the horizontal exchange coefficient (K_h), is used[16]

$$K_h = \frac{1}{2} k_o^2 h_e^2 ((\frac{\partial u}{\partial x} - \frac{\partial v}{\partial y})^2 + (\frac{\partial v}{\partial x} + \frac{\partial u}{\partial y})^2)^{1/2} \tag{16}$$

where h_e is the average element length and k_o is von Karman's constant. Vertical exchange coefficients for momentum, heat, and moisture are based on the Monin-Obukhov length, i.e.,

$$K_z = \frac{\kappa_o u^* z}{\phi_m(\zeta)} \tag{17}$$

where $\phi_m(\zeta)$ is the nondimensional wind profile and $\zeta = z/L$ (where L is the Monin-Obukhov length).

3. h-Adaptation

Bilinear, isoparametric quadrilateral elements are used in 2-D simulations; trilinear hexahedrals are used in 3-D. Quadrilaterals and hexahedrals, which are commonly used for fluid flow, create virtual, or "hanging" nodes when refined. These nodes must be tied back to the corner nodes, thus creating an element with one extra node on a face and causing a serious mesh compatibility problem. This problem is usually handled with some elaborate book-keeping. Although more computationally inhibiting as the number of nodes increases - especially when the problem is three-dimensional - element distortion is generally limited, and permits use of reduced integration. Another way to overcome the hanging node problem is to use an interface of triangles between regions of quadrilaterals to maintain element connectivity[17]. While easy to implement, the procedure leads to mixed elements within the overall array. In this study, we have elected to use the hanging node approach, coupled with averaging at the (parent) element corner nodes.

A simple adaptation scheme is shown in Fig. 1, where a 2-D array of quadrilaterals is refined. Element A is refined into 4 smaller elements; elements B, C, and D are checked. Element B is chosen for refinement and subdivided into 4 elements, similar to those in A. The process continues as element III in A is refined again, and its adjacent elements checked for refinement. Unrefinement follows similarly. Rules regarding the refinement and unrefinement of elements and nodal connectivity are discussed in Pepper and Stephenson[2].

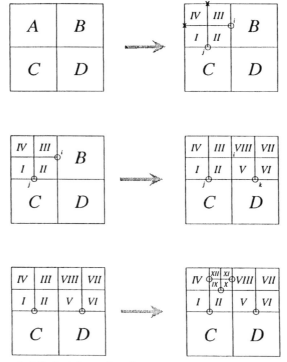

Figure 1. Mesh Adaptation for 2-D Bilinear Quadrilateral Element

4. Model Results

4.1 Subsurface Transport of Toxic Material

4.1.1 Saturated Case

Transient dispersion in saturated porous media is simulated using mesh adaptation for a buried leaking container (constant source). Figure 2 shows the final adapted mesh and concentration contours after 2096 days. The initial mesh contained 171 elements -which is coarse. Hydraulic conductivities typical of the strata found at the Savannah River Site were used[2,18], as shown in Table 1.

Table 1. Hydraulic Conductivities for Savannah River Site

Aquifer	K_{xx} (m/day)	K_{yy} (m/day)
1	1.10	6.0×10^{-3}
Aquitard (Clay)	1.0×10^{-7}	6.0×10^{-5}
2	1.2	6.0×10^{-5}

Several adaptations have occurred during the transient solution of the contaminant transport. The final mesh contains 684 elements. The concentration has slowed its vertical descent and is being transported horizontally along the upper surface of the aquitard.

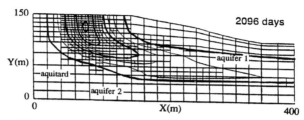

Figure 2. Solute Transport from a Buried, Leaking Container - Saturated Soil (Cross-section of Savannah River F-Area Basin, Aiken, SC)

4.1.2 Unsaturated Case

Groundwater flow and species transport within unsaturated porous media is calculated for a constant source[18] in the proposed Yucca Mountain Repository for nuclear waste. Figure 3 shows the final adapted mesh and concentration contours after 8000 years. The initial mesh contained 165 elements. Conductivities typical of the strata found at Yucca Mountain were used[18], e.g., $k_{xx} = k_{yy} = 7.0 \times 10^{-4}$. A constant infiltration rate of 0.1mm/yr was assumed at the top boundary. After several adaptations, the final mesh contained 495 elements. Model results indicate that the concentration begins to reach the water table (bottom boundary) after approximately 8000 years.

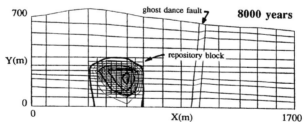

Figure 3. Solute Transport from a Leaking Container - Unsaturated Soil (Cross-section of Yucca Mountain Repository, NV)

4.2 Atmospheric Flow over a Ridge

Transient dispersion is simulated using mesh adaptation associated with the transport of contaminant from an elevated (z=50m), constant source (S=100gm/m³-s) upwind of two ridges[4]. A constant wind (u=5m/s) is assumed to flow into the problem domain from the left boundary. The vertical and lateral dimensions are 600m and 2000m, respectively; the height of the small ridge is 100m while the larger ridge is 300m. Figure 4(a,b) shows the final mesh and concentration pattern after 400 seconds. The original mesh consisted of 220 elements; the adapted mesh contained a total of 884 elements.

The concentration diffuses towards the surface between the ridges and continues to be transported horizontally over the surface of the higher ridge. Observation of the adapted mesh shows those regions where the concentration gradient is high, i.e., the added elements indicate where the most activity occurs. Knowledge of where such regions occur is essentially unknown prior to obtaining a complete solution; this typically requires *global* remeshing of the problem when using non-adapting finite element and finite difference methods.

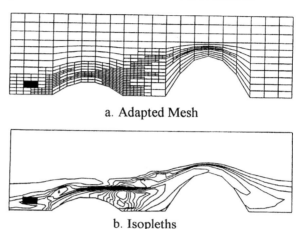

a. Adapted Mesh

b. Isopleths

Figure 4. Atmospheric Transport over Two Ridges from an Elevated Source

5. Conclusions

The development and implementation of an h-adaptive algorithm with a general finite element solver has been achieved for predicting species transport and dispersion for environmental problems associated with atmospheric or groundwater transport. Coupling h-adaptation with a Petrov-Galerkin finite element algorithm produces very accurate solutions with minimal computer time and storage demands. The C/C++ version of the model is easy to use, and runs under WINDOWS utilizing a mouse and pull-down menus; a mesh generator, which also runs under WINDOWS, allows the user to easily generate meshes using a mouse and point-and-click procedures, and establish boundary conditions.

Future research efforts will include the use of Delaunay Triangulation and h-p adaptation. In h-adaptation, convergence is algebraic; an h-p scheme should yield exponential convergence, thereby reducing the overall computation time and significantly reducing the number of degrees of freedom.

References

1. Devloo, P., Oden, J. T., and Pattani, P., 1988, "An h-p Adaptive Finite Element Method for the Numerical Simulation of Compressible Flow," *Comp. Meth. in Appl. Mech. and Engr.*, Vol. 70, pp. 203-235.

2. Pepper, D. W. and Stephenson, D. E., 1995, "An Adaptive Finite Element Model for Calculating Subsurface Transport of Contaminant," *Ground Water*, Vol. 33, No. 3, pp. 486-496.

3. Pepper, D. W. and Carrington, D. B., 1995, "An h-Adaptive Finite Element Model for 3-D Atmospheric Transport Prediction," *3rd. Int. Air Pollution Conf.*, Porto Carras, Greece, Sept. 26-28.

4. Pepper, D. W. and Carrington, D. B., 1995, "A Finite Element Model with h-Adaptation for Atmospheric Transport of Pollutants," *2nd UNAM-CRAY Supercomputing Conference, Numerical Simulations in the Environmental and Earth Sciences*, June 21-24, Mexico City, Mexico.

5. Oden, J. T., Liszka, T., and Wu, W., 1991, "An h-p Adaptive Finite Element Method for Incompressible Viscous Flows," *The Math. Of Finite Elements and Applications VII*, Academic Press, pp. 13-54.

6. Domenico, P. A. and Schwartz, F. W., 1990, *Physical and Chemical Hydrogeology*, J. Wiley and Sons, New York.

7. Bear, J., 1979, *Hydraulics of Groundwater*, McGraw-Hill, New York.

8. Yu, C.-C. and Heinrich, J. C., 1986, "Petrov-Galerkin Methods for the Time-Dependent Convective Transport Equation," *Int. J. Num. Meth. Engr.*, Vol. 23, pp. 883-901.

9. Istok, J., 1989, *Groundwater Modeling by the Finite Element Method*, AGU, Water Resources Monograph, 13.

10. Guymon, G. L., 1994, *Unsaturated Zone Hydrology*, Prentice Hall, NJ.

11. Anderson, M. P. and Woessner, W. W., 1992, *Applied Groundwater Modeling, Simulation of Flow and Advective Transport*, Academic Press, San Diego.

12. Pepper, D. W., 1990, "A 3-D Numerical Model for Predicting Mesoscale Windfields over Vandenberg Air Force Base," *Final Rept. No. USAF/AFSC F04701-89-C-0051*, Advanced Projects Research, Inc., Moorpark, CA.

13. Sherman, C. A. (1978), "A Mass-consistent Model for Wind Fields over

Complex Terrain," *J. Appl. Meteor.*, Vol. 17, pp. 312-319.

14. Pielke, R. A., 1984, *Mesoscale Meteorological Modeling*, Academic Press, Orlando, FL.

15. Zannetti, P., 1990, *Air Pollution Modeling*, Van Nostrand Reinhold and Computational Mechanics Pub., New York.

16. Smagorinsky, J., Manabe, S., and Holloway, J. L., 1965, "Numerical Results from a Nine-Level General Circulation Model of the Atmosphere," *Mon. Wea. Rev.*, Vol. 93, pp. 727-798.

17. Ramakrishnan, R., Bey, K. S., and Thornton, E. A., 1990, "Adaptive Quadrilateral and Triangular Finite-Element Scheme for Compressible Flows," *AIAA J. Thermophysics and Heat Transfer*, Vol. 28, No. 1, pp. 51-59.

18. Pepper, D. W. and Sethi, H. S. (1995), "An h-Adaptive Finite Element Model for Subsurface Contaminant Transport," presented at the *9th Int. Conf. on Finite Elements in Fluids*, Oct. 15-20, Venezia, Italy (also appears in the Proceedings of the conference).

Modelling depth-integrated contaminant dispersion in the Humber Estuary using a Lagrangian particle technique

R.V. Pearson, R.W. Barber
Water Resources Research Group, Telford Institute of Environmental Systems, University of Salford, M5 4WT, UK

Abstract

A two-dimensional Lagrangian particle model has been established to simulate depth-integrated contaminant dispersion in estuaries and coastal seas. The governing advection-diffusion equation is first rearranged into a Fokker-Planck equation allowing the implementation of a consistent random-walk method to simulate diffusion with a spatially variable diffusivity. The paper then describes a novel particle tracking algorithm which can be used with arbitrary non-orthogonal boundary-fitted coordinate meshes. By definition, the particle tracking method is perfectly conservative and free from numerical diffusion in the classical sense. The scheme is also able to preserve steep concentration fronts enabling accurate simulations of point sources.

1 Introduction

Increasing environmental awareness and the need to predict and improve the water quality in estuaries and coastal seas has led to significant developments in pollution transport modelling. Traditionally, contaminant dispersion has usually been simulated using finite-difference approximations of the standard advection-diffusion equation, expressed in depth-averaged form. However, Eulerian grid based finite-difference methods present a number of problems when applied to advection dominated flow regimes. In particular, grid based methodologies have difficulty modelling point sources of contaminant and steep concentration gradients. Furthermore, in simulations where the pollution does not occupy the whole flow domain, finite-difference schemes are often computationally inefficient compared to particle tracking methods.

The simulation of contaminant dispersion using a particle tracking technique is based upon the simple idea that the pollution may be represented by discrete particles which are subjected to advection, diffusion and decay. These processes are modelled by the combination of deterministic and stochastic numerical schemes; namely, translation of each particle using an advective velocity derived from the surrounding velocity field and diffusion of each particle using a random-walk technique.

2 Governing equations

The depth-averaged advection-diffusion equation representing the fate of a conservative pollutant can be expressed as

$$
\frac{\partial(cD)}{\partial t} + \frac{\partial(ucD)}{\partial x} + \frac{\partial(vcD)}{\partial y} = \frac{\partial}{\partial x}(D\Gamma_{xx}\frac{\partial c}{\partial x}) + \frac{\partial}{\partial x}(D\Gamma_{xy}\frac{\partial c}{\partial y}) +
$$
$$
\frac{\partial}{\partial y}(D\Gamma_{xy}\frac{\partial c}{\partial x}) + \frac{\partial}{\partial y}(D\Gamma_{yy}\frac{\partial c}{\partial y})
$$
(1)

where u and v are the depth-averaged velocity components in the x- and y-directions, D is the local water depth, c is the depth-averaged pollution concentration, and

$$
\Gamma = \begin{vmatrix} \Gamma_{xx} & \Gamma_{xy} \\ \Gamma_{yx} & \Gamma_{yy} \end{vmatrix} = \begin{vmatrix} \Gamma_L & 0 \\ 0 & \Gamma_T \end{vmatrix}
$$
(2)

is the diffusion tensor in the Cartesian and flow-aligned coordinate systems. If $\theta = \tan^{-1}(v/u)$ is the angle between the local flow direction and the x-axis, then the Cartesian diffusion coefficients may be evaluated as

$$
\Gamma_{xx} = \Gamma_L\cos^2\theta + \Gamma_T\sin^2\theta
$$
(3)

$$
\Gamma_{xy} = (\Gamma_L - \Gamma_T)\sin\theta\cos\theta
$$
(4)

$$
\Gamma_{yy} = \Gamma_L\sin^2\theta + \Gamma_T\cos^2\theta
$$
(5)

where Γ_L and Γ_T are the longitudinal and transverse diffusion coefficients measured parallel and perpendicular to the local velocity vector. In the present work, the diffusion coefficients are estimated using Elder's concept[1].

Following the method outlined by Józsa[2], the advection-diffusion equation is rewritten in a form which can be interpreted from a particle tracking perspective. This is accomplished by introducing a new concentration variable, $C = Dc$. Equation (1) can thus be rewritten after algebraic manipulation as

$$
\frac{\partial C}{\partial t} + \frac{\partial(UC)}{\partial x} + \frac{\partial(VC)}{\partial y} = \frac{\partial^2(\Gamma_{xx}C)}{\partial x^2} + 2\frac{\partial^2(\Gamma_{xy}C)}{\partial x\partial y} + \frac{\partial^2(\Gamma_{yy}C)}{\partial y^2}
$$
(6)

in which

$$U = u + \frac{\partial \Gamma_{xx}}{\partial x} + \frac{\partial \Gamma_{xy}}{\partial y} + \frac{\Gamma_{xx}}{D} \frac{\partial D}{\partial x} + \frac{\Gamma_{xy}}{D} \frac{\partial D}{\partial y} \qquad (7)$$

$$V = v + \frac{\partial \Gamma_{yy}}{\partial y} + \frac{\partial \Gamma_{xy}}{\partial x} + \frac{\Gamma_{yy}}{D} \frac{\partial D}{\partial y} + \frac{\Gamma_{xy}}{D} \frac{\partial D}{\partial x} . \qquad (8)$$

If C is considered as a probability density function, then eqn. (6) is identical to the Fokker-Planck equation. The modified advection velocities presented in eqns. (7) & (8) allow the particle tracking scheme to simulate diffusion using a consistent random-walk technique. Without this modification, the particles would accumulate in regions of low diffusivity (Hunter et al.[3]).

3 Numerical scheme

To facilitate an accurate representation of the complex flow domains found in coastal regions, the numerical model is based upon a boundary-fitted non-orthogonal grid. Following Thompson et al.[4], a smooth curvilinear grid is generated by solving a pair of elliptic Poisson equations:

$$\left. \begin{array}{l} \xi_{xx} + \xi_{yy} = P(\xi,\eta) \\ \\ \eta_{xx} + \eta_{yy} = Q(\xi,\eta) \end{array} \right\} \qquad (9)$$

relating the physical (x,y) coordinates to the transformed (ξ,η) coordinates. (Here, the subscripts denote the usual shorthand notation for partial differentiation). The functions P and Q are the so-called 'attraction operators' or 'control functions' which can be used to alter the internal structure of the curvilinear mesh. After interchanging the dependent and independent variables, eqn. (9) may be rewritten as a quasi-linear elliptic system:

$$\left. \begin{array}{l} \alpha x_{\xi\xi} - 2\beta x_{\xi\eta} + \gamma x_{\eta\eta} + J^2(Px_\xi + Qx_\eta) = 0 \\ \\ \alpha y_{\xi\xi} - 2\beta y_{\xi\eta} + \gamma y_{\eta\eta} + J^2(Py_\xi + Qy_\eta) = 0 \end{array} \right\} \qquad (10)$$

where

$$\alpha = x_\eta^2 + y_\eta^2 \quad , \quad \beta = x_\xi x_\eta + y_\xi y_\eta \quad , \quad \gamma = x_\xi^2 + y_\xi^2$$

and J is the Jacobian of the transformation, given by $J = x_\xi y_\eta - x_\eta y_\xi$. The mapping expressions shown in (10) are rewritten as finite-differences and solved using successive-over-relaxation to find a one-to-one mapping between the transformed (ξ,η) plane and the physical (x,y) plane.

Before solving the governing advection-diffusion equation, it is first necessary to decide upon a particular grid configuration on which to represent all constants and variables appertaining to the pollution transport model. The staggered computational grid used in the present work is shown in Figure 1.

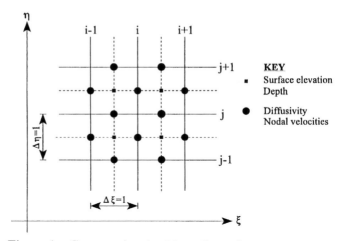

Figure 1: Computational grid configuration

Particle movement takes place under the action of both advection and diffusion each time step.   In order to translate the particles due to pure advection, it is necessary to calculate the modified advection velocities presented in eqns. (7) & (8) at each particle position.   Since the variation in diffusivity from cell to cell is generally quite small, then in the present scheme the derivative terms in eqns. (7) & (8) are assumed constant for all particles within the same cell during a time step.   These derivatives are with respect to the Cartesian coordinate system and before it is possible to calculate them on the non-orthogonal boundary-fitted grid they must be transformed and expressed in terms of the curvilinear coordinates (ξ, η).   The transformation is performed according to the numerical mapping formulae presented by Thompson et al.[4]:

$$\left.\begin{aligned}
f_x &= \frac{\partial f}{\partial x} = \frac{\partial(f,y)}{\partial(\xi,\eta)} \div \frac{\partial(x,y)}{\partial(\xi,\eta)} = \frac{1}{J}(y_\eta f_\xi - f_\eta y_\xi) \\
f_y &= \frac{\partial f}{\partial y} = \frac{\partial(x,f)}{\partial(\xi,\eta)} \div \frac{\partial(x,y)}{\partial(\xi,\eta)} = \frac{1}{J}(x_\xi f_\eta - f_\xi x_\eta)
\end{aligned}\right\} \qquad (11)$$

where f denotes a differentiable function of x and y.   Substitution of the above transformation formulae into all partial derivatives involving x or y in eqns. (7) & (8) leads to

$$\begin{aligned}
U = u + \frac{1}{JD}\Bigg[&y_\eta\left(\frac{\partial\Gamma_{xx}}{\partial\xi}D + \frac{\partial D}{\partial\xi}\Gamma_{xx}\right) - y_\xi\left(\frac{\partial\Gamma_{xx}}{\partial\eta}D + \frac{\partial D}{\partial\eta}\Gamma_{xx}\right) \\
&+ x_\xi\left(\frac{\partial\Gamma_{xy}}{\partial\eta}D + \frac{\partial D}{\partial\eta}\Gamma_{xy}\right) - x_\eta\left(\frac{\partial\Gamma_{xy}}{\partial\xi}D + \frac{\partial D}{\partial\xi}\Gamma_{xy}\right)\Bigg]
\end{aligned} \qquad (12)$$

and

$$V = v + \frac{1}{JD}\left[x_\xi \left(\frac{\partial \Gamma_{yy}}{\partial \eta} D + \frac{\partial D}{\partial \eta} \Gamma_{yy} \right) - x_\eta \left(\frac{\partial \Gamma_{yy}}{\partial \xi} D + \frac{\partial D}{\partial \xi} \Gamma_{yy} \right) \right.$$
$$\left. + y_\eta \left(\frac{\partial \Gamma_{xy}}{\partial \xi} D + \frac{\partial D}{\partial \xi} \Gamma_{xy} \right) - y_\xi \left(\frac{\partial \Gamma_{xy}}{\partial \eta} D + \frac{\partial D}{\partial \eta} \Gamma_{xy} \right) \right]. \tag{13}$$

The cell in which each particle lies is calculated using a method similar to that detailed by Milgram[5]. It is now possible to calculate all derivative terms in eqns. (12) & (13) using central differences about the cell centre. The values of the advective velocities u and v are computed using Taylor series expansions up to second order about the nearest non-zero velocity node in the cell containing the particle. If the distances in the x- and y-directions between the particle and the nearest non-zero velocity node are denoted by Δx and Δy and if the nodal velocities are represented as u_n and v_n, then the velocity components at the particle position can be expressed as

$$u = u_n + \Delta x \frac{\partial u}{\partial x}\bigg|_n + \Delta y \frac{\partial u}{\partial y}\bigg|_n + \frac{\Delta x^2}{2} \frac{\partial^2 u}{\partial x^2}\bigg|_n + \frac{\Delta y^2}{2} \frac{\partial^2 u}{\partial y^2}\bigg|_n + \Delta x \Delta y \frac{\partial^2 u}{\partial x \partial y}\bigg|_n \tag{14}$$

$$v = v_n + \Delta x \frac{\partial v}{\partial x}\bigg|_n + \Delta y \frac{\partial v}{\partial y}\bigg|_n + \frac{\Delta x^2}{2} \frac{\partial^2 v}{\partial x^2}\bigg|_n + \frac{\Delta y^2}{2} \frac{\partial^2 v}{\partial y^2}\bigg|_n + \Delta x \Delta y \frac{\partial^2 v}{\partial x \partial y}\bigg|_n . \tag{15}$$

Again it is necessary to transform the equations to curvilinear coordinates but this time second derivatives are also required. These are calculated by differentiating eqn. (11) with respect to x and y. The resulting equations are too long to be presented here, but full details can be found in Pearson & Barber[6].

Once the modified advection velocities are known, the new particle position due to pure advection can be calculated using a second-order accurate iterative technique which ensures that

$$\left. \begin{array}{l} \dfrac{x_a(t + \Delta t) - x_a(t)}{\Delta t} = U^* \\[3mm] \dfrac{y_a(t + \Delta t) - y_a(t)}{\Delta t} = V^* \end{array} \right\} \tag{16}$$

where,

$$\left. \begin{array}{l} U^* = \dfrac{U(x_a(t + \Delta t), y_a(t + \Delta t)) + U(x_a(t), y_a(t))}{2} \\[3mm] V^* = \dfrac{V(x_a(t + \Delta t), y_a(t + \Delta t)) + V(x_a(t), y_a(t))}{2} \end{array} \right\} \tag{17}$$

and x_a and y_a are the x- and y-coordinates of the advected particle position.

At the end of the advection calculation, the resulting particle distribution must undergo diffusion. This is accomplished by adding random velocity components to each particle with an appropriate standard deviation. The random longitudinal and transverse velocity components are generated as

$$u_L' = r_1 \sqrt{\frac{2\Gamma_L}{\Delta t}} \quad , \quad u_T' = r_2 \sqrt{\frac{2\Gamma_T}{\Delta t}} \tag{18}$$

where r_1 and r_2 are *independent* normally distributed random numbers, each with zero mean and standard deviation of unity. As shown by Hunter et al.[3], it is essential to have a random number generator which returns values that are sufficiently random so that diffusive processes may be simulated accurately. For this reason the routines RAN3 and GASDEV as described by Press et al.[7] are used to provide the normally distributed random numbers. After calculating the longitudinal and transverse diffusive velocities they can be expressed in the Cartesian coordinate system by transformation such that:

$$u' = u_L' \cos\theta - u_T' \sin\theta \tag{19}$$

$$v' = u_L' \sin\theta + u_T' \cos\theta . \tag{20}$$

The final x- and y-coordinates of the particle can now be expressed as

$$\left. \begin{array}{l} x = x_a(t + \Delta t) + u'\Delta t \\[2mm] y = y_a(t + \Delta t) + v'\Delta t . \end{array} \right\} \tag{21}$$

Any particles which cross a solid boundary are immediately reflected back into the flow domain thus maintaining mass conservation. The method is conservative, front preserving and unconditionally stable although the time step should be limited in order to obtain an accurate representation of the advective transport.

4 Results

The particle tracking algorithm was validated against an analytical test case of pure advection of a circular distribution of particles. The non-orthogonal boundary-fitted mesh used in the study is shown in Figure 2. The domain has a radius of 25 m and is 1 m deep. The fluid was given an anticlockwise angular velocity of 0.04 rads/s about the centre of the domain (0,0), giving a velocity of zero at the centre and 1 m/s at the perimeter. A circular distribution of 10000 particles and 5 m in radius was introduced, initially centred at (0,10). The particles were advected for 1 revolution using a model time step of 0.2 s. The time for one revolution is given by $2\pi/0.04$ which is approximately equal to 157 s. Theoretically, the circular distribution of particles should undergo no deformation. The results for 0.25, 0.5, 0.75 and 1.0 revolutions are given in Figure 2 and were indistinguishable from the analytical solution therefore implying that the advective scheme is phase preserving.

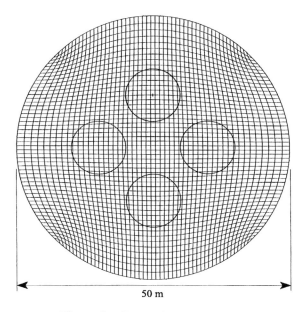

Figure 2: Pure advection test

The practical potential of the method is demonstrated by application of the model to a 50 km long section of the Humber Estuary on the east coast of England.   The Humber Estuary provides an outlet to the North Sea for the rivers Trent and Ouse, and shipping access to a number of ports including Hull, Immingham and Grimsby.   The present work has concentrated on the wider lower part of the estuary where the flow patterns are of a complex two-dimensional nature and tidal effects dominate.   Figure 3 shows the extent of the modelled area and the location of velocity observation sites used in model validation, whilst Figure 4 shows the boundary-fitted coordinate system representative of the estuary.   For the mesh illustrated, no grid line attraction was necessary and so the control functions, P and Q, in the grid generation equations were set to zero.   The depth of the bed below a fixed 'Chart Datum' was then calculated at all grid nodes using inverse power interpolation between scattered bathymetry points.

The hydrodynamic calculations providing the velocities to drive the pollution model were solved within the same computer code using the method described by Pearson & Barber[8].   The time step was identical for both models and equalled 12 s.   The purpose of the study was to examine the dispersion of pollution from a proposed sewer outfall north of Grimsby.   The position of the outfall is shown by the crosshair in Figure 5.   Pollution was discharged into the estuary at a rate of 1.25 m^3/s beginning 15 minutes after high water and ending 2 hours 15 minutes after high water.   The mean pollutant concentration was set to 1.0 kg/m³ and the particle mass was set to 1 kg. This led to a mass of 15 kg being introduced into the system each time step, equivalent to 15 particles. Thus, in total, 9000 particles were released into the estuary over the two hour period. The position of the plume on the ebb tide, 4.75 hours after high water, is presented in Figure 5 whilst the position of the plume on the flood time, 10 hours after high water, is shown in Figure 6. From the results, it is obvious that very little mixing occurs during the tidal excursion.

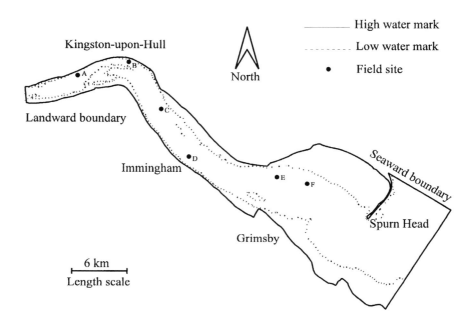

Figure 3: Map of the Humber Estuary

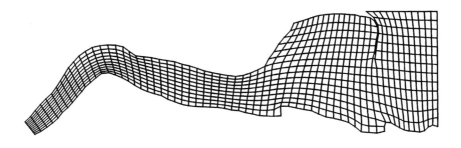

Figure 4: Boundary-fitted grid representative of the Humber Estuary

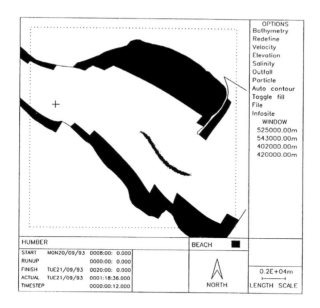

Figure 5: Plume transport on ebb tide

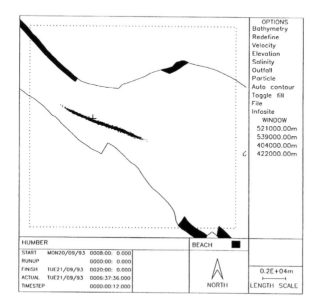

Figure 6: Plume transport on flood tide

5 Conclusions

A particle tracking model has been developed for use with non-orthogonal boundary-fitted grids to simulate depth-averaged pollution transport in well mixed estuaries and coastal seas. The method is conservative, front preserving and unconditionally stable although the time step should be limited in order to obtain accurate representation of the advective transport. In the case of practical estuary studies, the number of particles required to obtain accurate quantitative predictions may become prohibitive. However, the method is useful from the qualitative point of view as a means of assessing pollution transport trends and perhaps to assist in contingency planning.

References

1.   Elder, J.W. The dispersion of marked fluid in turbulent shear flow, *J. Fluid Mech.*, 1959, **5**, 544-560.
2.   Józsa, J. 2-D particle model for predicting depth-integrated pollutant and surface oil slick transport in rivers, *Proc. Int. Conf. on Hydraulic and Environmental Modelling of Coastal, Estuarine and River Waters*, 332-340, Bradford, U.K., 1989.
3.   Hunter, J.R., Craig, P.D. & Phillips, H.E. On the use of random walk models with spatially variable diffusivity, *J. Comp. Phys.*, 1993, **106**, 366-376.
4.   Thompson, J.F., Thames, F.C. & Mastin, C.W. Automatic numerical generation of body-fitted curvilinear coordinate system for field containing any number of arbitrary two-dimensional bodies, *J. Comp. Phys.*, 1974, **15**, 299-319.
5.   Milgram, M.S. Does a point lie inside a polygon?, *J. Comp. Phys.*, 1989, **84**, 134-144.
6.   Pearson, R.V. & Barber, R.W. Modelling depth-integrated contaminant dispersion in the Humber Estuary using a random-walk technique, *Proc. 9th Int. Conf. on Numerical Methods in Laminar and Turbulent Flow*, Part 2, 1088-1099, Atlanta, U.S.A., 1995.
7.   Press, W.H., Flannery, B.P., Teukolsky, S.A. & Vetterling, W.T. *Numerical Recipes: the Art of Scientific Computing*, Cambridge Univ. Press, Cambridge, U.K., 1986.
8.   Pearson, R.V. & Barber, R.W. Mathematical simulation of the Humber Estuary using a depth-averaged boundary-fitted tidal model, *Proc. 8th Int. Conf. on Numerical Methods in Laminar and Turbulent Flow*, Part 2, 1244-1255, Swansea, U.K., 1993.

A quintic interpolation scheme for semi-Lagrangian advection models

P. Holnicki

Systems Research Institute, Polish Academy of Sciences, 01-447 Warszawa, Newelska 6, Poland

1 Introduction

Algorithms of shape-preserving interpolation are now widely investigated in connection with important physical applications, for example those related to numerical weather prediction, analysis of global climate changes or modeling atmospheric transport processes. Used in most cases, effective semi-Lagrangian methods that follow wind characteristics backward in time, require interpolation of the initial profile at the upstream, departure point. It is then desired that the algorithm applied correctly reflects physical reality suggested by the data. Typical demand is, therefore, that interpolation scheme, except accuracy and low numerical diffusivity, should be positive definite and generate monotone interpolant in regions where the data are monotone. In some applications, also convexity (concavity) and conservation properties are important.

The commonly used piecewise polynomial interpolants, usually based on Hermite cubics [3, 4, 7] or quintics [3], give a compromise between accuracy and computational effort, but they do not automatically preserve the shape of the data, especially on steep gradient regions or in vicinity of local extrema. The desired positivity- or monotonicity-preserving properties are usually obtained by imposing the respective constraints on derivatives or by utilizing limiter function which generates shape preserving derivative estimates [7, 9, 10].

The paper presents a piecewise-quintic interpolation scheme, previously defined in [5] for advection problem and developed in [6]. The basic shape-preserving properties of the method as well as some numerical results are presented. The scheme considered is based on four-point grid stencil, defined on a uniform mesh. Four consecutive grid values and the first derivative estimates at the internal points are used to construct a quintic interpolation polynomial (compare [5] for details). The interpolant is the sixth-order accurate if the derivatives at the grid points are of the fifth order at least. It can also be combined with any algorithm of derivative estimation, but accuracy degenerates then according to monotonicity constraints applied.

In section 3 sufficient conditions for the scheme to be positive definite are formulated using the discrete maximum principle approach. The results obtained are slightly stronger than the respective formulations in [5]. More-over the technique applied in [7] for Hermite cubics is developed to formulate monotonicity conditions. The interpolant is combined with standard limiter functions for computing derivative estimates (Akima, Fritsch-Butland). To avoid accuracy degeneration by "clipping" effect, parabolic approximation [7] of derivatives in vicinity of extremum points is applied. Numerical results presented in section 4 refer to the application of the algorithm for 2D advection of standard test profiles. They show good accuracy and shape-preserving properties of the method.

2   Definition of interpolation polynomial

Consider a regular grid $\{x_i\}_{i=0}^{n+2}$ for $x_0 < x_1 < \ldots < x_{n+2}$, with the the mesh spacing $h = x_{i+1} - x_i$ and the corresponding data points $\{f_i\}$, which are samples of a piecewise smooth function f, such that $f_i = f(x_i)$. The slope of the piecewise linear interpolant (the first divided difference) will be denoted by $\Delta_i = (f_{i+1} - f_i)/h$. The respective set of derivative estimates at the interpolation points will be denoted by $\{d_i\}$.

To construct interpolation polynomial in subinterval $[x_i, x_{i+1}]$ we shall consider four-point stencil, based on the consecutive data values f_{i-1}, f_i, f_{i+1}, f_{i+2} and the derivative estimates at the internal points, d_i, d_{i+1}. The goal is to construct an interpolation polynomial $p(x)$, such that

$$p(x_{i+j}) = f_{i+j} \quad \text{for} \quad i = 1, \ldots, n, \quad j = -1, \ldots, 2,$$

$$\text{(1)}$$

$$\frac{dp}{dx}(x_i) = d_i \quad \text{for} \quad i = 1, \ldots, n.$$

Without loss of generality, one can assume a unit length of interpolation subinterval – $[0, 1]$, corresponding to linear transformation of independent variable $\alpha = (x - x_i)/h$, where $x \in [x_i, x_{i+1}]$. Then, as it was derived in [5], the respective interpolation polynomial has a form

$$p_i(\alpha) = c_{i-1} f_{i-1} + c_i f_i + c_{i+1} f_{i+1} + c_{i+2} f_{i+2} + \quad \text{(2)}$$

$$c_i' d_i + c_{i+1}' d_{i+1}, \quad (i = 1, \ldots, n),$$

where, for $\alpha \in [0, 1]$, coefficients are defined as follows:

$$c_{i-1} = \frac{1}{12} \alpha^2 (1 - \alpha)^2 (2 - \alpha),$$

$$c_i = 1 - \alpha^2 [1 + (1 - \alpha^2)(\frac{7}{4} - \frac{3}{4}\alpha)],$$

$$c_{i+1} = \frac{1}{4} \alpha^2 (1 + \alpha)[2 + (1 - \alpha)(8 - 3\alpha)],$$

$$c_{i+2} = \frac{1}{12} \alpha^2 (1 - \alpha^2)(1 - \alpha), \tag{3}$$

$$c_i' = \frac{1}{2} \alpha (1 + \alpha)(1 - \alpha)^2 (2 - \alpha),$$

$$c_{i+1}' = -\frac{1}{2} \alpha^2 (1 - \alpha^2)(2 - \alpha).$$

In most applications derivatives used in the interpolation scheme must be numerically approximated. The shape-preserving properties (nonnegativity, monotonicity) of interpolant depend then essentially on the estimation method applied. The approach is usually based on constraining that estimates to meet sufficient conditions of nonnegativity or monotonicity [3, 4, 7, 9]. Usually, derivative estimate satisfies the following constraint:

$$|d_i| \begin{cases} \leq \varrho \cdot \min(|\Delta_{i-1}|, |\Delta_i|), & \Delta_{i-1}\Delta_i > 0, \\ = 0, & \Delta_{i-1}\Delta_i \leq 0, \end{cases} \tag{4}$$

where $\varrho > 0$ is a given constant.

3   Nonnegativity and monotonicity

Positivity (nonnegativity) of interpolation scheme is one of the essential demands considered in most applications. Violation of this condition can lead to non-physical, negative values of the solution, overshooting the local maxima or to spurious oscillations, especially in vicinity of steep gradient. If the data represent, for example, the pressure or density of material, (e.g. in computational weather prediction or in modeling atmospheric pollution transport) – the negative or oscillatory values are non acceptable.

The sufficient conditions of nonnegativity of the interpolation scheme (2) – (4) can be formulated in terms of the discrete maximum principle. The following two propositions are a modified versions of the respective results presented in [5, 6].

PROPOSITION 1. *Consider interpolant (2) with the coefficients defined by (3). If the limiter constant in (4) is* $\varrho = 8/3$ *then the piecewise quintic interpolation function* $p(x)$ *satisfies the following maximum principle (we denote* $p_i = p_i(\alpha)$, *for* $i = 1 \ldots, n$, $\alpha \in [0, 1]$):

$$\min_{0 \leq j \leq n+2} f_j \leq p_i \leq \max_{0 \leq j \leq n+2} f_j. \tag{5}$$

PROPOSITION 2. *Let the assumptions of* Proposition 1 *hold. If* $\varrho = 3.5$, *then the interpolant* (2) *is positive definite for strictly monotone data and admits limited under- or overshootings in the neighborhood of extremum points.*

The respective proofs can be found in [6]. The formulated results state that the scheme is positive definite for $\varrho = 8/3$, but it can also be useful for the values of this coefficient up to about $\varrho = 3.5$, especially in regions with regular data or in flat gradient cases.

Preserving monotonicity of the input data is another important property of interpolation scheme. Following the known definitions [3, 4, 7], the data are *nondecreasing at* x_i if $f_{i-1} \leq f_i \leq f_{i+1}$ and they are *nondecreasing in* $[x_i, x_{i+1}]$ if they are nondecreasing at x_i and x_{i+1}. Analogous definitions hold for nonincreasing data. The data are *monotone at* x_i (or *in* $[x_i, x_{i+1}]$) if they are nondecreasing or nonincreasing at x_i (or *in* $[x_i, x_{i+1}]$). Consequently, the interpolant pf is *monotone in* $[x_i, x_{i+1}]$ if the values $(pf)(x)$ are monotone for all $x \in [x_i, x_{i+1}]$.

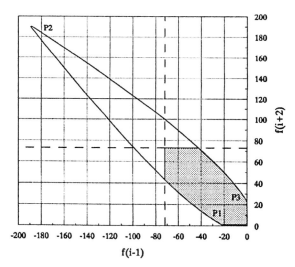

Figure 1. Monotonicity region for interpolant $p(x)$.

In general, monotonicity of (2) – (3) can be characterized as the respective subset of R^4 (compare [3]) depending on four parameters $(d_i, d_{i+1}, f_{i-1}, f_{i+2})$. In the case of quintic considered, there exists a relation between them due to definition of derivative estimates. Following the technique applied in [7] for cubics, four pairs of derivative estimates, $(d_i, d_{i+1}) = (0,0)$, $(8/3, 0)$, $(0, 8/3)$, $(8/3, 8/3)$ for $\varrho = 8/3$ can be considered. Then, to formulate monotonicity conditions, four auxiliary quintic polynomials $h_{(d_i,d_{i+1})}(\alpha; f_{i-1}, f_{i+2})$ which characterize the desired properties of (2) are constructed. If the auxiliary polynomials are monotone then any linear combination of them with positive coefficient is also monotone. Results of analysis [6] is summarized as the following

PROPOSITION 3. *Interpolant (2) with the coefficients defined by (3) and the limiter constant $\varrho = 8/3$ in (4) is monotone in $[x_i, x_{i+1}]$, provided that the grid values (f_{i-1}, f_{i+2}) are in the domain shown in Fig. 1.*

4   Implementation and numerical results

The interpolation method discussed was implemented for several algorithms of derivative estimates, known from the literature. Results of numerical tests presented in the following refer to Akima (AK) and Fritsch-Butland (FB) methods [3, 7, 9, 12], which are defined as follows:

$$\text{(AK):} \quad d_i = \begin{cases} \dfrac{\alpha\Delta_{i-1} + \beta\Delta_i}{\alpha + \beta}, & \alpha + \beta \neq 0, \\[2mm] \dfrac{\Delta_{i-1} + \Delta_i}{2}, & \alpha + \beta = 0, \end{cases}$$

$$\text{(FB):} \quad d_i = \begin{cases} \dfrac{3\Delta_{i-1}\Delta_i}{2|\Delta_{i-1}| + |\Delta_i|}, & (\Delta_{i-1}\Delta_i > 0) \wedge (|\Delta_{i-1}| \leq |\Delta_i|), \\[2mm] \dfrac{3\Delta_{i-1}\Delta_i}{|\Delta_{i-1}| + 2|\Delta_i|}, & (\Delta_{i-1}\Delta_i > 0) \wedge (|\Delta_{i-1}| > |\Delta_i|), \\[2mm] 0, & \Delta_{i-1}\Delta_i \leq 0, \end{cases}$$

where the discrete slope is $\Delta_i = (f_{i+1} - f_i)/h$ and the coefficients in the first formula are $\alpha = |\Delta_{i+1} + \Delta_i|$, $\beta = |\Delta_{i-1} + \Delta_{i-2}|$.

It is well known that the monotonicity constraints of a general form (4) cause loss of accuracy, mainly by generating "too flat" solutions near strict local extrema. This "clipping" effect follows from the property that the interpolant is still monotone in the neighborhood of extremum, even though the data are not. One of the method of overcoming this drawback and increasing the overall accuracy is to relax monotonicity constraints in vicinity of local extremum and replace them by higher order approximation. We consider here parabolic interpolation scheme discussed in [7]. The above defined limiters, with the additional parabolic approximation will be referred to as (AK2) and (FB2), respectively.

Interpolation scheme (2) with the above limiters has been applied for solving advection equation by the method of characteristics, combined with a uniform-grid spatial approximation (semi-Lagrangian method). In one-dimensional case a linear advection equation is considered

$$\frac{\partial f}{\partial t} + u\frac{\partial f}{\partial x} = 0, \qquad (6)$$

where $f(x,t)$ is a transported scalar quantity and $u(x,t)$ denotes x-component of the wind vector. Let τ denotes the time resolution step and h – the grid spacing. The method is based on integrating over the wind trajectory the profile that arrives at a grid point P at time $(n+1)\tau$

$$f_P^{n+1} = f_*^n,$$

where f_*^n is the value of the factor f at the departure point x_* at time $n\tau$ (see [5] for details of numerical scheme construction).

Estimation of the departure profile f_*^n is an essential step of this approach. It can be obtained by polynomial interpolation (2), where the interpolation parameter is determined as

$$\alpha = C - p, \qquad (0 \le \alpha \le 1),$$

for $C = u\tau/h$ – the Courant number. Here p – is a parameter (integer number) that indicates how many grid steps upstream from the arrival point lies the interpolation interval [1, 5, 8].

Shape-preserving properties of the interpolation method can be verified by advection experiment of standard test functions. Some results of one-dimensional advection can be found in [5, 6]. The method can be easily extended to two-dimensional case by use of directional split or tensor product [9, 12]. Two examples of rotating profiles considered below are commonly used in the literature [1, 2, 5, 8–12]. The first test function represents a unit amplitude cone rotating in an advection field of constant angular velocity. It is considered in the square domain $[0,1] \times [0.1]$ discretized with the step $h = 0.01$. The initial profile of radius $r_o = 15h$, centered at $(0.25, 0.5)$ is placed on a constant background equal to 25. One rotation of the cone is completed after 60 time steps.

Figure 2 shows results of simulation obtained after 6 revolutions, by (AK2) and (FB2) methods. The profile in both cases is accurately placed, with good shape preserving and very low clipping effect (the maximum values are 0.98 and 0.97, respectively). Regularity of the solution is remarkably good in case of (FB2) method.

The quantitative evaluation of accuracy in the case considered can be obtained by standard error measures [1, 8, 11]. Given a function f and its approximate solution F calculated on the set of K grid points, the mean-square error is defined as

$$\| f - F \|_h^2 = \frac{1}{K}\sum_k (f_k^n - F_k^n)^2,$$

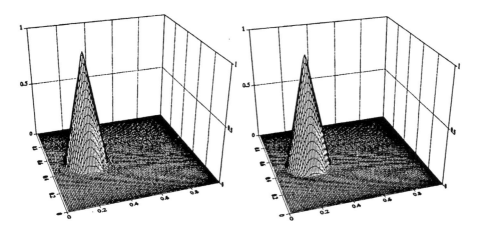

Figure 2. Rotating cone after six revolutions by AK2 (left) and FB2 (right) methods.

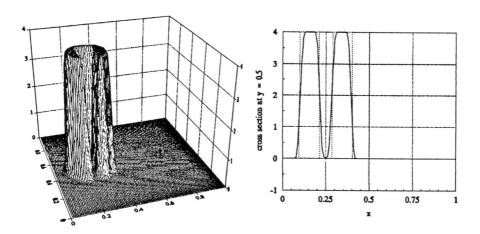

Figure 3. Slotted cylinder after six revolutions by FB2 method and a cross section at $y = 0.5$.

for the n-th time step. It can be shown [11] that the error is a sum of *dissipation error* and *dispersion error*

$$\| f - F \|_h^2 = E_{DISS} + E_{DISP},$$

where

$$E_{DISS} = [\sigma(f) - \sigma(F)]^2 + (\overline{f} - \overline{F})^2,$$
$$E_{DISP} = 2(1 - \rho)\sigma(f)\sigma(F).$$

Here $\overline{f}$ and $\sigma(f)$ denote the mean and the variance, respectively. The correlation coefficient between functions f and F on the set of grid points is denoted by ρ. Error measures for rotating cone experiment are presented in Table 1 and Table 2.

Table 1. Errors for rotaying cone experiment by AK2 method.

Step	$\int F / \int f_o$	$\int F^2 / \int f_o^2$	max(F)	min(F)	E_{DISS}	E_{DISP}
60	1.0001	1.0001	0.9662	0.0	0.621E-06	0.677E-05
120	1.0002	1.0007	0.9675	0.0	0.631E-06	0.919E-05
180	1.0002	1.0013	0.9695	0.0	0.632E-06	0.130E-04
240	1.0006	1.0018	0.9723	0.0	0.664E-06	0.178E-04
300	1.0010	1.0025	0.9753	0.0	0.702E-06	0.232E-04
360	1.0015	1.0031	0.9783	0.0	0.737E-06	0.293E-04

Table 2. Errors for rotating cone experiment by FB2 method.

Step	$\int F / \int f_o$	$\int F^2 / \int f_o^2$	max(F)	min(F)	E_{DISS}	E_{DISP}
60	1.0000	1.0002	0.9704	0.0	0.585E-06	0.718E-05
120	1.0000	1.0010	0.9720	0.0	0.580E-06	0.934E-05
180	1.0000	1.0017	0.9741	0.0	0.595E-06	0.128E-04
240	0.9999	1.0021	0.9734	0.0	0.604E-06	0.175E-04
300	0.9997	1.0024	0.9702	0.0	0.605E-06	0.235E-04
360	0.9995	1.0025	0.9689	0.0	0.605E-06	0.306E-04

Another two-dimensional computational example concerns rotation of a slotted cylinder (compare [1, 5, 8, 10]). The initial profile forms a cylinder with height equal to 4, radius $r_o = 15h$, and a center at $(0.25, 0.5)$. The slot has the width $6h$ and the length $22h$. The resulting profile and its cross-section at $y = 0.5$, obtained after six revolutions by (FB2) method are presented in Fig. 3. A similar result obtained by (AK2) algorithm can be found in [6].

Table 3. Errors for slotted cylinder experiment by FB2 method.

Step	$\int F / \int f_o$	$\int F^2 / \int f_o^2$	max(F)	min(F)	E_{DISS}	E_{DISP}
60	0.9996	0.8817	4.0000	0.0	3.511E-03	4.120E-02
120	0.9990	0.8683	4.0000	0.0	4.377E-03	4.768E-02
180	0.9986	0.8601	4.0000	0.0	4.955E-03	5.182E-02
240	0.9985	0.8544	4.0000	0.0	5.382E-03	5.499E-02
300	0.9984	0.8500	4.0000	0.0	5.726E-03	5.774E-02
360	0.9983	0.8464	4.0000	0.0	6.017E-03	6.026E-02

The solution obtained is positive, with no substantial diffusion and without overshootings. Both the upper face of the lobes and the depth of the slot are reproduced very well in comparison with other methods (compare Figure 3). The error measures presented in Table 3 are comparable or better than the similar results in references quoted. It can also be seen that the method has quite satisfactory conservation properties.

5   Summary

The paper contributes to the state-of-the-art in higher order interpolation methods, with emphasis on shape preserving properties [3, 5–7]. Nonnegativity conditions of two types are formulated as Proposition 1 and Proposition 2. They state that the scheme discussed is positive definite if the limiter constant in derivative estimate function is $\varrho \le 8/3$. However, for $\varrho \le 3.5$ the scheme is also positive definite for strictly monotone data, and admits very limited under- and overshootings of local extrema. Thus, it can be successfully applied for that value of ϱ, especially for smooth profiles and in flat gradient areas. Monotonicity conditions, which are characterized by Proposition 3, can be easily implemented as computational algorithm [6].

Numerical tests performed for two implementations of derivative estimates confirm good accuracy of interpolant and very good shape-preserving properties in simulation of advection process. An essential gain of the overall accuracy is due to relaxing the standard, first order monotonicity constraints in vicinity of local extrema, and replacing them by parabolic approximation [7]. This can be seen by comparing results presented in section 4 with those of [5] or by comparing accuracy of (AK) and (AK2) as well as (FB) and (FB2) algorithms.

Results of 2D advection tests can be compared with several other papers, where similar functions were considered [1, 2, 8–12]. In rotating cone example, both (AK2) and (FB2) methods show very good shape-preserving properties, low diffusivity and minor maximum clipping effect. Generally, the best results are obtained by combining interpolant (2) with Fritsch-Butland limiter and parabolic approximation near extremum (FB2).

References

1. Bermejo, R. & Staniforth, A. The conversion of semi-Lagrangian advection schemes to quasi-monotone schemes, *Mon. Wea. Rev.*, 1992, **120**, 2622 – 2632.

2. Bott, A. A positive definite scheme obtained by nonlinear renormalization of advection fluxes, *Mon. Wea. Rev.*, 1992, **117**, 1006 – 1015.

3. Dougerthy, J.R., Edelman, A. & Hyman, J.M. Nonnegativity-, monotonicity-, or convexity-preserving cubic and quintic Hermite interpolation, *Math. Comp.*, 1989, **52**, 471 –494.

4. Fritsch, F.N. & Butland, J. A method for constructing local monotone piecewise cubic interpolants, *SIAM J. Sci. Stat. Comput.*, 1984, **5**, 300 – 304.

5. Holnicki, P. A shape-preserving interpolation: applications to semi-Lagrangian advection, *Mon. Wea. Rev.*, 1995, **123**, 862 – 870.

6. Holnicki, P. *A piecewise-quintic interpolation scheme*, Rep. A30.82/95, pp. 2 – 43, Systems Research Institute, Warsaw 1995.

7. Huynh, H.T. Accurate monotone cubic interpolation, *SIAM J. Numer. Anal.*, 1993, **30**, 57 – 100.

8. Priestley, A. A quasi-conservative version of the semi-Lagrangian advection scheme, *Mon. Wea. Rev.*, 1983, **121**, 621 – 629.

9. Rasch, P.J. & Williamson, D.L. On shape-preserving interpolation and semi-Lagrangian transport, *SIAM J. Sci. Comput.*, 1990, **11**, 656 – 686.

10. Staniforth, A. & Côté, J. Semi-Lagrangian integration schemes for atmospheric models – a review, *Mon. Wea. Rev.*, 1991, **119**, 2206 – 2223.

11. Takacs, L.L. A two-step scheme for the advection equation with minimized dissipation and dispersion errors, *Mon. Wea. Rev.*, 1985, **113**, 1050 – 1065.

12. Williamson, D.L. & Rasch, P.J. Two-dimensional semi-Lagrangian transport with shape-preserving interpolation, *Mon. Wea. Rev.*, 1989, **117**, 102 – 129.

A space-time finite element formulation for the shallow water equations

F.L.B. Ribeiro,[a] A.C. Galeão,[b] L. Landau[a]

aPrograma de Engenharia Civil, COPPE / Universidade Federal do Rio de Janeiro, Caixa Postal 68506, Rio de Janeiro, RJ 21945970, Brasil
bLaboratório Nacional de Computação Científica, Rua Lauro Müller 455, Rio de Janeiro, RJ 22290-160, Brasil

Abstract

This paper presents a space-time finite element formulation for problems governed by the shallow water equations. A constant time-discontinuous approximation is adopted, while linear three node triangles are used for the spatial discretization. The streamline upwind Petrov-Galerkin (SUPG) method is applied in its equivalent variational form to fit the time discretization. Also, the correspondent semi-discrete SUPG version is established, and some numerical results are presented in order to compare the performance of these methods.

1 Introduction

As it is well known, the use of the classical Galerkin method to approximate convection-dominated phenomena leads to numerical solutions contaminated by spurious oscillations that are spread over the entire computational domain. In the context of weighted residual methods, a remarkable improvement on the numerical solution of such problems was provided by the consistent variational SUPG method proposed in by Brooks and Hughes[1]. Since then, many SUPG based methods have been used for multivariable systems of equations.. In Sharkib[2] and Almeida and Galeão[3], space-time Petrov-Galerkin (STPG) formulations were derived for the compressible Euler and Navier-Stokes equations, showing their inherent control of derivatives along the characteristics. This fact turns out to be an improvement over the classical semi-discrete SUPG formulation, where only streamline derivatives are controlled. The discontinuity capturing approach proposed in those references lead to stable and accurate methods to solve all details of sharp layers and/or shock discontinuities. If we realize that the mathematical structure of the above mentioned equations is identical to those that govern the shallow water problem, it can be immediately concluded that for such problems, P-G weighted residual methods will also perform well. This was done in Bova and Carey[4] and Saleri[5], where semi-discrete P-G (SDPG) finite elements were employed, and in Carbonel, Galeão and Loula[6], where a space-time P-G formulation was derived, using linear interpolation for both spatial and temporal discretization. In this paper, the continuous linear interpolation is retained for the spatial discretization, but constant time-discontinuous interpolation is adopted. For this choice of interpolation, the weighting function gives no contribution to the terms involving time derivatives. Even so, the resulting discrete space-time Galerkin (STG) equations coincide with the semi-discrete Galerkin approximation with an Euler backward difference scheme to approximate the first time derivatives. Nevertheless, this will not occur with the P-G weighting residual methods. The numerical examples that will be presented later will show that, under these circumstances, the STPG formulation will give more accurate results than the SDPG model.

2 Problem statement

Let $(x, y) \in \Omega \subset \Re^2$ define a set of points on an horizontal plane and let $z \in [-h, \eta]$ denote the vertical direction, where $h(x, y)$ represents the water depth and $\eta(x, y, z)$ is the water surface elevation, both measured from the undisturbed water surface. We start from the 3-D incompressible Navier-Stokes equations, after turbulent time-averaging, integrating these equations along the z direction using depth-averaged horizontal velocities. Under the simplifying assumption of a hydrostatic pressure distribution (negligible vertical acceleration), we arrive at the shallow water equations:

$$u_{,t} + uu_{,x} + vu_{,y} + (gH)_{,x} = gh_{,x} + fv + \frac{1}{H}(\sigma|\mathbf{w}|w^x + \frac{1}{C^2}g|\mathbf{u}|u) + \mu(u_{,xx} + u_{,yy})\quad(1a)$$

$$v_{,t} + uv_{,x} + vv_{,y} + (gH)_{,y} = gh_{,y} - fu + \frac{1}{H}(\sigma|\mathbf{w}|w^y + \frac{1}{C^2}g|\mathbf{u}|v) + \mu(v_{,xx} + v_{,yy})\quad(1b)$$

$$H_{,t} + Hu_{,x} + Hv_{,y} + uH_{,x} + vH_{,y} = 0 \quad\quad(1c)$$

In these equations, $H = h + \eta$ is the total water depth and $\mathbf{u}$ is the depth-averaged velocity, with components u and v in x and y directions respectively. The gravitational acceleration is given by g and f is the Coriolis parameter. The wind velocity is $\mathbf{w}$, with components w^x and w^y. σ and C are, respectively, the surface and Chezy friction coefficients, and μ is the eddy viscosity.

Multiplying the third equation by g and observing that,

$$(gH)_{,t} = \left(\left(\sqrt{gH}\right)^2\right)_{,t} = \sqrt{gH}\left(2\sqrt{gH}\right)_{,t} = c(2c)_{,t} \tag{2}$$

where $c = (gH)^{1/2}$, and considering similar expressions for $(gH)_{,x}$ and $(gH)_{,y}$, we obtain the shallow water equations in the velocity-celerity variables (see Saleri [5]) which, in matrix form, can be written as:

$$U_{,t} + A.\nabla U = F \; ; \qquad A.\nabla U = A^T \nabla U \tag{3a}$$

$$U^T = \begin{bmatrix} u & v & 2c \end{bmatrix}; \quad A^T = \begin{bmatrix} A_1^T & A_2^T \end{bmatrix}; \quad \nabla U^T = \begin{bmatrix} U_{,x}^T & U_{,y}^T \end{bmatrix} \tag{3b}$$

$$A_1 = \begin{bmatrix} u & 0 & c \\ 0 & u & 0 \\ c & 0 & u \end{bmatrix}; \qquad A_2 = \begin{bmatrix} v & 0 & 0 \\ 0 & v & c \\ 0 & c & v \end{bmatrix} \tag{3c}$$

$$F = \begin{bmatrix} gh_{,x} + fv + \dfrac{1}{H}\left(\sigma|w|w^x + \dfrac{1}{C^2}g|u|u\right) + \mu(u_{,xx} + u_{,yy}) \\[2mm] gh_{,y} - fu + \dfrac{1}{H}\left(\sigma|w|w^y + \dfrac{1}{C^2}g|u|v\right) + \mu(v_{,xx} + v_{,yy}) \\[2mm] 0 \end{bmatrix} \tag{3d}$$

Once an initial state $U_0(x, y)$ is specified at $t = 0$ and appropriate boundary conditions are prescribed, the system of equations above can be solved to give the unknown column vector U.

To obtain the space-time description of (3a-d) we introduce the variable change: $s = (1)t$, where (1) has units of velocity. Then if we define:

$$\overline{A}^T = \begin{bmatrix} I & A_1 & A_2 \end{bmatrix}; \quad \overline{\nabla} U^T = \begin{bmatrix} U_{,s}^T & U_{,x}^T & U_{,y}^T \end{bmatrix} \tag{4a}$$

where I is the (3x3) identity matrix, we will say that the space-time solution of the original problem (3a-d) is the (3x1) column vector U that satisfies the transformed equation:

$$\overline{A}.\overline{\nabla} U = F \quad in \quad S \equiv \Omega x(0, T) \tag{4b}$$

3 Petrov-Galerkin finite element model

In order to construct the space-time finite element subspace, let us consider partitions $0 = t_0 < t_1 < \ldots t_n < t_{n+1}$ of $\Re^+$ and denote by $I_n = (t_n, t_{n+1})$ the nth time interval. For each n the space-time integration domain is the "slab" $S_n = \Omega \times I_n$, with boundary $\overline{\Gamma} = \Gamma \times I_n$. If we define S_n^e as the eth element in S_n, $e = 1, \ldots (N_e)_n$, where $(N_e)_n$ is the total number of elements in S_n, then for $n = 0, 1, 2, \ldots$

(i) the space-time finite element partition $\Pi^{h, \Delta t}$ is such that:

$$S_n = \bigcup_{e=1}^{(N_e)_n} S_n^e \; ; \quad S_n^e = \Omega_e \times I_n \; ; \quad \Omega = \bigcup_{e=1}^{N_e} \Omega_e \; ; \quad \Omega_i \cap \Omega_j = \emptyset \quad for \; i \neq j \quad (5a)$$

(ii) the space-time finite element subspace consists of continuous piecewise polynomials on the slab S_n, and may be discontinuous in time across the time levels t_n, that is:

$$\hat{\mathcal{U}}_n^h \equiv \left\{ \hat{U}^h; \quad \hat{U}^h \in \left(C^0(S_n) \right)^3; \quad \hat{U}^h \big|_{S_n^e} \in \left(P^k(S_n^e) \right)^3; \quad \hat{U}^h \big|_{\overline{\Gamma}} = 0 \right\} \quad (5b)$$

where P^k is the set of polynomials of degree less than or equal to k.

According to the above definitions, the variational STPG formulation for the problem (4a-b) reads:

Find $U^h \in \mathcal{U}_n^h$ such that for $n = 0, 1, 2, \ldots$

$$\int_{S_n} \hat{U}^h \cdot \mathcal{L}^h d\Omega dt + \sum_{e=1}^{(N_e)_n} \int_{S_n^e} \left(\tau \, \overline{A} . \overline{\nabla} \hat{U}^h \right) . \mathcal{L}^h d\Omega dt + \quad (6a)$$

$$+ \int_{\Omega} \hat{U}^h(t_n^+) . \left(U^h(t_n^+) - U^h(t_n^-) \right) d\Omega = 0 \; ; \quad \forall \hat{U}^h \in \hat{\mathcal{U}}_n^h$$

where,

$$\mathcal{U}_n^h \equiv \left\{ U^h; \quad U^h \in \left(C^0(S_n) \right)^3; \quad U^h \big|_{S_n^e} \in \left(P^k(S_n^e) \right)^3; \quad U^h \big|_{\overline{\Gamma}} = G \right\} \quad (6b)$$

$$\mathcal{L}^h = \overline{A} . \overline{\nabla} U^h - F^h = U_{,t}^h + A(U^h) . \nabla U^h - F^h \quad (6c)$$

• *Remarks*

(1) If in (6a) the integrals are taken over Ω and Ω_e instead of S_n and S_n^e, respectively; the P-G weighting function $\left(\tau\,\overline{A}.\overline{\nabla}\hat{U}^h\right)$ is replaced by $\left(\tau\,A.\nabla\hat{U}^h\right)$; and in (5b) the finite element subspace is defined for all t (making the jumping term in (6a) to disappear); and finally, if we approximate the partial time-derivative by a time-differencing operator, we generate the semi-discrete SUPG method.

(2) If in (6a) we do not consider the added SUPG contribution represented by the second term under the summation symbol, we reproduce the time-discontinuous Galerkin method. If in this case, for instance, we use constant time-discontinuous interpolation, the resulting space-time finite element method will be identical to the backward Euler semi-discrete finite element procedure. Since constant time interpolation is used, $U_{,t}^h \equiv 0$, and therefore it is clear that the jumping condition, represented by the third integral in (6a), is the term responsible for this equivalence. Nevertheless, the STPG method and the correspondent SDPG method will be different.

(3) The definition of the matrix τ will be also different in these two formulations. This point is focused in the next section.

4 Purely hyperbolic problems

To simplify our analysis let us assume $F = 0$. Using the τ matrix definition found in Sharkib[2], we have:

• for the space-time formulation,

$$\tau_{st} = \left[\left(\frac{\partial\xi_0}{\partial x_0}\right)^2 \mathbf{I} + \sum_{i,l,j=1}^{2}\left(\frac{\partial\xi_i}{\partial x_l}\right)\left(\frac{\partial\xi_i}{\partial x_j}\right)A_l\,A_j\right]^{-1/2} \tag{7a}$$

where $x_0 = t$; $x_1 = x$; $x_2 = y$; ξ_k ($k = 0,1,2$) are the local coordinates of the parent element S_n^e; and A_1, A_2 are the Jacobian flux matrices defined in (3c).

• for the semi-discrete formulation,

$$\tau_{sd} = \left[\sum_{i,l,j=1}^{2}\left(\frac{\partial\xi_i}{\partial x_l}\right)\left(\frac{\partial\xi_i}{\partial x_j}\right)A_l\,A_j\right]^{-1/2} \tag{7b}$$

where ξ_i (i = 1,2) are the local coordinates of the parent element Ω_e. If, once again, for the sake of simplicity , we consider unidimensional problems and $\Pi^{h,\Delta t}$ partitions of equal $(h_e \times \Delta t) S_n^e$ elements, (7a-b) simplify to:

$$\tau_{st} = \left[\left(\frac{2}{\Delta t} \right)^2 \mathbf{I} + \left(\frac{2}{h_e} \right)^2 A_1^2 \right]^{-1/2} = \sum_{i=1}^{2} Z_i^{st} \left(\lambda_i^{st} \right)^{-1/2} \left(Z_i^{st} \right)^T \tag{8a}$$

$$\tau_{sd} = \frac{h_e}{2} \left(A_1^2 \right)^{-1/2} = \sum_{i=1}^{2} Z_i^{sd} \left(\lambda_i^{sd} \right)^{-1/2} \left(Z_i^{sd} \right)^T \tag{8b}$$

where,

$$\left(\lambda_{1,2}^{st} \right) = \frac{4}{\Delta t^2} + \frac{4}{h_e^2} (u \pm c)^2 ; \qquad \left(\lambda_{1,2}^{sd} \right) = (u \pm c)^2 \tag{9a}$$

$$\left(Z_{1,2}^{st} \right)^T = \left[-\frac{1}{\sqrt{2}} \quad \mp \frac{1}{\sqrt{2}} \right] = \left(Z_{1,2}^{sd} \right)^T \tag{9b}$$

are the eigenvalues and correspondent eigenvetors of $(\tau_{st})^{-2}$ and A_1^2 respectively. With these definitions we have,

$$\tau_{st} = \frac{1}{2\sqrt{\lambda_1^{st}} \sqrt{\lambda_2^{st}}} \begin{bmatrix} \sqrt{\lambda_2^{st}} + \sqrt{\lambda_1^{st}} & \sqrt{\lambda_2^{st}} - \sqrt{\lambda_1^{st}} \\ \sqrt{\lambda_2^{st}} - \sqrt{\lambda_1^{st}} & \sqrt{\lambda_2^{st}} + \sqrt{\lambda_1^{st}} \end{bmatrix} \tag{10a}$$

$$\tau_{sd} = \frac{h_e}{2|u^2 - c^2|} \frac{1}{2} \begin{bmatrix} |u+c| + |u-c| & |u-c| - |u+c| \\ |u-c| - |u+c| & |u+c| + |u-c| \end{bmatrix} ; \quad u \neq c \tag{10b}$$

Now let us introduce the non-dimensional factor $\alpha = \dfrac{h_e}{c\Delta t}$, or what is the same,

$CFL = \dfrac{1}{\alpha} \Rightarrow \alpha \geq 1$, where CFL denotes the well known Courant-Friedrichs-Levy number. Using this factor, the definition of τ_{st} becomes:

$$\tau_{st} = \frac{h_e}{2\delta} \begin{bmatrix} \frac{1}{2}(\gamma_1 + \gamma_2) & \frac{1}{2}(\gamma_2 - \gamma_1) \\ \frac{1}{2}(\gamma_2 - \gamma_1) & \frac{1}{2}(\gamma_1 + \gamma_2) \end{bmatrix} \tag{11a}$$

$$\delta = \left\{ \left[\alpha^2 c^2 + (u^2 - c^2) \right]^2 + 4\alpha^4 c^4 \right\}^{1/2} \tag{11b}$$

$$\gamma_1 = \left[\alpha^2 c^2 + (u + c)^2 \right]^{1/2}; \qquad \gamma_2 = \left[\alpha^2 c^2 + (u - c)^2 \right]^{1/2}; \tag{11c}$$

• *Remarks*

(i) Notice the intrinsic dependence of τ_{st} on the used time step Δt, which is not accounted for in τ_{sd}. In the limit, as $\Delta t \to 0$; $\tau_{st} \to 0$, in a consistent way. This does not occur with the semi-discrete formulation because τ_{sd} is independent on Δt.

(ii) Although not realizable from the practical point of view, $\tau_{st} \equiv \tau_{sd}$ if and only if $\alpha = 0$. Remind that α must be greater than one, in order to the CFL condition be attained. We restate this comment saying that τ_{st} approaches τ_{sd} as the time step Δt becomes larger, or, in other words, when accuracy decreases.

(iii) If we assume that $u \langle\langle c$, then τ_{st} and τ_{sd} become almost diagonal matrices, and can be replaced by,

$$\tau_{sd} = \frac{h_e}{2c} \mathbf{I}; \quad \tau_{st} = \frac{h_e}{2c} \left(\frac{\sqrt{1 + \alpha^2}}{\left\{ \left[1 - \alpha^2 \right]^2 + 4\alpha^4 \right\}^{1/2}} \right) \mathbf{I}; \quad \tau_{st}\big|_{\alpha=1} = \frac{h_e}{2c} \frac{\sqrt{2}}{2} \mathbf{I}$$

Because $\alpha \geq 1$, the above ratio between brackets is always less than one. Even in the most unfavorable situation, CFL $= 1 = \alpha$, τ_{st} introduces less dissipation than τ_{st}.

In order to get a deeper insight about the performance of the STPG and SDPG methods, some numerical experiments will be performed in the next section. For these examples, constant discontinuous time interpolation and piecewise linear continuous spatial interpolation will be adopted for the STPG method. For the correspondent SDPG formulation this same spatial discretization will be used, combined, with an implicit backward finite difference scheme for time discretization.

5 Numerical results

Our first example is the well known *dam break* problem, which consists of a wall separating two undisturbed water levels that is suddenly removed (Figure 1a). Friction effects are neglected and the spatial discretization is given by a 2x100 triangular elements mesh, as illustrated in Figure 1b. Figures 2-3 show the results for $t = 2.50$, respectively, comparing the solutions obtained with the Galerkin, the space-time and the semi-discrete formulations. For a time step $\Delta t = 0.10$ (Figure 2), Galerkin solutions exhibit oscillations in the entire computational domain. This does not occur for both, the semi-discrete and the space-time P-G solutions, which accurately approximate the high gradients between the three horizontal water levels. For this time step, the semi-discrete solution is sharper than that obtained with the space-time formulation. The effect of reducing the time-step is shown in Figure 3, where the results corresponding to $\Delta t = 0.05$ are plotted for $t = 2.50$. For the Galerkin solutions, the oscillations grow up. For the semi-discrete solutions some localized oscillations appear near the sharpest layer, while a sharper solution without oscillations is obtained with the space-time method.

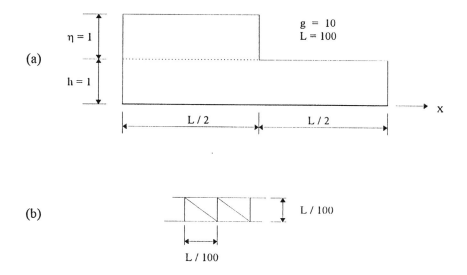

Figure 1: Dam break problem.

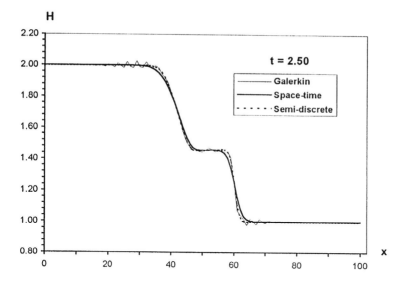

Figure 2: Solution for time t = 2.50, Δt = 0.1.

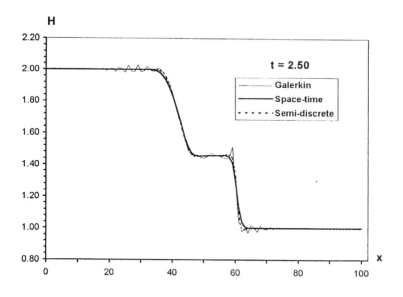

Figure 3: Solution for time t = 2.50, Δt = 0.05.

The second example, illustrated by Figure 4a, is the problem of a reflecting wave in a frictionless horizontal channel of length L = 5000, discretized with 2x10 elements, as shown in Figure 4b. The channel is open at the inflow boundary and closed at the opposed boundary. The system is subjected to a boundary condition at point A, raising the water level suddenly from the initial state of rest (H = 10) to H = 10.1, within one time step. The results can be seen in Figures 5-6. In these figures, the time-history responses for the water surface elevation at point B are depicted. For Δt = 10 (Figure 5), the curves of both, Galerkin and semi-discrete solutions, almost coincide. The STPG method presents an overdiffusive behavior for this time step, leading to a solution that progressively damps along time. However, a completely different behavior occurs when Δt = 1. For this time step, the STPG solution reaches the rectangular pulse form, while the Galerkin and SDPG solutions present some oscillations, as can be observed in Figure 6.

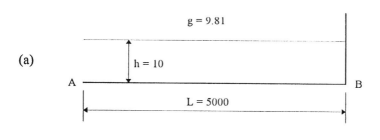

(a)

(b)

Figure 4: Reflecting wave in a frictionless channel.

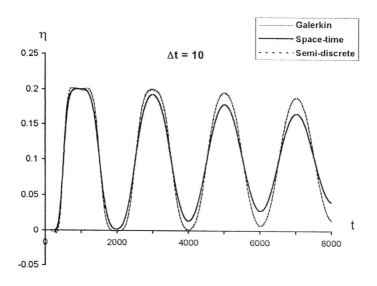

Figure 5: Solution at point B, with $\Delta t = 10$.

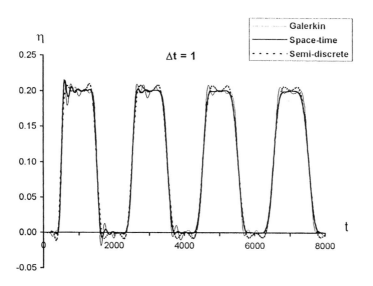

Figure 6: Solution at point B, with $\Delta t = 1$.

6 Conclusions

In this paper, a STPG finite element model was derived for problems governed by the shallow water equations. A piecewise linear continuous interpolation (in the space variables) was used, and piecewise constant discontinuous functions were adopted for time interpolation. In addition, the correspondent semi-discrete P-G model was also presented. The numerical results showed in this work indicate that the STPG method performs better   than the SDPG formulation. Finally we should comment about the necessity of using a consistent time-space shock-capturing PG weighting function such that, in the limit, when $\Delta t \to 0$, the approximated numerical characteristics approaches the true characteristics. This can not be actually provided by the SUPG method. The CAU generalized method proposed in Almeida and Galeão [3] for compressible flows, fulfill these requirements and, can therefore be successfully applied to shallow water problems.

References

1. Brooks, A. N., Hughes, T. J., Streamline Upwind Petrov-Galerkin Formulation for Convection-Dominated Flows with Particular Emphasis on the Incompressible Navier-Stokes Equations, *Comput. Meth. Appl. Mech. Engrg*, Vol. 32, pp 199-259, 1982.

2. Shakib, F., Finite Element Analysis of the Compressible Euler and Navier-Stokes Equations, *Ph.D. Thesis*, Stanford University, 1988.

3. Almeida, R. C., Galeão, A. C., An Adaptive Petrov-Galerkin Formulation for the Compressible Euler and Navier-Stokes Equations. *Comput. Meth. Appl. Mech. Engrg*, Vol. 129, pp 157-176, 1996.

4. Bova, S. W., Carey, G. F., An entropy Variable Formulation and Petrov-Galerkin Methods for the Shallow Water Equations, in: *Finite Element Modeling of Environmental Problems-Surface and Subsurface Flow and Transport*, ed. G. Carey, John Wiley, London, England, 1995.

5. Saleri, F., Some Stabilization Techniques in Computational Fluid Dynamics, *Proceedings of the 9th International Conference on Finite Elements in Fluids*, Venezia, 1995.

6. Carbonel, C., Galeão, A. C., Loula, A. D., A Two-dimensional Finite Element Model for Shallow Water Waves, *Proceedings of the 13th Brazilian Congress and 2nd Iberian American Congress of Mechanical Engineering*, Belo Horizonte, 1995.

SECTION 8:
SATELLITE DATA, IMAGE PROCESSING, PATTERN RECOGNITION AND REMOTE SENSING

Programming a Task Oriented Application of Spatial Information Systems Toolbox (TO-ASIST) - an example for the automated extraction of land cover and land use data from satellite imagery

N. Rollings

Department of Ecosystem Management, University of New England, Armidale, NSW 2351, Australia

Abstract

Spatial information technology has emerged as a major growth industry over the last 5 to 10 years. It is proffered that spatial information technology has much to offer the resource management community but is not widely utilised. It is suggested that current software is confusing to the lay-user, presenting a barrier to its operational use. Such persons invariably turn to system administrators who soon become swamped with data requests. The development of a Task Orientated - Application of Spatial Information Systems Toolbox (TO-ASIST) is offered as a possible solution to this problem. A TO-ASIST is an in-house developed piece of software dedicated to performing one task. Such systems may be developed using scripting languages offered by many vendors. Alternatively, the development of object oriented programming languages such as Visual Basic provide a simple environment which can tap into the rapidly expanding number of software custom controls. Major vendors are starting to release these custom controls providing direct access to their software functionality from within the design environment. Equally as important as demystifying spatial information technology, the TO-ASIST must also raise the technological skills of the lay-user via on-line documentation and theory lessons. Again using Visual Basic and custom controls, the HTML standard may be employed to develop a series of on-line instructions that new users may access. The concept of a TO-ASIST is demonstrated through Land Use Modeller, a system designed to map land use from multi-temporal satellite imagery. An application of Land use Modeller over the Shire of Melton, Victoria, Australia, is presented.

Introduction

Spatial Information Technology (SIT) has much to offer the professional and lay-resource manager. SIT includes such technologies as remote sensing and geographic information systems. Despite over 20 years of research and development, there are still surprisingly few true operational uses of SIT and large sectors of the resource management community are not taking advantage of its functionality.

Rollings[1] identified two major technical barriers to the wide-spread use of SIT by resource managers. First the complex nature of most SIT software is a deterrent for users to adopt the technology in an operational sense. Attempts have been made by vendors to overcome this problem For example, Earth Resource Mapping of Australia, developed a new concept of user interface while ESRI, arguably the largest GIS vendor in the world, have developed ARC/VIEW in order to make the technology more accessible. Despite these efforts SIT is still difficult to use for *non-experts*. Another barrier to the adoption of SIT is the lack of appropriately written technical material for new users to consult. Existing material appears to be overly technical and time consuming to digest.

Users of Spatial Information Technology

The implementation of a Spatial Information System (SIS) is driven largely from a technological perspective. Additionally, the impetus for such a system often comes from one individual, a section or a department within an organisation. Consequently, these systems are rarely designed with the whole organisation in mind and little attention is paid to end-users of the system. In resource management, the people who will make best use of the spatial database, the decision makers and their immediate staff, are not necessarily SIT literate and their computer skills may be low. Conversely, those who build and run the SIS are not directly involved in decision making, have very high computer literacy and are SIT experts.   Naturally, there will be exceptions but to build a link between these two groups can be very difficult and is critical to the successful operation of the SIS. Figure 1 shows a common scenario where a SIS may not perform optimally due to poor access to the system and data by end-users.

At the management level the system manager is in constant contact with other managers in order to promote the use of the SIS and gain feedback on management requirements. Similarly, there will be some interaction between the system administrator/technician and the end-users but the majority of queries should be conducted by the users themselves using tools that are developed by an applications developer. Unfortunately, end-users are rarely trained to an adequate standard to use the SIS software, even if it has undergone some customising. Often the tools are missing that enable end-users to access the GIS database and users are left to "fend" for themselves. In this situation, as depicted in figure 1, users will invariably request information from the system administrator/technician who soon becomes swamped with requests and is unable to perform their primary task (maintaining the GIS database) and cannot keep up with requests for data. Ultimately the system will collapse and cannot service the needs of its users effectively . A preferable situation is where a communication medium or similar is established for the end-users such that they can access the data they require directly. Under this scenario improved access to end-users can accomplished in one of three ways.

1. Improve technical skills of users such that they are familiar with and competent in the use of the SIS. While attractive to some, others will not want to (nor should they have to) become technically competent in SIS principles.
2. Employ project specialists or consultants to process data requests from end-users. This may be an attractive option for larger organisations who manage their activities on a project-by-project basis. A series of applications specialists may be employed on several projects at once to satisfy the requirements of the end-users.
3. Simplify the technology so that is easy enough for end-users to access it with minimal training.

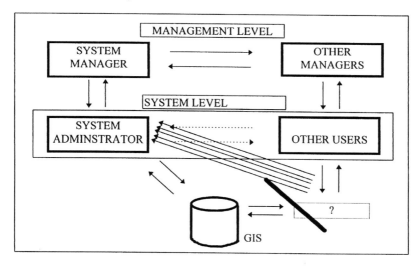

Figure 1. A common scenario when end-users attempt to access a SIS with little success

For those organisations who are not large enough to adopt solution 2, the answer lies in a combination of solutions 1 and 3. It is possible to simplify SIT to make it easier to access by end-users provided they have some degree of understanding and training in the use of SIT. In practice, a series of tools can be built to aid the end-user while an organisational, education programme can raise awareness of SIS and its capabilities. In figure 2 a series of tools have been built to allow end-users to access the GIS database directly thus freeing the system administrator to perform their essential role.

Rollings and Condurso[2] demonstrated the concept of a Task Orientated Image Processing System (TOIPS) that allowed land use information to be extracted from multi-temporal satellite imagery. This concept may be extended to SIS in general to produce the Task Orientated - Application of Spatial Information Systems Toolbox (TO- ASIST).

Task Oriented Application of a Spatial Information Systems Toolbox : TO-ASIST

A TO-ASIST may be defined as a *Customised software tool developed for a specific function using spatial information systems principles presented in an environment that promotes the understanding of SIT principles*. Notice that a TO-ASIST is dedicated to *one function*. If a more complex TO-ASIST was developed then we are returning to a user interface menu, the very thing we are trying to avoid. The development of a TO-ASIST must be driven from the end-user rather that a SIT specialist attempting to predict what is needed. From concept to operational product, the following procedure will ensure that a useable TO-ASIST is developed.

1. An information need is established by end-users. It is the end users who are in the position to describe what they need, and ultimately shape the final form of the TO-ASIST. To do otherwise would run the risk of the technology manipulating the process for technological outcomes rather than the management process manipulating the technology to achieve good management and planning decisions.
2. The task is divided into sub-tasks and a flow diagram created to depict each task and where it fits into the overall structure of the TO-ASIST.
3. All tasks are undertaken in a pilot project by an SIT expert to test the soundness of the methodology.
4. Technological barriers are identified against each sub-task.
5. Solutions to the technological barriers are devised (if possible).
6. Key SIT concepts are identified next to each task and theory lessons and informative practical tips are written.
7. The structure of the TO-ASIST is developed.

8. Based on the structure, an automated list of instructions are written for each step of each task. Theory lessons are linked to these steps.
9. Implementation options are considered. The system may be developed under the envelope of an existing system using a scripting language (eg. mapbasic in Mapinfo) or alternatively it may be developed using a lower level programming language such as Visual Basic.
10. The prototype system is developed and tested by the end-users.
11. Necessary modifications are made and the TO-ASIST is commissioned in the field.
12. The system is monitored for performance.

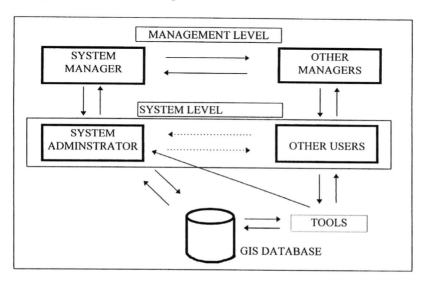

Figure 2. An ideal scenario for end-users to access a SIS.

Not all tasks will be suitable for customising into a TO-ASIST, particularly tasks that involved complex processing procedures with a high user-input based on experience. The use of expert systems may overcome this problem and is the focus of current research. Implementation of a TO-ASIST using a low level programming language versus a scripting language is a difficult decision. Tables 1 and 2 highlight some of the advantages and disadvantages of each method.

It is apparent that for simple systems such as IDRISI, the low level programming language is a better option while on more complex systems, such as Arc/Info, a scripting language may be the preferred option. This may change as software is implemented under the Windows environment and SIT software functionality may be called via Dynamic Link Libraries and through custom controls.

Table 1 Development Using of Low Level Programming Language.

Advantages	Disadvantages
1. Complete control over software functionality, system can be fully "open" 2. Distribution is easy, no need for expensive licenses in all offices	1. All functionality must be programmed from basics 2. Software development costs are high and need specialist staff 3. Many vendors use proprietary data structures which are not accessible to users.

Table 2 Development Using a macro or scripting Language.

Advantages	Disadvantages
1. Full power of system is available with less programming required. 2. Existing SIT experts can easily learn the language	1. Functionality may be restricted to software functionality 2. Interfaces can be clumsy 3. Licenses are often needed to run the software

The first signs of this are now evident as ESRI have announced a product called "Map Objects" which allow users to develop their own applications using Arc/Info tools via a custom control, designed for use with Visual Basic, C++, Delphi or a similar object oriented programming language.

Development of a TO-ASIST : Land Use Modeller

The need to map land use on a routine basis was established during a project concerned with water quality research Land cover and land use may be mapped via remote sensing but to make the application operational it was decided to develop a TO-ASIST to automate the land use extraction.

Concept

The purpose for which land is used is of interest both for planners and managers at all levels. Additionally, land use history can also be of interest when land use conflict issues arise, to model land degradation potential or when environmental degradation becomes evident. This is well illustrated by the current concern over toxic cyano-bacteria blooms in water storages and rivers whose catchments are dominated by agricultural land use. The extraction of land cover information over extended areas is feasible using satellite imagery. Current satellites can provide timely and repetitive information at a variety of spatial and temporal resolutions. While useful, the distribution of land cover is insufficient to provide reliable land use information. The ability to discriminate land use is dependent on resolving a time sequence of land cover types that correspond uniquely to a particular land use, eg a land parcel may be ploughed in month X but may be used for improved pasture, weed control or cropping. Only by observing that same paddock at a later date(s) can the land use be resolved. Hence a temporal database of land cover may be used to extract land use via the recognition of land cover patterns.

Development

The conceptual model as detailed by Rollings and Condurso[2] required several technical barriers to be overcome. The basis of Land Use Modeller (LUM), as the TO-ASIST was termed, was the classification of a sequence of satellite images over a yearly growth cycle. By careful selection of several images, land use may be interpreted from a sequence of land cover images. To facilitate this, each image is classified into one of four simple cover types, forest, active growth, inactive growth and bare. A land cover vs land use calendar can then be constructed for a 12 month period. A sequence of images, local knowledge and common sense were used to construct a land use calendar depicting the sequence of land cover types for each land use. This calendar can then be used to select images that can uniquely resolve the desired land use classes. For example, native pasture, grazing land and crops may exhibit the active growth cover-type during winter. During the spring however, native pasture may appear as inactive growth and in summer much of the vegetation will be grazed and the area may resemble bare ground. Table 3 shows the land use calendar compiled for this study.

Table3 Land use calendar as a function of land cover.

Land use	J	F	M	A	M	J	J	A	S	O	N	D
Improved Pasture	m	ml	ml	l	l	l	l	l	l	l	l	ml
Plough/fallow	mb	mb	m	m	m	m	b	b	b	m	m	m
Winter Plough/crop	mb	mb	mb	b	ml	l	l	l	lm	m	m	m
Summer plough/crop	bm	m	b	b	ml	l	l	l	lm	b	b	b
Native/Poor Pasture	mb	mb	m	l	l	l	l	m	m	m	m	m
Forest	f	f	f	f	f	f	f	f	f	f	f	f

m- Inactive growth   b - Bare   l -Active growth   f -Forest

Landsat MSS data was chosen for reasons of cost and the availability of archival data. Geometric integrity was assured by purchasing pre-rectified imagery from the data supplier. Once the data is received, LUM imports the data into the IDRISI data structure. The IDRISI structure was selected as it is a widely used system, the data structure is very simple and it is easy to manipulate IDRISI files using in-house developed software.

In order to make LUM easy to use it was decided that all image classification should be handled automatically with no user input. This required the use of an atmospheric compensation routine and a spectral library to classify the various cover types. Based on the image dates, entered during image import, LUM calibrates the images. Collection of spectral library was undertaken using the corrected imagery. Known cover types were identified in selected paddocks and their brightness values (B.V.) were recorded. This was undertaken over several years of imagery so an indication of variation could be quantified. For the final spectral library, averages were taken. These were later refined during an iterative classification of 8 images. In order to make LUM a stand-alone product, a standard minimum distance classification routine was written to apply the spectral library to the images. Classification files are stored in IDRISI format for further processing. To identify land use, the land cover patterns identified during the compilation of the land use calendar are applied. A new database layer of land use can now be extracted. This task is also automated and transparent to the user.

The development of LUM made the process of extracting land use from satellite imagery possible by the lay person. In order to raise the level of knowledge of the lay operator it was decided to include a series of help files that guide the user through the various steps in LUM. Further to this, a series of theory lessons are included for those users who wish to investigate the theory behind LUM.

Implementation

All software development for LUM was undertaken using Visual Basic Professional Version 3.0 (VB) and a series of VB add-ons (custom controls and DLL's developed by third party software specialists). This enabled full control over the code and its distribution. VB was selected because it uses the familiar BASIC programming language and graphical user interfaces may be built with little or no programming. This represents a considerable time saving. Coding was further simplified by using a third party coding assistant (VB Codemaster, Teltech, USA). VB Codemaster handles all code documentation, some dialogue box creation and all error handling. All image display was handled by a third party VB add-on (VisionTools, Evergreen Technologies, USA). This made image display much quicker than using VB code and will allow enhancement to be made to LUM in the future. All of the help files were created as Microsoft Help Files using a third party add-on (VB Helpwriter, Teltech, USA). VB Helpwriter reads a programmers code to generate the required help topics leaving the programmer to "fill in the blanks". Microsoft help files were kept to a minimum however in preference to hypertext markup language (HTML) documents. This allows the on-line help to be customised to users requirements at a future date. The additional theory files were developed in HTML within Microsoft Word 6.0 with a 3rd party add-on (Easyhelp, Eon Solution, UK). Easyhelp can generate Microsoft Help Files or HTML files from the same document. The HTML files are displayed from within LUM using another 3rd party add-on (Webster HTML OCX beta 914, Trafalmador, USA). The Webster custom control made it possible to write a HTML viewer and programatically change documents according to user input. Some video sequences are included as part of the theory lessons. These video sequences were captured using a video board (Videoblaster, Creative, Singapore). All video sequences were processed and edited on a PC (Adobe Capture and Premiere, Adobe, USA) and then passed to a dedicated video authoring package for hot-spot editing and compilation (MediaShop, Motion Works, USA).

New Visual Basic custom controls are appearing on the market everyday. Major vendors of SIT are recognising that 3rd party application development using their custom controls to provide access to sophisticated SIT processing is the way to proliferation of their technology without the need to make systems completely "open". As more of these products become available, the development of a TO-ASIST using programming languages such as Visual Basic will become more practical and attractive.

Application of a TO-ASIST : Land Use Modeller

The area selected for this study was the Shire of Melton situated 40km West of Melbourne. This area was chosen due to the variation in land use including cereal cropping, grazing of livestock, thoroughbred horse breeding, urban expansion and conservation. In order to test LUM under a variety of conditions, images were selected from years where diverse weather conditions prevailed. Figure 3a shows a Landsat MSS image for October 1987 while figure 3b shows the corresponding classification using LUM. Figure 3c shows one of the final land use image as determined by LUM based on 2 Landsat MSS images for 1987.

Discussion and Conclusion

Spatial information technology has much to offer the resource management community. Many agencies however are not benefiting from this technology. One reason for this is a technological barrier between managers who need information and the complexity of the systems that can provide it. Some vendors have attempted to address this problem by writing generic "front-ends" or user interfaces to their products. Unfortunately, these products often fall short of their original objectives and the software is still daunting to some people. The development of a series of Task Orientated - Application of Spatial Information Systems Toolboxes may be a possible solution. A TO-ASIST may bring the technology to level where lay-users may access it with minimal training. A critical part of a TO-ASIST is comprehensive on-line documentation and "theory lessons" designed to raise the awareness of the lay-user without necessarily expecting them to become SIT experts.

Visual Basic and the many 3rd party add-on modules available provide an easy to use environment for development of a TO-ASIST. The development of such a system requires an extensive knowledge of SIT in order to overcome the technical problems that will inevitably be encountered.

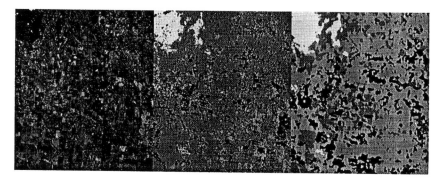

Figure 3a Landsat Image   b Land Cover   c Land Use

3b   Dark Grey - Active Growth   Black - Bare
      Light. Grey - Inactive Growth   White - Forest
3c   White - Forest Light Grey - Marginal Pasture Black - Winter Crops
      Medium Grey - Summer Crops Dark Grey - Fallow

References

1 Rollings, N.M. Developing a Task Oriented Application of Spatial Information Systems Toolbox (TO-ASIST) : Improving Access to Spatial Information Technology for Natural Resources Management. Unpublished PhD Thesis, in prep.

2 Rollings, N.M. and Condurso, C. Automated Land cover and Land use History Mapping from Satellite Imagery. *Proceeding of AURISA 94*, Sydney, 1994.

Remote sensing as a tool for 3D model of ecosystems: an application to lagoon of Venice

C. Solidoro, V.E. Brando, C. Dejak, D.Franco, R. Pastres, G. Pecenik

Department of Physical Chemistry, Sect. of Ecological Physical Chemistry, University of Venice, Dorsoduro 2137, 30123d Venice, Italy

Abstract

Three-dimensional models require a large amount of data for characterising the space and time evolution of the system and for testing the adequacy of the model. Remote sensing can greatly improve the results of these models because makes available space-time series of data hardly obtainable in other ways. Indeed, traditional kind of measurement, while more accurate, cannot provide data sampled as often, and on as a fine mesh, as the ones obtainable by remote sensing technique. Experimental observations derived from remote sensing images, has been coupled with a 3D-water-quality model, to analyse the effects of eutrophication in the lagoon of Venice, namely to define the modality of the spatial expansion of macroalgae community.

The model covers the central part of the lagoon of Venice, that includes the industrial area and the historic town of Venice. It combines transport and tropic processes, describing the seasonal dynamic of the main community of macroalgae (Ulva r.), phytoplankton and zooplankton densities, nutrients' concentration in water (NH_4^+, PO_4^{--}, NO_x), BOD and Dissolved Oxygen. The variation of light and temperature during the year is computed on the basis of meteoclimatic data combined with an energetic balance. The nutrients loading are interpolated from monthly averaged estimations and immitted as point sources.

Transport processes are simulated by 3-D anisotropic turbulent diffusion, which takes into account the average dispersive effects of the tidal agitation. All the state variables are considered as passive solutes, except for macroalgae colonies, which spread according with an empirical law (diffusive process but with different and empirical constant coefficient) and are always confined only into the bottom layer. Macroalgae surviving in the new colonised area will depend both upon rate of macroalgae spreading and bathimetryc conditions, as well as hydrodynamic and trophic conditions.

This is in agreement with the experimental observations which show that Ulva r. is present only in well defined areas. Comparison of the simulation outputs versus remote experimental observation gives the possibility of estimating the rate of spreading of macroalgae colony.

Even starting from homogeneous initial conditions, the model succeeds in describing the main features of the spatial pattern for Ulva r. and the seasonal evolution the other state variables, also as a consequence of the nutrient scenario input and transport phenomena. Results are markedly improved when initial conditions for Ulva, in the early spring, are defined according with a map of presence-absence, obtained by treating a remote sensing image. The observed spatial pattern is correctly described yielding sharp gradient of standing crops biomass.

These first results indicate that the simulations appear to be in reasonable agreement with experimental observations in this area, and therefore supports the plausibility of the Ulva spreading mechanism.

Introduction

In the last two decades the lagoon of Venice went through such modifications that macroalgae are no longer a negligible component of the ecosystem, but they actually are of major importance in the cycling of matter and energy[1].

Mathematical modelling, besides being an useful tool in planning management and recovering strategies, helps in the synthesis and interpretation of the data, acting as a theoretical framework in which one can fit most of the partial knowledge gathered by experimental scientists.

For this reason an eutrophication model, taking into account both phytoplankton and macroalgae dynamics, have been coupled to a three dimensional diffusion model of the central part of the lagoon of Venice[2,3,4,5,6,7]

Remote sensing

Mathematical model requires for a large amount of data, first to understand the physical processes which are going on and actually build a model for them, and then for estimating the parameters used in the formulation, for the calibration and for the validation of the model. This is all the more true for threedimensional model, whose validation goes through a comparison between simulations and experimental spatial distribution of the state variables.

Remote sensing techniques are amongst the more useful tools to this aim, since they are able to make available space-time series of data hardly obtainable in other way. Indeed traditional kind of measurement, while more accurate, cannot provide data sampled as often, and on as a fine mesh, as the ones obtainable by remote sensing technique. Coupling of remote sensing measurement with mathematical modelling are therefore a necessary step in designing a modern management and recovery strategies.

In this paper we used two different remote sensing images, Landsat-5/TM, referring to April 30th, and July 19th 1990 satellite pass.

Several algorithms are available in the specific literature for derive the presence of chlorophyll and suspended matter from the spectral signature of marine water. Nevertheless, the interpretation of remote sensing information collected for the Lagoon of Venice is by no means trivial, because of the many problems due to the particular environmental conditions, which present an high variability both from a geomorphologic point of view that for the water-quality one.[8]

The methodology here used for the recognition of macroalgae vegetation from Landsat-5/TM images[9,10] enable one to identify a spectral indicator, SI, based on the TM spectral bands correlated with dominant biochemical parameters of the physiological stages of macroalgae vegetation.

$$SI = (TM4-TM3)/(TM4+TM3) \qquad (1)$$

Figure 1. Remote sensing image for April 30 1990. Macroalgae presence is indicated by the white area

In Figure 1, referring to spectral indicator SI for April 30th 1990, macroalgae presence is associated to the white area in the very centre of the picture, just under the major island on which the city of Venice is built. From this one we have derived a map of presence-absence of ulva biomass, Figure 2, to be used as starting point for our simulations, by identifying the nine pixels of the landsat images corresponding to each node of the model bathimetry and checking whether or not at least half of them presented macroalgae vegetation.

Figure 3 refers again to the spectral indicator SI, but for July 19th 1990, and only for the area covered by the model.

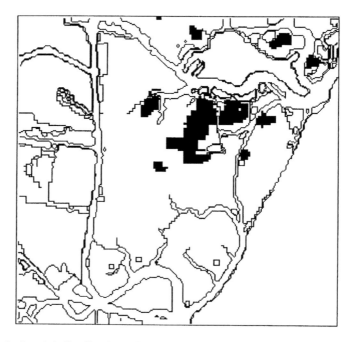

Figure 2. Spatial distribution of macroalgae derived from figure 1 and used as initial condition for our simulations.

Model of Ulva

In the venetian ecosystem *Ulva rigida* is, by far, the most abundant macroalgae species[1]. It colonises shallow environments because of its capacity to survive in adverse external condition like low irradiance or nutrient level. Nutrients are not assimilated from the sediment but from the water medium and there is the possibility for the algae to store part of the nitrogen but not immediately metabolised nitrogen, and to use it in a second time.

Figure 3. Remote sensing image for July 29 1990.   Macroalgae presence is indicated by the white area.

Therefore Ulva can survive for quite a long periods under limiting nitrogen condition as well as in absence of nutrients[11].

Ulva dynamics is described by the general equation

$$dB/dt = \nabla k \nabla B + f(x,t) \qquad (2)$$

the first term describes transport phenomena and second one represents the biological transformations.

The equations of the model for Ulva are summarised in the Table 1. Growth of Ulva follows a two steps kinetic, uncoupling growth from assimilation. Asymptotic formulations describe the influence of light and temperature, as well as respiration. Shading from phytoplankton and water, as well as selfshading, is considered.

A multiplicative model is used also for describing phytoplankton dynamic: nutrients limitation is modelled by Monod type kinetic while influences of the forcing functions are respectively simulated by the Steel function (for light influence) and a modified Lassiter-Kearns function for temperature. Steel formulation takes into account shading effects of water and phytoplankton itself along the water column. Zooplankton grazing follows a type II functional response, while linear terms described mortality losses.

Mortality losses go into the detritus compartment (labile fraction of dead organic matter), where the bacterial mediate demineralisation replenishes nutrients to the water medium. Detritus sinks at a constant rate and, eventually, reaches and accumulates on the bottom, where it is labelled as superficial sediment. The equations are described in more details in [4]

This model succeeds in describing the main features of the spatial pattern for Ulva r. and the seasonal evolution the other state variables: and differentiate the initially homogeneous spatial distribution of macroalgae density, giving rise to sharp gradient of biomass[2].

The second term of equation 1 accounts for the transport phenomena. Since macroalgae have a pretty large size and, even if weakly, are rooted to the bottom they can certainly not be treated as a passive tracer subjected to the eddy diffusion process, as is done for the other state variables. Nonetheless they do spread and give rise to a spatial distribution of the colonies which present also sharp gradient of density of biomass.

Result and discussion

Spreading rate is an important parameter in regulating the evolution of the colonies of macroalgae. The greater the spreading rate the greater the flux of biomass out of an area and, as a consequence, the loss in biomass for that area, which will therefore reach lower value of standing crop. On the other hand, the greater the spreading rate the larger the colonised area and the higher the amount total biomass globally present all over the lagoon.

There is not a clear idea on the modality following which the colonies of macroalgae expand in the lagoon of Venice, for the lack in experimental measurement from one side and of theoretical studies on the other one.

Our hypothesis is that macroalgae are confined in the bottom layer and that it is possible to model their expansion by the purely empirical assumption that diffusion along both horizontal directions is described by a single diffusivity, constant for all the model. Moreover we assume that a colony spreads only in spring and summer and only if its density is above a preassigned minimun threshold. These coefficients are empirical, and might be estimated by comparing visually the space distribution obtained by the model with the observed ones.

This mechanism does not take into account the possibility that floating algae, no longer anchored to the bottom but driven by the currents, will eventually accumulate in certain area. This phenomenon, which following several authors[12] should not be neglected, seems to be of minor relevance in the part of the lagoon covered by our model. Indeed, the hypothesis that residual current and negligible is the very assumption made on modelling transport phenomena as a eddy diffusion process[2].

The surviving of macroalgae in the new colonised area will therefore depends on the bathymetry, the trophic state in the new area and also on the spreading coefficients. Experimental observations confirm that *Ulva r.* is present only in well defined areas.

Several simulation have been performed varying the spreading coefficient. Bathymetry affect the spatial distribution of macroalgae, because the presence of channels or large area of 2 meters depth slowed down expansion acting as a natural barrier. On the other hand, the final distribution depends also upon the initial condition supplied to the model, as it has been seen by comparison (Figures 4 and 5).

For this reason an experimental map of presence-absence of ulva biomass, to be used as starting point for our simulation (Figure 2), has been derived from the remote sensing image referring to April, 30th 1990, as it was mentioned above.

Simulations have then be compared with the distribution, obtained by a second remote sensing image, Figure 3, referring to the end of July 1990. In the figure the colonies of macroalgae are coloured in white, and one can see as they are much larger than in the Figure 2. Our simulation, Figure 4, shows for the same day a similar distribution of macroalgae biomass: the area invaded by macroalgae is considerably larger than the starting one and also its shape resemble the experimental observation, as one can see noting, for example, as the simulated distribution presents the same peculiar shape of the left edge of the main colony (check its bottom-left corner) or the lower density level along the channel in the middle of the main colony.

Figure 4. Simulated spatial distribution of biomass

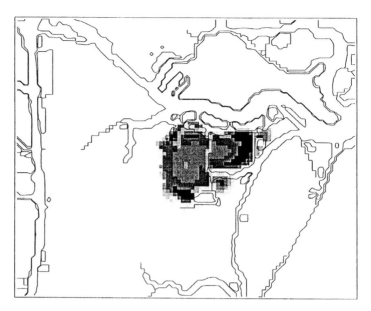

Figure 5. Spatial distribution of macroalgae biomass starting the simulation fron unrealistic initial condition

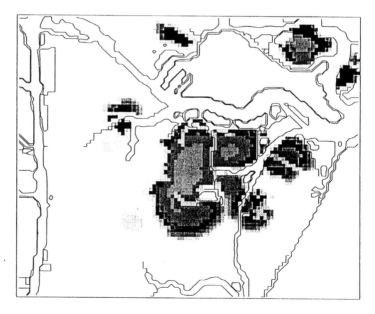

Figure 6. Spatial distribution of macroalgae biomass running the simulation with a lower spreading rate

Figures 5 and 6 respectively illustrate, for comparison, the simulated spatial distribution of biomass obtained starting the simulation from unrealistic initial condition and running the model with a smaller rate of spreading. The differences with the experimental distribution are self evident and indirectly confirm the estimate for spreading parameter used in the previous simulation.

Acknowledgement. The authors wish to thank dr. L. Alberotanza and dr. M. Pavanati for providing the Landsat/TM-5 images. The study was partially supported by ENEL (Ente Nazionale per l'Energia Elettrica) and by the CVN (Consorzio Venezia Nuova).

TABLE 1: MACROALGAE GROWTH MODEL:

Mass Balances

$$dB/dt = (\mu - f_{death}) \cdot B$$
$$dQ/dt = (V_{NH} + V_{NO}) - \mu \cdot Q$$
$$d[DO]/dt = [\varphi - f_{resp}] \cdot B$$
$$d[NH_4^+]/dt = -V_{NH} \cdot B$$
$$d[NO_x^-]/dt = -V_{NO} \cdot B$$
$$d[PO_4^{3-}]/dt = -PCR \cdot \mu \cdot B$$

Functional Expressions

$$\mu = \mu_{max} \cdot g_1(Q) \cdot g_2(P) \cdot g_3(I) \cdot g_4(T)$$
$$\varphi = \varphi_{max} \cdot g_1(Q) \cdot g_2(P) \cdot g_3(I) \cdot g_4(T)$$
$$g_1(Q) = (Q - Q_{min})/(Q - k_c)$$
$$g_2([P]) = [P]/(k_p + [P])$$
$$g_3(T) = \left\{ 1 + \exp\left[-\zeta_p \cdot (T - \vartheta_p)\right] \right\}^{-1}$$
$$g_4(I) = 1 - \exp(-I/I_o) \qquad I = I_{inc.} \cdot \exp\left[-\varepsilon_w - \varepsilon_B \cdot B\right]$$
$$V_{NX} = V_{mNX} \cdot f_1(NX) \cdot f_2(Q)$$
$$f_1([NO_x^-]) = [NO_x^-]/(k_{NO} + [NO_x^-])$$
$$f_1([NH_4^+]) = [NH_4^+]/(k_{NH} + [NH_4^+])$$
$$f_2(Q) = (Q_{max} - Q)/(Q_{max} - Q_{min})$$
$$f_{death} = \alpha B^\beta + k_t \cdot \frac{\max\left[(f_{resp} \cdot B - DO), 0\right]}{f_{resp} \cdot B}$$
$$f_{resp} = k_{resp} \left\{ 1 + \exp\left[-\zeta_{resp} \cdot (T - \vartheta_{resp})\right] \right\}^{-1}$$

References

1. Sfriso,A., Marcomini,A., Pavoni,B. Relationship between macroalgal biomass and nutrient concentrations in a hypertrophic area of the Venice Lagoon. *Mar. Envir. Res.*, 1987, 22: 297-312.
2. Dejak C. and Pecenik G. 1987. Special issue: Venice lagoon. *Ecol. Modelling* 37(1/2): 1-101.
3. Pastres R., Franco D., Pecenik G., Solidoro C., Dejak C. Using parallel computers in envinromental modelling: a working example. *Ecol. Modelling*, 1995 80(1): 69-86
4. Solidoro, C. Dejak, D. Franco, Pastres,R. and G. Pecenik, A growth model for macroalgae (*Ulva r.*) n the Lagoon of Venice. lModel structure identification and parameter estimation. *Ecol. Modelling.* (in press)
5. Solidoro, C. Dejak, D. Franco, Pastres,R. and G. Pecenik, 1995. A model for macroalgae and phytoplankton growth in the Venice Lagoon. *International Enviroment.* 21(5) :619-626
6. Solidoro C., Brando V.E., Dejak C., Franco D., Pastres R., and Pecenik G. Simulation of the seasonal evolution of macroalgae in the lagoon of Venice. *Environment Modelling and Assesment* (in press)
7. Solidoro C., Brando V. E., Franco D., Pastres R., Polenghi I. and Pecenik G. Simulation of macroalgae dynamic using a 3d water quality model of venice lagoon. *CINECA* (in press)
8. Alberotanza L., Pavanati M., Zibordi G., Zandonella A. Landsat-5/TM images used to idenify macroalgal vegetation in the Venice Lagoon, *Water Pollution: Modelling, Measuring and Prediction*, CMP and Elsevier pp 355-369, 1991
9. Alberotanza L., Zandonella A., 1995 Biochemical Information Extraction of Venice Lagoon Macroalgal Vegetation From TM Spectral Bands, *IGARSS '95* Florence, 1995
10. Alberotanza L., Zandonella A. Estrazione di informazioni biochimiche da bande TM a debole assorbimento per migliorare i modelli di riconoscimento della vegetazione macroalgale della laguna di Venezia. *Atti del V Convegno Nazionale AIT* pp 437-448, 1992
11. Fujita R.M. The role of nitrogen status in regulanting transient ammonium uptake and nitrogen storage by macroalgae. *J. Exp. Mar. Biol. Ecol.*, 1985, 92: 283-301.
12. Menesguen A. and Salamon J. Eutrophication modeling as a tools for fighting against Ulva coastal mass blooms. In: Schrefler B.A. (Ed.) *Computer modelling in ocean engineering.* Balkema, Rotterdam. pp 443-450. 1987

Studying vegetation distribution using ancillary and remote sensing data: a case study

M.T. Obaidat
Civil Engineering Department, Jordan University of Science and Technology (J.U.S.T.), Irbid, P.O.Box 3030, Jordan

Abstract

A Landsat Thematic Mapper image (TM) was used, in combination with ancillary topographic and topoclimatic data, to study the distribution of vegetation classes in the Niwot Ridge-Colorado, U.S.A. A logical channel approach; i.e. spectral and ancillary data, for digital classification of remote sensing data was used. The analysis was performed using the SPSS statistical package. The vegetation class was dependent on nine selected topoclimatic and topographic data variables. These variables include: topographic slope, aspect, albedo with and without slope/aspect consideration, Normalized Difference (ND) with and without slope/aspect consideration, convexity, Potential Solar Insolation (PSI), and Slope Aspect Index (SAI). Three random sampling techniques, to select vegetation classes samples from the map, were used: random samples from all the regions in the study area, samples from hilly areas only, and samples using a strip area along the map profile. The results of this study showed that the combination of TM data with the topographic and topoclimatic data variables is an efficient way to study the distribution of wet and dry vegetation classes. Minor effects were found for samples locations on the discrimination analysis process.

1 Introduction

Phillipson (1980) defined the remote sensing task as: it is used to collect, analyze, and convert remotely sensed data to useful information, for problem solving or decision making. Among other approaches, remote sensing data has been extensively used in the last two decades to study the classification of vegetation distributions and plant communities in various areas in the world. Becking (1959), and Jensen and Estes (1978) advocated the use of aerial photograph in vegetation and forestry in many fields such as mapping, interpretation of forest conditions, reconnaissance and management planning, and crop classification using digital Landsat data which proved superior to analysis of digitized high-altitude photography. Practically various techniques and strategies were developed to deal with digital classification of remote sensing data. Three approaches may be used for this purpose: pre-classification scene stratification, post-classification class sorting, and logical channel approach -classification modification- through increasing the number of observation channels (Hutchinson, 1982). Thus, ancillary data such as digitized maps and terrain data if combined with Landsat data is expected to improve the digital classification of vegetation classes. The resulted classification might be used in natural resource inventory, with condition that the analyst has a detailed understanding of the objects of interest and their relation with the ancillary data before using them. From this perspective, Cibula and Nyquist (1987) found that applying geographic models and topographic data such as solar insolation, snow cover, and wind besides spectral bands assisted the classification process, and high accuracy level was maintained.

Thematic Mapper (TM) data is normally used in classification studies, since there is a great sensitivity of a forest's apparent reflectance to the acquisition geometry in the middle infrared part of the spectrum of the vegetation (Leprieur *et al.*, 1988). Previous research indicated that using TM data besides topoclimatic and topographic variables could be used successfully to map dominant vegetation communities (Frank, 1988). Frank and Isard (1986) stated that accuracy measurement of topographic setting is necessary to distinguish between various alpine vegetation types in the rocky mountain, because the local topographic site factors influence the plant distribution.

This paper explored the vegetation distribution of Colorado Rocky Mountain Front Range according to the altitude and various topographic and topoclimatic data, and then compared the resulted classification to the map classification. In order to extract the samples of vegetation classes from the map, the study took three random sampling methods into consideration: samples from all locations of the map, samples from hilly areas only, and strip samples (profile). A discriminative statistical analysis to check the significance of the used variables and sampling methods was also performed

2 Study Site and Vegetation Distribution

The study area is located in the Niwot Ridge at Colorado Rocky Mountain enclosed within the Ward, which is part of the backbone of the Americas - a backbone that extends more than 10,000 miles from Alaska to Patagonia. The study area is extended from $105° 32' 30"W$ to $105° 37' 30"W$, and from $40° 02' 30"N$ to $40° 04' 00"N$. The area is characterized of high peaks (above 14,000ft). The difference between its summits and bases is about 5,000-7,000ft. The forests are extensive and great vistas showing spectacular landform with colorful rocks and forests. The vegetation of the Rocky Mountain is altitudinally distributed and zoned with a relationship with topography (Hunt, 1974). The vegetation classes on the mountain, starting from the top of the summits, include: Alpine Zone (A), Canadian Zone with Spruce-fir Forest (SF), Transition Zone (TZ), Mostly Yellow Pine (YP), Lodgepole Pine (LP), Upper Sonaran Zone with Pinyon-Juniper Woodland (PJ), Short Grass (SG), and Sagebrush (S).

The reasons behind the selection of this study site are due to the availability of the classified vegetation map (US/IBP Tundra Biome Program and A

Contribution to the US/MAB 6 Mountains and Tundra Project, 1976; and Frank, 1988), as well as the availability of the Landsat image of the area. Frank and Isard (1986) listed the NODA classification of vegetation for the Niwot Ridge area using six groups: dry meadow which includes other six classes (1 to 6), dry fellfield (2 to 5, and 7), moist shrub tundra (20), moist meadow (8-13), snowbed (14-17), and wet meadow (18-19). These classes were mainly used in the map to check the accuracy of our study. Other classifications, using hierarchical braun-blanquet classification system based on floristic-sociological principles, and dominant vegetation communities, were also used.

2.1 Factors Affecting the Vegetation Distribution

Many factors affect the distribution of the vegetation in the selected area including:
1. Snow cover and moisture content of soil.
2. Interactive relationship between insolation, snow, wind, soil moisture, relief, and vegetation distribution .
3. Atmospheric control which is presented by solar radiation.
The classification of vegetation communities was based on the fact that there is a correlation between these factors and the vegetation classes.

3 Material and Methodology

The image used in this study was taken by Landsat 5 for Niwot Ridge- Colorado on July 29, 1984 at 10 o'clock in the morning with a sun angle of 60 degrees. The image contains digital TM data which consists of seven bands including: Band 1 (Blue-Green) 0.45-0.52 μm, Band 2 (Green) 0.52-0.60 μm, Band 3 (Red) 0.63-0.69 μm, Band 4 (Near Infrared) 0.76-0.90 μm, Band 5 (Middle Infrared) 1.55-1.75 μm, Band 6 (Thermal) 10.3-12.5 μm, and Band 7 (Middle Infrared) 2.08-2.36 μm (Campbell, 1987). The image was rectified using ERDAS software with nearest neighbor resampling approach.

TM data besides ancillary data variables (including topographic and topoclimatic variables) were used to predict vegetation distribution of the study area. A combination of these variables was selected to discriminate between six vegetation communities using channel approach method; i.e. combination of spectral and ancillary data. This approach is used because it is expected to be efficient in the vegetation classification regions which have premafrost, snow accumulation, soil moisture growing season, and solar insolation effect.

All computer work was done using ERDAS software, whereas TERRAMAR software was used for three-dimensional (3-D) purposes.

3.1 Study Variables

The TM data was calibrated geometrically before quantifying the indexes in order to correct for latitude/longitude geographic location and time factor. Two cases were studied: calibration with and without slope and aspect consideration.

The following topoclimatic and topographic variables were used:
1. Slope: the magnitude of the elevation gradient.
2. Aspect (ASP): the direction of the slope ($0°$-$360°$).
3. Albedo (ALB): the outgoing radiation over incoming radiation, which is a measure of the biomass. This depends on the calibration of the TM data at a given time, the sun angle, the atmosphere, the slope and aspect of terrain, and the surface cover (Robinove, 1982). Two cases of albedo were studied: with and without slope-aspect effect (abbreviated as ALBT and ALB, respectively).
4. Normalized Difference Index (ND): a function of the near infrared band (NIR) and the red band (R). Equation 1 shows the ND formula (Frank, 1988):

$$ND=(NIR-R)/(NIR+R) \dotfill (1)$$

Normalized Difference Index was considered with and without slope-aspect effect (abbreviated as NDT and ND, respectively). Vegetation Index (VI) was also considered for preliminary study, and finally ignored due to similar characteristics with ND.
5. Convexity (CON): the second derivative of the slope.
6. Potential Solar Insolation (PSI): the direct beam potential solar insolation was computed from March to December for 24 hours a day.
7. Slope Aspect Index (SAI): a topoclimatic index used to study wind effect which is function of slope and aspect. Equation 2 shows the SAI formula (Frank, 1988):

$$SAI=(Sin(Slope)*Aspect)/(MaxSAI*k) \dotfill (2)$$

where: $max. SAI$ is the maximum index value, and k is a constant to convert to eight bit value.

The data variables, including the listed seven variables besides ALBT and NDT, were classified into nine variables, where each one of them was dependent on the Digital Elevation Model (DEM).

ERDAS software was used to produce the images of SAI, Slope, albedo, and ND. The important characteristics of the generated images are shown in Table 1. Layer combinations of bands, representing the combination between topographic and topoclimatic variables, using a SUBSET program were also generated. Combinations of these images include: convexity, ND, and aspect bands; slope, SAI, and ALB; PSI, SAI, and ND; Slope, aspect, and albedo; and convexity, PSI, and SAI.

TERRAMAR software was also used to construct 3-D images from various directions for the nine studied variables of ancillary data. The images represent a set of combination of 3-D images for: slope, PSI, and SAI; slope, aspect, and albedo; albedo, ND, and aspect; and ND, PSI, and convexity. The images clearly showed the accumulation of snow cover in East face which was associated with low SAI, while the west face was free of snow due to blow of the wind.

Table 1: Important characteristics of the image for the used nine variables.

Features	Characteristics			
	Minimum	Maximum	Mean	Standard Deviation
Slope	0	53	14.8	8.4
Aspect	0	359	156.9	106.4
ALB	15	98	39.3	21.8
ALBT	15	100	39.9	21.5
ND	0	255	127.5	63.3
NDT	0	255	127.5	63.3
Convexity	-10	96	10	13
PSI	0	1580	1315.1	203.2
SAI	0	132	27.2	18.9

3.2 Random Sampling

Three sampling techniques were used to represent the vegetation distribution on the used map. Around 30 points were digitized for each class from the map. This leads to at least 180 data points for each sampling case for the whole classes. The sampling points were kept approximately equally distributed for the specified samples. The three sampling cases used include:
1. All regions sampling including flat, hilly, and in between terrains.
2. Hilly areas sampling.
3. Strip (profile) sampling of about one tenth the width of the area.

4 Discriminant Analysis

SPSS statistical package was used to make discriminant statistical analysis to study the dependence of the vegetation classes on the nine used variables, using the three mentioned sampling methods. The variance-covariance matrices showed a strong correlation between ND and albedo, and slope and SAI, therefore, albedo and slope were not included in the final classification process. Thus, only five variables which represent the solar, topoclimatic, and topographic variables were used for the purpose of statistical analysis to discriminate between vegetation distribution for the study area. These variables include: PSI, convexity, normalized difference with topography consideration (NDT), aspect, and SAI. Minor correlation exists between these variables, but the SAI and convexity.

Only five vegetation classes were also used in the statistical analysis. Dry fellfield class was combined with dry meadow class to be referred as class 1, due to difficulties to discriminate between the two classes. The other four classes include: moist meadow (class 2), wet meadow (class 3), moist shrub tundra (class 4), and snowbed (class 5).

The discrimination analysis results for the three sampling cases are shown in Table 2. While the canonical discriminant functions evaluated at group means (group centroids) are shown in Table 3. Variables in the table are ordered by size of correlation within function. Four functions, for each sampling case, representing the discriminate function between vegetation distribution, were obtained as functions of the five used variables (NDT, aspect, convexity, PSI, and SAI). The functions were representing albedo (irradiance/radiance), snow and wind, solar insolation, and fourth functions with common characteristics of most of the variables. In the random sampling case the four function represent: albedo, topography, albedo, and potential solar energy, respectively. The strip sampling functions represent albedo, topography, solar energy, and a fourth function of common characteristics of the previous three functions, respectively. While, the hilly area sampling method functions represent albedo, potential solar energy combined with other factors, potential solar energy, and snow and wind, respectively.

From the previous functions, it is obvious that albedo, solar insolation, and snow accumulation and wind direction play as the major factors in the discrimination and distribution of the vegetation communities in the Niwot Ridge-Colorado. Albedo differentiated class 2 from other classes, potential solar insolation differentiate class 3 and class 5 from other classes, and snow and wind factors discriminate class 1 and 4 from other classes.

The trend for vegetation distribution is dependent on PSI by which the vegetation classes are going from dry to moist; i.e. from class 1 to class 5, as the PSI value increases. Except for classes 3 and 4, the vegetation distribution is becoming wet as the convexity increases; i.e. going down the crest. Similar observations could be drawn for ND, aspect, and SAI. SAI is a good discriminant of alpine vegetation, since the decrease of SAI is an indication of wind blowing and snow free. Thus, the vegetation distribution is mainly dependent on both topoclimatic and topographic variables.

Comparing the mean values of the three sampling cases for each variable, variances of less than 20% exist. This is due to the little effect of the way of selection of samples from the used map.

The classified percentages correct for the strip and random sampling cases were better than the hilly area sampling. They were 36.6%, 42.1%, and 42.5% for hilly areas, strip, and random sampling, respectively. Classes 2 and 4 were correctly classified up to 44.4% and 58.5%, respectively, using strip sampling. While hilly area sampling gave better classification results for Class 1 with 44.9% correct. Random sampling was so effective to discriminate Class 4 and Class 5 with percentages correct reached 73.3% and 63.3%, respectively. Whereas, Class

3 (wet meadow) was hardly discriminated using the three sampling techniques. It reached a maximum percentage correct of about 37.2% using hilly area sampling. Figure 1 depicts the correct percentages of classification for the five vegetation classes of the three sampling cases. As expected, taking comprehensive sampling for the hilly areas as well as flat areas; i.e. random sampling, is the best way to have a representative sample especially if it is well distributed all over the map area.

Table 2: Classification analysis results for the three sampling cases.

Actual Group	% Correct of Predicted Group Membership for Different Classes (Numbers in parentheses are the correct predicted classified samples)					
	No. of Cases	Class 1 (%)	Class 2 (%)	Class 3 (%)	Class 4 (%)	Class 5 (%)
A. Strip Sampling (% of classes correctly classified= 42.1%)						
Class 1	54	20.4% (11)	16.7% (9)	27.8% (15)	25.9% (14)	9.3% (5)
Class 2	27	3.7% (1)	44.4% (12)	18.5% (5)	3.7% (1)	29.6% (8)
Class 3	27	22.2% (6)	3.7% (1)	29.6% (8)	18.5% (5)	25.9% (7)
Class 4	41	4.9% (2)	2.4% (1)	24.4% (10)	58.5% (24)	9.8% (4)
Class 5	41	4.9% (2)	17.1% (7)	17.1% (7)	0% (0)	61.0%(25)
B. Hilly Area Sampling (% of classes correctly classified= 36.6%)						
Class 1	98	44.9% (44)	14.3% (14)	23.5% (23)	10.2% (10)	7.1% (7)
Class 2	55	12.7% (7)	38.2% (21)	16.4% (9)	9.1% (5)	23.6% (13)
Class 3	43	18.6% (8)	18.6% (8)	37.2% (16)	23.3% (10)	2.3% (1)
Class 4	32	31.3% (10)	9.4% (3)	25.0% (8)	25.0% (8)	9.4% (3)
Class 5	51	9.8% (5)	33.3% (17)	15.7% (8)	15.7% (8)	25.5%(13)
C. Random Sampling (% of classes correctly classified= 42.5%)						
Class 1	62	40.3% (25)	17.7% (11)	11.3% (7)	14.5% (9)	16.1% (10)
Class 2	29	20.7% (6)	31.0% (9)	0% (0)	6.9% (2)	41.4% (12)
Class 3	30	16.7% (5)	13.3% (4)	6.7% (2)	16.7% (5)	46.7% (14)
Class 4	30	6.7% (2)	16.7% (5)	0% (0)	73.3% (22)	3.3% (1)
Class 5	30	3.3% (1)	13.3% (4)	3.3% (1)	16.7% (5)	63.3%(19)

Table 3: Canonical discriminant functions evaluated at class centroids for the three sampling cases.

Class	Functions			
	Function 1	Function 2	Function 3	Function 4
Strip Sampling				
Class 1	0.24915	0.45501	0.01167	0.07238
Class 2	-0.56103	-0.00372	0.33797	-0.08824
Class 3	0.37043	0.14705	-0.20003	-0.17618
Class 4	0.82845	-0.48495	0.03244	0.03650
Class 5	-1.03108	-0.20871	-0.13865	0.0423
Hilly Area Sampling				
Class 1	0.35959	-0.35786	0.00665	-0.00346
Class 2	-0.55841	-0.01422	-0.23960	0.07627
Class 3	0.40840	0.50035	-0.21306	-0.07675
Class 4	0.28217	0.42499	0.32855	0.12162
Class 5	-0.61016	0.01446	0.21910	-0.08721
Random Sampling				
Class 1	0.02773	0.46304	-0.05657	-0.01214
Class 2	-0.53406	-0.25594	-0.38017	0.01462
Class 3	-0.07360	0.25102	0.28692	0.02652
Class 4	1.13765	-0.41396	-0.03007	0.00044
Class 5	-0.60511	-0.54660	0.22757	-0.01601

5 Discussion and Suggestions

Although the discrimination results, between the vegetation classes in the Niwot Ridge-Colorado using topoclimatic and topographic variables besides the TM data, were not absolutely effective, the study shows that using such variables is promising for discrimination purposes. According to NODA classification, it was found that it is extremely difficult to discriminate between dry fellfield and dry meadow. Other variables might be useful for the purpose of vegetation distribution prediction, such as: vegetation index (VI), elevation, reflectance absorption index

(R/A), relief, ND, aspect, and SAI.

Snow and wind, potential solar energy, and albedo had significant impact on the classification process. This is an indication of observed correlation between variables which had some sharing of common factors. Based on these indicators, as a generally speaking, the dry vegetation is located on the high altitude, while the moist vegetation is down the slope.

The effect of sample location on the classification process of vegetation distribution was minor. It might be due to the use of the same variables in the three cases and minimum number of points (about 30 for each class). Anyhow, the classification was enhanced of about 6% when using strip, or random sampling if compared to hilly area sampling.

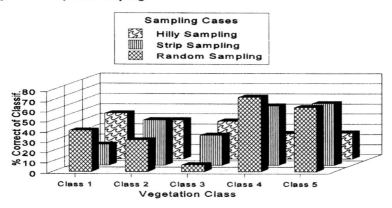

Figure 1: Classification percentage correct for five classes of all sampling cases.

For further future research, in order to enhance the classification process efficiency, it is recommended to do the following:

1. Add new variables to discriminate between class 1 and class 2. These variables might include VI, R/A, relief, and elevation.
2. Increase the number of samples for each class to assure that the sample will follow a normal distribution.
3. Use other classified map to include more detailed vegetation communities distribution.
4. Use approaches other than the logical channel approach to study the potential of remote sensing data on the vegetation distribution.

6 Conclusions

The logical channel approach is proven to be useful in feature characterization of vegetation communities. The classification accuracy of different types of vegetation was around 45%. This is due to the well known fact that spectral reflectance patterns are associated with certain variables such as slope, aspect, SAI, ND, PSI, albedo, and any combination of these variables. Using topoclimatic and topographic indexes besides the remote sensing data have the advantage of combining both spectral and information category classification. Samples for the vegetation classes should also be comprehensive and well distributed all over the vegetation classes in order to enhance the accuracy of the classification process of vegetation communities. Further, other parameters should be considered in order to practically adopt this approach of classification.

References

1. Becking, R. W. "Forestry Applications of Aerial Color Photography", *Photogrammetric Engineering and Remote Sensing (PE&RS)*, 1959, 25: 559-565.
2. Campbell, J. B. "Introduction to Remote Sensing", *The Guilford Press, New York*, 1987.
3. Cibula, W. G.; and Nyquist, M. O. "Use of Topographic and Climatological Models in A Geographical Data Base to Improve Landsat MSS Classification for Olympic National Park", *Photogrammetric Engineering and Remote Sensing (PE&RS)*, 1987, 53 (1), pp. 67-75.
4. Frank, T. D. "Mapping Dominant Vegetation Communities in the Colorado Rocky Mountain Front Range with Landsat Thematic Mapper and Digital Terrain Data", *Photogrammetric Engineering and Remote Sensing (PE&RS)*, 1988, 54 (12), pp. 1727-1734.
5. Frank, T. D.; and Isard, S. A. "Alpine Vegetation Classification using High Resolution Aerial Imagery and Topoclimatic Index Values", *Photogrammetric Engineering and Remote Sensing (PE&RS)*, 1986, 52 (3), pp. 381-388.
6. Hunt, C. B. "National Regions of the United States and Canada", *W. H. Freeman and Company, San Francisco*, 1974.
7. Hutchinson, C. F. "Techniques for Combining Landsat and Ancillary Data for Digital Classification Improvement", *Photogrammetric Engineering and Remote Sensing (PE&RS)*, 1982, 48 (1), pp. 123-130.
8. Jensen, J. R.; and Estes, J. E. "High-Altitude Versus Landsat Imagery for Digital Crop Identification", *Photogrammetric Engineering and Remote Sensing (PE&RS)*, 1978, 44 (6), pp. 723-733.
9. Leprieur, C. E.; Durand, J. M.; and Peyron, I. L. "Influence of Topography on Forest Reflectance Using Landsat Thematic Mapper and Digital Terrain Data", *Photogrammetric Engineering and Remote Sensing (PE&RS)*, 1988, 54 (4), pp. 491-496.
10. Philipson, W. R. "Problem-Solving with Remote Sensing", *Photogrammetric Engineering and Remote Sensing (PE&RS)*, 1980, 46 (10), pp. 1335-1338.
11. Robinove, J. C. "Computation with Physical Values from Landsat Digital Data", *Photogrammetric Engineering and Remote Sensing (PE&RS)*, 1982, 48 (5), pp. 781-784.

Computer aided scheming of large flood plains using remote sensing data

F.C.B. Mascarenhas, R.C.V. da Silva

COPPE, Federal University of Rio de Janeiro, UFRJ, Caixa Postal 68506, Rio de Janeiro 21945-970, Brazil

Abstract

The use of remote sensing data from satellite images is presently a wide world technique in the analysis of damage and dimensions of large flood plains. In large river basins with scarce hydrological and topographic data it is very difficult to accurately define the limits of flooded area from water levels and cartographic maps, as required to the mathematical modelling of flood propagation on a flood plain. This paper presents a methodology of analysis and scheming of flood plains using LANDSAT4 and 5 satellite images, which leads to the obtaining of the following information: the dimensions of the flooded area and their time evolution; the continuous distribution of the flood plain widths; the existence of non-flooded areas for different floods by superimposing two images; the potential lateral inflow/outflow using satellite sensor band that separates sediment laden flow (main channel) from clear water (flood plain flow). Those and other information were mathematically treated and used as input data for two kinds of flood plain models: a two-dimensional cell model and an one-dimensional one with lateral contribution. The methodology was applied to the basin of Upper Paraguay River, located in Brazil and presenting poor hydrological and topographic field data for modelling purposes.

1 Introduction

The accurate scheming of a flood plain is an important requirement for its mathematical modelling. This includes the correct delimitation of the flooded area and its classification as a storage plain and/or a dynamic one. In this situation, the remote sensing may be used as a very reliable complementary tool for field measurements at places where access conditions are difficult or impossible, for instance dense rain forest areas. In this paper we search to show how satellite data may improve the mathematical modelling of large flood plains, using as case study the upper Paraguay river wetland area, known as Pantanal Matogrossense which is an important environmental system located in the southwest of Brazil.

2 Imagery Analysis Methodology

The scheming of a flood plain usually requires the combination of detailed topographic surveys with a dense network of water level data, providing information about flooded areas and inundated widths. The density of level gauges is more important in dynamic flood plains due to the large level gradient observed in the flood propagation. In the upper Paraguay river basin the water level information is scarce and the best existing chart scale is of 1:100,000 m with topographic level curves spaced each 50 m.

The use of remote sense data is possible in this flood plain because of its dimensions (about 1,000 km long and widths running from 5 km up to 180 km) and its large average complete flow time (5 to 6 months). The case study area and its location in South America and Brazil is shown in figure 1. The above mentioned properties of the area under study are compatible with satellite image resolution and frequency. The figures 2 to 5 show the upper Paraguay river flood plain in the areas near to the river channel. The figure 2 is a combination of topographic chart with level meters and its drawing had been possible because of the better data acquisition system in the upper river stretch. The figures 3 to 5 were made from satellite overlay scanning. The analysis of these figures has provided useful information and data for the choice of the mathematical models to be used in the simulation.

3 Mathematical Modelling

According to Cunge [1], there are four types of flood plains scheming for modelling purposes:
. One-dimensional model with water flowing across the entire valley section, assumed as being a compound section.
.   One-dimensional   model with a main flow section and the flood plain simulated as a storage zone. The water level is assumed constant over the entire section.
. The same model above, but with different water elevations in the flood plain and in the main channel. The exchange process between them is simulated as a weir type link.

. Two-dimensional model, where river and flood plain are simulated as a box grid describing flow and storage areas. The connections between boxes may be of channel and/or weir type links (e.g. Zanobetti et al [3]).

The analysis of the images presented in the figures 2 and 3 shows a two-dimensional flood plain, and so a cell model derived from the model proposed by Zanobetti et al [3] has been used (e.g. Mascarenhas & Miguez [2]). After an analysis of the very great flood plain size of the area shown in the figure 4 we propose, as a fifth type, a reservoir mathematical model for its flood plain modelling. This model, based on the well-known SSARR model (U.S. Army Corps of Engineers [4]), is presently being developed by the authors and will be subjected to a careful calibration procedure to correctly simulate the associated flooded area. The figure 5 otherwise shows a nearly one-dimensional flood plain, and the Saint-Venant equations, with lateral inflow q_L representing the river-plain discharge exchanges, were applied to.

In the two-dimensional cell model the continuity equation is combined with the dynamic one, neglecting inertia terms, and is written for a generic cell i as:

$$A_{Si} \frac{dz_i}{dt} = P_i + \sum_k Q_{i,k} \tag{1}$$

where
z_i - water surface level at the center of cell i
A_{Si} - free surface area of the cell for the water level zi
P_i - rainfall flow over the cell i
$Q_{i,k}$ - discharge between cell i and its boundary cells j

The reservoir type model is based on the general balance equation that describes the water balance inside a wide area, involving inflow, outflow and volumetric storage variation:

$$\left(\frac{I_1 + I_2}{2}\right) \Delta t - \left(\frac{0_1 + 0_2}{2}\right) \Delta t = V_2 - V_1 \tag{2}$$

where
I_1, I_2 - inflow rate at the beginning and the end of time Δt
$0_1, 0_2$ - outflow rate at the beginning and the end of time Δt
V_1, V_2 - storage volume at the beginning and the end of time Δt

The Saint-Venant equations for the one-dimensional model, supposing a storage width B_s are:

$$B_s \frac{\partial h}{\partial t} + vB \frac{\partial h}{\partial x} + v\left(\frac{\partial A}{\partial t}\right)_{h=const} + A \frac{\partial v}{\partial x} - \frac{q_L}{B_s} = 0 \qquad (3)$$

$$\frac{\partial v}{\partial t} + v \frac{\partial v}{\partial x} + g \frac{\partial h}{\partial x} - g(S_0 - S_f) = 0 \qquad (4)$$

where
g - gravity acceleration
B, B_s - main channel and storage widths
h, v - dependent variables of flow depth and mean velocity
x, t - independent variables of space and time
S_0 - mean bottom slope
S_f - energy line slope
A - main channel wetted area
q_L - lateral inflow/outflow

The continuity equations of the type (1) for all the cells in the model and the system of Saint-Venant equations (3) and (4) are separately solved by finite differences. So the first model is two-dimensional, with hydraulic laws of channel and broad crested weir flow types for the discharge exchanges $Q_{i,k}$, and the third one is one-dimensional, while the second is a so-called hydrologic model. The water level results of the cell model for a selected flood are presented in the figure 6.

From the obtained spatial integration grid of the area shown in the figure 5, the Saint-Venant equations (3) and (4) were solved for all discrete sections and for each time step by the Preissmann implicit finite difference method. The figure 7 shows an example of the computed and measured discharge-time curves in the location named Fecho dos Morros. In that gauge section the flood is entirely confined between natural terrain elevations, and so this station can be considered as a very reliable water control for all the flood amount that actually leaves the inundation plain and an exact internal boundary condition for the mathematical simulation.

4 References

1. Cunge, J.A. Two-Dimensional Modelling of Flood Plains, Chapter 17, *Unsteady Flow in Open Channels*, eds K. Mahmood & V. Yevjevich, Vol. 2, pp 705-762, Water Resources Publications, Colorado, 1975.

2. Mascarenhas, F.C.B. & Miguez, M.G. Large Flood Plains Modeling by a Cell Scheme: Application to the Pantanal of Mato Grosso, in Engineering Hydrology (ed C.Y. Kuo), pp 1212-1217, *Proceedings of the ASCE International Symposium on Engineering Hydrology,* San Francisco, U.S.A., 1993.

3. Zanobetti, D., Lorgere, H., Preissmann A. & Cunge, J.A. Le Modele Mathematique du Delta du Mekong, *La Houille Blanche,* 1968, ns. 1,4 & 5.

4. U.S. Army Corps of Engineers *Application to the SSARR Model to the Upper Paraguay River Basin,* UNESCO Project Report, Portland, U.S.A., 1972.

Figure 1: Location Map in South America of the Case Study Area

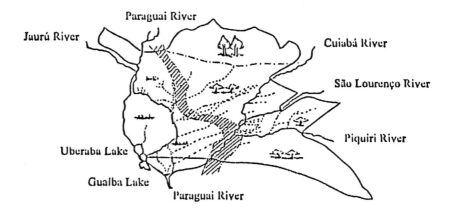

Figure 2: Upper Paraguay River Flood Plain - 1st stretch - From Cartographic Chart.

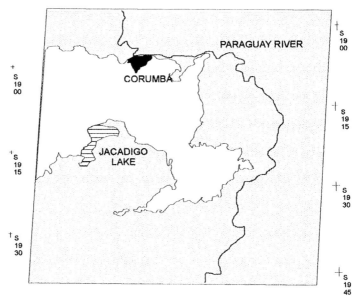

Figure 3: Upper Paraguay River Flood Plain - Partial Image of 2nd stretch - from Satellite Image of June, 1979.

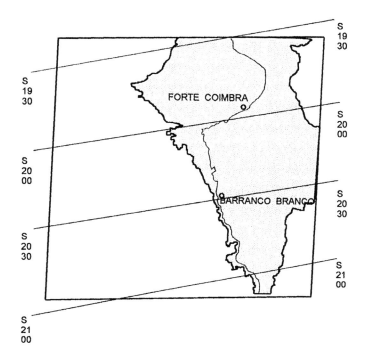

Figure 4: Upper Paraguay River Flood Plain - 3th stretch - From July, 1979.

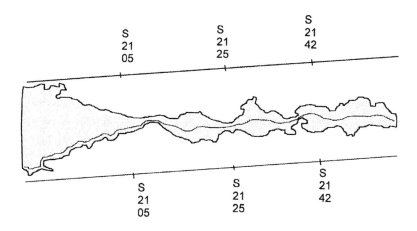

Figure 5: Upper Paraguay River Flood Plain - 4th stretch - From August, 1979.

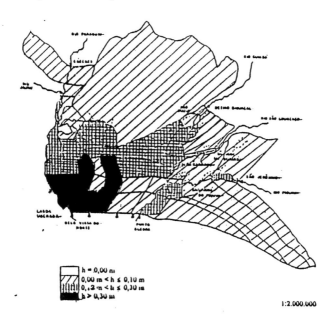

Figure 6: Computed Flood Plain from the Cell Model for 1970 Flood - 1st stretch.

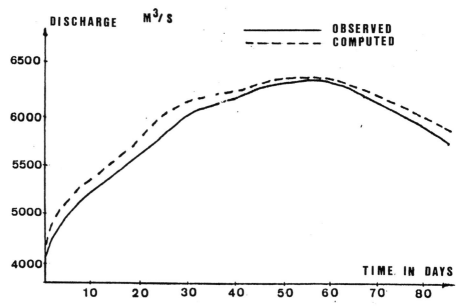

Figure 7: Computed and Observed Hydrographs of the 1979 Flood for the 4st Stretch from an One-Dimensional Model.

Methodic of sea bottom's shape modelling and objects situated on it according to echo analyses acquired from vertical echosounder and side looking sonar

G. Kuchariew,[a] St. Parczewski,[a] Z. Szymczak[b]

[a]*Autocomp Electronic Co. Ltd, ul. Wladyslawa IV nr1, PL 70-651, Szczecin, Poland*

[b]*Biuro Hydrograficzne Marynarki Wojennej, Gdynia, Poland*

Abstract

The article presents a methodic of 3D sea bottom pictures synthesis and algorithms used to find out object's coordinates based on real data. The described algorithms might be the basis for a computer system of marine oceanographical research.

1. Introduction

The work the measuring system which is used to pick up data from vertical echosounder and side looking sonar is shown on fig. 1.

The job of computer data work out system (acquired during echo sounding) is to synthetise the bottom shape, and create it's pictures: 3D or flat. The result of projection is the map of depth isolines.

The system of working out data acquired form side looking sonar works has a different task, which is to analyse the acoustic scheme of objects situated on the sea bottom - characteristic of reflected signals and acoustic shadows. Acquired description includes not only the 3D scheme of object but also it's coordinates.

The solutions for those problems are based on set assumptions about the shape of sea bottom, features of data, and on requirements of algorithms creating computer sea bottom picture. All those requirements are mentioned in articles[1-3] and in this one only the main ones are emphasised.

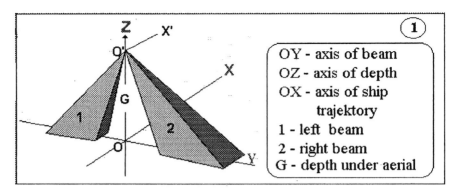

OY - axis of beam
OZ - axis of depth
OX - axis of ship
       trajektory
1 - left beam
2 - right beam
G - depth under aerial

1.1 The assumptions concerning sea bottom

1.1.1. The sea bottom is not flat but it is smooth - it has neither vertical upcasts nor uplifts.
1.1.2. Every point on the bottom level induces the position of associated point and all the other points. The location of one point interacts with location of the other points.
1.1.3. The influence of one point 's level of sea bottom on the other points levels is inversely proportional to a distance between them.

1.2. The assumptions for measurements

1.2.1. Initial data are modified and corrected - the influence of ship's oscillations are taken into account.
1.2.2. The data are modified with the seed of the sound in the water.
1.2.3. The main ray of antenna's directional characterictic is vertical to a direction of ship's move.

2. Mathematical description of 3D picture forming.

In the space with X,Y,Z coordinates there are N points $P_1(x_1, y_1, z_1)$, $P_2(x_2, y_2, z_2)$, . . . , $P_N(x_N, y_N, z_N)$, characterising xi,yi coordinates in surface XY and characterising depth measurements z_i in these points.

The depths are obtained from measurements in coordinates x_i, y_i
($\forall$ $i = \overline{1,N}$), *but these depths are not situated in the regular net on surface* XY.
Although to create bottom shape all the data must be collected in the table without any blank spaces and describing values of points $\{P\}$ with equally remote coordinates (in the regular net).
It shows, that it is essential to calculate value zql for new regular net with coordinates as follows: x_g, y_l, *where* $q \in \overline{1,Q}$ *i* $l \in \overline{1,L}$ *(under a condition that the new coordinates do not exceed the area of initially set coordinates).*

The scale of the coordinate net's can be varied and can be different for various coordinates X,Y.

2.1 Sea bottom picture forming. Solving the task.

The distance between two points is defined with following formula:

$$d_{12} = \left| (x_1 + (\sqrt{-1})y_1) - (x_2 + (\sqrt{-1})y_2) \right| ,$$

the coordinates have been described in compound form.

Additionally it might be assumed that every, separate pair of points $(x_i, y_i$ $\forall$ $i = \overline{1,N}$) shows different points in the field of acquired data on the surface XY. We define it as a condition accordance with which the coordinate values have not to be repeated. The problem is solved in two stages.

In the first one , based on acquired data x_i, y_i, z_i $\forall$ $i = \overline{1,N}$, we calculate interpolative model which is corresponding to this data.

In the second one we create interpolations on set net in XY surface.

2.1.1. First stage. Based on postulate 1.1.3 we define functions of interpolation f_i as depth dependence on x_i, y_i as follows:

$$f_i = \sum_{j=1}^{N} b_j * \left| k_i - k_j \right| , \quad \forall \ i \in \overline{1,N} \tag{1}$$

where:

f_i - function magnitude in coordinate i ;

b_j - factors , which influence the depth;

$k_i = x_i + (\sqrt{-1}) y_i$ - coordinate i ;

$k_j = x_j + (\sqrt{-1}) y_j$ - coordinate j;

$\left| k_i - k_j \right|$ - coordinate discrepancy module.

In (1) factors b_j (defining effect of influence on f_i) are inversely proportional to coordinate discrepancy module - this is the distance between i and j points. Formula (1) may be used only when condition of *not to be repeated* is satisfied.

$$\left| k_i - k_j \right| \neq 0 \quad \forall \quad i \neq j . \tag{2}$$

Factors b_j are defined when condition of identity in f_i and z_i is satisfied:

$$f_i \equiv z_i , \quad \forall \quad i = \overline{1,N} \tag{3}$$

2.1.2.The second stage.

Values z_{ql} (in the new net) are specified using already calculated factor vector b_j as follows:

$$z_{ql} = \sum_{j=1}^{N} b_j * |k_{ql} - k_j|,$$

(4)

where:

$$k_{ql} = x_q + (\sqrt{-1}\ )\ y_l\ ,\qquad q \in \overline{1,Q}\ \ i\ \ l \in \overline{1,N}\ ;$$

$$|k_{ql} - k_j|\ -\qquad \text{coordinate discrepancy module in surface } XY.$$

The set consisting of $Q*L$ calculated according to (4) value Zql gives the interpolated depths for set number of coordinates so it defines 3D shape of sea bottom.

2.2 Real data work-out.

The exemplifying trajectory of the model ship and a data histogram on analysed area shows fig.2.

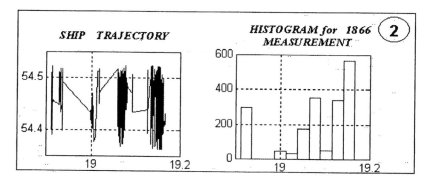

It must be added that in suggested method it is essential to solve the system of eguation with N index of the power, where N is the number of analyses. It is easy to see that N equals 1866. It appears, that to calculate b_j factors with satisfactory precision 300 measurements is quite enough [3]. So N should be the number equal or smaller to 300.

As the number of real measurements is much higher, only a representative part is selected from the set of the measurements (fig. 3) .

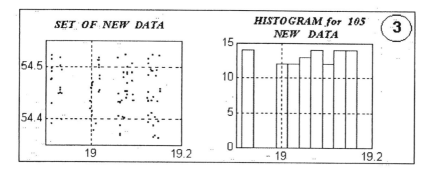

The selection is based on histogram analysis and picking up regularly diffused analysed points The selected points include also all limit coordinate values[2, 3].
Initial data acquired in this way are used to build the depth model and 3D model of bottom shape is formed, which is used as **Basic Sea Bottom Picture** (**BSBP**). It defines the bottom shape in whole analysed area. (Fig.4).

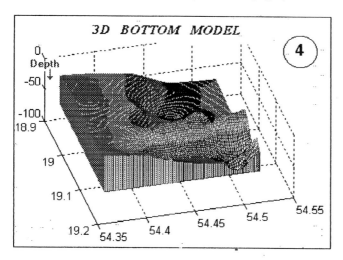

This **3D BSBP** may be recalculated (using interpolative formula (4)) on any optionally set net or it may be the base for further data analysis (depending on shape and corrugations of sea bottom).
All the other sea bottom pictures - flat picture of sea bottom and maps of depth isolines are created by operating with widely known methods of numerical pictures work out , including methods proposed by authors[4].
Every diagram showing analysed area, initial data and the result of interpolation (**3D BSBP**) is based on real data obtained from experiments realised in the Baltic sea.

3. Assumptions concerning initial data acquired during side looking observation.

3.1. The reflected signal is received by a side looking sonar's antenna system (in this case by "GBO-100"). It is recorded separately for each ship's side. Separate signal records are obtained with ship's move in direction parallel to axle OX (fig.1).

3.2 The coordinates of analysed water area and connected coordinates of hydrographical ship are given.(actually antenna's coordinates and depth of it's submersion under the water to be exact) and also the sea depth in the analysed area (under the antenna). *Presented assumptions allow to define the bottom shape in analysed area using already described methods.*

3.3. The directional characteristic of antenna is also given, but for side looking sonar (type "GBO-100") it is not possible to specify with it the distance between various objects on the bottom and the ship's trajectory.

4. Methodic of defining coordinates of objects on the sea bottom

Figure 5 presents the plan of completed experiments with side looking sonar. Each halls has been separated with continuous line, and the end of each halls trajectory has been marked with a number. Coordinates are given in metres.

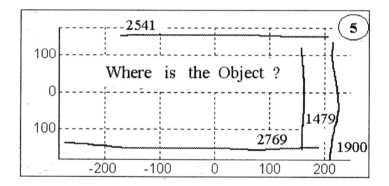

Based on one rectilinear halls the size of object connected with only one coordinate may be specified (e.g. width or length of object).

Coordinates (X and Y) of object's position on sea bottom may be obtained with minimum two halls measured by various angles. Defining coordinates of object situated on the bottom will be shown on sample of working out the data from halls 2541 and 1900.

Echogrames from both hallses are shown on pictures 6 and 7. Ship's trajectory has been indicated with arrows.

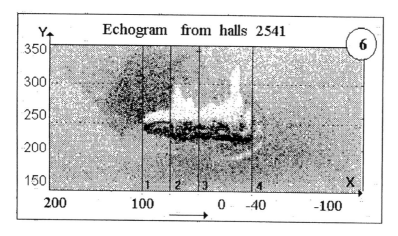

On the echogram from halls 2541 the outline of a ship wreck is clearly visible, as well as acoustic shadows from it's details. Echogram from halls 1900 does not show the wreck so clearly.

On figure 6 the way of finding out the wreck's gabarite in relation to X axel. Lines 1 and 4 indicate the length of wreck, line 2 the end of the bow and line 3 the middle of the ship.

In the same way the gabarites towards Axel X are specified. Shown on fig.7, but here the gabarites are indicated with dots.

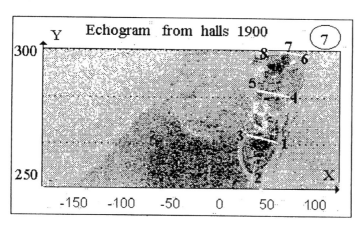

The result of crossing the gabarite lines from both hallses shows fig.8, where the approximate shape of object on the bottom is shown.

Figure 9 presents bottom shape based on real data and wreck's site on the bottom in set system of coordinates.

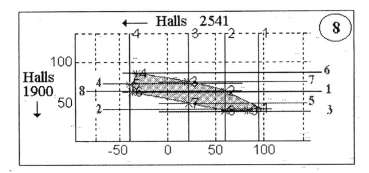

Presented effect is a result of combining both methods mentioned in the article. Specifying the sea bottom's shape in the area of discovered object may be used with the same methods.

To define the horizontal shape of an object the correlation between range of signal's echo and the length of shadow has been used:

$$H = G * R / L , \qquad (5)$$

where: R - shadow's length ;
            L - is a range of signal's echo.

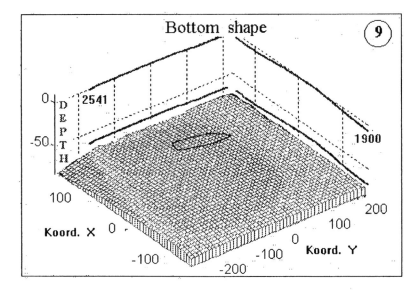

It must be stressed that the height of the relief's form does not depend on the angle of sea bottom's inclination. The result of specific calculations of bottom relief's shape around the wreck and reconstructed wreck's shape shows fig. 10.

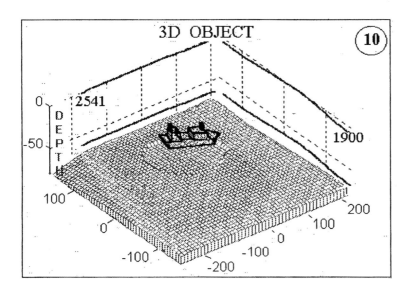

REFERENCE

1. Кухарев Г., Parczewski S. Алгоритмы автоматизированного формирования образов дна моря для систем информационных технологий в исследовании моря и океана. В кн. "Компьютерный анализ данных и моделирование" / сб. научных статей Международной конференции/, Минск, 1995, стр. 153 - 160.

2. Kuchariew G., Parczewski S. Method of Sea Bottom Picture Forming./ Proceeding of the Third International Conference "Pattern Recognition & Information Analysis", Szczecin, 1995, pp. 195-201.

3. Kuchariew G., Parczewski S.. Metoda i Algorytm Automatycznego Formowania Obrazu Dna Morza / Zeszyt naukowy Politechniki Rzeszowskiej, "Mechanika", z. 45 (Awionika, cz.1), s.141-149.

4. Bednarczyk D., Кухарев Г., Parczewski S. Алгоритмы вычисления и сглаживания контуров для срезов трехмерного изображения. В кн. "Современные методы обработки сигналов в системах измерения, контроля, диагностики и управления/ Материалы научно-технической конференции "ОС-95", Минск, 1995, стр. 18 - 22.

SECTION 9:
DATABASES

DataDelve client and EcoTrack server: a spatial data system for environmental warehousing

J.A. Bowers,[a] R.E. Osteen,[b] G. Rogers,[b] F.D. Martin[a]

[a]*Westinghouse Savannah River Company, Savannah River Technology Center, Environmental Sciences Section, Ecology Group, 773-42A Aiken, SC 29808, USA*

[b]*HOPS International Inc., 800 West Avenue, Miami Beach, FL 33139, USA*

Abstract

DataDelve Client and EcoTrack Server is an environmental data warehouse utilizing the Heuristic Optimized Processing System (HOPS), a client-server database management system. DataDelve's unique Spatial Browser Tool allows easy point and click data queries over georegistered map layers allowing access to virtually all types of environmental information within a spatial-temporal construct. Other features developed in DataDelve include a statistical toolbox for easy use of common parametric and nonparametric methods, a document retrieval system with search capabilities, and DDAdmin, a system and database administrators' toolbox to manage the warehouse.

1 Introduction

Environmental information exists in many more formats besides the classical "fields and records" that reside in a Relational Database Management System (RDMS). Information exists as aerial or satellite photographs, technical figures, written reports from past studies, georegistered spatial data in specialized Geographical Information Systems (GIS) and in relational database software using the Standard Query Language (SQL). Traditional solutions within large environmental organizations result in these varieties of information types being stored and distributed across an unweildly collection of software, hardware, Operating System (OS) environments. Environmental professionals needing to access all of this data have two choices, either learn all of applications and their OSs to directly access the data through networking desktop computers, or request software and OS professionals to download that information through a particular application-OS combination. The first choice is not generally found in the environmental sciences due to a lack of time for training and opportunities to experience this plethora of software and hardware. Therefore, direct access to the data is usually not possible. Our experiences at the Savannah River Site, a U. S. Department of Energy (DOE) facility and interviews with environmental scientists at other large U. S. government facilities led us to believe that data access for environmental professionals was difficult, slow, expensive and rarely directly accessible at the desktop level for synthesis, analyses and ultimately reporting.

What environmental scientists needed was an environmental data warehouse at the desktop level of computing. This warehouse would capture, store and distribute all of the above types of information and offer end-users direct access through an intuitive Graphical User Interface (GUI) requiring a minimum of training. Furthermore, the GUI would maximize the coupling of information to space and time because environmental information is, by definition, spatial and temporal in nature. Equally important, the information must be delivered in a timely fashion, minutes versus hours to fit the business tempo of large government and corporate institutions. Prior experience with traditional SQL-RDMS systems highlighted the need for a faster data engine and a query language easier for end-users. This requirement led to the use of the Heuristic Optimized Processing System (HOPS, HOPS International Inc.) because of its significant performance advantage and its dynamic query language obviating the need for the older SQL dialect. Perhaps equally important was an application that worked easily with other desktop software where most of the work transpires. The HOPS data engine provided the core of EcoTrack Server which contained responses to all of the custom client queries, system security controls over network access, the database administrators toolbox, DDAdmin, and client-server protocols in general. This application primarily functions as an environmental data warehouse in support of CERCLA wastesite remediation at DOE facilities [1, 2].

2 Graphical User Interface (GUI) Design

Our strategy for designing the application was to plan the basic component tools, develop a series of GUIs to access those tools and then furnish them to end-users for criticisms and suggestions. When a beta version was completed, end-user responses were solicited for further menu corrections and code debugging. Above all other factors in the GUI design was ease of use for end-users. Because environmental information is most often thought of in terms of space and time, information access occurs through a Spatial Browser Tool (Figure 1) pointing to all of the information except those datasets not referenced to any location or particular time period. The information itself is partitioned into a bi-level hierarchy of Themes and Topics which would begin any data browsing. Currently Data-Delve has six Themes each having their own list of second-level Topics. These Themes are Visual (the maps themselves, operating in both latitude-longitude and Universal Trans Mercator coordinate systems), Management (financial databases for projects), Surface (data files on soils, air, terrestrial plants and animals, wildlife, etc.), Subsurface (mainly soil cores, groundwater databases), Aquatic (water quality data and all aquatic animals and plants in streams, rivers, reservoirs and wetlands) and Documents (all textual data, reports, publications, procedures, references, etc.). This design also permits end-users to directly access information in situations where location and time are not pertinent.

Accessing data files within a space and time construct begins by selecting a Theme and Topic then clicking on the Set Focus command, dragging a rectangle over the chosen map layers and clicking on the Find button. A Bird's-eye window, which always illustrates the largest geographical context of the mapping, alleviates "getting lost" at magnified viewing levels. Distance and area tools are also provided. The Find command brings up another window for selecting a month-year time interval of the desired information (*See* inset window in Figure 1).

Figure 1. Spatial Browser Tool in DataDelve Client Application.

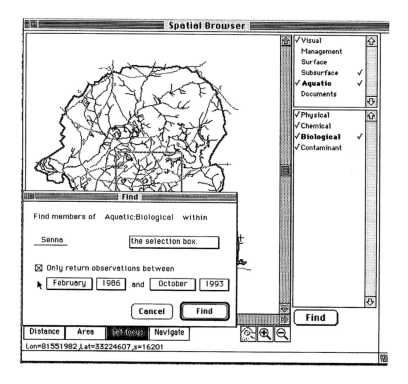

At this point end-users can choose the month and year time interval of the search and save the query as a file, if desired. Continuing the search results in a list of

HOPS data files satisfying the Theme, Topic and space and time interval conditions. Only those files meeting space and time conditions are retrieved. After selecting a file, end-users can then immediately view the file, see a list of the file's data fields, or continue and use the Make Seed command which allows further databasing with that file (joins to other files, filters, etc.).

Selecting the Documents Theme and Topics choices allows end-users to retrieve any document from the server using up to eight keywords and the boolean AND or OR query modes. After a list of document file names are retrieved, clicking on any filename returns an abstract. A retrieve button brings the document to the client in an Adobe Acrobat$^@$ .pdf file format. This format was chosen because of its platform independence and its ease of use for further queries within the document and cut, copy and paste abilities.

Besides the Spatial Browser Tool, the client application offers end-users additional utilities for savings map preferences, cut, copy and paste to other currently running applications, password changes, and up and downloading of data files into EcoTrack Server or end-user drive volumes. Data import and export uses the ASCII file format to accommodate most other applications.

File security and access was an important consideration in designing DataDelve Client/EcoTrack Server because of the complicated quality assurance protocols required of environmental information submitted for federal and state regulatory compliance standards. Therefore DataDelve Client incorporates four security levels from a minimum view-only privilege to full access to all information for the Database Administrator and application developers. Additionally, EcoTrack Server allocates each end-user a workspace (approximately 30-50 Mb) where an end-user can freely import and export data files. All shared data on all server volumes are loaded into EcoTrack Server by the Database Administrator who is responsible for the system's data dictionary and quality assurance protocols.

Another utility feature, a Statistical Toolbox, was developed for immediate data analyses without end-users having to move the data to a statistical application. It was purposefully kept small in the belief that many other applications currently exist that perform these sorts of tasks in a user-friendly environment, but that their powerful features are only applied occasionally. Rather those tests were provided that end-users noted they used most frequently when beginning to analyze raw data sets. Those tests are: Descriptive Statistics for one variable (Mean, Mode, Median, Range, Maximum, Minimum and Quantile Distribution), Student's T-Test, Kolmogorov-Smirnov Test, Linear Regression and Correlation (AOV Table), Moving Average (Smoothing), Auto (Cross) Correlation, and a spline-algorithm contouring package. Data visualization is accomplished with scatter and line plots and histograms, all with end-user chosen axis scaling features.

A fully developed context-sensitive help system was also developed for end-users. The default start-up window (optional in preferences), launched after of the start-up application screen, points out all of the major commands in the Spatial Browser Tool. Using the Option key with the "?/" key launches the help system from anywhere in the application and is command specific.

3 Graphical Users Interface (GUI) Development

The HOPS Application Programmer's Interface (API) was used to create Data-Delve Client and EcoTrack Server. This API is a suite of tools written in C, compiled in Macintosh Programmers Workshop (MPW)$^@$ and is executed as a compiled task. Source Bug$^@$ was the main debugger used with MPW although other debuggers can be applied. Programmers have the ability to write data queries to the server in addition to a complete set of GUI forms (windows, scroll bars, buttons, menus, lists, icons, etc.) for the client application. Two important issues during the creation of DataDelve was the intuitiveness of the client application and client-server loads. Where possible, much of the work was kept on the client CPU such as the Statistical Toolbox, graphics routines and GUI views. This permits EcoTrack Server to mainly accommodate data queries.

While the system can display coordinates in Latitude/Longitude or Universal Trans Mercator (UTM) they are stored internally in latitude/longitude. Each measurement is kept in an 8 digit integer representing millionths of degrees. An 8 digit geokey is created using the 4 most significant digits of the latitude followed by the 4 most significant digits of the longitude. This allows for fast retrieval given a geographical area.

Most of the data in the system is represented by a three-tier object hierarchy consisting of Themes which contain Topics and within Topics, Members. Themes are overall areas of interest such as visual layers for maps that are displayed or scientific layers such as aquatic data. Topics are specific types of entities to be found in a theme. For instance roads are a specific type of entity under the visual theme and biological data is a specific type of aquatic data gathered. Members are objects that can be mapped and or described. A member might be a physical object such as a monitoring well that can be drawn on the screen or a data file that consists of data records gathered from several monitoring wells. This configuration allows us to quickly locate data. For instance sampling stations are mapped to all technical data files they have collected data for. This information is kept in a metadata file shown in Table 1. Since each sampling station is georegistered, when a geographic search for any type of data is initiated, the stations are quickly retrieved along with the information about what files are linked to this station and the data contained in each file.

Table 1: Theme-Topic-Member scheme in DataDelve.

StationMemberID	TechDataFileIndex	Theme	Topic	MemberID	Count
0	35	4	2	21	39.00
0	36	4	2	22	30.00

There are two metadata files kept on the server for the Theme Documents, a keywords file and an abstracts file. The keywords file contains each document name along with up to eight keywords, while the abstract file contains document name, abstract, and location of document's file on the EcoTrack server. When a document is requested, it is retrieved from the server and, through the use of Apple Events[@], the Acrobat[@] file reader is launched with the requested document.

The statistical algorithms used in DataDelve are modifications from Press et al. [3]. These were chosen because they have all been validated independently. While the data file is read from the server, all calculations are performed on the client computer. This keeps the server free for data and map queries. The client then posts an appropriate graph or table. These can be saved, printed, cut, and pasted for use in reports.

The context-sensitive help system in DataDelve Client uses the freeware library Help on Wheels for Apple Macintosh programs. It functions through a series of context links that are added to the program. The help system is really a separate program communicating through Apple Events[@].

The administration program, called DDAdmin, resides in the EcoTrack server application and allows the System and Database administrators, to manage all normal duties such as adding, modifying, and deleting user accounts. Users are provided with disk quotas and access privileges to files. It also provides the ability for the Database Administrator to add data to the system without shutting the system down.

4 Database Administration within DataDelve using DDAdmin

Management of environmental data requires a level of understanding of environmental chemistry and biology beyond what most business database managers currently hold. For example, data to be used for calculating a biotic index must be capable of being split out to species or subspecies levels at one stage of calculation while later it may require being lumped by higher taxa, ecological guilds, developmental stages, or sampling method. In addition, as biologists redefine taxa, taxonomic synonyms must be dealt with. Data structures must take these facts into account. For DataDelve/Ecotrack Server most of the administrative details for this level of handling data structures are easily handled within HOPS itself.

When the person using data gathered that data, special information attached as metadata is not usually critical, but in large organizations and for data going back over a number of years, it is often necessary to examine the original field data books to assess the quality and usability of the data. The database administrator, if he or she is a subject matter specialist, can reduce the need for most end-users to go through this effort. DataDelve approaches this problem through metadata associated with each data file. This metadata allows the user to view the file structure and a prose description of the data contained in the file without having to view the data. The next version of DataDelve will also have the data dictionary associated with the file so that, when viewing the file structure, field definitions can be seen as needed. Entry of these metadata and georegisteering objects for inclusion in the database is automated through DDAdmin.

From experience at SRS, the most difficult to deal with maintenance issues are related to synonyms within chemical and biological data. There are several systems of chemical nomenclature and biological taxonomy is changing rapidly as organisms are studied with newly developed methods. All files being entered into the system must be brought into uniform nomenclatural system. The other serious problem is caused by the lack of precision of some of the Global Positioning Systems (GPS) used to georegister data points. As information comes in to increase the precision of these data, files must be updated frequently. This problem is addressed through tools in DDAdmin.

5 Conclusions

Implementing Data Delve and functioning in a production mode for approximately a year and a half has provided the development team with valuable insights for environmental warehousing within a spatial design. Besides the normal issues encountered in installing a new application, an environmental warehouse of georegistered information offered new challenges not common to the typical database system. Georegistered data files required new functionality in the administrators' toolbox for data entry and metadata editing. Although end-users had unlimited access to large volumes of shared data, end-user requests led to functionality for easy up(down)loading of data files into each end-user's secured workspace on the server and their own desktop volumes. In this case having a good API facilitated end-user needs. It cannot be overemphasized that data quality has been crucial to our success with DataDelve. Because EcoTrack Server was designed with restricted data entry menus, managing quality issues was made easier.

Currently DataDelve Client and EcoTrack Server are production applications on the Macintosh OS System 7.X. A beta version of DataDelve 2.0 is being tested using the Digital DEC Alpha 2100A operating in DEC/UNIX with cross-platform DataDelve Clients (MacOS 7.5.X native for PowerPC computers and MS WindowsNT Workstation 3.51 OS). The principle reason for this system is the ability to distribute clients to virtually all desktop computers. Moreover, this code port significantly increases both server and client performances plus adding the ability to accommodate increased end-users concurrently on-line. Functional enhancements include ODBC connectivity to other database engines using SQL protocols, Theme and Topic hierarchies capable of n levels instead of the current bi-level system, RAID Level-5 disk drive storage and data files georegistered to points, lines and polygons.

Acknowledgments

The authors are indebted to Dr. John B. Gladden for his continued and enthusiastic support for the acquisition of the HOPS system and development of DataDelve Client and EcoTrack Server. Funding from the Department of Energy's Office of Technology Development permitted the Savannah River Technology Center of Westinghouse Savannah River Company to team with HOPS International Inc. to create DataDelve Client and EcoTrack Server, an application in HOPS to function as an environmental warehouse, specifically as a Decision Support System (DSS). This work was funded by U.S. Department of Energy under contract no. DE-AC09-89SR18035 to the Westinghouse Savannah River Company.

References

1. Department of Energy. Five-year cleanup plan released for public comment. *DOE This Month*, 1993, 16(2), 22.

2. Norton, S.B., D.J. Rodier, J.H. Gentile, W.H. VanderSchalie, W.P. Wood, & M.W. Slimak. A framework for ecological risk assessment at the EPA. *Env. T ox. Chem.*, 1992, 11, 1663-1672.

3. Press, W.H., Flannery, B.P., Teukolsky, S.A. & Vetterlin, W.T. (ed). *Numerical recipes in C: The art of scientific computing.* Cambridge University Press, Cambridge, United Kingdom, 1988.

Development of a database for small hydropower technology projects and their environmental attributes

E.C. Kalkani

Department of Civil Engineering, National Technical University of Athens, Patission 42, Athens 10682, Greece

Abstract

A Microsoft Access database is presented developed in Microsoft Windows environment. The database refers to small hydropower technology projects and includes information in engineering, economics, development and environmental attributes. The main aspects of environmental attributes included in the database are those on environmental values, protected environment, pollution, hazards and health-safety.

1 Introduction

Hydropower is based on well established technology, while technological improvements are introduced into the design for commercial operation of hydro powerplants. The exploitation of plant sites depends on the water flow, the available head, the civil engineering works, the turbine-generator system installed, and the environmental impact of the project. The Thermie program of the European Union, wanting to sponsor innovative small hydro powerplants, invited proposals by the private and the public sector. The financially supported proposals are included in the database presented here.

The database is developed by using Microsoft Access in Microsoft Windows environment, with the goal to process information on small hydropower technology projects, as presented in European Union publications, and describe the environmental attributes of the projects. The database is called Eu(ropean) Ther(mie) Hy(dro technology projects). The number of project records available is small, and most of the information available is in engineering, economics, development and environmental attributes. The main aspects of environmental attributes referenced in the database are the attributes on environmental values, protected environment, wildlife refuge and preserve, pollution and hazards, as well as health and safety attributes.

2 Overall design of the database

The database described here is a collection of data, which are grouped in categories. The data refer to small hydropower technology projects described in Thermie (1993). An example regarding hydro power categories may be a group of general information, a group of plant characteristics and development stage, and a group of environmental attributes.

Using the Microsoft Access (1994) database, a relational database is created which stores related data in one place. As an example, one builds the relational database to store the data related to powerplant information in one place, the data related to development stage in another, and so on.

The advantage of relational databases is that it is easier to find, analyze maintain and protect the data, since the data stored in one place is easier to protect.

The number of different objects that the Microsoft Access database can contain are the following six:
- the tables that store data,
- the queries that gather data which are requested from one or more tables to view, edit or print,
- the forms that display data from tables or queries used to view or edit data, and enter data,
- the reports that summarize and present data from tables and queries to print or analyze it,
- the macros that perform actions without the need of programming,
- the modules that customize, enhance, and extend the database by using the Access Basic Language.

3 Designing the tables

A table is the collection of data regarding a specialized region, ie. information on small hydropower projects specific topics on economics or development. A presentation of the four tables used in the EuTherHy database is shown in Fig. 1. Each table organizes data into columns called fields and into rows called records.

Each field contains information regarding each small hydro power project, concerning the title, the position, the contractor, the address, phone and name of the contact person. Each record contains the information about the small hydro powerplant project with its specific project identification.

The two views of each table is the design and the datasheet view. The design view is used to build or modify the structure of the table, where the different kinds of data to be held in the table are specified, ie. text, number, date/time, etc. The datasheet view is used to add, edit or analyze the data recorded on a table.

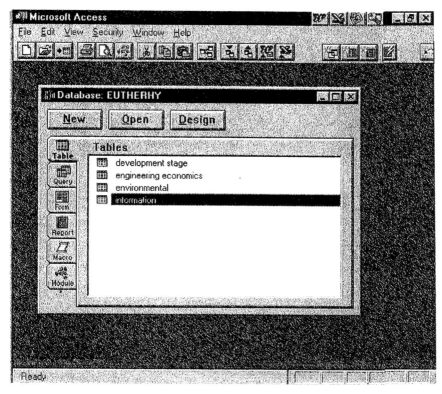

Fig. 1.   Presentation of the four tables in Microsoft Access environment, which are used in the database EuTherHy.

The tables of a database can be built by using cue cards from a file drawer.   In this case information on small hydropower projects is in different drawers and by different names.   The organization of the information requires first the creation of subjects, categories and groups.

For instance, information regarding the projects, such as the title of the project, the identification code (part of a larger coding system), and even a picture of the site can be included.

After creating the table one can keep track of additional information entered on hydropower technology projects, such as new activities and the completion of the project, the date of information aquisition and the date of initiation or completion of an activity on the site. The table can be used to copy data from spreadsheet files, sequential file databases and make these part of the database described here.

The Microsoft Access database gives the possibility of link to a Paradox (Ansa Software, a Borland Co.) database, to a dBASE (Borland Int., Inc.) file or network, and the possibility to work dynamically with data.

When working with data on a table, in case one needs to add lines of

Table: engineering economics

Field Name	Data Type
head (m)	Number
flow (m3/sec)	Number
type of turbine	Text
number of units	Number
type of generator	Text
total capacity (MW)	Number
energy output MWh/year	Number
currency	Text
cost per kWh	Number
total cost	Number
investment period years	Number

Table: information

Field Name	Data Type
ID	Counter
id code number	Text
title	Text
contractor	Text
address	Text
telephone, fax	Text
responsible person	Text

Table: development stage

Field Name	Data Type
ID	Counter
existing plant	Text
refurbishment	Text
new plant	Text
clients	Text
stage of development	Text
completion date	Date/Time

Table: environmental

Field Name	Data Type
ID	Counter
recreational value	Text
cultural value	Text
wildlife, bird and fish value	Text
geologic value	Text
wildlife refuge or preserve	Text
reservoir pollution, hazards	Text
protected environment, land	Text
health and safety	Text

Fig. 2.   Contents of the Tables on Information, Engineering-Economics, Development Stage and Environmental.

new projects into the projects file, one can edit the names and addresses of responsible persons at all or some of the sites. It may be possible to hide some columns that are not interesting at the moment, such as fax or telephone numbers when a list of mailing addresses is necessary, or put side by side interesting information such as project name and responsible person for contact, and phone and fax numbers.

Sorting information is possible for each column of the table, regarding names, dates and values recorded on the table. Also, one can view projects of specific interest, just by sorting the value for the specific interest by larger to lower values and printing that information. A table datasheet can be given on specific information that can be viewed at a glance. And as well, one can replace very easily recordings that have to be renamed with the new names at a very easy way.

The fields in each of the four tables (information, engineering-economics, development-stage, environmental) are shown in Fig. 2, along with the data type necessary for the recording. Further, some entries recorded in the Information Table (already 118 entries) and a print-out view are shown in Fig. 3, indicating the identification and code number, the title of the project, the contractor's name and address, phone number and contact person for each project.

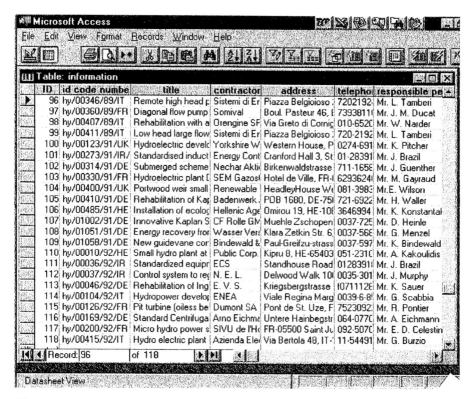

ID	id code numbe	title	contractor	address	telepho	responsible pe
96	hy/00346/89/IT	Remote high head p	Sistemi di Er	Piazza Belgioioso ;	7202192	Mr. L. Tamberi
97	hy/00360/89/FR	Diagonal flow pump	Somival	Boul. Pasteur 46, [	7393811I	Mr. J. M. Ducat
98	hy/00407/89/IT	Rehabilitation with a	Orengine SF	Via Greto di Cornic	010-6520	Mr. W. Narder
99	hy/00411/89/IT	Low head large flow	Sistemi di Er	Piazza Belgioioso ;	720-2192	Mr. L. Tamberi
100	hy/00123/91/UK	Hydroelectric develc	Yorkshire W	Western House, P	0274-691	Mr. K. Pitcher
101	hy/00273/91/IR/	Standardised induct	Energy Cont	Cranford Hall 3, St	01-28391	Mr. J. Brazil
102	hy/00314/91/DE	Submerged scheme	Nechar Akti	Birkenwaldstrasse	711-1658	Mr. J. Guenther
103	hy/00330/91/FR	Hydroelectric plant [	SEM Gazosl	Hotel de Ville, FR-I	6293624I	Mr. M. Gayraud
104	hy/00400/91/UK	Portwood weir small	Renewable	HeadleyHouse We	081-398	Mr.E. Wilson
105	hy/00410/91/DE	Rehabilitation of Kar	Badenwerk.	POB 1680, DE-75I	721-6922	Mr. H. Waller
106	hy/00485/91/HE	Installation of ecoloc	Omirou 19, HE-10E	3646994	Mr. K. Konstantal	
107	hy/01002/91/DE	Innovative Kaplan S	CF Rolle GN	Muehle Zschopen	0037-725	Mr. D. Heinle
108	hy/01051/91/DE	Energy recovery fror	Wasser Vers	Klara Zetkin Str. 6.	0037-56E	Mr. G. Menzel
109	hy/01058/91/DE	New guidevane cor	Bindewald &	Paul-Greifzu-strass	0037-597	Mr. K. Bindewald
110	hy/00010/92/HE	Small hydro plant at	Public Corp.	Kipru 8, HE-65403	051-231C	Mr. A. Kakoulidis
111	hy/00036/92/IR	Standardized equipr	ECS	Standhouse Road	0128391I	Mr. J. Brazil
112	hy/00037/92/IR	Control system to rec	N. E. L.	Delwood Walk 10	0035-301	Mr. J. Murphy
113	hy/00046/92/DE	Rehabilitation of Ing	E. V. S.	Kriegsbergstrasse	I071112E	Mr. K. Sauer
114	hy/00104/92/IT	Hydropower develor	ENEA	Viale Regina Marg	0039-6-8!	Mr. G. Scabbia
115	hy/00126/92/FR	Pit turbine (oiless he	Dumont SA	Pont de St. Uze, F	7523092	Mr. R. Pontier
116	hy/00169/92/DE	Standard Centrifuga	Arno Eichme	Untere Hainbegstr	064-077C	Mr. A. Eichmann
117	hy/00200/92/FR	Micro hydro power s	SIVU de l'Hc	FR-05500 Saint Ju	092-507C	Mr. E. D. Celestin
118	hy/00415/92/IT	Hydro electric plant	Azienda Elei	Via Bertola 48, IT-'	11-54491	Mr. G. Burzio

Record: 96   of 118

Datasheet View

Fig. 3. Outline of the Information Table and the recorded information on hydropower technology projects (118 entries).

4 Designing the queries

A query is the question one wants to ask about the data recorded in the database. As an example one can ask:
- which hydropower projects have environmental values regarding wildlife, bird and fish, cultural, recreational, or geologic,
- which hydropower projects are within areas of wildlife refuge or preserve,
- which hydropower projects are subject to reservoir pollution and natural hazards,
- which of the powerplant sites have problems with protected land and protected environment,
- which hydropower projects have health and safety problems.

With the query the information is gathered from one or more tables and it is put on a dynaset that can be edited or on a snapshot that cannot be edited.

The query is a permanent request saved in the database, so that each time the query is run the latest information appears on the dynaset or on the snapshot to see, or it performs an action on it, for instance deleting or updating data.

The available views in a query are the design and the datasheet views. The design view is used to build or modify the layout of the query. On the design view one asks questions about the data in order to specify what data are needed and how the data should be arranged. The queries may be developed by using SQL commands (Structured Query Language). The datasheet view can be used to add, edit and analyze the data itself which are contained in a dynaset or recordset, which answers the questions posed.

To design a query cue cards are used, in order to:
- collect and organize related data from more than one table, such as recreational and geologic values of a site, along with contact person's name and phone number,
- present a dynaset displaying only the largest powerplants regarding installed capacity or produced annual energy,
- indicate total attributes of powerplants (capacities, energy, total cost),
- assign a problem factor to hydropower projects that have long term delays.

The great advantage of the query is that one can work with data and perform specific tasks, such as to:
- view related data at a glance from several tables, such as development stage and reservoir pollution or hazards,
- find and replace entries with a specific name with a more explicatory expression,
- edit values in the query dynaset and update directly on the dynaset, such as telephone numbers and names of responsible persons (the changes are updated in the relevant table),
- hide uninteresting columns such as hydropower project identification and move interesting columns side by side, such as recreational value and health-safety attributes.

5 Designing the forms

A form is a way to view and edit information in the database record-by-record. The form displays the information we want to see in the way we want to see it.

The form uses familiar controls such as text boxes and check boxes used in windows, which make entering data and viewing them an easy procedure. The forms can be colorful and distinctive since we have control over the size and appearance of every element on it.

We can work on the design view and the form view. The design view is used to build or modify the structure of the form, by adding controls which are bound to fields in a table or query, including text boxes, option bottons, graphs and pictures. In the form view one can add, edit and analyze the data itself record-by-record. In the datasheet view one can see a row-and-column presentation of the data.

To design a form by using cue cards one may:

   - create a form for the projects displaying useful fields such as
     name of the project and the closest environmentally protected area,
     but not the name of the contact person and his phone number,
   - add controls, for instance to a pamphlet to the management,
     regarding an easy to use check box for information to be added,
   - add a command button to the form that prints the current record
     when clicking on the button,
   - add a graph on the form or update the graph without leaving
     Microsoft Access.
   To work with data on a form one uses cue cards to perform different
activities, such as to:
   - add or edit data from a hydropower projects information form,
   - view development in the last year with certain characteristics of
     projects completed,
   - sort the projects by the size of plant (installed capacity) or by the
     name of the country in which the projects are found,
   - illustrate the different projects with the type of turbine available by
     using a series of pictures,
   - find entries in the same country and replace the name with a
     complex one, indicating for instance the region of the project.

6 Designing the reports

The reports are used to view and print information from the database. With
the report, information is displayed in the way we want to see it. Also a
grouping in many levels of the records can be established, so that totals and
averages can be computed by checking values from many records at once.
One can have control of the report, its size and appearance, and of all the
information on it.
   The report can be viewed in the design view and the print preview. The
design view is used to build or modify the structure of a report. One can
add controls that are bound to fields in a table or query, or unbound
controls that calculate totals or averages. The print preview is used to
analyze the data itself, or open a report in sample preview, so that the layout
is checked.
   To design a report cue cards are used, in order to:
   - group the hydropower projects with the wildlife high values,
   - group a list of projects by type of dam and sort them by years of
     completion,
   - group the projects with problems of reservoir pollution and
     hazards,
   - count the number of projects in each country and calculate the
     total capacity installed or the total produced energy per year,
   - add a graph to the list of the hydropower projects indicating the
     type of dam or the position on the map, and update it without
     leaving the Microsoft Access.

7 Designing the macros

A macro is a set of commands for specific actions. It may open a form or print a report, and in general it performs automation of specific tasks with the goal to work smarter and save time.

Specific things made faster by using macros are to:
- set up and print reports,
- open the everyday working forms automatically,
- check data to make sure they are correct,
- open a second form at a click of a botton to find related records.

To write a macro cue cards are used. The different actions may be to:
- set up and print reports automatically when one opens the database starting his work,
- open a variety of forms one works on every day just by pressing a key,
- open a second form at the click of a botton and find records related to the first form,
- check data on a form to make sure all information entered is valid.

8 Designing the modules

Modules are units of codes written in basic (Access Basic Language). Modules are useful to automate and customize the database in some sophisticated ways.

One needs to be comfortably programming in basic, and modules are more powerful than macros to use, although they are more complex to write.

9 Conclusions

A database is developed with the use of Microsoft Access in Microsoft Windows environment. The database includes most of the information available in engineering, economics, development and environmental attributes of small hydropower technology projects. The main aspects of environmental attributes referenced in the database are the attributes on environmental values, wildlife refuge or preserve, protected environment, pollution, hazards, health and safety.

References

1. Thermie Project, *Community Energy Technology Projects in the sector of Hydro Electric Energy - Small Hydro Power Plants*, Commission of the European Communities, Directorate General for Energy, DG XVII, Oct. 1993, p. 212.
2.   Microsoft Access, *Building Applications, Relational Database Management System for Windows*, Microsoft Office, Microsoft Corporation, Document DB57111-1294, 1994, p. 428.

SECTION 10:
ENVIRONMENTAL MANAGEMENT AND DECISION ANALYSIS

Developing environmental model bases

G. Guariso, A. Rizzoli
Department of Electronics and Information, Politecnico di Milano, Milan, Italy

Abstract

Though a systematic method for the design of model bases is not yet available, the paper revises a number of recent experiences. They show that database techniques must be integrated with expert systems and hypermedia to support model selection and application by inexperienced users efficiently and correctly. In particular, it appears that a database approach is preferred when dealing with a very large and heterogeneous set of models, while rule base systems better support model selection in a restricted knowledge domain and hypermedia can be the best tool to supply experience on model use. In most cases also object-oriented techniques are very useful.

1 Introduction

In a recent survey among environmental analysts, the availability of computerised bases of environmental models was ranked as a first priority by 23% of the interviewed. Indeed, a routinely application of mathematical models and related software for the simulation and the management of natural resources is still prevented by the difficulties encountered in selecting and properly use the available tools.

The storing and retrieval of information related to a model is more complex than that traditionally used, for instance, to search for a book or to register population data. Models are active components that must in turn provide significant answers to users' problems and thus their management requires specific tools. The problem is even more relevant if the user wants to find models that describe different compartments of a complex environmental system and wants to link them together to explore alternative actions or decisions on that system. For instance, in a model of lake eutrophication, one sub-model can describe limnological processes and another ecological processes, the latter can also be decomposed into sub-models and these models may in turn be used within a mathematical programming program to find optimal treatment decisions according to a given performance criterion.

When a modeller starts working on a new environmental problem different approaches may be selected:

- pre-packaged routine modelling;
- innovative modelling:
- multi-domain model integration.

The first case is the most common: the problem is well structured and widely studied. Other modellers in the past have produced standard methodolgies and implementations to solve the modelling problem. The problem is, first, to find out the model(s) suitable for that problem and then to convert the data in the format required by the specific model implementation. These problems are solved, at various levels of user friendliness, by *model directories* and *model libraries*, which are characterised by the ability of storing and retrieving a variety of programs each with specific characteristics and a fairly rigid structure.

The second case, despite being the hardest from the modelling point of view, is the most welcome by model definition and integration theorists. They are in fact free to ask the modeller to write the model using their preferred tools and languages that allow for perfect integration and re-use of the newly created pieces of work. Only rarely, however, the amount of really innovative work written in the model would justify its complete rewriting. All models are based on many years of previous work and therefore they also must re-use older components. It is only a limit of present model development packages the fact that it is often easier to rewrite everything rather that to adapt existing models.

In the third case, the modeller has already browsed through model libraries and found all the relevant ingredients for the job. The problem is thus to assemble these models, sometimes written in different programming languages or operating systems, imposing different requirements on data format, and to make them work together.

As noted by Dolk [1], research in the last ten or fifteen years has mainly concentrated on the development of model management systems (MMS) to better support the creation of new models and their use. In fact, a summary of the desirable features of MMS, recently given by Potter et al. [2] on the basis of several previous works, lists the following points:

1) The MMS should be able to create new models quickly and easily.
2) The model building-blocks of MMSs should be cognitively meaningful chunks of knowledge to the user.
3) MMSs should be able to inter-relate models with appropriate linkages, thus providing the functions of model integration and model decomposition.
4) MMSs should be able to manage the model base with functions analogous to data base management.

5) MMSs must have a meta-level encyclopaedia, analogous to a DBMS's data dictionary, which includes a repository of data, heuristics, tasks, models, and the relationships between them.

Some packages implementing these ideas have already become commercially available (e.g., STELLA™ or Simulink™). However, to support the third kind of modelling listed above, MMSs should have another important feature: the ability to incorporate executable models written by other modellers and to connect them in a seamless way. This is essential in the environmental domain that has a long tradition of model development and some programs have been applied in hundreds of studies, thus becoming a legacy of primary importance. Research in this area is only at the beginning and aims at making model re-use more convenient than model re-write.

2 Pre-packaged models

The literature on environmental models has grown in recent years to considerable dimensions. Recent surveys about available software implementing such models quote numbers of the order of several hundreds for groundwater problems [3], more than hundred for air pollution [4,5], few tenths in areas like ecology or surface water management [6]. The problem of finding the one(s) suitable for a given application is thus becoming very difficult.

2.1 Management of models with a fixed structure

Lists of available models have been compiled, mainly in paper form, since long ago. The Environmental Software Directory [7] and Elsevier's Report on Environmental Software [8] are examples of this approach. Only since very recently, however, these libraries are being supplied in a digital form on high volume storage devices such as CD-ROM's [6]. These applications have been mainly developed for modelling experts and thus provide only moderate support to the model screening phase, when the user is looking around for a model that fits his/her needs.

Designing a database for model selection has a number of different and sometimes conflicting objectives. The information should be highly structured so that the computer could easily access it and should be highly flexible to allow easy interaction with the non computer-oriented user. The model must be well described so that the user can get at least some idea of its suitability for his/her specific application, but should be short enough to allow a fast browsing among a number of similar models.

The LIVIA modelbase [9] was built for a wide target audience with quite different, and usually rather basic computer backgrounds. It was implemented on a personal computer together with facilities for model validation and comparison. At present, the model base consists of about 100 well tested and documented programs for which the executable code is also available. They were acquired from various private and public environmental agencies and cover the following main environmental fields: ground water (17 models), air quality (31), dynamics of natural ecosystems (12), hydrology (10).

The information about models in the LIVIA base was structured into four different levels. The first level corresponds to a short description of the model characteristics by the use of a number of *keywords*. These represent a strongly condensed form of description, but they allow a fast search operation and the possibility of displaying the basic information about a model on a single computer screen. For instance, to search a model via keywords, the user can specify one or more values referring to one or more of the following aspects: model name, model objectives, application field, techniques used, type of computer. This structure allows all the normal database operations such as the insertion of new models/values as well as the retrieve and deletion of existing ones.

At the second level, one can find the information organised in a sequential way, that is, a short manual is available that summarise model input and output data format, program installation, price and availability, etc.

The third level is more exhaustive and presents the description of the main features of the models in hypertextual form. The user can browse the information using a set of words that are highlighted on the computer screen without following a sequential order. This technique allows a fast and direct access to the desired information.

The final and more complete level presents the complete model manuals in a hypertextual way. The manuals are stored on a CD-ROM and can be consulted directly on the computer screen even during program execution (figure 1).

2.2 Management of models with a variable structure

All the models referred to in the previous section may be thought of as basic models, i.e., as a mathematical formulation associated with a real world object that has a *flat* structure. For instance, a water reservoir can not be considered a flat domain object when it is seen as an ecosystem composed of fish, zooplankton, and phytoplankton. Conversely, when it is used to describe a simple storage of water in a more complex water management system, neglecting the ecological components, it can be characterised as a basic domain object.

A compound model, on the contrary, represents the aggregation of basic models, i.e., it defines a much more complex structure. Environmental analysts understood long ago that most environmental problems were too varied to be solvable by a fixed model structure. The development history of a river quality model like QUAL2 [10] can be seen as the progressive inclusion of a number of sub-models to better cover a variety of practical situations. The structure of these sub-models was fixed and their number for each compartment was limited. Though QUAL2 has hundreds of options, the user is requested to

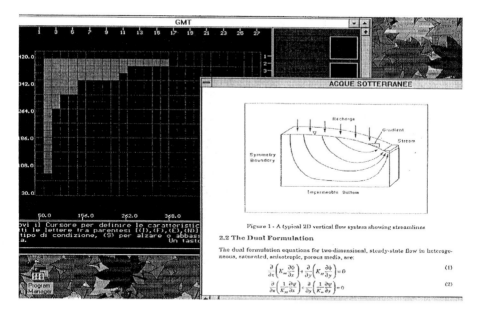

Figure 1: Accessing a specific page of the model manual, during model execution.

select in a simple menu the few that relate, for instance, to flow speed or light penetration.

Modern software engineering tools, particularly object-oriented programming, allowed to further extend the flexibility of similar programs by giving more freedom also to the overall model structure beside the choice of the components. For instance, MASAS [11] and Lakemaker [12] are recent examples of this approach to study lake eutrophication. They exploit the object-oriented approach to systematically support the model definition and execution phase. Each model, each component, each process, and even each individual variable, is represented as a separate object so that its interface with other objects is well defined. Any modification or adjustment of its data attributes may be performed without affecting the rest of the model. For instance, a specific model editor allows, in the Lakemaker package, to select the required model components, attributes, processes and parameter values (figure 2 shows the set of choices required to define the evaluation of a specific subprocess - light limiting factor).

This structure forces a useful and traditional top-down approach to model building. The user is progressively brought down from the most significant choices to the finest details with the possibility of visualising, at any time, the model structure defined up to that point.

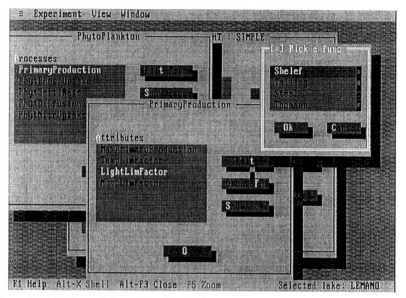

Figure 2: Selection of alternative descriptions of the same subprocess (light limiting factor).

2.3 Tools for supporting model selection

As already mentioned, assisting the user in finding a correct model is a complex task. Two modern computer tools seem particularly suitable to support this task: expert systems and hypermedia.

Expert systems have been used to support model development since much longer than hypermedia. A prototype expert system was developed to help users with the selection of the parameters needed to run the HSPF hydrologic model back in 1981 [13]. With the advent of commercial shells, many other expert systems were developed (see [14] for a survey) which were all efforts to improve the usability and effectiveness of well-known environmental models. The expert system was used to support the definition of a model with flexible structure and was thus dealing with a very narrow problem domain.

FRAME [5] has been an attempt at addressing the wider problem of selecting a legacy model in the domain of air pollution. It comprises two basic modules: a relational database and an expert system shell.

The relational database stores the possible attributes of all air quality models (more than hundred) in a large table. Each row of the table corresponds to a specific model with the last column containing the name of the file with the complete model description, its application conditions, and related bibliographical references. A consultation with FRAME starts with the description of a specific environmental situation. The user is then presented with a list of all the attributes valid for models that satisfy the first query. The consultation proceeds by defining the characteristics of the desired models more and more specifically.

The expert system helps the user in consulting the database, which includes a reference to the set of rules that defines each attribute. Say, for instance, that the user is uncertain whether the value of the "plume" attribute is "buoyant" or "jet". He/she can invoke the expert system, which accesses the set of rules required to define these values. The user is then asked about some meteorological and physical parameters that are necessary to distinguish between the two types of plumes. Using the information provided by the user together with the rules in the knowledge base specific to that attribute, the inference engine computes the suggested value for the parameter under discussion and submits it to the user, that can use it to refine the search.

FRAME structure has the advantage of addressing a large problem domain by decomposing it into a set of small knowledge bases, which allow an easier debugging and maintenance.

Hypermedia technologies allow a more sophisticated and effective presentation of material in a given domain than the traditional, sequential printed matter. The two major features that characterise hypermedia are in fact the dominance of multimedia data (free text, pictures, graphics, video, sound, animation) over traditional formatted data and the emphasis on interactive navigation, i.e., traversing explicit connections among pieces of information.

HYPERMODEL [15] is a project based on the use of hypemedia techniques for browsing through a model base. It is based on four different entities:

- "Model" which accesses a technical description of each program in the base similar to the keywords in the database described above;
- "Theory" that provides some explanation of the theoretical basis of the models,
- "Manual" that allows to flip through the complete model documentation and, finally,
- "Documents" that shows the conditions and the results of the application of a model to specific cases.

This last entity is particularly important because it allows to insert in the system any type of information useful to describe the application of models to real cases. By looking at maps, pictures or short videos, besides traditional explanation text, the user may have a better understanding of the conditions under which the model was applied or about the simplifications that have been necessary in the modelling process; two aspects of paramount importance particularly for teaching purposes.

3 Development of new models

To further expand the generality of MMSs and to allow for the creation of really new models (not simply a new combination of existing options), developers have been forced to specify a structure for the models and their internal components. A number of systems have been proposed based on an Operational Research perspective (decision variables, constraints, objective functions) [17, 18, 19] and others on a System Theory perspective (input, output, state variables varying in time) [20, 21].

All these packages share, at different levels, the assumption that the user is able to translate his/her problem into a mathematical formulation and then translate back a sequence of numbers in something meaningful for understanding and/or decision making.

Another feature of these systems is that the overall model must be built by connecting new or existing components with a typical bottom-up approach, quite different from the traditional environmental software, built for a user who perfectly knows the problem domain and wants to progressively refine its description. For these reasons and despite their conceptual value, these packages do not seem to represent viable tools for environmental analysts, at least in the near future. In the long term, on the contrary, some of them may become a kind of "standard" modelling language and this would justify the costs and difficulties of re-writing available models in a compatible form.

4 Integration of models from different domains

Model integration is needed when the modeller wants to intersect the competencies of various models to solve a complex problem, a common practice in environmental modelling. Model integration has been always done by hand. The modeller studies the whole problem, decomposes its domain in subdomains where he/she can apply existing models, collects the needed data (modifying them in order to comply with format specifications), and finally runs the model implementations, often feeding the output data of one model to the input of another. This activity is time consuming and very hard to replicate each time there is a change in the modelled scenario.

The purpose of model integration is to devise tools and techniques to facilitate these tasks and operations. While current research is active in studying the problems of model integration [1, 22], very few are the case studies in which real model integration took place [23].

If a model is designed from scratch, clearly the modeller has to use abstraction principles in order to represent the objects he/she has previously identified as components of the problem domain. On the contrary, when an existing model has to be integrated in a new environment, the model integrator is not free to construct his/her own representation of the problem domain: the models already contain their own (hidden) representations.

Thus, major problems when integrating legacy models are:
- mapping the legacy model to the problem domain;
- linking together legacy models.

Classification, aggregation and *inheritance* principles, inspired by object-oriented analysis and design methodologies, can be used for the purpose of integrating existing models.

The model integrator must extract the hidden representation of the problem domain contained in the model and define accordingly a set of abstract data structures (problem domain classes) which constitute its explicit representation. Then, links between the interface of the legacy model and the relative class must be established. This is not a trivial problem. The problem domain classes may present a different organisation from the data structures in the legacy model. A particular attention must be given to the mapping of similar concepts into the problem domain, in order to avoid useless duplication or unwanted simplifications.

Another difficulty that arises in integrating existing models, is to make them "communicate". There is no standard procedure. When the source code is unavailable and there is no way of writing "drivers" accessing the inner model mechanisms, the most common approach is to set up communications through data files. "Embedding" the model in the problem domain classes can overcome this problem since inter-model communications are filtered through class interfaces (data attributes and methods) which also provide a way of synchronising and transforming data items which do not have the same spatial and temporal resolution.

Figure 3 illustrates how these principles may be applied in practice [2 4] . Two legacy models are embedded in two separate classes (class LandSys and class StreamSys) and then integrated in a unique class Catchment.

The local data attribute of the class Catchment are re-routed to the import data attributes of the subclasses and the outputs of subclass StreamSys to the export data attributes of class Catchment. The class structure also allows to connect, via the export and import data attributes, the outputs of model LandSys to the inputs of model StreamSys.

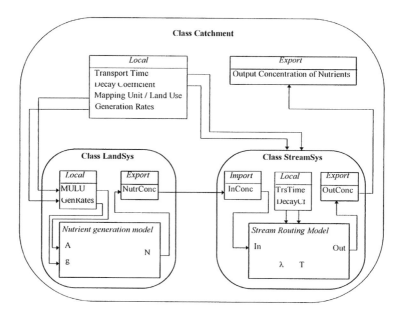

Figure 3: An example of model class for a river quality problem.

5 Concluding remarks

The development of efficient model bases and model management systems is becoming essential to support a wider utilisation of environmental models. Modern computer tools allow an improved interaction with the user and thus make this objective feasible. In this paper, we have presented some ideas and realisations of model bases for different purposes and using different computer tools. A summary of this brief survey is presented in Table 1. It shows that object-oriented techniques are probably going to have a major impact on the development of environmental modelling, while expert systems may be important whenever dealing with a sufficiently narrow knowledge domain and hypermedia when the scope of the base is more directed toward providing practical experience to the user.

Table 1: a summary of model characteristics and proposed computer tecniques.

Model characteristics ------------------------ *Computer tecniques*	Fixed structure	Variable structure	Entirely new	Integration of existing models
database	very important	useful	useful	useful
expert systems	useful	important		useful
hypermedia	useful	useful		useful
object-oriented approach		very important	very important	very important

References

1. Dolk, D.R. & Kottemann J.E., Model Integration and a Theory of Models, *Decision Support Systems*, **9**, 1993.
2. Potter, W.D., Byrd, T.A., Miller, J.A. & Kochut, K.J. Extending Decision Support Systems: the Integration of Data, Knowledge, and Model Management, *Annals of Operations Research*, **38**, 1992.
3. International Ground Water Modeling Center (IGWMC), *Software Catalog*, Boulder, CO, 1995.
4. Zannetti, P. *Air Pollution Modeling.* Computational Mechanics Publications, Southampton, UK, 1990.
5. Calori, G., Colombo, F. & Finzi, G. FRAME: a Knowledge-Based Tool to Support the Choice of the Right Air Pollution Model, in *Computer Support for Environmental Impact Assessment* (eds. G. Guariso, B. Page), pp. 211-222, North Holland, Amsterdam, 1994.
6. Dodson & Associates, Inc., *HydroCD*, Huston, Texas, 1996.
7. Donley Technologies, *Environmental Model Directory*, 1995.
8. *Environmental Software. A strategic study of the environmental software market*, Elsevier, Amsterdam, 1992.
9. Del Furia, L. & Petrucci, M. LIVIA Project: a computer laboratory for environmental impact assessment, , in *Computer Support for Environmental Impact Assessment* (eds. G. Guariso, B. Page), pp. 311-320, North Holland, Amsterdam, 1994
10. Barnwell, T.O., Brown, L.C. & Marek W. Development of a prototype expert advisor for the enhanced stream water quality model QUAL2E. Internal Report, US EPA, 1986, Athens, GA.
11. Ulrich, M.M., Imboden, D.M. & Scwarzenbach, R.P. MASAS-A user-friendly simulation tool for modeling the fate of anthropogenic substances in lakes, *Environmental Software*, 1995, 10, 43-64.
12. Del Furia, Rizzoli, A. & Arditi, R. Lakemaker: a general object-oriented software tool for modeling the eutrophication process in lakes. *Environmental Software*, 1995, **10**, 43-64.

13.Gashing, J., Reboh, R. & Reiter, J. Development of a Knowledge-based Expert System for Water Resources problems, SRI Project 1619, 1981, SRI International, Palo Alto, CA.

14.Davis, J. R. & Guariso, G. Expert systems support for environmental decisions, *Environmental Modelling Vol. II,* ed. P. Zannetti, pp. 325-350, Computational Mechanics Publications, Southampton, 1994.

15.Guariso, G. & Tracanella, E. Hypermodel: An hypermedia guide to environmental models, in *Proc. Int. Conf. on Environment and Informatics* (Z. Harnos ed.), pp. 12-19, Budapest, 1995.

16.Muhanna, W.A., SYMMS: a Model Management System that Supports Model Reuse, Sharing, and Integration, *European Journal of Operational Research*, 1994, **72**, 214-242.

17.Bhargava, H.K. & Kimbrough S.O. Model Management: an Embedded Languages Approach, *Decision Support Systems*, **10**, 1993.

18.Dolk D.R. & Kottemann J.E, Model Integration and Modeling Languages: a Process Perspective, *Information Systems Research*, **3**, 1992.

19.Hong, S.N., Mannino, M.V. & Greenberg, B, Measurement Theoretic Representation of Large, Diverse Model Bases: the Unified Modeling Language L_U, *Decision Support Systems*, **10**, 1993.

20.Rozenblit, J.W., Jankowski, P.L., An Integrated Framework for Knowledge-Based Modeling and Simulation of Natural Systems, *Simulation*, **57**, 1991.

21.Guariso, G., Hitz, M. & Werthner, H. An Integrated Simulation and Optimization Modelling Environment for Decision Support, *Decision Support Systems*, 1996, **16**, pp. 103-117.

22.Rizzoli, A. A software architecture for model management and integration: theoretical background. *Technical Memorandum 94.10. CSIRO Division of Water Resources.* ISBN 0 643 05591 6. 1994

23.Davis, J.R, Abel, D.J., Zhou, D., Rizzoli, A. & Kilby, P. HYDRA: a Generic Design for Integrating Catchment Models, Presented at the *American Society of Civil Engineers 21st Annual Conference on Water Resources Planning and Management Division*, Denver, Co, 22-26 June 1994.

24.Rizzoli, A., Davis, J.R., Reed M. &Farley T. 1996. A DSS for catchment management. To appear in: "Environmental Software, Vol. III"; P. Zannetti (ed).

VIM: a screen model for the evaluation of the environmental impact from a multisource area

T. Tirabassi,[a] U. Rizza,[b] C. Mangia,[b] L. Lepore,[c] U. Poli[c]

[a]Institute FISBAT of C.N.R., Via Gobetti 101, Bologna, Italy
[b]Institute ISIAtA of C.N.R., Via Arnesano, Lecce, Italy
[c]ISPESL, Via Urbana 167, Roma, Italy

Abstract

A screening model (VIM: Valutazione dell'Impatto da Multisorgenti) is presented as a method for estimating maximum ground-level concentrations in an area with many emission sources, as a function of stability, wind speed and wind direction. It is designed for a low-cost, detailed screening of a multi-source area in order to determine maximum hourly concentrations and decide whether or not the use of a more sophisticated model is required. The model is considered to be a useful tool for screen analysis as it constitutes a relatively simple evaluation technique that provides a conservative estimate of the air quality impact of a specific multi-source area.

1. Introduction

The availability of preliminary estimates provided by a simple model can be extremely useful in evaluating air quality and the impact produced by a number of polluting sources. Such estimates are also useful in the planning of new emissions to ensure that projects conform to legal standards. A preliminary evaluation model is of considerable use when it is able to estimate concentrations also in the presence of only a few or even no meteorological measurements.

The VIM (Valutazione dell'Impatto da Multisorgenti) model is in fact able to provide estimates of maximum concentrations produced in a specific area, also in the absence of meteorological data. It is a model that calculates the maximum concentrations of a pollutant for all meteorological scenarios possible in a given area.

The VIM packet is made up of various codes. The VIM_GRID calculus code provides the concentration of a pollutant at all grid points for each of the meteorological scenarios envisaged by the VIM model. It is therefore a model that writes on the output all concentrations and not only the maxima. The results obtained with VIM_GRID can subsequently be employed to build maps.

2. Meteorological scenarios

In reference to the atmospheric dispersion of materials, a meteorological scenario is characterised by wind speed and direction and by atmospheric turbulence intensity responsible for the dispersion of the pollutants in the atmosphere. Thus, the scenarios are given by all the combinations of wind speed, wind direction and atmospheric turbulent intensity. This way of proceeding is the one indicated by the United States Environmental Protection Agency (U.S.EPA), which identified 49 meteorological scenarios (derived from realistic combinations of 6 classes of turbulence and 14 of wind speed), among which are sub-divided the various situations of atmospheric dispersion (Pierce et al., 1982). Added to these are 8 different wind directions, giving a total of 392 scenarios considered by VIM.

Atmospheric turbulence can be divided into 6 classes according to the Pasquill-Gifford scheme (Pasquill and Smith, 1983). Such classes are usually referred to with the first six letters of the alphabet, from the most unstable to the most stable class, respectively. Wind speed can be reasonably divided into 14 classes (Pierce et all., 1982) represented by the following speeds in m/s.: 0.5, 0.8, 1, 1.5, 2, 2.5, 3, 4, 5, 7, 10, 12, 15, 20.

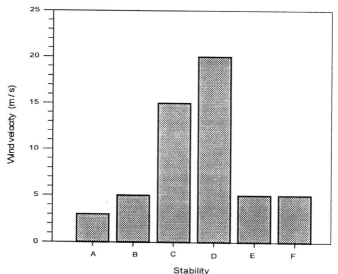

Figure 1. The combinations of stability and wind speed

The combinations of stability classes with the wind speeds and wind directions give rise to the various meteorological scenarios. Taking account that not all combinations are feasible in real situations, the six stability classes, the 14 wind speed classes (Figure 1 shows the combinations of stability and wind speed) and the 8 wind direction classes add up to the above total of 392 scenarios.

3. Characteristics of the model

VIM is a steady-state Gaussian plume model for flat terrain derived from the single source model of the U.S.EPA PTPLU2 (Pierce et al., 1982). In fact, VIM calculates the concentrations using the Gaussian equation (Pasquill and Smith, 1983), which is an exact solution of the advection-diffusion equation in the case when both the wind and diffusion coefficients are constant. Since, however, the latter conditions are unrealistic, the Gaussian solution is forced to represent real situations through empirical parameters, the so-called "sigma". In particular, within a reference system where x is in the along-wind direction, y is perpendicular to wind direction and z is the height, the model makes use of the formulas below.

In stable conditions (classes E and F) or for an atmospheric boundary layer taken to be infinite:

$$C = Q \ 1/u \ g1/(\sqrt{2\pi} \ \sigma_y) \ g2/(\sqrt{2\pi} \ \sigma_z)$$

where:

$$g1 = \exp(-0.5 \ y^2/\sigma_y^2)$$
$$g2 = \exp[-0.5 \ (z-H)^2/\sigma_z^2] + \exp[0.5 \ (z+H)^2/\sigma_z^2]$$

Q is the emission flux of the pollutant
u is the wind speed
H is the effective height of the source, or the height reached by emissions due to Archimedes forcing, if hot, or by inertia, if cold
σ_y and σ_z are the horizontal and vertical diffusion parameters.

In unstable or neutral conditions, when σ_z is greater than 1.6 L, where L is the mixing layer height:

$$C = Q \ 1/u \ g1/(\sqrt{2\pi} \ \sigma_y) \ 1/L$$

In unstable or neutral conditions, when both H and z are less than 1.6 L and, if σ_z is less or equal 1.6 L:

$$C = Q \ 1/u \ g1/(\sqrt{2\pi} \ \sigma_y) \ g3/(\sqrt{2\pi} \ \sigma_z)$$

where:

$$g3 = \sum_{n=-\infty}^{\infty} \left\{ \exp\left[-0.5(z - H + 2nL)^2 / \sigma_z^2\right] + \exp\left[-0.5(z + H + 2nL)^2 / \sigma_z^2\right] \right\}$$

The infinite series rapidly converges and the calculation is limited to values of n between -4 and +4.

Atmospheric stability is evaluated using the Pasquill-Gifford classes (Pasquill and smith, 1983). The wind direction is divided into 8 sectors, each of 45 degrees: the x axis is rotated 90 degrees in a clockwise direction with respect to the y axis and the wind within the sector 1 blows along y, in sector 2 between x and y, in sector 3 along x and so on.

For the evaluation of atmospheric turbulence and diffusion parameters, VIM follows the method proposed by Pasquill (1961). In open country conditions, vertical diffusion is described by:

$$\sigma_z = a\, x^b$$

where a and b are functions of atmospheric stability and the distance from the source. The vertical diffusion parameter   is taken to be equal to 5000 m in stability A for distances greater than 3.11 km and in stability B for those greater than 35 km.

Lateral dispersion is estimated by the model using the function:

$$\sigma_y = 1000 \ \times\ \tan(\theta)/2.15$$

where x is the distance from the source in kilometres and   is obtained by logarithmic interpolation between two values corresponding to the two distance x = 0.1 km and x = 100 km.

For urban conditions, the VIM model follows the parameterization of Briggs (Gifford, 1976). The diffusion parameters are presented in Table I.

Table I. Diffusion parameters in urban areas (Briggs, 1973)

Stability	σ_y (m.)	σ_z (m.)
A and B	$\sigma_y = .32\, x / (1+0.4\, x)^{1/2}$	$\sigma_z = .24\, x / (1+ x)^{1/2}$
C	$\sigma_y = .22\, x / (1+0.4\, x)^{1/2}$	$\sigma_z = .2\, x$
D	$\sigma_y = .16\, x / (1+0.4\, x)^{1/2}$	$\sigma_z = .14\, x / (1+ 0.3x)^{1/2}$
E and F	$\sigma_y = .11\, x / (1+0.4\, x)^{1/2}$	$\sigma_z = .08\, x / (1+ 1.5x)^{1/2}$

Plumes from a point source, because they have a higher temperature than the ambient air and/or the speed of inertia, proceed to rise after emission in to the atmosphere to reach a height that is referred to a effective height. This height is calculated in the VIM model according to the method of Briggs (Briggs, 1969; Briggs, 1971).

The VIM model offers the options of calculating the stack downwash (i.e. the effect of the wind field perturbation due to the source itself), the buoyancy-induced dispersion and the gradual plume rise.

If the latter option is adopted, the effective height of the source H is a function of the distance x from the source up to the point where the effective height is reached. If distance x from the source of the point of concentration calculation is less than the distance at which the plume ceases to rise and reaches the effective height, plume height is calculated using the following formula:

$$H = h + (160 \ F^{1/3} \ x^{2/3}) / u$$

where H is the height of the plume axis, h is the physical height of the source, u the wind speed at height h and F is the buoyancy parameter (Briggs, 1969).

If the stack downwash option is adopted, the model modifies the physical source height on the basis of the following formula (Briggs, 1973):

$$h' = h + 2 \ [\ (V_s \ / u) - 1.5 \ ] \qquad \text{for } V_s \ < 1.5 \ u$$
$$h' = h \qquad \text{for } V_s \ \geq 1.5 \ u$$

where h' is the source height modified for downwash, V_s is the plume exit speed and u is the wind speed at height h.

For strongly buoyant plumes, entrainment as the plume ascends through the ambient air contributes to both vertical and horizontal spread. If the buoyancy-induced dispersion option is adopted, the effective dispersion (σ_{ye} and σ_{ze}) is determined by adding the contribution due to the entrainment. Since in the initial growth phases of release, the plume is nearly symmetrical about its centerline, buoyancy-induced dispersion in the horizontal direction equal to that in the vertical is used:

$$\sigma_{ye} = (\sigma_y^2 + \sigma_p^2)^{1/2}$$
$$\sigma_{ze} = (\sigma_z^2 + \sigma_p^2)^{1/2}$$

where:

$$\sigma_p = (H - h) / 3.5$$

The increase of wind speed with height is described by a power low profile:

$$u = u_0 \ (z/z_0)^p$$

where u_0 is the wind speed at the height z_0, while the model uses default values of the exponent p from Irwin (1979). The exponent p used as default values are reported in table II.

Table II. Wind profile exponents used as default values.

Stability	A	B	C	D	E	F
urban condition	0.15	0.15	0.20	0.25	0.30	0.30
country conditions	0.07	0.07	0.10	0.15	0.35	0.55

4. The VIM code

The VIM code is written in FORTRAN and calculates the concentrations produced by all the emission sources considered. On a file called VIM.OUT, it writes the maximum concentrations (different from 0) of each meteorological scenario, the co-ordinates of the positions of the maximum, the downwind direction sector, atmospheric stability and wind speed. The data relating to a meteorological scenario are printed on the same line. The meteorological scenarios are ordered and numbered (in decreasing order) according to the maximum concentration values to allow a rapid analysis of the results. In figure 2 an example of a part of VIM.OUT is shown.

5. The VIM_GRID code

The VIM_GRID code is written in FORTRAN and calculates and records in the VIM.DAT file the concentrations at all grid points of a given scenario selected by the user. In VIM_DAT the results are written in columns. In particular, the first column gives the value of co-ordinate x in meters, the second that of Y and the third the concentrations in $\mu g / m^3$. The values are reported without comment so that they can be easily used by a graphics programme for the mapping of concentration isolines.

The user types in the number given in VIM.OUT for the scenario to be mapped. He types 1 if he wishes to map the scenario responsible for the maximum concentrations, number 2 if he is interested in the scenario responsible for the second highest concentrations, and so on.

6. Conclusions

A Gaussian model (VIM) has been presented that is able to determine the maximum hourly concentrations produced by a multi-source emission area for 392 meteorological scenarios, corresponding to 392 combinations of wind direction, wind speed and atmospheric stability class.

VIM is a screening model that can provide estimates of maximum concentrations, also in the absence of meteorological measurements and is therefore useful in providing a conservative estimate of the air quality impact of a specific multi-source area.

```
· · · · · · · · · · · · · · · · · · · · · · · · · · · · · · · · · · · · · · · · · · · · ·
MODEL OPTIONS
iopt(1)= 0 (1: gradual plume rise, 0: no gradual plume rise )
iopt(2)= 0 (1:stack downwash, 0: no stack downwash)
iopt(3)= 0 (1:buoyancy-induced disp., 0:no buoyancy-induced disp.)
idflt=  1 (1:  default exponents of wind profile, 0: no default exponents)
muor=  2 (1: urban conditions, 2: country conditions)
DATA
mixing layer height =1500. (m)
air temperature =  293.0 (⁰K)
receptor height =  .0 (m)
```

exponenents of power-low wind profile
.07   .07   .10   .15   .35   .55

number of sources = 5

num.	conc. (ug/m^3)	x (m)	y (m)	sect.	stab.	wind (m/s)
1	.840E+02	400.	600.	4	1	1.00
2	.840E+02	1400.	400.	2	1	1.00
3	.840E+02	600.	400.	8	1	1.00
4	.840E+02	0.	1600.	1	1	.80
5	.840E+02	1600.	0.	3	1	.80
6	.840E+02	400.	0.	7	1	.80
7	.840E+02	0.	400.	5	1	.80
8	.836E+02	600.	400.	8	1	.80
9	.836E+02	400.	600.	4	1	.80
10	.836E+02	1400.	400.	2	1	.80
11	.805E+02	1600.	0.	3	1	1.00
12	.805E+02	0.	1600.	1	1	1.00
13	.805E+02	400.	0.	7	1	1.00
14	.805E+02	0.	400.	5	1	1.00
15	.770E+02	0.	1400.	1	1	2.00
16	.770E+02	1400.	0.	3	1	2.00
17	.770E+02	600.	0.	7	1	2.00
18	.770E+02	0.	600.	5	1	2.00
19	.748E+02	1400.	0.	3	1	2.50
20	.748E+02	0.	1400.	1	1	2.50
21	.747E+02	600.	0.	7	1	2.50
22	.747E+02	0.	600.	5	1	2.50
23	.723E+02	400.	600.	4	1	1.50
24	.723E+02	600.	400.	8	1	1.50
25	.723E+02	1400.	400.	2	1	1.50

Figure 2. Part of the output of code VIM.

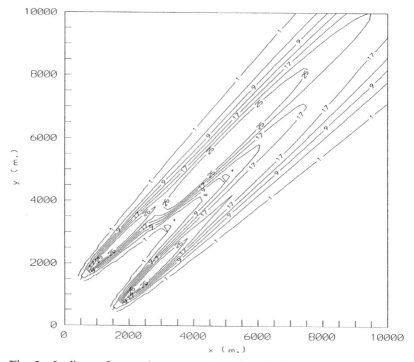

Fig. 3 - Isolines of ground concentrations (μ g/m3) due to emissions from 5 sources located at co-ordinates (Km): 0,1; 1,0; 2,2.5; 2,3; 2.5,1.5. Meteorological scenario: stability = D, wind velocity = 3 m/s, sector = 2.

By way of example, Figures 3 shows the map of ground level concentrations obtained by means of applying the SURFER code (Golden Software Inc.) to the results of VIM_GRID.

7. References

Briggs G.A. (1973). Diffusion estimation for small emissions. *Rapporto NOAA n. 79*, Oak Ridge, TN (U.S.A.).

Briggs G.A. (1969). Plume Rise. *Rapporto USAEC n. TID-25075*, National Technical Information Service, Springfield (VA), U.S.A.

Briggs G.A: (1971). Some recent analyses of plume rise observation. Atti del Second *International Clean Air Congress,* a cura di Englund H.M. e Beery W.T., Academic Press, New York, pp. 1029-1032.

Briggs G.A., Diffusion estimation for small emissions. Rapporto NOAA n. 79, Oak Ridge, TN (U.S.A.), 1973.

Gifford F. (1976).Turbulent diffusion - typing schemes: a review. *Nuclear Safety*, Vol. 17, pp. 68-85.

Pasquill F. e Smith F.B. (1983). *Atmospheric Diffusion.* Ellis Horwood, Londra, 1983

Pasquill F. (1961).The estimation of dispersion of windborne material. *Meteorol. Magazine,* Vol. 90, pp. 33-49.

Pierce T.E., Turner D.B., Catalano J.A. e Hale F.V. (1982). PTPLU - A single source Gaussian dispersion algorithm. *Rapporto EPA-600/8-82-014,* U.S.EPA, Research Triangle Park (USA).

IrwinJ.S. (1979). A theoretical variation of the wind profile power-low exponent as a function of surface roughness and stability. *Atmos. Environm.,* Vol. 13, pp. 191-194.

Matrix for Evaluation of Sustainability Achievement (MESA): determining the sustainability of development

N.S. Fleming,[a] T.M. Daniell[b]
aConsulting Engineer, RUST PPK Pty Ltd
bDepartment of Civil and Environmental Engineering, The University of Adelaide, Adelaide, SA 5005, Australia

Abstract

Decisions generally made by planners, designers, economists, engineers, and politicians occur in a multi-objective environment and are based upon non-standardised and imprecise data. In the past, projects were justified almost solely on economic grounds, with insufficient consideration of environmental and social issues. The shifting of societal values has placed greater emphasis upon careful management of the resources. Techniques are therefore required to deal quantitative and qualitative information. The MESA methodology has been developed to promote 'sustainable' planning decisions. Decisions are based upon a formal rigorous decision strategy involving Fuzzy Set Theory allowing quantitative and qualitative values to be evaluated. The MESA method is able to incorporate public opinion with expert knowledge, which is vital in the development and assessment of environmental policy.

1 Introduction

Most development decisions are multi-objective in nature, involving economic, political, social and environmental issues. This reality is often unsatisfactorily addressed as most development decisions are justified almost solely on economic grounds (employing cost-benefit or cost-effectiveness analysis), with little or no consideration of other issues. A holistic review of development proposals is necessary to avoid unsustainable development that benefits only a few to the detriment of many others. Procedures for quantifying environmental and social impacts are time consuming and occur late (if at all) in the planning process because there are often conflicting objectives that cannot be expressed in commensurable units [1]. Attempts at expressing environmental, social and political impacts in monetary terms have been made with limited success [2].

Many problems occur within cost-benefit analysis when attempting to include social and environmental issues. The imputation of market prices to intangible items is questionable; while equity problems are typically ignored [3]. There has been an acceptance that parameters which are non-quantifiable in monetary terms must be considered in the evaluation of alternatives. A wide range of multi-objective decision-making techniques have been developed to facilitate the involvement of the public in planning efforts. In fact, there are so many multi-objective techniques available that methods have been developed to assist in the selection of the most suitable technique [4],[5],[6]. Multi-objective methods should try to meet the following criteria [1], [7]:

- address the prediction of futures, both with and without the project;
- allow consideration of alternatives both separately and in combination;
- quantify probabilistic nature of economic, environmental and social impacts;
- be based on principles and assumptions that are valid and easily illustrated;
- employ qualitative and quantitative information in a sound way;
- make explicit subjective and value judgments;
- communicate the predicted impacts with respect to their spatial and temporal distribution;
- involve inter-agency coordination and public involvement;
- yield results that are understandable to decision makers and the public;
- should be completed within a reasonable time frame; and
- require minimal expenditure for project evaluation.

Goicoechea et al. [8] identify the shifting of societal values, relating to the use and management of natural resources, as a prime motivation in the development of multi-objective analysis. Shafer and Davis [9] state that it is essential to include all the perceived costs and benefits in project evaluation because the quality of life and of environments in the future are shaped by the long-term planning priorities, decisions and activities undertaken today.

The expansion of population in almost every region of the world has led to increased use of the Earth's resources [10] and a focus on sustainable development. A realistic method is needed that includes and evaluates quantitative and qualitative values.

2 Development of the MESA method

The Matrix for the Evaluation of Sustainability Achievement (MESA) has been developed to concurrently address economic, environmental and social issues, indicating whether a project assists or retards the achievement of sustainability. MESA aims to ensure:
- public involvement in the establishment of priorities;
- evaluation of qualitative and quantitative information on costs and benefits;
- assessment and comparison of multiple projects or alternatives;

- consideration of the consequences of development relative to current conditions (i.e. 'with' versus 'without' the project);
- combination of impact assessments in a methodologically sound way without resorting to estimating monetary values of intangible items; and
- a simple, logical technique that allows a relatively quick appraisal.

A matrix style of decision-making was selected because it is a simple yet effective medium for surveying development consequences in environmental impact analysis [11]. In an attempt to eliminate the need to place monetary values on costs and benefits, coupled with the desire to use the unavoidable subjective responses to some criteria, use of *fuzzy set theory* was adopted. Fuzzy set theory has the ability to represent multi-objective decision problems involving vague or fuzzy objectives or constraints [12]. Fuzzy sets combine quantitative and qualitative information by assigning to each of the criteria (for a given alternative) a number, in the range 0 to 1. The number 0 indicates that the criterion is not achieved, while 1 refers to satisfaction of the criterion. Some early attempts to include fuzzy set theory into engineering project decision-making were Ragade et al. [13], Daniell [14], Znotinas and Hipel [15], and Alley et al. [16]. In recent years fuzzy set theory has re-emerged and found increasing acceptance in natural resource management [8], as well as civil engineering and water resources applications [17], [18], [19].

3 Undertaking the MESA method

The MESA technique combines information on the pre-development condition, the potential impacts of a range of development alternatives and the importance attached to each evaluation criterion.

Evaluation by the MESA method adopts the following procedure:

1. Define the problem which has initiated the development investigations, and outline the objectives that alternative solutions should meet.
2. Determine the criteria and values (in addition to those presented in the MESA) that are important in the selection of the best alternative.
3. Develop alternative schemes to meet these objectives.
4. Assemble a multi-disciplinary team possessing knowledge and expertise in areas such as economics, engineering, law, the natural environment and social issues. Assess the importance of each of the assessment criteria to develop a weight factor to be attached to each criteria.
5. The members of the multi-disciplinary team should assess the degree to which the present conditions satisfy the designated *Goals and Constraints*. The alternatives should then also be evaluated.
6. Combine these assessments mathematically to provide an indication of the performance of each alternative relative to the criteria under consideration:

$$WP_{ij} = (PC_j \times PP_{ij})^{\alpha_j} \qquad (1)$$

where
   i = the alternative number,

j = the sustainability goal or objective,

a_j = weight attached to criterion j,

PC_j = Present Condition rating for criterion j,

PP_{ij} = Project Performance rating for alternative i and criterion j, and

WP_{ij} = Weighted Performance of alternative i relative to criterion j.

7. The Fuzzy Set Combination is the minimum WP_{ij} for each alternative; that is, the minimum value in the column corresponding to each alternative.

8. A Dominance Matrix is constructed to examine the cumulative benefits of one alternative over another. The Dominance Factor is calculated as an effectiveness measure of one alternative relative to all other alternatives.

9. The Weighted Performance and Dominance Factor product determines the final ranking of alternatives. The closer the product is to 1, the greater the satisfaction of the criteria and provision of benefits over other alternatives.

Perhaps the most important task is the specification of criteria against which alternatives are assessed. An evaluation by practitioners in South Australia of key criteria of a sustainable water resources management strategy for a peri-urban area of Adelaide has been undertaken [20]. Deason and White [21] note that despite the growth in number of multi-objective methods, they are yet to enjoy wide success in solving real problems.

Table 1   Sustainability evaluation criteria

Maintenance of habitat and ecosystems	Local infrastructure compatibility
Preservation of native plant species	Involvement of the community
Preservation of native animal species	Use of renewable energy sources
Preservation of areas of landscape/amenity value	Airborne disposal within assimilative capacity
Preservation of areas of cultural value	Energy efficiency
Reclamation & reuse of wastewater	Public acceptability
Wastewater disposal within assimilative capacity	Groundwater extraction within sustainable yield
Minimisation of greenhouse gas emission	Improved access to public open spaces
Improvement in surface water quality	Improved recreational opportunities
Improvement in groundwater quality	Full cost recovery for good or service
Productive use of fertile soils	Annual equivalent cost-benefit ratio
Prevention of erosion	Costs borne by consumers
Application of clean technology	Equitable cost-benefit distribution
Waste recycling or use	Unit cost for good or service
Material utilisation allowing recycling	Increased in employment opportunities
Increased use of metal substitutes	Capital cost funding capability
Compatibility with existing operations	

3.1 Selection of criteria for project evaluation

The specification of an appropriate and operational set of objectives is as difficult as it is important. The MESA Table 1 lists 33 goals and constraints, and includes criteria relating to the economic, environmental and social aspects of development. Many modifications to this list were required to achieve the final list. It is considered that the list contains the fundamental objectives that would constitute a strategy for the achievement of sustainability, while maintaining sufficient generality to be applicable to all large project conditions. Any issues or objectives which are project specific should also be included.

3.2 Assigning the relative importance to each criterion

Many different methods have been developed for investigating the relative importance of different criteria or objectives. These methods range from simple ordinal ranking of objectives, to more sophisticated mathematical methods. The method proposed for use with the MESA is that developed by Saaty [22]. This technique determines a ratio scale for the group of criteria or objectives, based upon a pairwise comparison of the individual criterion. The importance assigned to each criterion is calculated from each member's responses. The weights determined by each member are then averaged across each criterion. Assume that a comparison is being made between criterion j and criterion j+1 to determine which of these two criteria are more important. A value is assigned based on this judgment and then combined to construct an n×n matrix A, so that the following rules are satisfied:

1. if criterion *j* is more important than *j+1*, assign a number to a_{jj+1} (Table 2);
2. $a_{jj} = 1$; and 3. $a_{j+1j} = 1/a_{jj+1}$.

Table 2   Importance scale and its description

Intensity of importance	Definition	Explanation
1	Equal importance	Two activities contribute equally
3	Weak importance of one over the other	Experience and judgment slightly favour one activity over another
5	Essential or strong importance	Experience and judgment strongly favour one activity over another
7	Demonstrated importance	Activity is strongly favoured and its dominance is demonstrated in practice
9	Absolute importance	The evidence favouring one activity over another is of the highest possible order of affirmation
2, 4, 6, 8	Intermediate values between adjacent judgments	When compromise is needed

Saaty [23] showed that the eigenvector corresponding to the maximum eigenvalue of **A** is a ratio scale for the criteria compared. This eigenvector is further modified to produce a vector where the average importance is unity. There can be a large number of comparisons to be made in the process of evaluating the weights. Given that there are *n* criteria, then the number of comparisons to be made can be calculated using the equation:

$$Comparisons, \; f = \frac{n(n-1)}{2} \qquad (2)$$

It is recommended that the number of items for comparison be limited to between 7 and 12. Where the number of criteria exceeds 10, then a hierarchical structure could be used to segregate the criteria into categories, thereby reducing the number of comparisons in any group. The pairwise method does help reduce the decision-maker to two alternatives in any one decision task.

3.3 Assessing criteria satisfaction

The members of the project evaluation team should be retained to assess the satisfaction of the criteria under existing conditions, and the potential beneficial or adverse impacts of each alternative. Each member of the team would undertake an assessment relevant only to their field of expertise.

The alternatives are assessed to determine the satisfaction of each of the listed criteria. A value in the range of 0 to 1 is assigned to represent this level of satisfaction, where 0 is analogous to no satisfaction, and 1 represents complete satisfaction. A value between 0 and 1 represents an intermediate condition. Where quantitative information exists, such as minimum or maximum acceptable conditions, these can be assigned 0 and 1. However, in some cases the assessment must be subjective, in which case the value 0 can be considered to correspond with the linguistic classification of 'very bad condition', whereas 1 corresponds to 'very good condition'.

The assessment of present conditions and criteria satisfaction for each alternative project are combined by taking the mathematical product, since this corresponds to the linguistic 'and' condition. This accounts for the impact of an alternative with respect to the current circumstances, i.e. would acceptance of an alternative result in an adverse impact and compound an existing problem area where the satisfaction of the criteria is already low?

Raising this product to the power of the criterion weight serves to further modify the grade of membership in the overall fuzzy set combination for each alternative. The cumulative satisfaction of all criteria for each alternative is determined to be the minimum membership value in each column. This reflects the minimum level of satisfaction of any criteria for each alternative. This is referred to as the 'minimal aggregation' since it assumes that the worst evaluation of any feature for an alternative should be taken into consideration. Fuzzy set theory suggests that the alternative corresponding to the largest of these minimum membership values is the best alternative.

3.4 Dominance Matrix

The use of a Dominance Matrix was first suggested by Alley et al.[16], and was used with similar evaluations of criteria satisfaction to rank a set of alternatives. An alternative was defined to be superior to a second alternative if it dominates the second alternative in more features than the number of features in which the second dominates the first.

In order to display the dominance structure between all possible pairs of alternatives, a square matrix D of order p is used (where p is the number of alternatives or projects) and is called the Dominance Matrix. In the matrix [16] the element d_{ij} was the number of criteria for which the membership value of alternative j dominates or is greater than alternative i. A dash is entered for the diagonal elements. If the kth column is summed, the total number of dominances of alternative k over all other alternatives is obtained. Similarly, if the kth row is summed, the number of times the kth alternative is dominated by all the other alternatives is determined. A modification to this technique is suggested. The situation existing with the original dominance matrix was that the dominance of one alternative over another for a low importance criterion possessed equal weight to that corresponding to a criterion of high importance. To improve this situation, the element d_{ij} is redefined; d_{ij} becomes the sum of the weights corresponding to each criterion for which alternative j dominates alternative i. The Dominance Factor, DF has been devised as an overall measure of the superiority of an alternative, and is defined as:

$$DF = \frac{MD - RT_k}{MD} \times \frac{CT_k}{MD} = \frac{(MD - RT_k).CT_k}{MD^2} \qquad (3)$$

$$\text{Maximum Dominance, MD} = n(p\text{-}1) \qquad (4)$$

where
   MD = the maximum possible dominance, which is equivalent to one
          alternative dominating all other alternatives in all criteria;
    n = the number of criteria used in the assessment of alternatives;
    p = the number of alternatives being considered;
  RT_k = the row total for alternative k; and
  CT_k = the column total for alternative k.

The more favourable alternatives have higher column sums and lower row sums, that is reflected in the way the DF is formulated. A high column total and low row total will produce a DF close to 1. Similarly, a low column total and high row total will produce a DF close to zero. Therefore, a DF near 1 indicates a favourable alternative while a DF near zero is a poor alternative. Taking the product of the DF with the weighted performance measure, tempers the pessimistic aggregation produced using fuzzy set theory, to identify the alternative which most effectively meets all criteria, and is consistent with the priorities assigned via the allocation of weights. The results should be checked for consistency and errors, and to ensure that no unacceptable impacts exist with the preferred solution.

4 Benefits and limitations of the MESA method

The MESA technique provides a methodology for dealing with multi-objective development problems at the planning level. The numerous advantages of the MESA technique are:
- The method is logically sound, standardised and repeatable.
- Quantitative and qualitative data can be incorporated into the decision-making process without resorting to estimations of the monetary value of costs and benefits.
- Available quantitative data is valuable and easily incorporated into the technique. The data can be used to set limits of acceptability (e.g. total

available funds, allowable levels of pollutants, etc.) against which the performance of alternative projects can be assessed.

- It provides a decision-making 'audit trail', a result of the formal structure and the method of combining the evaluations of dissimilar criterion.
- The use of weights to reflect the importance of the assessment criteria, and the assignment of impact ratings through membership values, inherently reflects people's attitudes to risk and uncertainty.
- A major shortcoming of many forms of public participation is that pressure or loud groups are heard rather than a balanced public cross-section [24]. MESA allows the coordinator to place a weight on the relative importance of the input of groups thereby partially overcoming this problem.
- The use of pairwise comparisons in the evaluation of criteria importance is a simple yet effective technique to compare a large number of criteria.
- The use of fuzzy set theory and the pessimistic aggregation method ensures a risk averse selection of the 'best' alternative.
- There is considerable flexibility and scope for sensitivity analysis of the fuzziness of membership values and criteria weights. The potential impact of differing political viewpoints can be examined through application of criteria weights offered by each member of the evaluation panel.
- The numerical evaluation of project alternatives and combination of assessments is mathematically simple.
- The method accommodates the selective comparison of dissimilar criteria and consideration of the potential trade-offs between alternatives.
- The spatial scale of alternative projects and the implications for sustainability can be tested quickly, as membership values (of criteria such as cost, use of fertile land, increase in salinity, etc.) could be constructed as functions of spatial size.

There are some limitations to the MESA method which include:

- Results depend upon the subjective evaluations or approximation of social, economic, political and environmental values (improvements can be made as better information comes available).
- Results reflect only the knowledge and judgment of the panel participants.
- The method is new and unfamiliar to decision makers.
- The issue of sustainability is the subject of ongoing debate, therefore the criteria listed for evaluation of alternatives is a good basis for discussion.
- There is a requirement that decision-makers overlook short-term interests to the long-term interest of the community as a whole. The discount rate used in economic evaluation reflects this preference, and so it is important that longer-term goals be evaluated using the MESA.
- Bias in selection of members for the evaluation committee by the views and nature of the assessment agency, thereby influencing the results obtained.
- There are a large number of comparisons to be made in the process of evaluating the importance of criteria. In most instances computer software would be required to combine and determine the criteria weights.

5 Conclusions

The MESA methodology attempts to encourage development that is 'sustainable'. It accounts for subjective judgment and contains a formal rigorous decision strategy that takes the place of intuition when quantitative and qualitative values of environmental activities need to be evaluated. The MESA method is capable of incorporating public opinion, a vital component in the development and assessment of environmental policy. Use of the multi-disciplinary approach enables public opinion to be combined with expert knowledge for the successful preparation and promotion of environmental policies and development plans.

The effectiveness of any multi-objective approach in practice will depend in part on how easy it is to understand, how well it addresses the appropriate issues or problems. The MESA approach does not eliminate subjectivity, but rather makes it explicit, spelling out the basis of the judgment and facilitating discussion of that assessment. If an honest appraisal of a project is required and there is the will to carry out an evaluation of the development proposals using the MESA methodology, then the process will enable a greater comprehension of the problem and a method for determining the 'best' alternative.

6 References

1. Loucks D.P. Analytical methods for multiobjective planning, In: Viessman W., Schilling K.E. (Ed.) *Social and environmental objectives in water resources planning and management,* American Society of Civil Engineers, New York, 1986, pp. 169-183.
2. Ahmad Y.J. (Ed.) *Analysing the Options,* United Nations Environment Program, Studies 5, Nairobi, Kenya,1982.
3. Nijkamp P., Rietveld P., Voogd H *Multicriteria Analysis in Physical Planning,* North-Holland, Amsterdam, 219 pages, 1990.
4. Gershon M., Duckstein L. A procedure for selection of a multi-objective technique with application to water and mineral resources, *Applied Mathematics and Computation,* 1984, Vol. 14 No. 3, pp. 245-271.
5. Hobbs B.F. What can we learn from experiments in multi-objective decision analysis? *IEEE Transactions on Systems, Man and Cybernetics,* Vol. SMC-16 No. 3, 1986, pp. 384-394.
6. Karni R., Sanchez P., Rao Tummala V.M. A comparative study of multi-attribute decision-making methodologies, *Theory and Decision,* Vol. 29, 1990, pp. 203-222
7   Smith P.G.R., Theberge J.B. Evaluating natural areas using multiple criteria: theory and practice, *Environmental Management,* Vol. 11, 1987, No. 4, pp. 447-460.
8. Goicoechea A., Hansen D.R., Duckstein L. *Multiobjective Decision Analysis with Engineering and Business Applications,* John Wiley, N Y., 1982.

9. Shafer E.L., Davis J.B. Making decisions about environmental management when conventional economic analysis cannot be used, *Environmental Management,* Vol. 13 No. 2, pp. 189-197, 1989.

10. Hipel K.W. Multiple objective decision making in water resources, *Water Resources Bulletin,* Vol. 28, 1992, No. 1 (February), pp. 3-12.

11. Biswas A.T., Geping Q. (Ed.) *Environmental impact assessment for developing countries,* Vol. 19, 1987, Natural Resources and the Environment Series. Tycooly International, London. 232 pages.

12. Yager R.R. Multiple objective decision-making using fuzzy sets, *Int. Journal of Man-Machine Studies,* 1977, Vol 9 No. 4, pp. 375-382

13. Ragade R.K., Hipel K.W., Unny T.E. Nonquantitative methods in water resources management, *Journ. of Wat Resources Planning and Management, A S C E,* Vol. 102 No. WR2 (November), pp. 297-309, 1976.

14. Daniell T.M. *Concepts and methods of water resource planning,* Master of Administrative Studies Thesis, ANU, Canberra. 58 pages, 1978.

15. Znotinas N.M., Hipel K.W. Comparison of alternative engineering designs, *Water Resources Bulletin,* Vol. 15 No. 1 (February), pp. 43-58, 1979.

16. Alley H., Bacinello C.P., Hipel K.W. Fuzzy set approaches to planning in the Grand River basin, *Adv. in Water Res.,* Vol. 2 March 1979, pp. 3-12.

17. Ayyub B.M. Systems framework for fuzzy sets in civil engineering, *Fuzzy Sets and Systems,* Vol. 40, 1991, pp. 491-508.

18. Eldukair Z.A., Ayyub B.M. Multi-attribute fuzzy decisions in construction strategies, *Fuzzy Sets and Systems,* Vol. 46, 1992, pp. 155-165.

19. Esogbue A.O., Theologidu M., Guo K. On the application of fuzzy sets theory to the optimal flood control problem arising in water resources systems, *Fuzzy Sets and Systems,* Vol. 48 p. 1992.

20. Fleming N.S., Daniell, T.M. Water Resources Management For Sustainability: A Practitioners' Perspective. *Results of Workshop on Water Resources Management,* Department of Civil and Environmental Engineering, University of Adelaide, June 1995.

21. Deason J.P., White K.P. Specification of objectives by group processes in multi-objective water resources planning, *Water Resources Research,* Vol. 20 No. 2, 1984, pp. 189-196

22. Saaty T.L. *An eigenvalue allocation model in contingency planning,* The University of Pennsylvania, Philadelphia, 1972.

23. Saaty T.L. A scaling method for priorities in hierarchical structures, *Journal of Mathematical Psychology,* Vol. 15, 1977, pp. 234-281

24. Kiely-Brocato K.A., Buhyoff G.J., Leuschner W.A. An attitude matrix scaling system with relevance for resource management, *Journal of Environmental Management,* Vol. 10, 1980, pp. 71-81.

Advanced systems for environmental assessment and control

N. Quaranta
Department of Environment, Lombardy Region, Milan, Italy

abstract

Advanced systems information staff at the Department of Environment of Lombardy regional Administration have been developing two packages for on line decision and analytical processing.

ARIANNA is a powerful, stand-alone, PC based tool for the geographical navigation on air quality monitoring data, that allows the operator to acquire data via modem and use them locally for both a spatial evaluation and a time analysis.

AURORA is a package providing a software architecture for the environmental analysis and assessment at regional level that guarantees the largest opening and flexibility to integrations and future expansions in terms of technological improvements, new functionalities for the user and integration with other systems.

The central elements of this system are two: the man-machine interface that is based on an ingenious geographical navigation system; the environmental "services" that furnish the information and the analytical processes (about different topics), which are distributed on different server machines in the network.

1.   Introduction

The department of the environment of the Lombardy Region has an own e.d.p. center that manages the systems of acquisition of environment monitoring data and the files and databases with the territorial information.

Traditional databases are surely useful for the management of the administrative transactions, but they have come out unsuitable for the on line analysis oriented in line to the territorial planning.

The use of a G.I.S package is usually considered from the experts to be the most effective solution for such a kind of problem; but in practice this solution has shown a set of operational difficulties:

- the available GIS tools on the market are very complex, difficult to configure and use in a distributed software environment;

- is very difficult to connect effectively in a net of distributed services the GIS workstations with remote files and databases;

- to insert the "intelligent" applications (like mathematical models, simulation models, expert systems, neural nets, multimedia documents) in a G.I.S preconstituted software environment is a very hard job;

- the man-machine interface that usually these systems provide is valid for an experienced computer operator but not adequate to the mentality and to the requirements of a territorial operator;

- the cost of license and management of a GIS is usually very high.

Another usual technique in the "d.s.s." (decision support systems) circle and betwen the so-called "OLAP" (on-line analytical processing) systems is that of the "datawarehousing", that allows the access to huge files containing the transversal information about all domains and topics that could interest planning activity.
Unfortunaly, to tell the truth, this technique suits well those companies that have consolidated methodologies for the own "business process" analysis (like the financial societies, banks, the manufacturing industries), and therefore in the field where have outcome the so-called "e.i.s." (executive information system) tools, while for what concerns the analysis of environment evaluation and planning doesn't exist standard methodologies or consolidated technologies.

For these reasons already for years the information systems staff of the Department Environment has started the study and the realization of any tools expressly conceived for the support of the environmental operators at every level of activity; two lines of products particularly have been identified:

♦ "stand alone" packages for a remote user that must analyze and visualize a particular situation in real time, or a series of local data already elaborated at synthetic level;

♦ systems for the user that needs to assess the state of the environment, in a certain geographical context, weighing the conditions of the different environmental variables (air, water, ground, subsoil) for estimating the "stress" of the territory either from the point of view of the human settlement either of the animal and vegetable one.

2.   Arianna

Arianna (v. 3.0) is a program carried out to solve the problem of the acquisition and the knowlegde in real time of the data concerning the distribution of the atmospheric pollutants in all the Lombardy region.

This tool doesn't have comparison of hits kind either in terms of performances than of versatility and simplicity of use.
Arianna is a program "plug & play" self configuring on personal computer, that is composed of two separate modules:

• a communication module that acquires the data in real time from a remote server;
• an integrated module for the management, presentation and elaboration of the acquired data.

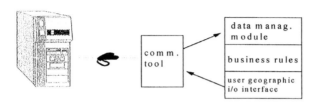

The user interface is provided of a powerful geographical navigation system in the territorial regional context, with information about each monitoring station and its characteristics; it is possible to select a parameter (pollutant or meteo), a type of elaboration, the required period of sampling, and to get an output of spatial type (thematic map) or of temporal type (linear or bar graph).
User has a large number of available functions and he could shape either the scale of the graphic either the background or representation colors.

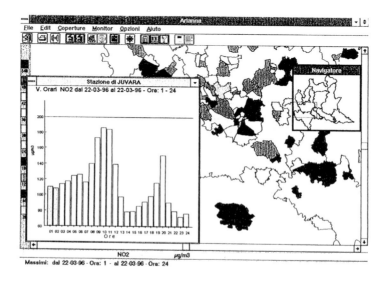

By this way from any site it is possible to connect (by means of a modem and a telephone line) to the regional center, acquire the data of the day and check the level of reached pollution.

Arianna is a program developed in C++ that requires at least a 486 PC with 8 MB RAM.

3.   Aurora

In the field of the environment analysis the early question of the lack of data has been replaced from that of the organization of the data: in fact, if people are not able to define precise environment indicators oriented to the planning of the choices and of the interventions, it is not possible to realize a data model and some "business rules" for the evaluation of the environment condition.

The control of the environment and the environmental impact assessment require the availability of many different tools like: DBMS, simulation models, image processing, GIS, expert systems, graphic and statistics libraries, forecasting models et cetera.
The need of the territorial operators is to have an integrated tool that allows a 'crosswise' analysis of the different variables and environmental indicators.

The EDP staff of the Department Environment has therefore projected and realized an integrated system of analysis, control and planning of the environment, that allows to get synthetic reports on the state of the territory, either in tabular form either on graphics, cartographic (thematic maps) or like images.
The starting objectives for the plan were:

- user interface very simple, provided with "intelligent" software functions able to manage very complex processes for the user;
- same methodology for the access to the more heterogeneous applications also, in such a way to furnish an homogeneous base in the investigation from different point of view;
- access to the applications and databases both local and remote by means of the implementation of "services" for the final user. "Services" are either single data elaboration either homogeneous application fields either a set of homogeneous output forms concerning an environment topic.

The system, called Aurora, is provided of a graphic interface that is a real G.I.S., able to manage both vectorial and raster maps for the user "navigation" on the territory.
The man-machine interface was on purpose projected to guarantee the greatest simplicity with the most efficiency as possible; the functionalities of the interface have been expressly designed for the needs of people that must analyze environmental problems and of impact on the territory.

Software architecture is of type 'three tier' client-server using a full "object oriented" environment:

client   application server   d.b. server

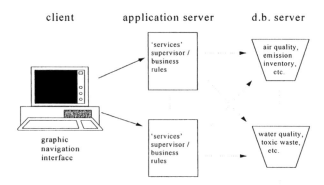

- client console supports the man-machine interface and manages the windows of textual output, graphic, multimidia; besides it receives every request for the more or less complex data elaborations (called "services") and information to furnish with reference to the selected objects (an "object" could be a city, a Commune, a Province, an industry, a river, a road, and therefore an entity geographically definable).

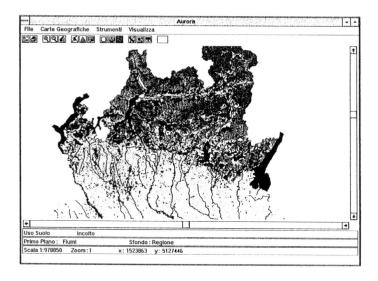

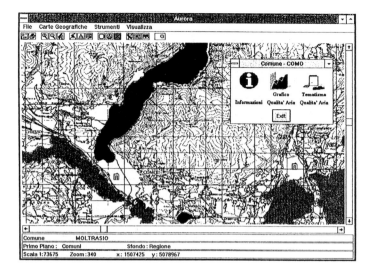

- the application server(s) is able to connect the user to the different software applications on the phenomena concerning the environmental pollution (atmospheric pollution, waste management, evaluation of the industrial risk, water quality, simulation models, etc.), and also to communicate with data files and distributed database.

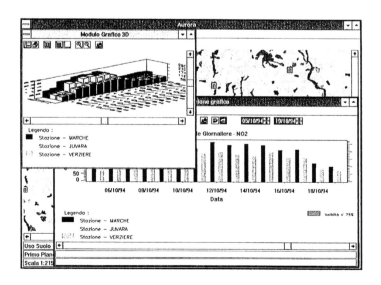

- the database server(s) contain the data necessary to do the elaborations, but also information of various kinds, like statistic data, meteorological data, data about the state of the parks, on the roads and generally on the structure of the territory and on environment condition.

Using the paradigm of the "services" several applications have been implemented on different topics, like:
- thematic maps on the distribution of socioeconomic variable;
- graphic and thematic maps on the quality of the air;
- reports and thematic maps on the distribution of the emissions;
- expert system on the analysis of the traffic emissions;
- diffusion models of the atmospheric pollution (in phase of development);
- analysis of the industrial risk;
- statistic analysis of the production and distribution of the industrial and urban waste;
- characteristics of the natural parks (in phase of study);
- analysis and multimedia presentation of the alpine glaciers (in phase of study).

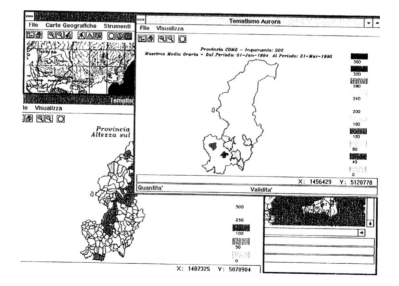

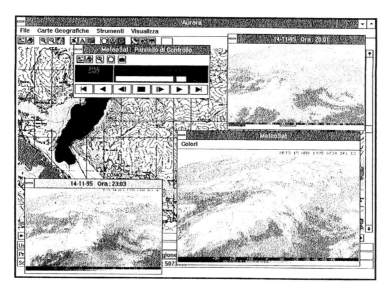

At the present the architecture of the system seems very strong, and a phase of experimentation and evaluation has started beside the users of the Department Environment.

4.   Nebula project

Currently many regional Administrations are developing and/or using their own system for environment planning and control.
These systems are not compatible each other, so that for getting an overview on data coming from different administrations it must to implement specific, more or less complicated, program interfaces.

The Lombardy Region and the Emilia-Romagna Region have reached an agreement on a common project (called "Nebula") that aims to promote and realize a whole of network services necessary to the public authorities and Administrations involved in the management of the environment.

For this purpose the system AURORA will constitute software architecture and connectivity tool between "clients" and "services".

By this way, to give an example, a user will be able to analize air quality across several regions, doesn't matter wich site the data come from (the system will combine data acquired on line from different servers); or a user could to use a mathematical model running on a remote server, combining meteo data provided by a specialized center with emission data coming from another db server, to evaluate air pollution over selected territory.

While the client-server software is always the same of Aurora, the novelty is the attempt to create, on a virtual network, several centers specialized to furnish thematic information accessed and displaied in a 'transparent' way by Aurora.

NEBULA - AURORA architecture

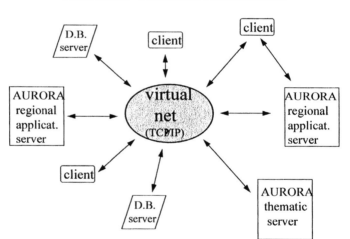

An expert system approach to identification keys

K. Gregson

Department of Physiology and Environmental Science, University of Nottingham, Nottingham, UK

Abstract

There are many environmental applications which rely on the accurate identification of insects, plants, etc. The availability of personnel with sufficient identification and taxonomic skill is often a limitation and the production of identification keys has been one means of alleviating this problem. It is the purpose of this paper to show that expert system techniques can be employed to provide computer based keys that are both simpler to use and easier to construct and maintain.

This paper describes the structure, development and use of such an expert system based identification key. Experience indicates that quicker and more accurate identifications are possible and that there are other less obvious benefits. In particular the system can be both easily modified and maintained and that the addition of new knowledge is a relatively simple operation.

Introduction

The system described in this paper (MIV) came about in response to the problem of identifying macro-invertebrates in samples of river water. The identifications were relevant to a project within the University of Nottingham concerned with monitoring water quality and the recognition of pollution incidents and their possible causes. It soon became apparent that the identification process was a major bottleneck.

As a result of developing the system at Nottingham I became involved with similar work in the Ecological and Monitoring Assessment Program undertaken by the United States Environmental Protection Agency which is a major task involving the collection of data samples in both spatial and temporal domains. In this case the necessary taxonomic identifications are performed (at the Oregon State University site) within the Stream Team Laboratory and again it was clear that the Laboratory was under considerable pressure and unable, due to time considerations, to provide the full analyses required. The same kind of problem will occur in many other investigations and at many other sites.

The intent of the following work is to show how expert systems may be used to expedite some areas of the identification process and hence to facilitate the work involved in environmental impact and monitoring programmes.

Expert Systems

The phrase "expert system" is a generic term used by computer and other specialists to describe any computer program that behaves in some way like a human *expert*. There are many definitions of varying complexity but the above loose description will suffice for our purpose. The development of expert systems within academic departments of psychology, and their association with artificial intelligence, has lead to the general perception that expert systems are large and extremely complicated computer systems which are of little use in ordinary applications. In fact expert systems need not be complicated and can make useful contributions in a wide area of applications.

An expert system consists of a data-base made up from the known facts and rules within the specific problem area (*knowledge domain*) together with some form of inference mechanism. The knowledge base is elicited from *real* experts and the inference procedures are usually provided by a language like Prolog or as part of an expert system shell.

Identification of Macro-invertebrates (MIV)

The expert system, MIV, developed at Nottingham in order to help with the identification of macro-invertebrates was based upon the key constructed by AIDGAP [1]. Even in the very early stages of its development we found the computer version to be beneficial in terms of both speed and accuracy, especially when used by relatively inexperienced researchers. The main reason for this is that the mechanics of manipulating the path through a key are taken care of by the computer, so that the frustrations of backtracking, losing place-markers, etc. are avoided.

Development carried out at Oregon State University restructured the program in terms of separate modules corresponding to each stage in the identification. Thus the initial module identifies the *class* of the organism, then if required passes control to the next module which enables identification at the *order* level within the *class*. Identification may then proceed until no more detailed knowledge base module is available, at which point the process is terminated and the identification presented.

Currently the knowledge modules available are:

class	order
bivalvia	
crustacea	
gastropoda	
hirudinae	
insecta	*trichoptera*
	diptera
	coleoptera
	hemiptera
	ephemeroptera

together with identification of *trichoptera* larvae to family level (see later).

It is perhaps worthwhile considering the problem of knowledge representation at this stage. I suspect that the computer-phobe, of which there are many - and I have respect for them, will be wondering how difficult it is to translate the written knowledge into some sort of computer representation. I shall illustrate this by giving the following computer representation for the order *ephemeroptera*:

'ephemeroptera' if
            (lives = free_living and
            jointed_legs = 3 and
            not true_wings and
            (tails = 3 or tails = 5) and
            tail_termination = hair_like and
            abdominal_gills = lateral)

The above is written in the language of Advisor2 but is typical of many expert system representations. You will see that it is not too far removed from normal language which is important for validation and checking purposes.

Each knowledge module will consist of many such rules describing the different taxa. In general the position of the rules within the knowledge base is not important, though it is logical and helpful to keep related rules together. However it is possible to manipulate the order in which the expert system asks questions by changing the order of the rules in the knowledge base. This can be utilised in several ways. It is possible for example to force a consultation in a particular way, this may be useful as a teaching exercise. It is also possible to optimise the search process and there are techniques available for achieving this.

The resulting knowledge base can be easily maintained and extended to include additional *classes* or to extend the identification to *families*, etc.

It is important to remember that the elicitation and engineering of the knowledge within each module is a time-consuming and difficult task requiring the co-operation and time of a classification expert. However the provision of further modules is a task that could be distributed over different working groups, each concentrating on one closed knowledge domain (*sub-family* or even *species*). This would allow the construction of a very large and comprehensive knowledge base.

Currently MIV is based on the Advisor2 [2] expert system shell. It can be run on any IBM-PC compatible machine that has an 80286 (or faster) processor, a hard disc and monochrome VGA (or better) monitor.

MIV operates by asking the user a sequence of questions concerning the morphological and behavioural characteristics of the individual creature. Using a process of logical inference the program makes an identification based upon the rules and facts held within the knowledge base. As the identification progresses the current knowledge can be reviewed at any stage and erroneous observations removed. It is also possible to seek additional help in the form of diagrams or explanations at various points in the identification. When the program has received enough information to make a positive identification a picture of the invertebrate is produced which allows the user to confirm the identification. If the identity is wrong or unclear the process can be continued by removing properties that are doubtful, the program will then resume the search along different lines until a positive identification is made.

During development and testing of MIV it was found that there were a number of distinct advantages of computer based keys over their paper based counterparts;

1 The question sequence is taken care of by the inference procedures within the program. This provides a faster and less error prone identification. If an identification is incorrect it is possible to discard some of the answers and renew the identification without discarding the remaining observations. This form of backtracking is difficult with a paper based key.

2 It is portable, user friendly, and provides on-line help facilities throughout a consultation to identify difficult features.

3 We found the computer based system to be faster than paper keys, on average it used 35% fewer questions to carry out an identification. One important feature is that questions are not restricted to the dichotomous nature of paper keys.

4 Occasionally the user may find a question that is difficult to answer. An expert system should allow the user to defer the consultation and present the possible alternatives to the user. This can be helpful whilst identifying damaged specimens or ambiguous morphological features.

North American *Trichoptera* Larvae

The identifications carried out at OSU make extensive use of the key for North American Insects by Merrit and Cummins [3]. In order to demonstrate the versatility of MIV we constructed an additional knowledge module for the identification of *trichoptera* larvae based upon this key. This part of the work was carried out with much help from Bill Girth in the Stream Team Laboratory whose expertise, together with that of Merrit and Cummins [3] is the basis for this key.

Discussion

The collection and storage of environmental and ecological data is a major, and growing, task which makes large demands upon resources. Whilst the amount of data analysed and stored could be reduced by performing less rigorous identification and storing biotic indices rather than the raw data, this would be an unsatisfactory solution. The only other option therefore is to provide some means of help in order to facilitate the identification process. MIV is one such tool that can provide some immediate help and which could be extended to include both a wider range and a more detailed identification procedure.

The program is now in a form which is easily maintained and can be extended as new knowledge is coded.

It should be remembered however that this knowledge is very specialised. Even though it is widely available in book form it will be necessary to consult with taxonomic experts to clarify and rationalise some of the details. Their co-operation and enthusiasm are imperative for the success of such a system.

One important aspect of building such a system is that the keys are often **not** the best place from which to extract the knowledge. In general it will be more accurate to base the knowledge upon the original taxonomic definitions. Indeed **all** that is necessary is to provide the taxonomic details, the intricacies of how the program reaches its identification can largely be left to the internal inference procedures. This is an important point of great relevance to taxonomists: the provision of easy to use paper based key is an extremely complicated and time consuming process that could largely be eradicated if keys were to be produced electronically in the way that has been described.

The production of easy to use paper-based keys forces the taxonomist to compromise between ease of use and accuracy. In particular the identifications of obscure and unusual taxa cause extreme problems, especially if they are treated as "special cases." One advantage of the expert system approach is that each taxon is considered as its own special case, and can therefore be defined within the knowledge base in exactly the same way as any other. The procedure for reaching the identification can be left to the program without the need to worry about the ordering of questions or the order in which tests are carried out.

There are still some tricks that could be incorporated into the search process in order to speed up the identification. These are problems that should be the concern of the computer scientist rather than the taxonomist however.

Conclusion

The availability of systems like MIV would have obvious benefits to ecological surveys. In particular, they could be used by relatively inexperienced personnel as tools to help with the classification. They may also be used on-site provided they could be based on a sufficiently robust portable computer. Such machines are becoming widely available.

The development of such systems would give the environmental scientist the ability to collect and use data which currently can not be handled. As I have tried to show, the development of these tools is not a difficult computational task and is well within the limits of available technology. It is my hope that the taxonomists will make their skills available by helping to provide the knowledge relevant to the value of such software.

Acknowledgements

I would like to acknowledge the work of Michael Renshaw whose MSc project at the University of Nottingham laid the foundations for the MIV system. I would also like to thank the Environmental Protection Agency, the Center for the Analysis of Environmental Change and the Department of Statistics at Oregon State University for support during this work, part of which was carried out during sabbatical leave from the University of Nottingham, England.

References

[1] Croft P S. AIDGAP - "A Key to the Major Groups of British Freshwater Macro-invertebrates", Field Studies, Vol 6 No 3 (1986).
[2] Advisor2 - An expert system shell developed by Expert Systems International, Oxford, England.
[3] Merrit & Cummins *Aquatic Insects of North America*, Kendall/Hunt.

Ergonomic presentation of results given by ASTRAL (a nuclear accident consequences assessment software)

J.C. Bernié, Ph. Renaud, H. Maubert, J.M. Métivier
Nuclear Protection and Safety Institute, CEA Cadarache, France

Abstract

Although every precaution is taken about the use of nuclear energy, the occurrence of a major accident cannot be totally ruled out. The Institute of Protection and Nuclear Safety set up a Technical Crisis Centre to face such an eventuality. This centre is often tested through crisis exercises. In order to meet the needs in environmental consequences assessment of nuclear accidents a software called « ASTRAL » (technical assistance in post-accidental radioprotection) is under development. The software relies on a calculation module integrating the most recent advances in radioecology, a data base, and a Geographical Information System. It allows to : characterise the radioactive state of the environment, forecast its evolution in time and space, estimate the efficacy of rehabilitation actions and countermeasures and compare the calculated values with regulatory levels.

In order to reach such a goal it is necessary to manipulate a vast amount of data, giving in turn numerous results. In addition, it is not always easy for a non specialist to get familiar with the nuclear units such as Becquerels, Sieverts and to relate the announced results to intervention levels.

This paper will be focused on the ergonomic presentation of the system output. It shows how the complexity of the results may be simplified under the form of graphics and maps in order to give to decision making people an accurate, complete and easy to catch information.

1 Introduction

In the nuclear field, assessment in emergency situations should, in the case of a major accident, allow one to meet the needs of the public authorities, to have expert advice available so as to estimate the consequences of this accident and prepare decisions in matters of protection of the environment and population.

The ASTRAL software (technical assistance in post-accidental radioprotection), developed by the IPSN, is an assessment and decision-aid tool. Its aim is to appraise the consequences of the accident, notably on foodstuff resources and on the population, and to estimate effectiveness of the various measures designed to reduce these consequences.

These assessment figures are based on the analysis of a great number of data and information ; meteorological, agro-climatic, radioecological, geo-economic, and concerning of course the accident itself. Although complex and evolving in space and time, they must be transmitted to decision-makers in a form that allows rapid understanding of the situation and its evolutions, the opinion of experts as well as the main constituents of this judgment.

Particular care has been given to the ergonomics of result presentation in the design of ASTRAL, and the availability of main explanatory and contextual data or information which are implied.

2 Principal types of ASTRAL results

The ASTRAL software, from a basis of radioactivity deposited on the ground following an accidental release from a nuclear installation, allows one to assess :

• concentrations in radioactive elements in main vegetal or animal agricultural productions,

• surfaces and quantities concerned by these radioactivity levels for different administrative sectors,

• radioactivity incorporated by the population of the areas affected, and the expectancy of these populations expressed in terms of dose,

• the numbers of these population groups.

These results are subjected to various processes : graphics, mapping, index calculations, etc., which give them presentation ergonomics susceptible to satisfying experts and decision-makers, and the demands of emergency situations.

3 Cartographic representations

31 The need to adapt the representation of certain results to space conditions

Radioactive deposit at ground level is more often than not made up of several radioactive elements in different proportions. These proportions, as indeed the overall importance of the deposit, sometimes vary significantly from one spot to another depending on the meteorological conditions and distances. In other respects, the radioactive elements of different chemical and nuclear characteristics do not all behave in the same way in the environment, do not have the same future, and do not present the same potential dangers.

In ASTRAL, assessments are therefore made for each radioactive element and for different surfaces on which the deposit is considered as homogeneous. These surfaces can correspond to one or many districts, or even counties or regions affected by the accident.

In this way it appears that only a spatial representation allows a rapid grasp of the importance and the spread of radioactive fallout over a given territory. Figure 1 shows the spread of caesium 137 deposits following an accident simulation on the Polyarnye Zori reactor in the Kola peninsula in Russia.

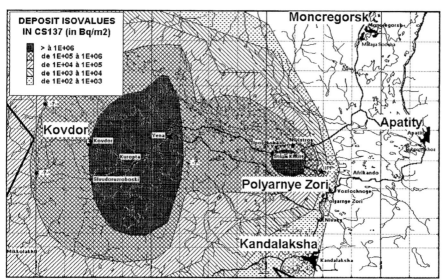

Figure 1: Cartographic representation of isovalues

32 Thematic cartographic representations

321 Index calculations - Radioactivity levels in agricultural productions are expressed by their « content » or « concentration » in each of the radioactive elements, in Becquerel per kilogram (Bq/kg). This unit is not highly significant to non-specialists. However, it is to be noted that this unit is small and that one often has recourse to multiples (k Bq/kg or even M Bq/kg), and lastly that many people are in the habit of using the old unit, the Curie per kilogram (Ci/kg). If one wishes to announce a result of this type, one should therefore be prepared in advance for the following questions - What does this represent ? How does this result stand when compared to what is usually tolerated ? How does it compare to natural radioactivity ? Or more simply, is it acceptable ?

It is the comparison of concentrations with different criteria - standards, intervention levels or even average natural radioactivity - which allows a rapid grasp of the importance of the impact on agricultural productions.

In ASTRAL, depending on whether the results are in graphic or cartographic presentation, it is possible either to show these standards on the presentation itself of results, or to calculate, then represent a « concentration index » (1) which corresponds to a ratio of the calculated concentration of the selected criteria.

$$ I = \frac{Concentration \ (Bq/kg)}{Criteria \ (Bq/kg)} \tag{1} $$

As this index is calculated for each agricultural product and each radionucleide, the terms of the result « Wheat at harvest time will have a caesium 137 concentration of 2500 Bq/kg » is replaced by « Wheat at harvest time will be as twice as high as the European Standard for caesium 137 ». This index furthermore allows on to classify the couples (agrocultural product, radionucleide) from the most penalizing to the last penalizing so as to show up rapidly the most important problems.

322 Spatial representation of the concentration index classifications - The index previously calculated also allows specifying the classifications which make appraisal by the expert easier and which can be represented cartographically. Figure 2 shows that it is thus possible to show up on which areas the concentrations of a product are ten or hundred times lower that the intervention level (white = areas without problem), roughly the same (pale grey = specific dispositions required), or exceeds it considerably (dark grey = consumption prohibited).

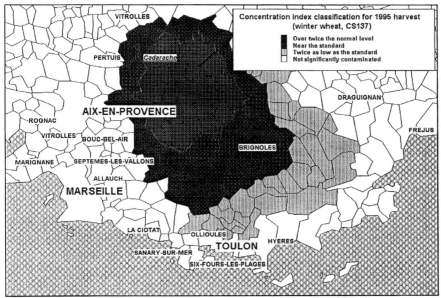

Figure 2: Cartographic representation of index classifications

323 Cartographic representation of impact on populations - On the basis of concentrations in food products and food rations, it is possible to calculate the radioactivity ingested daily by the inhabitants, then the dose expressed in Sieverts (Sv) that results from this incorporation. At this stage, one can go deeper into the impact on the body as a whole or a sensitive organ (dose on a organ). The results may be represented by mapping, either relatively with index calculations, or absolutely with for example proportional symbols. Figure 3 shows doses on the thyroid gland which result from ingestion of Iodine 131 in the various localities fictitiously affected during the accident simulation exercise of Kola in Russia.

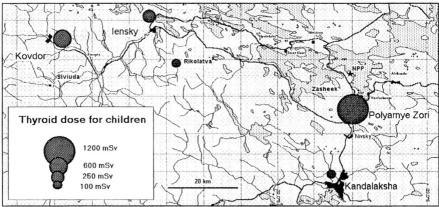

Figure 3: Cartographic representation by proportional symbols

4 Graphic representations

41 The need for dynamic representation of certain results

In a post-accident situation, the instantaneous image of the situation is not sufficient. All the results evolve in time and sometimes rapidly. The transfer phenomena to be taken into account are not always the same, their intensity varies, the agro-climatic data changes throughout the year, and the radioactive decay of certain radionuclides can lead to their disappearance in a few months. The effect of measures taken to reduce the impact on the environment and the populations referred to as « countermeasures » can be added to these evolutions which are not associated with any particular disposition. In this way, when experts announce an estimation of this type - « milk produced in this district is not fit for consumption », they should be prepared for such questions as « for how long ? Can one do something to remedy this in the less long term ? ».

Graphs on evolution, concentration and the daily ingestion of radioactivity, for example, which take into account dynamic parameters and which are susceptible to simulate the implementation of various countermeasures, answer these questions.

42 Dynamic evolution curves showing concentrations

Milk is certainly one of the products where radioactivity concentration is the most likely to vary significantly. This is mainly due to zootechnical practicalities of animal feed which vary throughout the year (feeding schedules). What is more this production allows the application of particularly efficient countermeasures.

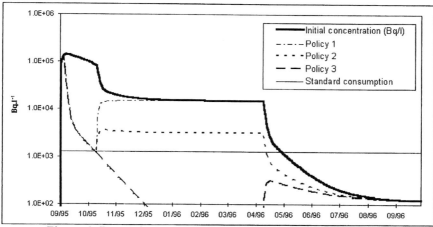

Figure 4: Dynamic concentration graphic - the example of milk

Figure 4 shows the evolution of concentrations in Bq/l of caesium 137 in milk during the year following an imaginary radioactive deposit which occurred in September 1995. The feeding schedule used is the one shown in Figure 6.

The bold line corresponds to the evolution of the concentration in milk if no protection measures have been taken. Policy 1 corresponds to a withdrawal of cows from pastures 5 days after the deposit. Policy 2 consists in administering to the animals a product designed to reduce the contamination of their milk. Lastly the third policy consists, when the animals are is stable over the winter period, in feeding them with a fodder of maize from non-contaminated regions. It seems that only this method reduces concentration below the standard level.

Dynamic representation of index classification (Figure 5) is another way of appreciating the impact of this third policy.

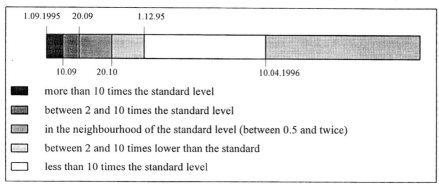

Figure 5: Dynamic graphic representation of index classifications

43 Graphic representations of impact on populations

Different graphics can allow one to represent the impact on populations. The radioactivity ingested daily by consuming contaminated food can, for example, be explained by a dynamic curve. When it is a question of showing the various contributions of the elements of a result expressed in terms of dose, graphics of the « pie graph » type are particularly well suited.

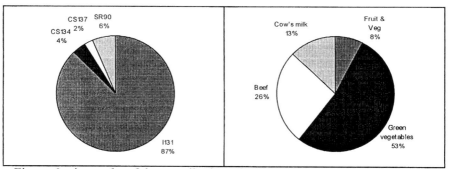

Figure 6: pie graphs of the contribution of radionuclides and food products in terms of dose per ingestion

Figure 6 shows two examples of pie graphs. The first shows the contribution of radionuclides in terms of dose per ingestion of contaminated products in the first thirty days after deposit. It appears that Iodine 131 contributes the essential part (87%) of this dose. Given the rapid decay of this radionuclide, for the following days the same graph would show a lower percentage, then its complete disappearance.

The second graph shows, equally in this period of the first thirty days, that it is the green vegetables (salad, cabbages and leeks) which represent the major part of the ingested radioactivity.

5 Representations of contextual data

Certain data or information which imply results must equally be represented. This is the case of feeding schedules (Figure 7) which allow one to understand evolutions in concentrations of animal products.

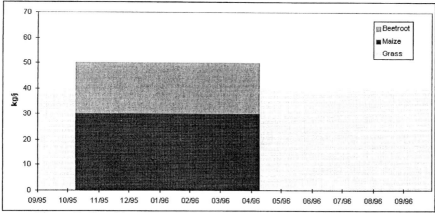

Figure 7: Graphic representation of milking cows' feeding schedule

Generally speaking, it is important that all data which allow placing it in its context appears on the representation of the result itself. This is the role of title blocks (Figure 8) designed not only as a heading or a comment, but equally to guide the user of such and such a document towards further information. Reference for further documentation by their title blocks must allow the user to extrapolate the result that he has in front of him for almost all the data and parameters which have been used to obtain it.

It is also with this objective to traceability of results, data and actions that the « decision chronology » is to be found (Figure 9) ; emergency situation logbooks which must be designed to allow recapitulation and the dating of all decisions, simulations, choice of experts and data modifications which have marked the post-accidental phase.

GENERAL CHARACTERISTICS OF THE ACCIDENT		
SITE	Name Polyarnye Zori Plant	Geographical coordinates lat. 00.00, long. 00.00
DATES	Date of the accident 25/04/95	Date of the deposit 01/05/95
USER SELECTIONS	Agro-climatic region Northern Russia Annual Incorporation Limit EC	Type of soil Clay Consumption Standard EC
STUDY ZONE	Average deposit 10^6 Bq/m²	Reference radionuclide CS137

Figure 8: example of title block

AGRONOMIC COUNTERMEASURES				
COUNTERMEASURE	APPLICATION PRODUCT	EFFECTIVENESS	START	END
Pasture withdrawal	Cow's milk	(-)	05/09/95	15/10/95
Prussian Blue	Beef	5	15/10/95	15/04/96
Prussian Blue	Cow's milk	5	15/10/95	15/04/96
Fertilizer addition	Winter wheat	2	20/08/95	(-)

HEALTH COUNTERMEASURES				
COUNTERMEASURE	LIFE STYLE	PRODUCT CONSUMED	START	END
Consumption prohibited	Children in rural areas	MILK	01/09/95	10/09/95
Consumption prohibited	Adults in rural areas	MILK	01/09/95	10/09/95

Figure 9: example of decision chronology

6 Conclusion

The main difficulty encountered for the presentation ergonomics was to categorize, indeed to explain the volume of data generated by ASTRAL so as to guide the Authorities in the decisions to be taken immediately or in the longer term.

In the short term, that is to say during the hours following the accident, it is the maps and dose contributions that allow replying to essential questions such as which are the areas the most contaminated ? Do they correspond to inhabited areas ? If so, what are the immediate measures to be taken ?

At a later stage, one will try to reduce the economic and environmental impact of the accident on a particular area and specific production. The graphic representations then allow one to assess and compare, case by case, the long-term effectiveness of each rehabilitation policy.

Decision support and idea processing systems: challenges and opportunities for aquaculture research, management and education

F. Stagnitti,[a] C. Austin,[b] S.S. De Silva,[b] R. Carr,[a] P. Jones,[b] D. Lockington[c]

[a]School of Computing and Mathematics, and Centre for Dynamical Systems & Environmental Modelling (CADSEM), Faculty of Science and Technology, Deakin University, P.O.Box 423, Warrnambool, Victoria 3280, Australia

[b]School of Aquatic Science and Natural Resources Management, Deakin University, P.O.Box 423, Warrnambool, Victoria 3280, Australia

[c]Department of Civil Engineering, University of Queensland, St Lucia 4072, Australia

Abstract

Environmental Decision Support Systems (EDSS) and Idea Processing Systems (IPS) are now being developed to facilitate the collection, manipulation and analysis of a variety of kinds of environmental data in a variety of fields. EDSS/IPS's typically consist of integrated hardware and software packages that facilitate operational decision making via user-friendly, visually interactive interfaces. To date, few EDSS/IPS applications for aquaculture have been developed. Therefore, the purpose of this paper is to address how EDSS/IPS technology may benefit aquaculture research, management, planning, knowledge generation and education. This paper discusses the functional units of an EDSS/IPS, the system's life-cycle approach to system development, structured and semi-structured decision processes and decision support for aquaculture applications. It is the contention of this paper that computer systems development for aquaculture can be greatly improved by applying the EDSS/IPS technology to system's analysis and design.

1 Introduction

Advances in computer technology have changed the way many things are done in society including the management of aquaculture facilities. Aquaculture, particularly in the developing world, is increasingly moving from extensive to semi-intensive and intensive operations and the increasing complexity of the production process demands better management in order to sustain high production. As management decisions have considerable effects on both the economics of production and on environmental pollution, decision making can be enhanced by increasing the access to and improving the handling and presentation of data. An efficient, cost-effective means of achieving this is the application of EDSS/IPS technology. The objectives of this paper is to (a) identify the decision-making and knowledge requirements for aquaculture professionals, (b) to compare and contrast existing EDSS/IPS systems, and (c) develop a comprehensive framework for the conceptual design of EDSS/IPS suitable for research and management of aquaculture facilities, and for education and training.

2   A Conceptual Framework For Applying EDSS/IPS Technology To Aquaculture

Decision Support Systems and Idea Processing Systems (DSS/IPS) are a class of computer-based information systems that provides interactive, user-friendly, decision support for managers and planners for solving structured and semi-structured problems Sprague[1]. EDSS/IPS are a sub-class of DSS/IPS that operates with environmental data to solve environmental problems. Computer scientists, eg. Guariso & Werthner[2], more rigorously define EDSS/IPS's as a class of DSS's that possess some or all of the following characteristics: (a) Dynamics. (Decisions are significantly affected by past phenomenon), (b) Spatial coverage. (Environmental data are collected from various geographical locations - the spatial component may be part of the modelling process), (c) Periodicity. (Environmental variables that are affected by seasonal variations), (d) Randomness (due to natural and experimental variation), and (e) Significant data requirement (particularly for spatially referenced data). Although these five characteristics do not exclusively define or are necessarily unique to EDSS/IPS, they are useful in distinguishing EDSS/IPS from other classes of information systems. The presence (or absence) of these characteristics and the relative importance of each characteristic in relation to others will depend on the application and decision requirements. In aquaculture, all these characteristics are operative and need to be accounted for in the design and construction of a suitable EDSS/IPS. EDSS/IPS's are also distinguished from other environmental computer systems in their emphasis on assisting management decision processes in an environment where the decisions may be complex and/or fuzzy. The emphasis is on decision support rather than environmental control. EDSS/IPS's are not designed to replace human cognitive processes but to enhance them. To achieve this, it is imperative that two conditions are satisfied: The manager or planner must (a) have direct access to the data or information without the need to go through an intermediary (eg. a database administrator or programmer), and (b) have the ability to conduct "what-if" analyses and simulation. A strong feature of EDSS/IPS is management by perception. Consequently the user-machine interface must be designed in a manner which permits the user to remain in control and unstressed throughout the problem-solving process.

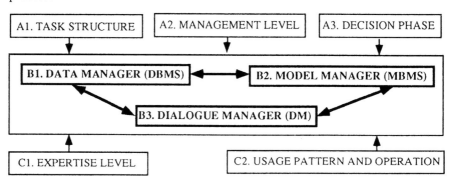

Figure 1: Conceptual framework for EDSS/IPS design and factors influencing components of the EDSS/IPS.

In the past decade there has been considerable debate as to what constitutes a DSS. Much of this debate was generated by computer professionals dealing with management information systems (MIS) for commercial and financial applications. There now appears to be a general consensus on DSS technology. However, the literature on EDSS/IPS is virtually non-existent. Therefore, in this paper, a conceptual model for EDSS/IPS is developed using generally agreed principles of DSS technology and illustrated on several examples in the aquaculture industry.

Figure 1 presents a conceptual model of an EDSS/IPS. Various components and functions are now discussed. EDSS/IPS consists of three components: a Dialogue Manager (DM), a Data-Base Management System (DBMS) and a Model-Based Management System (MBMS) (Pearson & Shim[3]). The interface between the manager and the machine is called the Dialogue Manager. The DM has the responsibility of translating the user requests into machine operable functions and returning the requested information in a user-friendly format. An important feature of the DM is the provision of user-controlled, human-directed dialogues that facilitate structured and semi-structured decision support. To achieve this human-machine synergy, the DM must interface directly with the data management system (DBMS) and the model management system (MBMS). The primary task of the DBMS is for the capture, retrieval and manipulation of environmental data and to translate user-requests obtained by the DM into a form that is tractable for use by the underlying models in the MBMS. The level of sophistication of the DBMS will depend on the application and user requirements for information and may range from a simple software program that sorts and extracts data from flat-files to high-powered database systems that handle operations on complex data models. Programs offering structured-query language support (SQL) appear to be popular in DBMS development.

The primary task of the MBMS is to provide analytical tools to augment the decision-making process. These tools might include statistical and graphical tools for data description, summary and analysis (eg. ANOVA, PCA, cluster analysis, linear and nonlinear regression, MDA, etc), simple mathematical and numerical techniques (eg., linear programming, dynamic programming, differential growth functions, nonlinear optimisation techniques, etc) and decision, idea processing and judgement simulation models (eg., analytic hierarchy process (AHP)). Depending on the nature of the decision process, the MBMS may be required to have considerable flexibility in order to handle qualitative as well as quantitative variables. If the EDSS/IPS includes decision, idea processing and judgement simulation then the user should be able to assign relative importance values to the various environmental parameters and management practices. The EDSS/IPS should then be able to provide an assessment of the potential success or otherwise of a certain management strategy. The outcomes of the modelling process are often expressed as probabilistic assessments associated with a set of user-specified criteria.

Each component of the EDSS/IPS (ie. the DM, MBMS, DBMS) to various degrees provides a set of capabilities to the operational manager or planner and improves the overall effectiveness of the decision making process. These components constitutes the core of the EDSS/IPS. Associated with the core are a set of functional units as indicated in figure 1. These functional units can be described in two broad dimensions; (1) task characteristics and (2) access patterns (Ariav & Ginzberg[4]). Task characteristics include: (a) task structure, (b) management level, and (c) decision phase. Access patterns include: (a) computer skill level and knowledge requirement of user, and (b) usage pattern and mode of operation. The task structure is concerned with the level of problem structure and definition required by the system to operate effectively. If all environmental parameters and decision variables are known and can be defined by equations or unambiguous statements, then the task is well structured. The decision models required to implement well-structured tasks are usually easy to implement in an EDSS/IPS, even though the decision or mathematical models might be complex. In semi-structured decision-making contexts, environmental parameters may be qualitative as well as quantitative. The decision environment may be loosely defined around desirable objectives or outcomes. Semi-structured environments often rely on heuristics as well as known scientific laws. The user generally is required to play a greater role in the decision process. Semi-structured tasks are generally more difficult to implement in an EDSS/IPS. The nature and style of modelling also depends on the task structure and decision phase. Semi-structured tasks rely more on "what-if" analysis, objective and goal generation, judgement simulation and probability assessments of outcomes. Well structured tasks rely more on traditional quantitative modelling practices such as operations research, numerical and statistical analysis.

The pattern of usage may also influence system design and operation. Environmental data is increasingly being generated from many different sources (eg. governmental, statutory and non-statutory bodies) and this is also true for aquaculture. Better management may result from increased accessibility to data from multiple external sources. Indeed future EDSS/IPS developments will, by necessity, need to integrate with distributed databases in order to improve environmental decision making. Another consideration of pattern of usage is the number of simultaneous users. Most EDSS/IPS are designed for single users. However, group-EDSS/IPS (multiple-simultaneous users) can also be important particularly for long-term, strategic environmental planning. The timeliness of information is another important factor. Technicians and operational managers are often required to make decisions daily or more frequently and are mainly concerned with control. Often rapid guesstimates are sufficient to arrive at appropriate actions for environmental control of aquaculture facilities. Accuracy can be sacrificed for speed. On the other hand, strategic management occurs on a longer time frame and there may be less requirement for timely information but more requirement for accuracy. Forecasting may be more important for strategic managers than control. The framework presented here is not only useful for systems design and implementation but can also be used for comparing and contrasting systems capabilities. This is illustrated in the next section.

3   EDSS/IPS's For Aquaculture: Systems' Description And Comparison

3.1 Aquaculture Research and Monitoring System (ARMS)

ARMS was one of the first EDSS/IPS developed for aquaculture. It was developed by the Aquaculture Research Group at Deakin University in 1991 (Bourke, Stagnitti & Mitchell[5, 6]). ARMS is a flexible software system that aids in the collection, monitoring and analysis of aquaculture experiments. It was specifically designed to capture and analyse physio-chemical and biological data generated in the research on culture of freshwater crustaceans. ARMS has implemented many features of a EDSS/IPS including a capacity for semi-structured decision support and what-if analysis. A key-feature of ARMS is a probabilistic estimate of specific experimental outcomes using Saaty's[7] Analytic Hierarchy Process (AHP) for simultaneously evaluating qualitative and quantitative variables. ARMS may be used to examine the validity of the assumptions used in past aquaculture experiments and, more importantly, to simulate new experiments before actual implementation. It also has a demonstration capacity that may be used to train other managers, planners or students. The design of ARMS stresses modular independence and thus the decision process and database management are not necessarily coupled to the experimental data and maybe modified to suit the requirements of other organisations or applications.

ARMS interfaces to WQMS, a water quality management system that provides on-line monitoring of real-time data on pond temperature, conductivity, pH and dissolved oxygen (see figure 2). Using ARMS, the user can also store and retrieve textual descriptions of each pond experiment in flat files and match these descriptions to various previously stored historical data sets. It provides a range of statistical and graphical functions, and allows decision modelling. The MBMS of ARMS is a simple engine that interfaces into a number of existing commercial packages (Minitab, Microsoft Excel and Microsoft Word) to provide basic statistical support, spreadsheet functions and word processing requirements. ARMS uses a menu-driven, graphical front-end program written in Borland's Turbo Pascal 4.0. Many of ARMS functions have been implemented using Borland's Turbo Toolboxes. ARMS uses a hierarchical file structure to store information and is managed by routines written in Turbo Pascal. It operates under DOS. ARMS contains nearly 8000 lines of code and consists of three major components: a Statistical Summaries Module, a Decision Support and Judgement Simulation

Module, and a Display and Calculation Module (see figure 2). These modules are briefly described below.

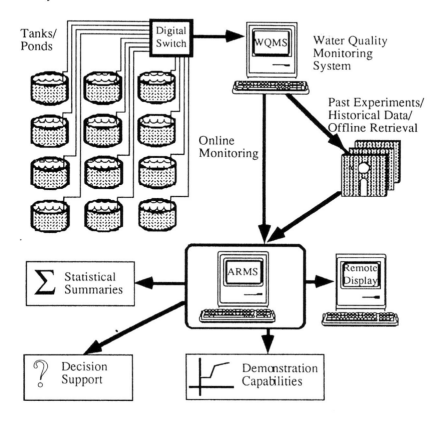

Figure 2: Components of the ARMS system. (Modified from Bourke, Stagnitti & Mitchell[6]).

ARMS provides simple on-screen graphing and descriptive statistics. However, the major statistical analyses are executed in Minitab. By choosing the Statistical Module option in the ARMS menu, ARMS will reformat the requested data into a file which is input directly into Minitab. The user then has a number of specially written Minitab macros available to perform regularly requested operations on the data (eg. regressions, correlations and ANOVA) or can use the full range of generic functions offered by Minitab. Minitab was chosen as the base module for statistical calculations not only because it is easy to use and very efficient, but also covers a wide range of desired analyses required for all current applications in aquaculture operations. It is also inexpensive and has good support and upgrade paths. Graphical displays can be generated in two ways - directly within ARMS or via MS Excel. The on-screen graphs are low resolution but quick to generate and contain cubic spline smoothing. They are useful for on-line monitoring and demonstration purposes. High resolution graphs for use in reports can be generated in MS Excel by custom-made macros. Other macros perform frequently required database operations and mathematical calculations. As with Minitab, ARMS reformats the selected data and directly transfers it into an Excel spreadsheet. When in Excel, the user also has the flexibility of developing their own personal macros by modifying those provided or using generic functions.

Conducting individual experiments, particularly those which might continue for a year or more, is a very costly enterprise. Therefore, ARMS has implemented a decision support tool that allows the user to make decisions about planned experiments by assigning importance values to a number of qualitative and quantitative variables. This approach has been used successfully in other environmental studies (eg. see Itami & Stagnitti[8]). The underlying decision model used in ARMS is AHP and was developed by Saaty[7]. AHP is a multicriteria, multiobjective decision model that uses a hierarchical or network structure to represent the decision problem and then develops priorities for alternatives based on judgements made by the decision maker. The operation of AHP basically involves the creation of a decision hierarchy by decomposing the decision problem into levels consisting of interrelated decision elements. ARMS has implemented the decision problem as a three-level hierarchy in which the top level of the hierarchy contains the most macroscopic decision objective; ie. the aquaculture experiment. The next level of the hierarchy contains detailed attributes of the experiment (eg., stocking density, substrate, pond temperature, etc) and the third level of the hierarchy contains desired experimental outcomes such as survival rate, production failure and biomass increase. The user inputs relative importance ratings and pairwise comparisons to all variables at the various levels of the decision tree. These importance ratings and pairwise comparisons provide the user with a basis for assessing the utility of the judgements made in setting up the experiment. This method can rapidly produce a scenario of "predicted" outcomes without the need to individually conduct each experiment. The set of conditions which leads to the "best-case" scenario can then be implemented at less risk and cost.

3.2 Decision Support For Pond Aquaculture (POND)

POND is a recent EDSS/IPS developed for the management of an entire aquaculture pond facility including species/facility customisation for particular aquaculture operations, economic and biomass forecasting, operational lot management, water quality and sediment management, and estimation of feed and fertiliser requirements. The main focus of POND is to provide a view of the pond dynamics at both the individual pond level as well as the facility level (Nath, Bolte & Ernst[9]). It is being developed by the Biosystems Analysis Group in the Department of Bioresource Engineering at Oregon State University and is supported by the Pond Dynamics/Aquaculture Collaborative Research Support Program (PD/A CRSP) funded in part by the U.S. Agency for International Development. POND grew around an earlier decision tool called PONDCLASS which provided analytical tools for fertilisation and liming requirements for aquaculture ponds. POND now encompasses four general functional areas: (a) estimation of fertilisation and liming requirements for individual ponds, (b) pond simulation models, (c) economic analysis, and (d) parameter estimation (see figure 3).

The fertilisation recommendation schedules are based on linear programming models developed earlier by PONDCLASS and tested on several sites in Asia and Honduras. The objective of this module is to provide the least-cost combination of fertilisers that meet nutrient requirements of a pond. The simulation modules of POND are a collection of analytical, deterministic formulae that address various pond dynamics. These formulae are expressed in terms of ordinary differential equations and are solved numerically. The simulation capability of POND is powerful and can be used to perform what-if analysis. POND's simulation models are organised into a 3-level hierarchy as shown in figure 3. Level 1 models are concerned with predicting fish growth based on a simple view of the pond including water and heat sources and sinks, fertiliser application, feed, pond temperature and volume. Level 2 models includes prediction of phytoplankton, zooplankton and nutrient dynamics (nitrogen, phosphorus and total carbon). Level 3 models incorporate the functionality of the first two levels and includes additional capabilities for simulating bacterial kinetics and detailed water quality and sediment dynamics where relevant for certain culture practices. User-specified feeding and fertilisation schedules are matched with predicted nutrient mass balances to estimate nutrient consumption and production rates.

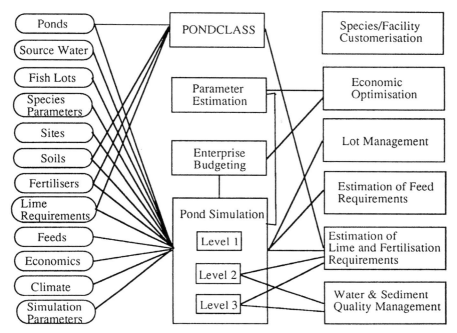

Figure 3: Architecture of POND showing databases, functional modules and applications for decision support. (Source: Nath, Bolte & Ernst[9])

POND includes economic forecasting in the form of enterprise budgets. The user inputs various interest rates and income sources and when the simulation is complete, POND generates a report of costs based on an areal, unit of production, and facility basis. This module permits the user to quickly generate income/cost strategies based on various facility configurations. The parameter estimation module is used to calibrate best-fit model parameters for different species, locations or management strategies. The calibration process permits the user to adjust model parameters based on their experience or by matching actual growth data of other species. The functional modules of POND might be thought of as the MBMS of the system. The dialogue manager of POND is windows-based and very flexible. Graphical displays of information are easily generated in a variety of formats. The DBMS is simple but adequate for retrieving ASCII data stored in flat files. The purpose of the DBMS is to supply information for the simulation models. The databases (files) in POND are managed by a consistent dialogue interface that allows the user to browse, add, delete, copy and save records. A free download of the software and user manual is available at the following internet address: http://biosys.bre.orst.edu/pond/pond.htm.

3.3 Decision Support Tool For Rating Soil and Water Information For Aquaculture Pond Location (DESTA)

DESTA is a decision support tool for rating soil and water characteristics for new aquaculture facilities. It was recently developed by Stagnitti[10]. The software is based on a conceptual framework developed earlier by Hajek & Boyd[11]. The user of DESTA is requested to enter values for some 50 environmental parameters that potentially effect the development or operation of new ponds (see figure 4). DESTA produces limitation ratings for each environmental variable supplied by the user. User responses are classed into slight, moderate or severe limitation ratings. Slight ratings indicate that in-situ soils or water supplies have properties favourable to the design, construction, operational management and/or maintenance of the aquaculture facility. Moderate ratings indicate that the environmental parameter has properties that require attention. The degree of limitation can be reduced only by special planning, design, management or maintenance. Severe limitations indicate one or more properties of the soil or water supplies are likely to produce a significant effect on the successful operation of the pond or facility. In this case, significant cost and effort are required to overcome severe limitations. Severe limitations consequently may force a potential site to become economically non-viable. User inputs are required in three main sections: (a) soil limitation ratings for excavated ponds, or soil limitation ratings for pond embankments, dikes and levees, (b) source-water limitations, and (c) watershed features and water availability (see figure 4).

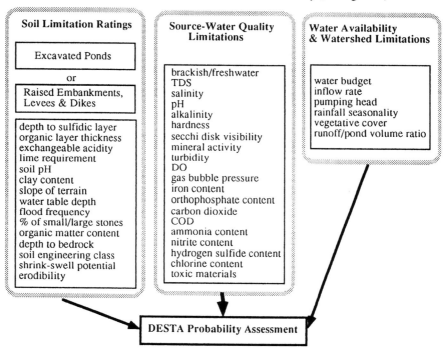

Figure 4: DESTA system overview.

DESTA provides a tabular report, grading all environmental variables into limitation classes and provided that sufficient information has been entered by the user, DESTA will make a probability estimate of the potential success of a new venture. If very few severe or moderate limitations in the environmental parameters are found, then a high probability assessment will result. If there is insufficient information supplied by the user, then DESTA will report only the limitation ratings for specified environmental parameters and no probability assessment is made. At this stage, the user cannot not adjust the importance ratings on the environmental parameters, nor can s/he adjust the ranges of values which determine which limitation class a particular environmental parameter value might lie in. However, DESTA will be modified soon to permit this. A beta 1 version of DESTA is freely available at the following internet site: http://www.cm.deakin.edu.au/~frankst/.

4   Systems Comparison and Classification

A conceptual model of a generic EDSS/IPS was developed earlier in this paper. The knowledge and decision requirements for aquaculture were outlined as well. The conceptual model for EDSS/IPS development and the decision environment for aquaculture can serve as invaluable tools for potential designers of new decision support systems for aquaculture. The framework outlined in this paper is also useful for comparing and contrasting existing systems' functionality. Three specific EDSS/IPS for aquaculture have been discussed. Using the criteria developed in section 2, a summary of their capabilities is presented in table 1.

Table 1: Systems comparisons using the conceptual framework in figure 1.

System	POND	ARMS	DESTA
A1. Task structure			
Structured	Yes	Yes	No
Semi-Structured	No	Yes	Yes
Unstructured	No	No	No
A2. Management-level supported			
Technicians	Yes	Yes	Yes
Pond managers	Yes	Yes	Yes
Executive/ Financial/ Administrative	Yes	No	No
A3. Decision phase supported			
Identify & define problems	Limited	Yes	No
Identify opportunities and outcomes	Limited	Yes	No
What-if scenarios	Yes	Yes	No
Probabilistic assessment of outcomes	No	Yes	Yes
Generates summaries and reports	Yes	Yes	Yes
Generates graphical and tabular summaries	Yes	Yes	Yes
Support for strategic decision making	Limited	Limited	No
Support for tactical decision making	Yes	Yes	Yes
B1. Database management			
Interacts with database/s	Yes	Yes	No
Database/s exclusive to EDSS/IPS	Yes	Yes	No
Distributive databases	No	No	No
DBMS queries and other functions	Simple	Simple	No
Extraction of data from several sources	Yes	Yes	No
B2. Model management			
Quantitative modelling & statistical analysis	Yes	Yes	No
Decision analysis & modelling	Limited	Yes	Limited
Models interact with DBMS	Yes	Yes	No
Simulation & Forecasting	Yes	Yes	No
B3. User Interface/ Dialogue Management			
Command line/ Power-user	No	No	No
Interactive/ Menu-driven	Yes	Yes	Yes
On-line help	Yes	Yes	Yes
C1. Required expertise level of user			
Computer skills (High, Moderate Low)	Low	Low	Low
Quantitative skills (High, Moderate Low)	Moderate	Low	Low
Discipline knowledge level (aquaculture)	Moderate	Moderate	Moderate
C2. Usage pattern and operation			
Single direct use	Yes	Yes	Yes
Group DSS	No	No	No
Links to external CBIS's	No	Limited	No
Implementation			
Operating system	Windows 3.1	DOS 4.0	Windows 3.1
Minimum platform	386 IBM	286 IBM	386 IBM
Minimum storage requirements (Mb)	1.5	0.32	0.4
Minimum memory requirements (Mb)	4	2	4

Conclusion

We have introduced a framework for the application of EDSS/IPS technology in aquaculture and have shown that it may contribute to improved design and management of environmental data and models. Using this framework several current EDSS/IPS's developed for the aquaculture industry have been contrasted and reviewed. Common features of these systems are powerful information processing and modeling routines that are supported by user-friendly, visually-interactive interfaces or dialogue managers. Using advanced simulation modules, prediction tools and decision analysis, EDSS/IPS's for aquaculture significantly reduce the time currently spent by managers, planners and researchers in manipulating data and enhances the quality of experimental reports. The ability to anticipate experimental outcomes by calibrating or simulating various management or production practices without the need to physically conduct the experiment is extremely cost and time effective and reduces the risk of production failure and poor economic performance. Indeed the costs associated with development of EDSS/IPS's are likely to be minuscule in comparison to the costs of constructing an aquaculture facility. EDSS/IPS's are therefore an important innovation to aquaculture. The challenge for future EDSS/IPS development is the need to better integrate with on-line distributed data sources such as satellite imagery and weather data.

References

1.   Sprague, R.H., Framework for the development of decision support systems, in *Decision Support Systems: Putting Theory Into Practice*, R. H. Sprague and Watson H.J., Eds.: Prentice Hall, 1989, pp. 9 - 35.

2.   Guariso, G. & Werthner, H., *Environmental Decision Support Systems*. Chichester: Ellis Horwood Ltd, 1989.

3.   Pearson, M.J. & Shim, J.P., An empirical investigation into DSS structures and environments, *Decision Support Systems*, vol. 13, pp. 141 - 158, 1995.

4.   Ariav, G. & Ginzberg, M.J., DSS design: A systemic view of decision support, *Communications of the ACM*, vol. 28, pp. 1045 - 1052, 1985.

5.   Bourke, G., Jones, T., Stagnitti, F., Mitchell, B., & Collins, B., The design of an environmental decision support system for aquaculture research, *Australian Society of Operations Research, Bulletin*, vol. 10, pp. 7 - 15, 1991.

6.   Bourke, G., Stagnitti, F., & Mitchell, B., A decision support system for aquaculture research and management, *Aquacultural Engineering*, vol. 12, pp. 111 - 123, 1993.

7.   Saaty, T.L., *The Analytical Hierarchy Process*. New York: Mc Graw Hill, 1980.

8.   Itami, R.M. & Stagnitti, F., New software for user controlled variable combination methods, *Land Use Modelling Quarterly*, vol. 2, pp. 2 - 9, 1980.

9.   Nath, S.S., Bolte, J.P., & Ernst, D.H., Decision support for pond aquaculture, Internet reprint document of a paper presented at the Sustainable Aquaculture Conference at PACON International, Honolulu, Hawaii. (Internet address: http://biosys.bre.orst.edu/pond/pond.htm), 1995, pp. 11pp.

10.   Stagnitti, F., DESTA. A decision support tool for optimally locating new aquaculture facilities based on soil and water information, Centre for Dynamical Systems and Environmental Modelling, Geelong, Technical Report Series (in prep.), 1996.

11.   Hajek, B.F. & Boyd, C.E., Rating soil and water information for aquaculture, *Aquacultural Engineering*, vol. 13, pp. 115-128, 1994.

Operational control for a multireservoir system - multiobjective approach

J.J. Napiórkowski, T.S. Terlikowski
Institute of Geophysics, Polish Academy of Sciences, ul. Ks. Janusza 64, 01-452 Warsaw, Poland

Abstract

The real case of a complex multireservoir and multiobjective water reservoir system is presented. The basic elements of the used Two-Level Optimization Method is discussed in details. It is shown that the introduced optimization concept improves considerably the system performance in comparison with Standard Operation Rule for 90 year long historical data record.

1. Introduction

Technique for determining the yield of multireservoir water supply has been developed and applied to the system serving the industrial region of the Upper Vistula River. The major objectives of this particular system are to supply water for the industrial and municipal water-users, the steel works, the chemical plan and the fish farms. At the same time, concentration of pollutants in the river should be maintained at the levels compatible with water quality requirements. The unified methodology that enables to comprise a large class of acceptable solutions and to cover the wide range of specific conceptual approaches is presented. It enables the inclusion of the operator's preferences, intuition and experience. The presented technique may be reduced to the following conjunct parts: the optimization of a simplified quantitative model of the actual system and the multiobjective verification/comparison through simulation. The first part consists in constructing a relatively wide class of control structures based on the two-layer optimization technique method (Terlikowski[1]). The second part is based on the simulation performed for historical data over a long time horizon (ninety years). This simulation is an active research and consists in testing and adapting the control rules by computation of many objective values. Several control schemes have been proposed in the form of computer programmes for the Upper Vistula Reservoir System. They have been compared for a large number of simulated years and for many objectives. One can see the ambiguity of different unified, aggregated evaluation methods in such a problem. The proposed control schemes are compared with the so called Standard Decision Rule, to present their undoubted advantages. The theoretical case of perfectly known future inflows is also tested to show the quality of the proposed control structure.

2. Description of the case system model

The water resources system concerned consists of two aggregated reservoirs located on two rivers (Soła River and Vistula River) and of five water users. The scheme of the system is shown in Figure 1.

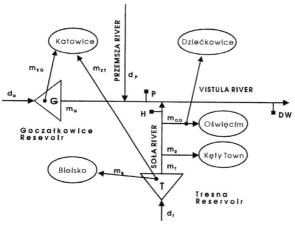

The major objectives of the system are to secure the water supply for the industrial and municipal water users, namely Katowice and Bielsko; to supply the steel works "Katowice" via the Dziećkowice Reservoir, and to supply water to the chemical plant Oświęcim and fish farms around the town of Kęty. At the same time, concentration of pollutants which are discharged mainly to the Vistula River downstream of the outlet of the Przemsza River should be maintained at the levels compatible with water quality requirements. The basic hydrological and reservoir characteristics are given in Table 1.

Figure 1: The Upper Vistula River System.

The purpose of the model is to describe relationships between flow rates in the rivers and in the conduits delivering water to users over a long time horizon (one year) with the discretization period of one decade (10 days). Therefore only the dynamics of the storage reservoir is considered, while effects of dynamics of flow in the river channels are neglected.

Reservoir	Tresna	Goczałkowice

historical inflows:
    the lowest
    the highest
    average
	1.18 m³/s
	1469 m³/s
	19.5 m³/s
		0.36 m³/s
		581 m³/s
		7.75 m³/s

catchment area	1095 km²	522 km²
total storage capacity V_{max}	139.7 mln m³	202.8 mln m³
dead storage V_{min}	13.6 mln m³	20.0 mln m³
flood control zone	27.0 mln m³	30.4 mln m³
max. outflow m_{max}	1730 m³/s	935 m³/s
min. outflow (biol. crit.) m_{min}	0.93 m³/s	0.45 m³/s

For brevity, the following notation is used:

j	- number of decade
V^j	- state of the reservoir at time j
d^j	- natural inflow to the reservoir or to the river at time j
u^j	- flow in a given cross-section
z^j	- water demands at time j
m^j	- outflow from the reservoir or water supply to a user, a control variable at time j
S^j	- pollutant load discharge at time j (kg/m³)
C^j	- admissible pollutant concentration at time j
T, G	- denote the Tresna and Goczałkowice reservoirs
H	- control cross-section at Soła River
P	- control cross-section at Vistula River (down Przemsza River)
DW	- control cross-section at Vistula River

and the following subscripts are introduced:

B	refers to Bielsko
T	refers to Tresna
O	refers to Oświęcim
R	refers to fish farms;
D	refers to Dziećkowice

According to introduced notation, we are able to write state equations for the system of reservoirs and flow balance equations formulated for the selected cross-sections (H, P, DW). State equations are:

$$V^{j+1} = V^j - B*m^j + C*d^j \tag{1}$$

where

$$V = [V_T, V_G]; \quad m = [m_T, m_G, m_B, m_{KT}, m_{KG}]; \quad d = [d_T, d_G]$$

$$\tag{2}$$

and

$$B = \begin{bmatrix} 1 & 0 & 1 & 1 & 0 \\ 0 & 1 & 0 & 0 & 1 \end{bmatrix}; \quad C = \begin{bmatrix} 1 & 0 \\ 0 & 1 \end{bmatrix} \tag{3}$$

The flow balance equations for the considered cross-sections are as follows:

$$u_P^j = m_G^j + d_P^j \tag{4}$$

$$u_H^j = m_T^j - m_R^j - m_{OD}^j \tag{5}$$

$$u_{DW}^j = u_P^j + u_H^j \tag{6}$$

3. Formulation of the optimization problem

The objective function of the optimization problem under consideration for any time instant (for any decade) and for annual time horizon (T=36) can be written in the form of a penalty function:

$$Q(m,V) = \sum_{j=k}^{k+T} [a_B^{+j}(m_B^j - z_B^j)^2 + a_R^{+j}(m_R^j - z_R^j)^2 + a_{OD}^{+j}(m_{OD}^j - z_{OD}^j)^2$$

$$+ a_K^{+j}(m_{KT}^j + m_{KG} - z_K^j)^2 + a_P^{+j}(u_P^j - S_P/C_P)^2 + a_{DW}^{+j}(u_{DW}^j - S_{DW}/C_{DW})^2$$

$$+ a_H^{+j}(u_H^j - z_H)^2 + b_T^j(V_T^j - V_T^{*j})^2 + b_G^j(V_G^j - V_G^{*j})^2] \tag{7a}$$

In equation (7), symbols a and b with respective subscripts denote weighting coefficients, while CP and C_{DW} denote values of pollutant concentration which should not be exceeded at the cross-sections P and DW. Inserting equations (4,5,6) into equation (7) the performance index Q can be expressed explicitly on controls m^j and the state trajectory V^j (reservoir contains). Other quantities which occur in its formulation are treated as parameters.

$$Q(m,V) = \sum_{j=k}^{k+T} Q(m^j, V^j) \tag{7b}$$

Required retention trajectory V^{*j}

It is assumed that the operation of the reservoir system is carried out on annual bases in the following way:

* by late December, the reservoirs normally are returned to low level to prepare the system for the next flood season completing the annual cycle.

* the storage reservation for flood control on January 1 was determined for controlling the maximum probable flood. During the normal filling period, January-April, the reservoirs should be filled up completely.

* during the May-November period the water stored in and released from the reservoirs is used for municipal, industrial and fish farms needs.

Weighting coefficients a^{j+} and b^j

According to the general objective of the control problem, which is aimed at the rational protection against water deficits and at reaching the desired state at the end of April, the following values of weighting coefficients in the optimization problem are used: $a^{j+}=1$ if demands are greater than supply and $a^{j+}=0.01$ otherwise, for k=[1,36]. As far as the second coefficient is concerned, in order to avoid a good performance in one year followed by a very poor performance in the next year $b^j=.001$ for j=[1,30] (May - February), $b^j=.004$ for j=[31,33] (March) and $b^j=.01$ in April, for j=[34,36].

The objective function during each decade is subject to the constraints on the state of the system, controls and flows in given profiles

$$V^j_{min} \le V^j \le V^j_{max} \; ; \; m^j_{min} \le m^j \le m^j_{max} \; ; \; u^j_{min} \le u^j \le u^j_{max}$$

$$(8)$$

2. TWO-LEVEL OPTIMIZATION TECHNIQUE

To solve the aforementioned problem we adjoin the equality constraints (1) with Lagrange multiplier sequence λ (prices). The Lagrangian function has the form:

$$L(m,V,\lambda)=\sum_{j=k}^{k+T} [Q(m^j,V^j)+\lambda^j(V^{j+1}-V^j+B*m^j-C*d^j)] \qquad (9)$$

To include the state-variable and outflow constraints the above problem is solved by means of the two-level optimization method and solved in a decentralized (coordinated) fashion. At this stage we make use of the additivity of the Lagrangian function (9) and the possibility of separation of the decision variables.

The Lagrangian function has a saddle point which can be assigned by minimizing $L(\lambda,V,m)$ with respect to V and m, and then maximizing with respect to λ. Finally, the optimization problem can be expressed in the form:

$$\max_{\lambda} \; [\min_{S \; u} \; L(\lambda, \; V, \; m)] \tag{10}$$

with inequality constrains on state and control and no constrains on Lagrange multipliers. Figure 2 illustrates how the two-layer optimal control method works.

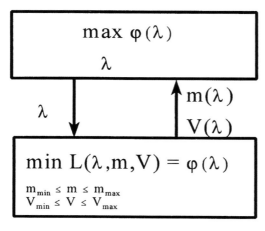

At **lower level** for given values of the Lagrange multipliers we look for the minimum of the Lagrange function. The necessary condition is the zero value of the gradient with respect to **m** and **V**. The task of **the upper level** is to adjust the prices λ in such a way, that the direct control of the reservoir, affected by λ, results in the desired balance of the system (the mass balance equation (1) is fulfilled satisfactorily). On the upper layer, in the maximization of the Lagrange function with respect to λ the standard conjugate gradient technique is used.

Figure 2: Two-level optimization method

In the applied Two-Layer optimization control method (TLM) the solution of the two-level optimization problem (10) is the essential "upper layer part". Note, that this planning layer "proposes" the sequence of T control variables $\{m^k,...m^{k+T}\}$ for one year long time horizon. At the current decade they are taking into account by the lower layer that tries to apply them in the real conditions and eventually subject to some additional operator's interventions.

4. Comparison of control methods through simulation

The simulation of some chosen control methods has been carried out over the long time horizon of 90 years, with the real, historical data of natural inflows to the system. The methods under investigation have been partially discussed in the previous sections. Let us mention here once again those of them, which - after an initial stage of synthesis consisting in adjusting their parameter values - have been thoroughly compared by simulation.

1.   SDR -The natural, standard decision rule realizing mainly the following principle: „take as much as you need if you can" for any particular water user.

2.   TLM - Two-layer optimization method with:
            1) the complex, long-term planning aiming at the optimization of all the particular goals in a compromise manner.

2) the realization of the planned decisions (water supplies and discharges) in the real, current conditions.

3.   SS - The method TLM supplemented by some elements of standard, direct decision rules.

In the last two methods, requiring solution of the optimization problem (10), the average values of real historical data for the period 1901-1990 have been taken as the long-term prediction of inflows. Furthermore, to compare and investigate the „power" of optimizing methods, the variant denoted REAL has been considered, which differs from the optimizing methods only in the fact that real current values of inflows are put in place of predicted values. It is worth to note that it is possible to make use of a more precise knowledge of future inflows only in the methods including a long-term planning.

Each method is evaluated through many different **performance indices -** equivalent, in a way, to degree of realization of conflicting goals. Hence, the indices reflect only the partial, not global, effects of system performance. In our model, each performance index is represented as a function of time:

- For water users (B, R, OD, K and the cross-sections H, P, DW) this is the **deficit** function expressed with respective time unit (decades).

- For the reservoirs this is the function of **storage level** (we are interested mainly in its average value in the summer period).

At the same time, each index is evaluated through many **scalar criteria.** In order to define them precisely, let us consider the deficit in meeting the needs of a given water user, e.g. R (fish farms) in a period of 1 year. The function m_R^j, where j corresponds to decade - together with z_R^j (representing the needs of fish farms) - characterize this one particular index in the most complete manner. However, in order to compare in a clear, well ordered manner the results of different controls m_R^j, and furthermore with the results of the others controls, we introduce some scalar criteria depending on these functions.

The following criteria have been proposed for the functions which represent the **water users** performance (i, j, k are included in a given period of 1 year, i.e. of 36 decades):

- global deficit time TD:

$$TD = Card(\{ i: \quad m_R^j < z_R^j\}) \tag{11}$$

- maximal continuous deficit time TDc:

$$TDc = \max(\{k \le l|: \quad k \le l \ \wedge \ [\forall \ k \le j \le l; \quad m_R^j < z_R^j]\}) \tag{12}$$

- average relative deficit AvD:

$$AvD = \sum_{j=k}^{k+T} \frac{(z_R^j - m_R^j)_+}{z_R^j} \frac{100\%}{T} \qquad (13)$$

- maximal average relative continuous deficit AvDc;

$$AvDc = \max(\mid \sum_{j=k}^{l} \frac{z_R^j - m_R^j}{z_R^j} \frac{100\%}{T} : \forall\ k \le j \le l\ m_R^j < z_R^j\ \mid) \qquad (14)$$

- maximum relative deficit MxD;

$$MxD = \max\ (\mid \frac{(z_R^j - m_R^j)_+}{z_R^j} \frac{100\%}{T} : k \le j \le k+T\ \mid) \qquad (15)$$

Any function defining particular water user supply (m_B, m_R, m_{OD}, m_K), as well as the flow in the cross section (H, P, DW) is characterized for a given one year period by 5 numbers as defined above.

At the same time the trajectories V_T^j, V_G^j are described by 2 criteria. For example average water content in the Summer period for Goczałkowice Reservoir is:

$$V_G Av = \sum_{j=1}^{12} \frac{V_G^j}{12} \qquad (16)$$

As a result, we obtain for each user/goal the sequence of 5 numbers, characterizing a given performance index function in a synthetic way. This could be sufficient to evaluate and compare the different functions for one, fixed index, e.g. with the aid of any multiobjective optimization method. However, it is more complicated, because we have to compare the control effects for 9 "users" and not for particular year, but for 90 years long historical record.

To solve such a problem it is necessary to use a specific approach, which is arbitral to some extent and makes use of intuition. To obtain the final comparison results we analyze the diagrams of s.c. **frequency (reliability) criteria**, calculated on the basis of simulation for each of nine „users" and for each of 5 or 2 scalar criteria (11-16).

Those frequency criteria are also functions, but defined over the set of values of respective scalar criteria TD, ..MxD, ..etc. Their values represent the number of years, for which the respective scalar criterium has its values in a given range. Formally, e.g. for MxD we have:

$$f_{MxD}(x) = Card(\mid I:\ x - \Delta\ \le MxD^I \le\ x \mid) \qquad (17)$$

$$F_{MxD}(x) = Card(\mid I:\ MxD^I \le\ x \mid) \qquad (18)$$

where MxD^I denotes the value of criterium MxD (15) for the year I, and Δ is the

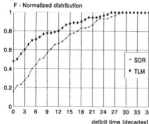

Figure 3: TD reliability criterion for Katowice

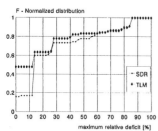

Figure 4: MxD reliability criterion for Katowice.

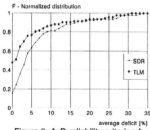

Figure 5: AvD reliability criterion for Katowice

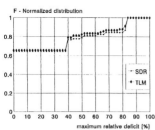

Figure 6: MxD reliability criterion for Bielsko

step of discretization of values of MxD (e.g. 2 per cent). As it is seen, f corresponds to the notion of density function and F - of commutative distribution function of the „random variable" MxD^I, when I is treated as representing the elementary events.

5. Results and conclusions

Some of the simulation results for the considered control methods, namely SDR, TLM, SS and REAL are presented below by means of the reliability criterium F.

Figures (3-5) show the diagrams of normalized distribution F corresponding to the criteria TD, MxD and AvD for the deficit in Katowice. Two methods of control are compared: Standard Decision Rule (SDR) and Two-Layer optimization Method (TLM). For the water user Katowice, the advantage of TLM is evident in the sense of all the considered scalar criteria. This results from the character of this particular user, taking water from the both parts of the system: Tresna Reservoir and Goczałkowice Reservoir. The TLM takes into account the cooperation of the whole system and coordinates the partial decisions. It gives distinctively better results.

It appears that for the users taking the water directly from Soła River (R, OD and B) the optimizing method also gives better results than SDR. The difference is, however, less evident (see Figure 6). Generally, the obtained results show that the TLM improves the control quality in comparison to SDR for any "real" water user (K, R, OD, B), especially for the MxD criterium of maximum deficit value.

In the case of flows required in the chosen control cross-sections the comparison of SDR with TLM and SS methods does not give so univocal conclusion. For the H cross-section at Soła River the SDR method gives better results than TLM or SS in most cases. This may be explained by the local character of this "user". Nevertheless, if we apply the TLM technique with the perfect knowledge of future inflows - as in REAL method - we get considerable improvements in all indices. This is shown

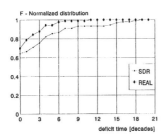

Figure 7: TD reliability criterion for the H cross-section (Sola River).

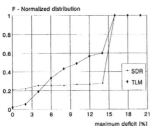

Figure 8: MxD criterion for the P cross-section (Vistula River).

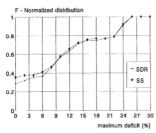

Figure 9: MxD criterion for the WP cross-section (Vistula River).

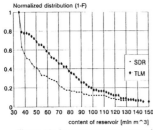

Figure 10: Average water content of Goczalkowice Reservoir in Summer.

in Figure (7), which presents the diagrams of F_{TD} criterion.

For the P cross-section at Visula River the TLM appears to be better than SDR even in real situation (uncertainty in future inflows) but not for any scalar criterion. For the F_{TD} criterium SDR produces generally worse results while for F_{mxD} - maximum deficit criterium SDR is often more advantageous as is shown in Figure (8).

For the DW cross-section (see Figure (9)) the TLM technique occurs to be better than SDR with respect to any scalar criterium. The above conclusion reflects one more fact that the optimizing method improves particularly those partial effects of control which is more structurally connected to the system as a whole and in consequence depend more on a proper coordination of the system.

Finally, Figure (10) shows the exemplary diagrams of reliability criteria for reservoirs. The idea of these criteria is, in a way, inverse to (18), because we prefer possibly large values of reservoir contents. Namely we have, e.g. for $A_G Av$ (see (16)):

$$F_{V_G Av}(x) = Card(\mathbf{I}\; I:\quad V_G Av^{\,I} \geq x\;\mathbf{I})$$

One can see that the TLM allows to keep the given water contents more frequently, even for relatively high values (90-120 mln m³).

Acknowledgments

This study has been supported by Polish Committee for Scientific Research, Grant No.PB0021/S4/94/06.

References

1. Terlikowski T. Comparison of two methods of water reservoir control - the case of the Goczalkowice Reservoir, *Archives of Hydroengineering*, 1993, **1-2**, 115-134.

WATER_SOFT: optimization of water use and reuse in power plants

F. Sigon,[a] C. Zagano,[a] M. Rovaglio[b]

[a]*ENEL Spa, R&D Division, Environment and Materials Research Center, Via Rubattino 54, 20134 Milano, Italy*
[b]*Dipartimento di Chimica Industriale ed Ingegneria Chimica, "G. Natta" Politecnico di Milano, Piazza Leonardo da Vinci 32, 20133 Milano, Italy*

Abstract

The most challenging target in the industrialized countries will be the development of an effective management of the environmental impact due to industrial activities. At the RIO Conference a precise statement, on the optimization and prioritization of the water uses, was approved in the Agenda 21: human needs being to be preserved. A new aptitude in designing industrial processes and plants is required. ENEL therefore developed the WATER_SOFT code to analyze and to foresee the performances of existing "water system" of its power plant. This code is providing the capability to assess the impact of alternative technologies and processes and to re-design advanced configurations with equivalent performances and better environmental impact (lower water requirements and minimum waste production). Comparison between predicted and experimental data has been reported for a complex industrial application. Such results confirm the reliability of the simulation model here presented. Extension of this code to study alternative process configurations are also investigated.

1 Introduction

Water is the most used fluid for energy production all over the world. The water need can be estimated around 0.5-2 l/kWh for power plants based on cooling tower technology, while for applications using "through-condenser" the water requirement grows up to 100-110 l/kWh. However, such amount cannot be considered a real consumption since the water, after use and minor chemical treatments, is entirely released with only a negligible temperature increase. Along the cycle water consumption can be generally estimated in 0.17 l/kWh for conventional plants but such value can be strongly reduced to 0.025-0.05 l/kWh by improving the technology related to the adopted physicochemical treatments (Queirazza et al., 1995). Such decrease allows to reduce the kWh operating costs and to follow the environmental politics about consumption minimization of natural utilities for industrial production.

From these considerations, the greatest interest for such investigating tools is expected to analyze, improve and optimize the adoption of water related technologies in the power industry and generally in the process industry.

With particular reference to power plants, the "water-system" can be defined as the complex of fluxes and treatments finalized to the production of operating fluids (industrial water and demineralized water), to the use of such fluids (cleaning, conservation treatments, etc.) and to the waste water disposal. The sequence of the process treatments is generally based on an "open-cycle" configuration where the inlet water is completely returned after a single use in the water system. Such a structure can be adopted only when the following conditions are satisfied:

- total and unlimited water sources of low cost;
- outlet streams easy to clean before their draining off;

moreover, the following features must be also taken into account:

- high consumption of water and chemical reactants;
- high sludge production.

The water system optimization can be achieved by different topological choices which can lead progressively to a configuration named "closed-cycle". For such purpose the following aspects are of a specific interest:

- optimization of the water treatments plant management;
- reuse of process water and/or use of alternative sources (e.g.: rainwater);
- new technologies for water treatments to be added or substituted to the existing ones.

Since the number of alternative plant configurations to be evaluated becomes very high and moreover there is a large number of data to be considered together with a complex plant structure, related to the presence of internal recycles, a suitable simulation tool is strongly recommended.

WATER_SOFT is a specific software designed and developed to simulate the steady-state conditions of any plant configuration with the aim of finding out the optimal solution related to the minimization of water need, reactants consumption and amount of waste and sludge produced. The analysis through such a simulation tool has the following advantages:

- to compare several process configurations on the basis of the prediction of their performances;
- to reduce time and developing costs for new process configurations supporting the experimental validation through pilot plant only for the most promising technologies;
- to solve complex plant structure where the presence of an high degree of recycles requires a robust and reliable numerical algorithm for the evaluation of each single unit capability together with the global plant performance.

2 Solution approach

Before any description of the unit models adopted, it is important to specify the objectives which must be reached through the simulation model itself. Namely, three are interesting for a specific mention:

- model capability to represent existing apparatuses and new solutions proposed by the end-user;
- choice of the correct description scale which allows to verify the unit performances in terms of outlet composition streams without introducing detailed phenomenologies related to the internal system behavior;
- input/output models which can be easily introduced in a more general program able to simulate, in a reasonable computing time, a complex scheme.

The achievement of such objectives is strictly related to the "solution approach" adopted to numerically solve the problem here examined.

In literature two main approaches are often mentioned: "Sequential Modular" and "Equation Oriented". They have been analyzed, discussed and compared in several application examples typical for chemical plant flowsheeting. For details see reviews of Pierucci (1978), Westerberg (1979), Evans (1984).

WATER_SOFT is based on a sequential modular approach where in an open functional architecture, "modules" and "connecting lines" can be used to represent a water system through simple block schemes. Modules can be easily assembled to build any plant configuration.

Each module represents one apparatus related to the water treatments: filters, deareators, clarifiers, reverse osmosis, etc.. The connecting lines correspond to inlet and outlet fluxes represented by information vectors (flowrate, composition, pH, temperature, enthalpy, etc.).

Table 1: WATER_SOFT unit modules categorized on the species treated/removed

TOTAL DISSOLVED SOLIDS	TOTAL SUSPENDED SOLIDS
reverse osmosis	gravity separation
mechanical evaporation	coagulation, flocculation, clarification
staged evaporation	granular media filtration
ion exchange	membrane filtration
DISSOLVED ORGANICS	HEAVY METALS
aerobic biological oxidation	chemical precipitation
anaerobic biological oxidation	ion exchange
chemical oxidation	reverse osmosis
DISSOLVED GASES	
air stripping	
steam stripping	
ion exchange	
biological nitrification	

The simulation model for the different units have been developed on both theoretical base and information derived on experience and the technical specifications supplied by vendors. With the aim of reducing the computing time, the model has been generally derived from simple equations which allow the evaluation of the whole system performances (reactants and energy consumption, outlet composition, operating costs, etc.). However, some algorithms are more complex and consequently the corresponding models developed require specific attention (e.g.: pH and precipitation which involves the evaluation of chemical equilibria). A list of modules (representing unit operations) and their use is summarized in Table 1.

Since the global algorithm is based on a sequential modular approach, each single module can be updated, modified or changed without compromising or influencing the basic code structure.

For the solution of the recycle problems, WATER_SOFT adopts a sequential solution strategy using a convergence algorithm based on evolutionary models (Boston 1980, Rosen 1980, Westerberg 1981). They enable an approximated representation of the overall block scheme through reduced simulation models which contain evolutionary parameters updatable along the convergence loop. This solution technique has been chosen for its greater efficiency and convergence velocity with respect to alternative methodologies (Pierucci, 1982).

During a simulation, the operating conditions of each unit are verified and automatically compared with the nominal design conditions. Warnings and other messages highlight when process constraints are violated (minimum or maximum pH, maximum temperature, maximum pressure drops, etc.).

Unit and plant performances are summarized through three different indexes:

- Merit function which is related to the functional condition of the unit assuming as optimal the average point equidistant from process constraints. A weighted sum of the units merit function allows to define the global plant merit function.
- Water quality given in terms of composition and compared with operating specifications or law limits.
- Operating costs corresponding to unit and plant costs evaluated on the basis of reactants and energy consumption.

3 Experimental Validation

The effectiveness of the developed software and its applicability as a reliable and useful tool in correctly simulating the performances of a real plant are illustrated by the analysis of a typical water system of a thermoelectrical power plant located on a river and of some alternatives for its configurations.

The complete plant, sketched in fig. 1, can be schematically divided in five subsystems:

1. Industrial water production. A clarifier/softener allows a first water hardness reduction then followed by sand filters which retain most of the suspended solids.

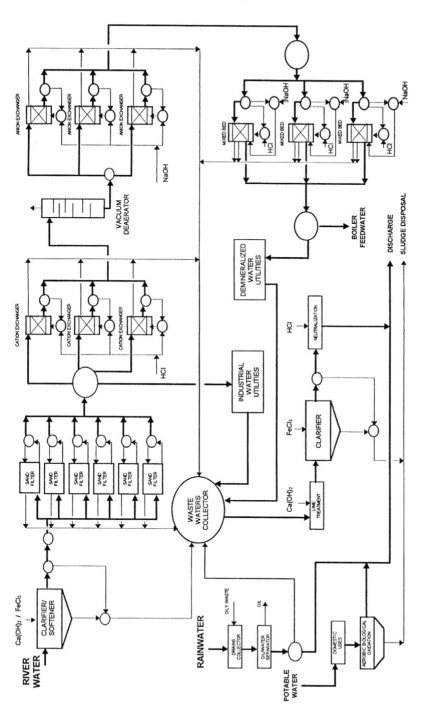

Fig. 1 : "Water system" for a thermoelectrical power plant located on a river

2. Demineralized water production. A train of different ion exchanger resins (cation exchangers, anion exchangers and mixed beds) and a vacuum deaerator drastically lower the silica and ion presence inside the main water stream.
3. Industrial water utilities especially required for washing most of filters and devices.
4. Demineralized water utilities, used at 90% to make up the boiler feed water. In this section the washing water from the filters of the boiler system is added. Before, the washing water is purified by a cake filter and a mixed bed ion exchanger.
5. Waste water treatment. All the water streams from the utilities and process units are collected and treated to match the law constraints before their discharge.

The complete scheme to be solved results of a very large dimension:
- 90 different units, considering mixers and splitters;
- more than 200 streams containing about 30 different species;

The difficulties in the numerical solution are increased by the presence of recycling streams.

Another problem to be faced is the correct definition of all the inlet streams, not only in terms of quantity, but mainly of composition. As a matter of facts, together with the river water, whose average composition can be considered sufficiently constant, and the known chemical reactants introduced by the operators, there is the need of identifying the species coming from the different utilities (both from plant and domestic uses). This last contribution to the total amount of water treatment must be carefully managed, changing periodically in a drastic way. Boiler or filter washing occurs only sometimes and for short time. It is then necessary to individuate an appropriate operation average time to average these conditions. In our analysis one year average has been chosen. Of course, this approximation can give wrong information in different local situations, particularly when strongly non linear models are involved (like chemical equilibrium). Table 2 shows a schematic comparison of the predicted and measured values for the water and sludge released to the environment from an hipothetic power plant. Despite the necessary approximations, assumptions and the difficulties in measuring all the inlet compositions, a fine agreement has been obtained. It is worthwhile remembering that, as above discussed, reactant species (for instance $Ca(OH)_2$, $FeCl_3$, HCl) are averaged on the year consumption and that the salt precipitation equilibrium is often very sensitive to the solution pH: small variations in the operative conditions can bring to very different equilibrium conditions. Moreover, it is important to predict the relative presence of each specie and the trends of their amounts with respect to different operating conditions or process alternatives.

It must be underlined, even though there is an high number of equations and recycles, a single simulation of the whole plant can be performed in about 300 s CPU on an Alpha Vax workstation.

Tab.2:   Experimental and WATER_SOFT predicted composition
of the water and sludge released from an hipothetic plant

Chemical species	Experimental data:		WATER_SOFT results:	
	Water (%)	Sludge (%)	Water (%)	Sludge (%)
Na^+	17.43	0.05	18.41	0.10
K^+	0.49	< 0.01	0.44	< 0.01
Ca^{++}	20.72	30.39	17.13	34.76
Mg^{++}	2.51	1.07	1.72	1.90
Fe^{3+}	0.03	7.85	0.02	8.45
Cu^{2+}	< 0.01	0.06	< 0.01	0.05
Ni^{2+}	< 0.01	0.41	< 0.01	0.38
Zn^{2+}	< 0.01	0.11	< 0.01	0.10
Co^{2+}	< 0.01	0.01	< 0.01	0.01
Mn^{2+}	< 0.01	0.04	< 0.01	0.04
Pb^{2+}	< 0.01	0.01	< 0.01	0.01
V^v	0.01	0.39	0.01	0.37
Al^{3+}	< 0.01	0.61	0.11	0.37
Cr^{3+}	< 0.01	0.01	< 0.01	0.01
Cl^-	33.74	0.09	38.13	0.19
$CO_2+HCO_3^-+CO_3^-$	3.37	54.34	2.50	52.02
NO_3^-	1.92	0.01	1.69	0.01
$PO4^{3-}$	0.04	0.07	0.02	0.08
SO_4^-	17.66	3.11	17.77	0.09
NH_4^+	1.22	< 0.01	1.23	0.01
SiO_2	0.85	1.37	0.82	1.05
TOTAL	100.00	100.00	100.00	100.00

4 Evaluation of Process Alternatives

WATER_SOFT, allows to discriminate among different process alternatives during the design phase or to address new operating solutions and revamps of existing plants.

An example of these possibilities is discussed and the results of three alternative configurations are compared in terms of water consumption, chemicals needs, operative costs and water quality. These new process designs refer to the introduction of updated technologies (membrane systems) or internal recycle for an efficient reuse of water.

Alternative 1: Reverse Osmosis
The actual configuration of the plant, above discussed, is modified by the introduction of a membrane system (reverse osmosis) in substitution of ion exchange resins and vacuum deaerator. This new layout allows to substantially reduce the chemical reactants consumption with a corresponding increase of the electrical energy consumption.

By the simulations, we derive three evidences:
- 30% increase of inlet river water
- 20% decrease of discharged salts
- 15% increase of sludge disposal.

The increase of water need is due to the larger amount of water to be treated by the reverse osmosis treatment to get the same output flowrate. Nevertheless, the discharge stream contains a significant salt reduction because of the lesser use of chemical reactants.

Alternative 2: Water Reuse

To reduce the water inlet from the river, the 90% of the stream sent to the neutralization (fig. 1) is recycled back to the initial clarifier /softener.By the simulation we derive the following:
- 60% decrease of inlet river water
- 280% increase of discharged salts
- 20% decrease of sludge disposal.

The sludge reduction is a direct consequence of the lower need of river water. The substantial salt increase in the discharge stream is determined by the chemical reactants for the ion exchange resin regeneration. As a matter of facts, the river water presents about 320 ppm of salts inside, while the recycled stream contains more than 20,000 ppm of ions in solution. That means a reduction of ion exchanger run time and a consequent more frequent regeneration.

Alternative 3: Reverse Osmosis and Water Reuse

This configuration is the sum of the previous ones. The rationale of this choice is that the recycle stream can decrease the river water feed and compensate for the water need of the membrane system. Moreover, the lack of chemical reactants in the reverse osmosis allows to avoid the growth of the salt presence inside the plant. The benefit of the introduction of the water reuse is not any more damaged by the need of large amounts of chemical reactants. The new input/output conditions, in respect with the actual configuration, are then evaluated by the simulation results:
- 45% decrease of inlet river water
- 70% decrease of discharged salts
- 30% decrease of sludge disposal.

The advantages of alternatives 1 and 2 are maintained, while disadvantages are almost canceled by their combined use. Only membrane exhaustion is more rapid in comparison with the simple introduction of reverse osmosis: the recycle stream in the present configuration contains about 2000 ppm of salts in solution (about 10 times higher than river water).

Simulation results of the different process alternatives is summarized in Tables 3-5.
Table 3 shows the unit and plant merit index (Queirazza et al., 1996) scaled between 0 (bad operating conditions) and 1 (optimal operating conditions). From these results it can be underlined that the plant merit index does not change significantly if all the units work within the correct operating range. In

particular, the merit index of alternative 2 and 3 shows that the process design is flexible enough to face a large amount of internal water reuse.

Table 3 : Units and Plant Merit functions

UNIT	Actual Configuration	Alternative 1	Alternative 2	Alternative 3
Clarifier /softener	1.0	1.0	1.0	1.0
Sand filters	0.63	0.63	0.63	0.63
Cation exchangers	0.84	-	0.60	-
Vacuum deaerator	0.71	-	0.71	-
Anion exchangers	0.74	-	0.60	-
Mixed bed exchan.	0.67	0.78	0.67	0.65
Cake filters	0.61	0.61	0.61	0.61
Clarifier	1.0	1.0	1.0	1.0
Reverse Osmosis	-	0.43	-	0.41
Plant	0.72	0.71	0.67	0.67

The quality index of the different water types are reported in table 4. This index is related to the check between the water composition and the operating or law limits and rapidly approaches zero when one or more constraints are violated. WATER_SOFT simulations indicate that the primary clarify/softener is not able to correctly operate in presence of a high recycle stream. As a matter of facts, alternatives 2 and 3 show a reduction of the industrial water quality index. This suggests either to reduce and optimize the recycle amount or to substitute the clarifier with a different unit (e.g. electrodialysis).

Table 4: Water quality index

Water type	Actual Configuration	Alternative 1	Alternative 2	Alternative 3
industrial	0.81	0.81	0.58	0.58
demineralized	0.99	0.95	0.96	0.97
waste	0.96	0.97	0.91	0.97

Finally, table 5 compares the operating costs (reactants and energy) and clearly indicate the significant reduction obtained through the use of reverse osmosis even in presence of high internal recycles.

Table 5 : Plant Operating Cost Estimations

Operating costs	Actual Configuration	Alternative 1	Alternative 2	Alternative 3
deminer. water (Lit/m^3)	390.	185.	2390.	145.

5 Conclusions

The optimization of the water system in a power plant is a very challenging problem because it needs to proceed by multiple objectives (water quality, system operability, process costs) and to choose among multiple choices (processes, technologies, operating conditions). Therefore, the availability of a suitable tool becomes necessary to analyze and to identify the best process alternative.

As reported above, WATER_SOFT seems to be a reliable simulation tool which allows to represent qualitatively and quantitatively a complex water system.

Moreover, synthetic information as the merit function and the water quality index or the operating costs evaluation allow to compare complex process structure in a very simple and easy way. Simulation results clearly show that, with reference to a real application, significant improvements can be obtained through the use of the WATER_SOFT analysis so reducing the environmental impact of the plant and the process operating costs.

References

- Queirazza G., F. Sigon, C. Zagano "L'esperienza straniera nel riuso delle acque nelle centrali termoelettriche", ATI - Sezione Lombarda, Camogli (I), 31 Maggio - 2 Giugno 1995
- Westerberg A. W. "Optimization in Computer-Aided Design", Foundation of Computer Aided Process Design, AIChE, New York, 1981
- Evans L. B. "Proceedings Symposium: Simulation et Conception Assistées par Ordinateur des Procédés Chimiques", Institut du Genie Chimique, Toulouse (F), 1984
- Pierucci S., G. Biardi, E. Ranzi, M. Dente "Proceedings Symposium: Contribution des calculateurs électroniques au development du génie chimique et de la chemie industrielle", Paris (F), 1978
- Pierucci S., E. Ranzi, G. Biardi "Solution of Recycle Problems in a Sequential Amodular Approach", AIChE journal, 28, 820, 1982
- Boston J. F. " Inside-out Algorithms for Multicomponent Separation calculations", Computer Application to Chemical Engineering, ACS symposium Series No. 124, 1980
- Rosen E. M. "Steady-state Chemical Process simulation: state of the art review", Computer Application to Chemical Engineering, ACS symposium Series No. 124, 1980
- Westerberg A. W., H. P. Hutchison, R.L. Motard, P. Winter " Process Flowsheeting", Cambridge, (UK), 1979

Reducing the environmental and economic costs of handling iron ore

J.E. Everett

Department of Information Management & Marketing, University of Western Australia, Australia

Abstract

Iron ore is shipped out in large quantities from northern West Australian ports. The ore is railed to the port, stock piled and then loaded on to ships. Customers require that successive shipments are uniform in composition, with respect to several elements, including iron, silicon, aluminium and calcium. Variability in composition can be smoothed out by stacking the arriving ore onto large stock piles, which are recovered for shipment. With random stacking and recovery from a single stock pile, improved variability requires larger stock piles, and/or increased handling of the ore. Both these involve environmental costs as well as economic costs. The larger the stock piles and the greater the amount of rehandling, the greater the resulting dust pollution and land degradation.

This paper shows how the size of stock piles and the amount of rehandling can be reduced by intelligent stacking and recovery procedures. The decision model discussed uses a combination of computer simulation and analysis to develop ore handling rules. Forecasts of the incoming ore composition combined with a monitoring of stockpile composition can be used to decide where ore should be stored. Similar rules are developed to choose the stockpiles from which to recover ore when ships are being loaded. We find that direct loading a proportion of the ore can further reduce environmental and economic costs, without loss of composition uniformity.

The computer simulation model is used on a microcomputer to evaluate and demonstrate graphically the considerable reduction in land use and dust pollution that can be achieved by intelligent stacking and recovery procedures. The computer simulation model not only helps evaluate alternative policies using real data, but also provides a graphical interface which has enhanced communication between managers, operators and policy advisers.

1 Introduction

Iron ore provides 5% of Australia's exports, totalling over 100 million tonnes, or over US$2 billion dollars value, per year[1]. Most is mined in the north west of Western Australia and railed to a few coastal ports, then crushed, stock piled and shipped out, to customers in Japan, China, Southeast Asia and Europe.

Customers are concerned about variability of composition in the iron ore. Their blast furnaces are tuned to ore of particular composition, and require considerable adjustment if composition varies between shipments. The major quality control for iron ore is therefore to minimise variability in composition, not just in the percentage content of iron (Fe), but also of silica (SiO_2), alumina (Al_2O_3) and calcium oxide (CaO). From some mines, the phosphorus content can also be important, though not in the example to be considered here. Customers are very aware of the variability offered by suppliers, from Australia and elsewhere. For example, the Japanese industry publishes an annual comparison of suppliers' ship-to-ship variability[4]. Improving the composition variability can therefore provide a substantial market advantage to an iron ore producer.

Figure 1 shows schematically the port facilities for a typical iron ore producer. The mine can plan the average composition of ore that it produces, but there will still be appreciable variation in what is mined from day to day. The ore arrives by train, is crushed, sampled for assay and stacked on stock piles. These stock piles are then reclaimed and loaded on ships for export. It is also possible to direct load some of the ore, bypassing the stock piles.

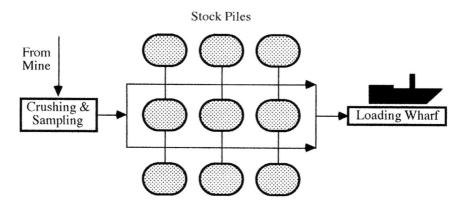

Figure 1: Layout of the ore handling facility.

Stock piles provide an effective means of reducing ore variability. Ore is stacked by laying the ore back and forth in one direction. Operationally, a stock pile cannot be reclaimed until it has been filled. The ore is reclaimed by cutting slices at right angles to the direction of stacking. So it can be assumed with reasonable confidence that ore reclaimed from a stock pile is of uniform composition. Clearly, the larger the stock pile, the greater its averaging or smoothing effect on composition. In effect, the stock pile is a low pass filter.

There are environmental as well as economic costs incurred in the port handling of iron ore. The stock piles occupy (but do not beautify) a large area of prime coastal land. The handling process, stacking and reclaiming of stock piles, creates dust which is environmentally deleterious.

This paper will show how computational procedures can be used to enhance the handling process to achieve as good or better quality control while using less volume of stock pile. It is also shown that a portion of the ore can be direct loaded, bypassing the stock piles, with no loss of quality. Both improvements can be expected to reduce the environmental and economic costs, by reducing the volume of the stock piles used, and by reducing the amount of ore handling.

2 Measure of Performance

The objective is to ship out ore of minimum variability. For each mineral there is a target percentage, and a tolerable range. The tolerable range was established by discussion with the marketing staff. For example, if the $\{Fe, SiO_2, Al_2O_3, CaO\}$ targets were $\{57.29\%, 5.35\%, 2.64\%, 0.56\%\}$, and the tolerable ranges were $\{0.40\%, 0.30\%, 0.14\%, 0.30\%\}$, then an Fe content of 56.89% or 57.69%, or an SiO_2 content of 5.05% or 5.65% would all be equally distasteful to the customer. It is convenient to define a "stress" for each mineral:

$$\text{Stress} = (\text{Actual \% - Target \%}) \,/\, \text{Tolerance \%} \qquad (1)$$

The aggregate stress (A) can then be defined by:

$$A^2 = \text{Stress}^2(Fe) + \text{Stress}^2(SiO_2) + \text{Stress}^2(Al_2O_3) + \text{Stress}^2(CaO) \qquad (2)$$

The individual mineral stresses can be positive or negative, but the aggregate stress must be positive. The aggregate stress provides a suitable measure of the quality of a shipment of iron ore. The objective is to despatch shipments of iron ore for which the root mean square (RMS) aggregate stress is minimised.

3 Data Source

This study was carried out in cooperation with Robe River Iron Associates, who export 30 million tonnes of iron ore yearly through their port facility at Cape Lambert., on the northwest coast of Western Australia.

Incoming iron ore is sampled for each four-hour period of production. The incoming tonnage and assays for iron, silica, alumina and calcium oxide are recorded. A data file of 2,014 such records was supplied, spanning about a year of operation of the port facility, and was used as the basis for this study.

Each four-hour period of incoming ore averaged close to 12,500 tonnes. The median ship capacity was about 150,000 tonnes. For the purpose of the simulation studies to be reported here, it will be assumed that each four-hour production is exactly 12,500 tonnes, and each ship has capacity 150,000 tonnes.

Table 1 shows statistics for the year's production. The four mineral stresses have quite strong cross correlations, and each exhibits strong serial correlation.

Table 1. Analysis of a year's production sampled at 4-hour intervals.

	Fe	SiO_2	Al_2O_3	CaO	Aggregate
Mean	0.000	0.000	0.000	0.000	6.863
Standard Deviation	1.360	1.667	1.179	0.922	2.620
Cross Correlation with: Fe		-0.761	-0.375	-0.511	-0.271
SiO_2			0.340	0.042	0.270
Al_2O_3				-0.175	0.108
CaO					0.162
Auto Correlation, Lag = 1	0.719	0.764	0.673	0.728	0.560
2	0.552	0.621	0.511	0.554	0.371
3	0.447	0.521	0.420	0.448	0.269
4	0.386	0.463	0.371	0.345	0.198
5	0.323	0.403	0.300	0.297	0.135
6	0.285	0.365	0.263	0.264	0.119
7	0.254	0.329	0.251	0.233	0.099
8	0.232	0.295	0.239	0.196	0.100

4 Simulation Models

To investigate the effectiveness of alternative ore handling procedures, the system of Figure 1 was modelled using the graphical simulation package Extend[2].

The model allowed up to three stock piles to be built concurrently from incoming ore, and for ships then to be filled from up to three stock piles simultaneously. We will discuss various procedures for determining how to build the stock piles, and how to recover ore from them to load the ships.

5 Single Stock Piles

The simplest procedure is to build a single stock pile to completion before the next is started, and to recover from a single completed stock pile to fill each ship. If the incoming stresses were not serially correlated, then the RMS aggregate stress of the completed stock piles would be inversely proportional to the square root of the stock pile size. Because of the strong serial correlation, the stress does not fall off quite so fast. Figure 2 shows how the root mean square aggregate stress decreases with increasing stock pile size. For comparison, the broken line shows the behaviour that would occur if the stresses had no serial correlation. Since ships are being filled each from a single stock pile, the RMS aggregate stress applies both to completed stock piles and to filled ships.

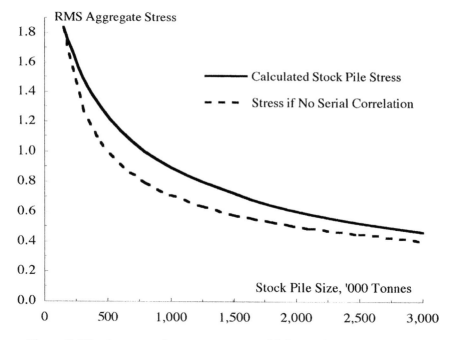

Figure 2: The decrease of aggregate stress with increasing stock pile size.

Larger stock piles occupy more land, implying greater environmental and economic costs. If more sophisticated ore handling methods can use a lesser stock pile area, then some environmental and economic benefit is obtained.

In the following sections we will consider a range of alternative ore handling methods, and compare the stock pile capacity needed with that required to give the same level of RMS aggregate stress using the single stock pile configuration.

6 Multiple Stock Piles Built in Turn

We have seen that the RMS aggregate stress does not fall off as quickly with stock pile size as it would if there were no serial correlation. Serial correlation can be reduced by building multiple stock piles in turn, to increase the building time each stock pile spans. Consider building simultaneously three stock piles, each of 300,000 tonnes. Since it would be operationally infeasible to change the destination stock pile every four hours, we shall assume that the destination is changed three times per stock pile. This corresponds to changing the destination stock pile each 100,000 tonnes, or 32 hours, of production.

Building the three stock piles could be done in a number of different sequences. Table 2 shows the sequence used, which ensures that each stock pile spans nine changes, the maximum possible. Maximising the building span of each stock pile reduces the serial correlation. Pile "3" is started when pile "1" is two-thirds full and pile "2" is one third full. The next 100,000 tonnes goes to pile "2"; then pile "1" is completed; pile "4" is commenced, and so on.

Table 2. Stock pile building sequence to maximise span ("three in turn").

Stock Pile	1	2	3	4	5	6	7
	{ 9	12	15	18	21	24	27
Production Period	{ 5	8	11	14	17	20	23
	{ 1	4	7	10	13	16	19

Table 3 compares building 300,000 tonne stock piles singly, with building three in turn (as in Table 2). Building three stock piles in turn gives considerable reduction in aggregate stress, equivalent to single stock piles of 706,000 tonnes. However, the single stock pile system averages two stock piles half full. If we build three stock piles in turn, and fill ships from the last stock pile completed, on average four stock piles will be half full. Taking this into account, building the three stock piles in turn gives 15% saving in storage space.

Table 3. RMS stress for stock piles built singly, and for three in sequence.

Stock Pile Sequence	Aggregate Pile Stress	Equivalent '000 Tonne	Piles in use	Stock Pile Saving
Single stock pile at a time	1.490	300	2/2	0%
Three in turn (see Table 2)	1.062	706	4/2	15%

7 Multiple Stock Piles Built Intelligently

If we know the stress of the incoming ore, then we can modify the sequence in which the stock piles are built, using the assay information to reduce the stress of the completed piles. We shall refer to procedures using the assay information as "intelligent" procedures. Consider a stock pile "j" containing ore of weight W_j, with stress components $\{S_{ij}\}$, where i goes from 1 to 4, for the four mineral stresses. We can define the "pain" P_j for stock pile "j" as:

$$P_j = W_j A_j^2 = W_j \Sigma_i S_{ij}^2 \qquad (3)$$

For incoming ore of weight w and stress $\{s_i\}$, we need to choose which stock pile (j = 1, 2, ...) to add it to. It is sensibleto choose the stock pile for which the rate of increase in "pain" per tonne added $(\partial P_j / \partial w)$ is smallest.

$$\partial P_j / \partial w = Lt_{w \to 0} [\partial \Sigma_i (W_j S_{ij} + w\, s_i)^2 / \partial w] = 2W_j \Sigma_i S_{ij} s_i \qquad (4)$$

So an appropriate criterion in choosing a destination stock pile "j" is to select the one which minimises the criterion $W_j \Sigma_i S_{ij} s_i$. With a delay of several hours in processing assays of incoming ore, the stress $\{s_i\}$ is not accurately known. However, an unbiassed forecast of the stress, based on previous assays, can be used instead. An exponentially smoothed forecast was used. Analysis using the SPSS Trends[3] package found the best fit with an alpha value of 0.7.

Table 4 shows the RMS aggregate stress that would be obtained if the 300,000 tonne stock piles were built using accurate assay data, and if forecast data were used. Although the accurate assay data would not be available, the forecast data gives almost as good a result. We see that the exponentially smoothed forecast is better than a naive forecast (the previous assay). The decrease in RMS aggregate stress, obtained with intelligent sequencing using the exponentially smoothed forecasts, gives a 30% saving in stock pile space.

Table 4. RMS stress for stock piles built using assay information.

Three Pile Sequence Controlled by:-	Aggregate Pile Stress	Equivalent '000 Tonne	Piles in use	Stock Pile Saving
Accurate assays	0.936	918	4/2	35%
Forecast based on:				
Exponential smoothing	0.966	861	4/2	30%
Previous assay	1.005	789	4/2	24%

8 Intelligent Ship Loading

We have so far assumed each ship is fully loaded from a single completed stock pile. Further improvement in variability can be achieved by intelligently filling each ship from multiple stock piles, taking account of the assay stresses of the stock piles. Filling a ship from "k" available stock piles, the proportion of load taken from each stock pile is the vector $\{p_j\}$, for $j = 1 \dots k$. The stress of stock pile "j" is given by the vector $\{s_{ij}\}$, where $i = 1 \dots 4$, represents the four minerals. The ship may already contain a proportion p_0 with stress $\{s_{i0}\}$. The aggregate stress of the filled ship is to be minimised, subject to the constraint that $\sum_j p_j = 1 - p_0$. Using the Lagrange multiplier μ, we can minimise a modified objective function "Y" with respect to each of $p_1 \dots p_k$ and μ:

$$Y = \sum_i (p_0 s_{i0} + \sum_j p_j s_{ij})^2 - \mu(p_0 + \sum_j p_j - 1) \qquad (5)$$

Solving the resulting equations gives the proportion p_j to take from each stock pile. If more is demanded from a stock pile than it contains, the stock pile is emptied to the ship, and the calculation repeated, including the next stock pile. Also, some p_j may be negative. If only one p_j is negative, the optimum is found by repeating the calculation omitting that stock pile. In the rare case of two or more p_j values being negative, then little (if any) optimality is lost by omitting the stock pile with the most negative p_j and repeating the calculation.

Both these problems could be dealt with by quadratic programming and the simplex algorithm, with the non-negativity and stock pile tonnage constraints included. However, this would require considerable extra computational effort with little better performance than the branch and bound method just described.

Table 5. RMS stress for ships loaded using stock pile assay information.

Stock Piles Built Using	Aggregate Ship Stress	Equivalent '000 Tonne	Piles in use	Stock Pile Saving
Three in turn (see Table 2)	0.590	2,086	6/2	57%
4-hour exponential forecast	0.564	2,248	6/2	60%
Accurate assays	0.575	2,178	6/2	59%

Table 5 shows that the intelligent ship loading procedure greatly improves the RMS aggregate stress, using stock piles created by "three in turn" and by intelligent methods. Even though the system uses six stock piles that are on average half full (three being built and three being recovered to the ship), the saving in stock pile space is 60%. Although building the stock piles intelligently gave less stock pile variability than did the "three in turn" method, we find that intelligent ship loading gives almost as good a result when it uses stock piles built "three in turn" as when it uses stock piles built intelligently.

9 Direct Loading

Referring back to Figure 1, it is possible to load some of the incoming ore directly to the ship, bypassing the stock piles. If a proportion p_0 is direct loaded, the rest can be intelligently selected from stock piles as described in the previous section. The simulation was run again, for a range of proportions of direct loading. Figure 3 shows that 20% or more of the ore can be direct loaded without significant increase in variability. Indeed, direct loading about 10% of the ore gives a small improvement in performance. This can be explained by the fact that direct loading increases the span of arrival time for ore going into a ship, and thus decreases the serial correlation problems. In the limit, as direct loading approaches 100%, the performance is no better than building single stock piles of 150,000 tonnes each. Being able to direct load a portion of the ore reduces the amount of handling and therefore the amount of dust pollution generated. This provides both an environmental and an economic benefit.

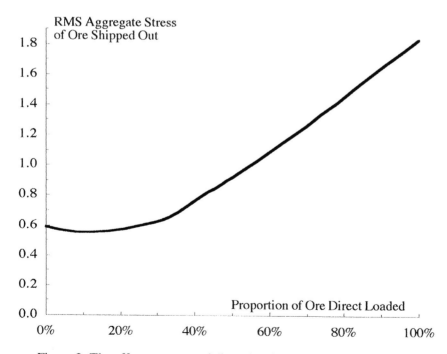

Figure 3: The effect on stress of direct loading a proportion of the ore.

10 Conclusion

The study has demonstrated how some fairly simple software, easily run on a microcomputer, can be used to reduce the variability in ore composition by intelligently controlling the building of stock piles from incoming ore, and the loading of ships from multiple stock piles. The computer simulation not only helps evaluate alternative policies using real data: the graphical interface has also enhanced communication between managers, operators and policy advisers.

We have seen that the software can be used to save about 60% of the space that would otherwise have to be dedicated to stock piles. This saving in stock pile space represents a considerable environmental as well as economic benefit. The software can also be used to control the direct loading of up to 20% of the ore, without increasing the variability of the shipments. Direct loading a fifth of the ore considerably reduces the amount of ore handling and resultant dust pollution.

One great advantage of the two forms of environmental benefit - saving in stock pile space and reduction in dust pollution - is that they are both accompanied by economic benefits. Environmental savings that can be achieved with a decrease rather than an increase in economic costs are clearly far more likely to be adopted, especially in a highly competitive industry.

The software described here is now being used by Robe River Associates to guide their iron ore handling operations at Cape Lambert. In the words of their Manager of Marketing - Technical:

> "... the recommendations from the work have been implemented, resulting in a measurable reduction in ship to ship variability compared to the base period. Robe has adopted your finding that preparing three simultaneous piles over a nine day period will optimise blending efficiency and the procedure has been incorporated into our Quality Management System. It was also presented to customers in Japan during our recent Technical Presentations to the Japanese Steel Mills and was well received. ... In essence, a good news story like this should not be in any way confidential."

References

1. Australian Bureau of Statistics, 1995, *Year Book Australia,* Australian Bureau of Statistics, Canberra.

2. Diamond, B. and Lamperti, S. 1995, *Extend™: Performance modelling for decision support.* Imagine That Inc., San José.

3. SPSS, 1988, *SPSS^X Trends.* SPSS Inc., Chicago.

4. TEX, 1995, *Iron Ore Manual 1994-95,* The TEX Report Co., Tokyo.

Creation of a concept for advanced use of ecological risk assessment in environmental management

G. Lefebvre, L. Houde

Environment Branch, Hydro-Québec, 75 W. René-Lévesque Blvd, 16th Floor, Montréal, Québec, Canada

Abstract

In order to manage the environmental risks related to its installations, Hydro-Québec has done many studies and built a number of computing tools. Most of the software has been developed in a shell which allows easy data input and graphical representation of the results, which can be georeferenced and superimposed on maps. The shell also allows comparison and reuse of results between models.

The most recent software is designed for ecological risk assessment for which Hydro-Québec is developing a framework involving the use of mathematical models. The ultimate aim is to provide a tool which can estimate the fate of pollutants in some compartments of an ecosystem and evaluate the potential effects of chemicals at levels of biological organizations.

1 Introduction

Hydro-Québec is a government-owned electrical utility in the province of Québec, Canada. It generates electricity mainly from hydro sources (96%) with minor contributions from nuclear and conventional thermal plants. The utility's generating capacity in 1994 was close to 30 GW.

The Hydro-Québec transmission system is unusual in that most of the energy is produced in northern areas whereas the customers are mainly concentrated hundreds kilometers to the south. The territory served covers roughly 1.54 million km^2 with more than 160,000 ha of rights-of-way to be managed and vegetation control to be done on 75% of these.

Hydro-Québec also distributes electricity, operating an extensive network of lines and more than 500 substations. With approximately 2.2 million wood poles, most of them PCP-treated, the utility has to manage wood pole storage sites as well as dangerous-waste storage sites.

2 An ecosystemic approach

Assessing the ecological risks inherent in Hydro-Québec's operations is important. For this reason the utility has developed a general framework allowing this exercice to be done systematically. The framework is intended to serve as the guideline for an iterative approach to each of the specific activities identified. For each case, the goal in applying the framework is to help specialists characterize specific threats to ecological resources and select effective solutions for mitigating unacceptable risks.

With respect to Hydro-Québec operations, we have identified seven specific installations and activities where an ecological risk assessment could be usefully performed:
- Yard storage for PCP-treated poles
- Potential use of soil moderately or slightly contaminated with PCP or oil
- Hazardous-material storage sites
- Rural and urban substations
- Thermal generating stations
- Herbicide spraying
- Mercury contamination in hydroelectric reservoirs.

In most of these cases, the primary contaminated media are site soils, nearby surface waters, and groundwater. The main pathways of contaminant transport are infiltration of soils and groundwater, and water runoff to receiving surface waters. Quantifying exposure entails measuring or modeling. Since sampling and chemical analysis are very expensive, computer simulation models of contaminant transport and distribution can be used to estimate contaminant concentrations in soils, surface waters, sediments and groundwater at lower cost.

The sequential assessment of risks could proceed through the use of more detailed, process-level ecological models which focusing on population projections based on demographic or bioenergetics models. These predictive models offer support to assessors who have to deal with very complex problems when predicting the probability of adverse effects resulting from exposure to contaminants.

To conclude this section, we present one of the problems mentioned above, the potential reuse of soil moderately or slightly contaminated with hydrocarbons. When land is transferred or reassigned, its purpose or use often changes. In some cases, Hydro-Québec wants to reuse the excavated soil at other sites. An ecological risk assessment (ERA) study could ensure that the use of contaminated soil at a specific site does not pose a threat to the environment. Such study depends on mathematical modeling.

3 Computing tools

In order to help specialists in their work, Hydro-Québec, in cooperation with different partners, has developed a number of computing tools in the 1980s. These are composed mainly of mathematical models, expert system, artificial-intelligence system, decision support systems, and a wide variety of databases. On the whole, this software is based on the use of a development shell.

3.1 Simulation system

The architecture of the simulation system is presented at figure 1. It is in fact a set of software tools which allow users to structure and pilot a problem-solving process in the environmental field. The system facilitates the acquisition, analysis, synthesis and transmission of environmental information required to assess impacts on the environment. It includes several models with a common user-friendly interface. The simulation system offers high-performance visual tools and a data management system which allows easy comparison and reuse of results between models. Description of hydrodynamic characteristics of a river, for instance, which might represent a major time-consuming step, can be used either for the hydrodynamic model and the mass pollutant dispersion in surface water model.

Data acquisition can be done in various ways: graphical information systems (GIS), laser videodisk, satellite images, topographical maps, etc. An automatic finite-element mesh builder is available for all the computing tools, which substantially reducer's the user's task in the discretization process.

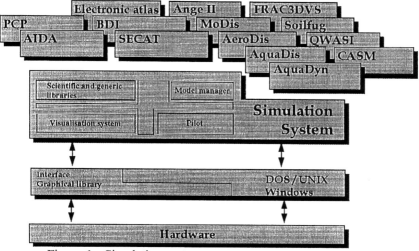

Figure 1 - Simulation system

3.2 Mathematical models

A firm believer in the concept of sustainable development, Hydro-Québec has committed itself to pursuing a more efficient approach to managing the different sources of risk inherent in the operation of its facilities and equipment. Better risk management calls for greater understanding and more accurate assessment of the sources of danger. In order to reach this objective and allow the potential impact mentioned in section 2 to be adequately assessed, the utility has developed or incorporated a number of mathematical models for contaminant dispersal in air, water, and soil. It has also developed a model for simulating the hydrodynamic characteristics of a body of water (lake, river, etc.). Some of these models will be described in more detail in section 4.

3.3 Expert system and Artificial Intelligence system

An expert system has been developed to provide a learning tool for those responsible for taking action when accidental spills occur at a substation. In order to simulate the expert's behavior as closely as possible, the intervention process was broken down into four stages:
- characterization of the site and the spill conditions and identification of the possible paths that the contaminants could follow;
- visual reconnoitering of the probable contamination sites;
- prioritization of the potential intervention points;
- determination of the most suitable means of containing and recovering the contaminants at each site.

Another system, which has been developed in collaboration with Carnegie-Melon University concerns impact assessment by automated learning.

3.4 Decision support systems

In order to help managers in their decision-making task in critical situations or to better manage existing facilities and develop environmental standards, a number of administrative units at Hydro-Québec have jointly developed several computerized tools using decision support systems.

These decision support systems are designed specifically for the management of nuclear crises and assessment of the repercussions of technological accidents at certain installations, e.g. explosion of an oil tank, failure of a turbine in operation at a thermal generating station, etc. Another decision support system was developed for assessing the consequences of environmental accidents that could occur at or near hazardous-waste storage sites. To improve the management and location of PCP-treated wood poles, the utility is now in the process of developing another system comprising four mathematical modules specifically for that purpose. The most recent decision support system is for ecological risk assessment. The remainder of this paper will focus on this system.

4 Decision support system

In order to resolve some of the problems identified above, we needed mathematical models that could assess problems different from those we could assess with our existing models. We also needed models that could complete those already available in the shell. Many models are neither user-friendly nor available in a same platform. We therefore decided to integrate four new models in our shell and to transfer our simulation system to an OLE 2.0 Windows-based format.

4.1 Choice of the models

We developed a number of model selection criteria which included:
- Accepted physical equations: the models had to be based on accepted physical equations.
- Input data requirements: a model might be very interesting but, if it requires a lot of complex input data, it will be neither useful nor easy to use.
- Validated/recommended models: as we have to use the models in our studies, we looked for standard models. Discussions with authorities and the scientific community often oriented us to some specific models.
- Number of product classes: the more product classes (organic compounds, metals, etc.) the models can take into account, the more advantageous they are.
- Uncertainty analysis: Monte Carlo simulation or other uncertainty analysis was a primary characteristic sought.

Other criteria such as cost, access to source codes and availability of a designer to modify the model were also considered.

In order to establish a preliminary list, we based our research on different software databases and conducted a bibliographical review. This first step allowed us to identify a list of 76 potential models which we then classified into four groups according to the area of application: atmospheric dispersion, surface water dispersion, soil and groundwater dispersion, multimedia and ecological risk. Then, according to the information we were able to gather on these models and by consulting recognized specialists, we reduced the list from 76 to 27. Since most of our production is hydroelectric, atmospheric dispersion is not a major stake for Hydro-Québec and, in any case, we already had our own model, we decided to focus on the other groups.

Evaluation of the models and many discussions with specialists led us to identify four models of potential use.

4.1.1 Qwasi Model (Mackay[1,9])

The Qwasi (Quantitative Water Air Sediment Interaction) model is typically based on the concept of fugacity, which expresses the tendency of a chemical to escape from one chemical phase to another. This model was developed by Mackay from the University of Trent (Ontario).

The model treats four bulk compartments: air, water, soil and sediment. Each of these is considered to consist of subcompartments of varying fractions of air, water and organic matter. Equilibrium is assumed to exist within, but not between, compartments. The model calculates the steady-state distribution and concentration of chemicals and the persistence in the environment. The accumulation media and dominant pathways of transfer are determined.

4.1.2 FRAC3DVS Model (Therrien[2,5], MacQuarrie[3], Schäfer[4])

This model provides a three-dimensional solution of the flow equations for groundwater and mass transport in porous media. The program was developed by Therrien at the Laval University in Québec city. The solution can be obtained using either the finite-element or the finite-difference approach; both steady-state and transient situations can be simulated. The porous medium can be completely or partially saturated, which means that the zone above the water table can be simulated explicitly.

Three-dimensional simulation of the transport of one or several dissolved substances is also possible. The transport processes considered are: advection, dispersion, adsorption and degradation, represented by a first-order law of kinetics. Specific or diffused sources of pollution, whose intensity can vary over time, can be considered for the simulation.

The use of numerical solution approaches allows heteregeneous porous media to be taken into account. The model has already been used in a probabilistic approach (Monte Carlo) where the hydraulic conductivity of the medium varies from one element to another.

Another feature of this model is that it can simulate the presence of pumping or injection boreholes, observation shafts, surfaces allowing seepage above the water table and recharge by infiltration. The model has already been used and validated by researchers at universities across North America.

4.1.3 SOILFUG Model (Di Guardo[6,7])

This model is a fugacity-based model developed by Di Guardo *et al.* from the Institute for Environmental Studies at the University of Toronto, to describe the unsteady-state behavior following application of agricultural chemicals. It evaluates the importance of different phenomena (degradation, volatilization, runoff/leaching) and the possibility of a build-up of concentrations after prolonged use of chemicals. It is an unsteady-state but equilibrium event model.

The simulation is performed by computing the amount of organic matter present in the soil over a period of time according to slow and fast degrading fractions. The model can take into account different soil types and different rainfall scenarios (steady rainfall or sporadic storms).

Many validation exercices, in the form of field experiments, for instance, have been performed with pesticides such as chlorobenzene, hexachlorobenzene, atrazine, carbofuran, lindane, simazine, etc. in Canada, the United States and the United Kingdom. Most produced conclusive results.

4.1.4 CASM Model (Bartell[8], Suter[10])

The Comprehensive Aquatic Simulation Model (CASM), developed by Bartell from SENES Oak Ridge (USA), simulates daily changes in the biomass of aquatic food web populations. The aquatic food web consists of ten functionally defined populations of phytoplankton, five zooplankton populations, three planktivorous fish populations and a single population of piscivores. The model uses difference equations to simulate daily changes in biomasss of the food web populations. Input requirements include daily values of light, temperature, and nutrients. This model requires good knowledge of the effects of toxic substances in order to estimate changes in biomass that result from exposure to a constant chemical concentration.

The model itself has been subjected to detailed numerical sensitivity analyses. Predicted toxic effects on selected model components have been compared to effects measured for phenolic compounds in experimental ponds.

4.2 Results of the project

The integration of recognized models in the same environment allows us to simulate the behavior of pollutants in different compartments of an ecosystem. As shown in Figure 2, users can build their own set of models.

In the case of aerial application of herbicides in rights-of-way, for instance, the user can identify the area to be sprayed by defining the herbicide characteristics, soil characteristics, etc. which completes the description of the preprocessing. The user can then choose the compartments to be simulated, which corresponds to choosing different solvers. Lastly, the user selects the way the results should be presented. The input data, e.g. preprocessing inputs and results of one model, are then passed from one model to the other automatically.

Oil leakage from a transformer represents another potential use of the system. For this simulation, the user can define the area to be simulated, using the GIS (Geographical Information System) component, and then define preprocessing inputs. The user can choose the problems to be assessed by selecting the different models. In this case, the oil leakage can leach into this soil or migrate at the soil surface toward a small pond located beside the substation. The user should then select FRAC3DVS, Soilfug, Qwasi and Casm.

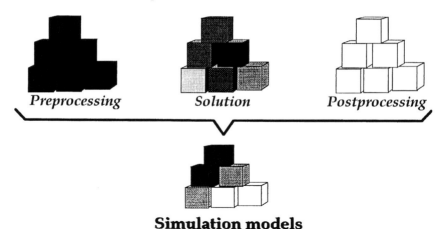

Simulation models

Figure 2 - Composition of simulation by user

The foregoing examples represent just a few of the possibilities for the different models. The next two figures show the preprocessing module for the QWASI model (Figure 3) and an example of postprocessing options for the QWASI-Fish model (Figure 4).

5 Conclusion and future work

As mentioned, Hydro-Québec has to manage facilities and activities that represent environmental risks and has therefore developed software to help specialists and managers in their work. The latest DSS represents a new approach, since it allows the user to choose a specific problem and then build up the expected application in order to assess the fate and transport of a pollutant in different compartments of the environment.

The system described represents phase 1 of the project. In the near future, Phase 2 of the DSS will begin. This consists in integrating at least one other model and adapting the previously integrated ones to Hydro-Québec's specific problems and needs, in collaboration with model designers. Interfaces with databases on specific products, fish species native to Québec and also on Hydro-Québec's installations, to name but a few, should be developed shortly. This kind of modifications can be done relatively easily and at reasonable cost with the architecture already developed.

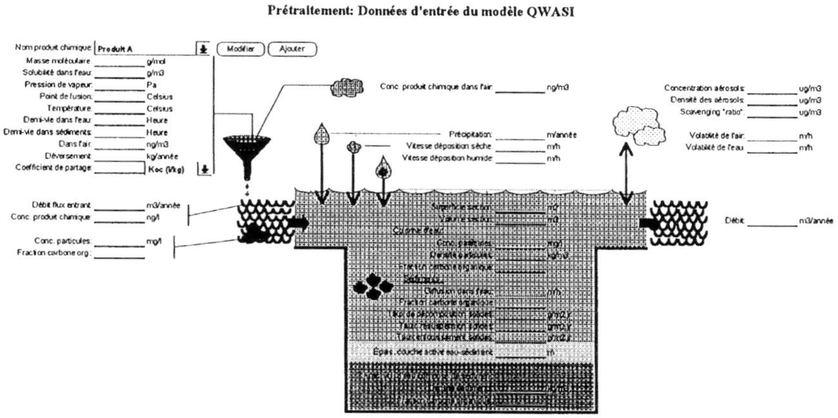

Prétraitement: Données d'entrée du modèle QWASI

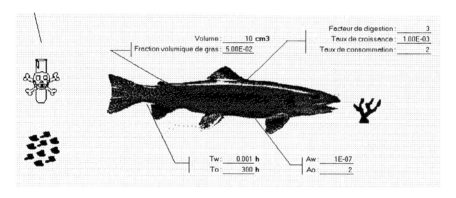

Figure 3 - QWASI model (preprocessing)

Figure 4 - QWASI-Fish model (postprocessing)

6 References

1. Mackay, D., *Multimedia Environmental Models - The fugacity approach*, Lewis publishers, Michigan, 1991.

2. Therrien, R., Sudicky, E.A., Three-Dimensional Analysis of Variably-Saturated Flow and Solute Transport in Discretely-Fractured Porous Media, *Journal of Contaminant Hydrology*, May 1995.

3. MacQuarrie, K.T.B., Sudicky, E.A., On the Incorporation of Drains into Three-Dimensional Variably-Saturated Groundwater Flow Models, *Water Resources Research*, May 1995.

592 Computer Techniques in Environmental Studies

4. Schäfer, W., Therrien, R., Simulating transport and removal of xylene during remediation of a sandy aquifer, *Journal of Contaminant Hydrology*, February 1995.

5. Therrien, R., Sudicky, E.A., McLaren, R.G., User's Guide for FRAC3DVS : An Efficient Simulator for Three-dimensional, Saturated-Unsaturated Groundwater Flow and Chain-Decay Solute Transport in Porous or Discretely-fractured Porous Formations, September 1994.

6. Di Guardo, A., Mackay, D., Cowan, C.E., Calamari, D., A Fugacity Model of Chemical Fate in Soil: Application to Amendment Associated Chemicals, *Poster presented at SETAC - 15th Annual Meeting*, Denver, Colorado, U.S.A., October 1994.

7. Di Guardo, A., Williams, R.J., Matthiessen, P., Brooke, D.N., Calamari, D., Simulation of Pesticide Runoff at Rosemaund Farm (UK) Using the Soilfug Model, *Environmental Science and Pollution Resources* , 1994, 1 (3) 151-160.

8. Bartell, S.M., Gardner, R.H., O'Neill, R.V., *Ecological Risk Estimation*, Lewis Publishers, Michigan, 1992.

9. Mackay, D., Paterson, S., Mathematical models of transport and fate, Chapter 5, *Ecological risk assessment*, ed. Glenn W. Suter II, pp 129-152, Lewis Publishers, Michigan, 1993.

10. Suter, G., Bartell, S., Ecosystem-Level Effects, Chapter 9, *Ecological risk assessment*, ed. Glenn W. Suter II, pp 275-308, Lewis Publishers, Michigan, 1993.

Universal information models for environmental management

B.N. Rossiter,[a] M. Heather[b]

[a]*Computing Science, Newcastle University, Newcastle NE1 7RU, UK*

[b]*Sutherland Building, University of Northumbria, Newcastle NE1 8ST, UK*

Abstract

Software to manage the environment not only needs the usual structural requirements of the standard database systems but also very high–level global consistency. Database management systems to implement any schema based on the familiar network, hierarchical or relational models have rather too many limitations for complex heterogeneous data involving environmental biodiversity. Newer categorical models offer more promise because they are constructive, integrative and based on naturalness and universals. The constructions in category theory are very similar to those of the new theoretical models based on natural systems, enabling limits and chaos to be modelled. Work is continuing at Newcastle on developing a generalized categorical model which can be adapted to handle a variety of complex ecosystems.

1   Database Management Systems

There are a number of database models available for environmental software such as the hierarchical, network and relational. The network model is based on directed graph theory and the relational on functions and relations. The hierarchical is a special case of the network model with a restriction on the linking mechanism.

All these models are aimed at real–world data where there will be a small number of classes of data, each with many instances. However in practice there may not be very many exhibits in any one category. For the evidence in the example of an investigation or trial of a chemical disaster would be quite heterogeneous. Data might consist of items like shipping, aircraft, personnel, cargo, itineraries, charts, legal documents, technical checks, safety records, telephone logs, bank records, photographs, satellite readings, forged papers and so on. Each probably has too few examples to make use of the power of relational query languages.

Furthermore these basic models do not provide enough control to represent real–world data. Much of real–world data has a very complex structure as, for example, in text and graphical data. For these, it is necessary to develop a more generalized model often termed semantic models in contrast to the standard syntactic models. Some useful ideas for handling this kind of data is available from an object-oriented approach which offers the user a number of powerful abstractions to assist in rationalizing complex situations such as the ability to construct generalizations and compositions.

However, this approach at present lacks a formal model. This raises uncertainties about consistency and therefore reliability. The object–oriented paradigm must accordingly have a cloud of doubt hanging over its use for environmental software until the day it can be converted into some formal model.

1.1   Needs of Environmental Data

We need to examine the structural requirements and semantic factors of ecological systems in the context of generalized database management systems. Some needs are satisfied by standard classical database features like views, integrity and security, display formats, temporal management and retrieval languages, but which may still need to be greatly enhanced and developed in the ecological context.

Controlling ecological features requires the availability of timely accurate information on as many variables as possible. What is needed is a portfolio of software tools for monitoring the planet in real time both for long term legal procedures as well as to provide high quality immediate decision making for action to be taken by emergency forces of international environmental agencies. The requirement is to integrate information from the different sources in such a way that deductions and inferences can be made to give a true overall picture of objects such as the forest at any time, including future projections.

Some data is short term, some long term. The methodology of the data capture needs to be invariant of the times and conditions and of particular technological developments; however the methods themselves may change. The term dynabase has been used [1] for those data systems where all the information recorded or generated over a long period of time is to be seamlessly integrated. There is a major technical challenge to capture the large amount of knowledge in each domain and the dependencies between domains. Since data is likely to be held on different systems, the question of interoperability arises between different databases, where each has its own data model and inferencing mechanisms.

It will also be necessary in environmental information systems of the future to be able to model limits. This concept concerns a series of boundaries within which periodic transitions occur. Some boundaries will be very local, others more global, indicating the extent to which an entity may change its status or behaviour. If a system remains within such limits, environmental changes will be manageable by gradual adaption [2] and our system is relatively stable. However, environmental disasters are typified by a lack of stability caused by systems going off–limits into chaos [3]. In effect, therefore we require an information model that can handle both periodic oscillation and chaos.

1.2   Formal Database Methods

The universal nature of the problems involving the environment means that the information methods need to be universal. Universal means formal but we are searching for formal methods appropriate to environmental data rather than the simpler data constructions often illustrated in formal specification languages such as Z and VDM. These two languages are based on mathematical logic applied to sets as in functional programming.

However, a mathematics which is more expressive of the complex data types [4] and limits is necessary for environmental software. In theoretical computer science, the new subject of category theory [5,6] seems the best current method to work with. Instead of the set being thought of as the basic building block, category theory is founded on the morphism usually expressed by the concept of an arrow. Category theory also has a well-established notion of limit so that stability mechanisms can be investigated and conditions for the onset of chaos [7] predicted.

The structure of the rest of this paper is first to introduce the concepts of category theory with explanations of category, functor, natural transformation and adjointness; then to discuss the architecture for an integrative approach; and finally to describe two examples to illustrate the modelling power of the categorical approach in its own right on specific environmental problems, followed by a discussion of the results.

2   Categorical Models

Category theory is based not on the set as a fundamental but on the concept of a morphism, generally thought of as an arrow and represented by $\longrightarrow$ [5]. The morphism can be regarded as an imperative arrow or as a relationship in computing. The arrow represents any dynamic operation or static condition and can cope therefore with descriptive/ prescriptive equivalent views.

The arrow is an effective representation of real–world phenomena. $A \longrightarrow B$ can represent an action from a state A to a state B, an interaction of A with B, for example a product of A with B, or a type change from type A to type B. $A \longrightarrow B$ may be a descriptive action or a prescriptive one. Alternatively it may be a probabilistic relationship. There may be any number of different arrows between the same objects.

The arrow can represent a more general relationship than the set-theoretic function. Much of scientific modelling is taken up with handling general relationships which exist between real world data. For example where there is more than one polluter for a given pollution, *polluter* is not a function of *pollutant*. The arrow can relate objects that are not sets like for instance bags and lists.

These simple categories are promising for modelling global real–world dynamic events as category theory is based on principles of naturalness that all the time relate universals with particulars [8,9]. Physical processes (interpreted widely to include chemistry and biology) are those that exist because they can be constructed. They are likewise universal in the sense that the laws of physics are invariant in all reference frames.

2.1   Functors and Transformations

An arrow between categories is termed a functor as shown in Figure 1. A functor provides the facility for transforming from one type of structure defined by one category to another type of structure defined by another category.

Functors are structure–preserving morphisms from one category to another. In Figure 1, the functor K assigns from each source object A, in category **A**, a target object K(A) to the object C in category **C**, and from each source arrow f, in category **A**, a target arrow K(f) to the arrow g in category **C**. Note that categories are given names in bold capitals. Functors really map structures. They carry across the high–level relationship as well as dealing fully in an integrated fashion with any lower–level relationship which needs to be constituted consistently within any higher–level mapping. This functorial character preserves the detailed information within transforming structures. In many models, the user has to flatten the structure when operating across different levels. Stochastic models are an example of this phenomenon where a structure may be collapsed into some statistical parameter.

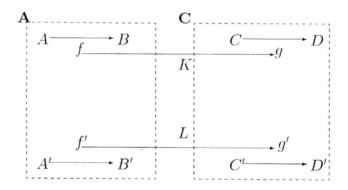

Figure 1: Functors compare Categories

A geophysical example could be the generation of earthquakes in fault regions. The aggregate activity is composed of large area faults in dynamic stability with smaller regional ones. A large earthquake could be initiated by local activity, or alternatively by large movements, or indeed by the interaction between small movements at both levels which in themselves would not be sufficient to cause a significant activity. Constructing functors to map from current inter–level geophysical structures to potential target geophysical structures is the first step in producing a model for investigating the limits and stability of target states as described later under the concept of adjointness.

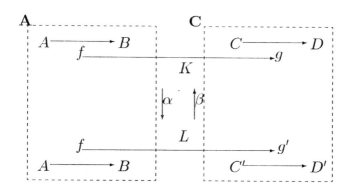

Figure 2: Natural Transformations compare Functors

An arrow between functors is termed a natural morphism (or transformation) as shown in Figure 2 where there is a natural transformation α from K to L, written $\alpha : K \longrightarrow L$

Natural transformations correspond to metalevel control mechanisms such as policy, political principles and ideals, regulators, audit programs, the principles in documents like the EMAS (Eco–Management and Audit Scheme) and the ISO standards (e.g. the British BS7750).

2.2   Adjointness

In dealing with the environment and biodiversity, we need to deal in universal properties. A very important universal concept which has emerged in the second half of the twentieth century is the principal of adjointness which can be expressed naturally in categorical models. Adjointness expresses an essential concept for formally handling complex environmental processes. Trying to integrate across levels is difficult with set theory and the more natural approach of category theory to express adjunctions is sufficient justification, in our view, for resorting to this new branch of mathematics.

Adjointness is particularly relevant to the environment for it relates to relative ordering which is the basis of natural balances and of the way that equilibria are controlled. Ormerod [10] shows how the order in nature is the result of very complex interactions, is very easily destabilized and need not be harmonious at all. By induction, he considers the same principles should apply to economics. Such apparent complexity can be modelled naturally by adjointness.

Adjointness between two categories **A** and **B**:

$$F \dashv U : \mathbf{A} \longrightarrow \mathbf{B}$$

has left and right components which specify how an arrow in category **A** is related to an arrow in category **B**. The left component is the free functor $F : \mathbf{A} \longrightarrow \mathbf{B}$ and the right component the underlying functor $U : \mathbf{B} \longrightarrow \mathbf{A}$. F is left adjoint to U and U is right adjoint to F; F may preserve colimits (sums) and U may preserve limits (products). There is a natural bijection between arrows which holds subject to the condition for all objects $A \in \mathbf{A}$ and all $B \in \mathbf{B}$ such that:

$$F(A) \longrightarrow B \text{ implies and is implied by } A \longrightarrow U(B)$$

F is a generalization of natural growth processes and evolution in an open–ended environment while U is the underlying genetic codes, laws of continuity, conservation, chemistry, thermodynamics, etc. With this condition there are two natural transformations or unit of adjunction:

$$\eta : 1_{\mathbf{A}} \longrightarrow UF, \quad \epsilon : FU \longrightarrow 1_{\mathbf{B}}$$

Adjunctions are universal descriptors for any kind of correspondence between systems whether in space or time, for example: thermodynamics stability, chemical equilibrium, biodiversity, radioactivity, and body temperature regulation.

3   Databases in Category Theory

The extra 'dynabase' advantage of the arrow in geometric logic can be seen by examining our preliminary database architecture, shown in Figure 3, with a dynamic categorical representation for use with real–world data. Each box represents a category and each arrow between categories is a functor. There is a pair of adjoint functors between each category. For ease of representation, only a functor in one direction is named.

This model provides the ability to integrate diverse models in a dynabase fashion. In effect it provides the 'glue' for linking together seamlessly the various models whether they are semantic (e.g. object–oriented, extended entity–relationship, Taxis, SDM, functional, etc), syntactic (e.g. relational, network, hierarchical) or simply stored files. The 'intelligence' in the integration comes from the functorial mappings MMt and MMt' between the categorical definition and the other models. The categorical definition then acts at the meta-meta level relating concepts and values in one model to those in another. The role of the categorical model can be compared to that of the ISO–standard Reference Model [11] which provides a standard reference point against which other models can be compared.

From the environmental data viewpoint, all components of the architecture can be heterogeneous and of arbitrary complexity to represent any data available on the biosphere. However, we envisage that a categorical system can do much more than just act as 'glue' between other models. Some components of the environmental model could be modelled and implemented directly in a categorical system, a prototype of which has been produced at Newcastle [12], as an extension of a functional database model.

We conclude with two examples which show what individual components of the model might look like with categorical modelling. Each would be one component of the semantic modelling level in Figure 3. The first example looks at high–level balances in biology in geophysics, biology and humans; the second at a more detailed example of balances in carbon dioxide and nitrogen between vegetation, atmosphere and soil,

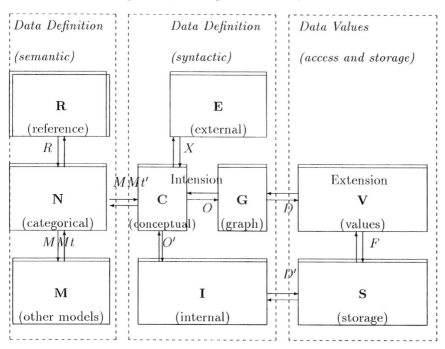

Figure 3: Integrative 'Dynabase' Architecture in Functorial Adjunction Terms

4   Global Categories

The environment is a very fine example of the adjunction between global categories. The environment is an arrow. The biosphere consists of a balance or equilibrium between the three categories shown below in Figure 4(a).

The adjointness between **B** and **H** represents the balance between the

category **B** of biological species and their interrelationships and the category **H** of human activity and policy. The human category includes the axiom of choice in that we can choose to some extent the course of circumstances. The adjointness between **B** and **G** represents the balance between the category of biological species and their interrelationships **B** and the category of geophysics **G** including climate, minerals and weather and their interrelationships. This adjunction can represent the impact of climate change on food supplies [13]. The adjointness between **G** and **H** represents the balance between the category of geophysics **G** and the category of human activity and policy **H**. This adjunction can represent assessments of hazards by extra terrestrial bodies [14].

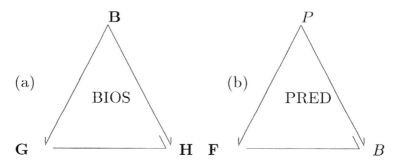

Figure 4: Global Balances between (a) Geophysics (**G**), Biology (**B**) and Humans (**H**); (b) Birds of Prey (P), Other Birds (B) and Food in General (**F**)

These adjunctions are shown in the Figure 4. The whole of this figure can be viewed as a category of categories **BIOS** representing the biosphere. This type of construction is termed a topos in category theory.

The categories **B, G** and **H** have their own complex internal structure, involving a number of local balances in addition to the top–level ones in Figure 4(a). For example, category **B** could be defined as a collection of adjoint triangles, each the shape of Figure 4 and representing a particular balance in nature. For instance, consider the relation between birds of prey (represented by the object P), birds in general (the object B) and food in general (the category **F**, as shown below in Figure 4(b). Note that this is a fractal of the more general diagram $BIOS$.

5   Categorical Models for Natural Systems

Another example of natural equilibrium, to augment those of Figure 4, is shown below in conventional terms in Figure 5(a) as reported by [15]. This is a more detailed example looking at balances in carbon dioxide and nitrogen between vegetation, atmosphere and soil,

Our equivalent categorical model is a direct representation of the balances with categories for local systems, functors relating local systems either by structure preserving or by structure transformation, and adjoints maintaining an equilibrium between two categories connected by functors in both directions. This model is shown in Figure 5(b). Note the adjointness between the atmospheric carbon dioxide category $\mathbf{A}$ (for ACD) and the vegetation category $\mathbf{V}$ (for VEG). Here we see represented two separate equilibria: $G : \mathbf{A} \longrightarrow \mathbf{V}$ and $R : \mathbf{V} \longrightarrow \mathbf{A}$.

The first is a free functor G (for GPP) assigning objects in the category $\mathbf{A}$ to objects $C_V \in \mathbf{V}$. The second is a forgetful functor R (for R_A) assigning part of the structure in $\mathbf{V}$, that is C_V, to the category $\mathbf{A}$. There is another intra–category adjointness between objects N_{VS} and N_{VL} in $\mathbf{V}$ where N_{VS} is nitrogen in vegetation in the structural pool and N_{VL} in the labile pool.

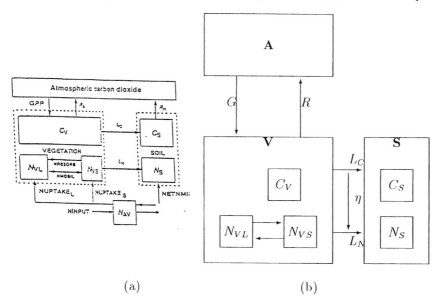

(a)   (b)

Figure 5: Global Balances for Carbon Dioxide and Nitrogen in
(a) conventional terms, (b) Categorical Terms

If the adjointness $G \dashv R : \mathbf{A} \longrightarrow \mathbf{B}$ holds, in the sense that local limits and colimits are preserved by the functors, there is a balance in nature. If it does not hold, examination of the morphisms involved may establish stability within wider limits enabling the system to be viewed as an oscillating dynamism within broad boundaries. Total failure to find any global limits within the model would mean that the system was in a chaotic region. There are then a very large number of possibilities as regards transitions in the system: optimal handling of this problem is part of current work at Newcastle University on handling uncertainty in information systems.

We also see two functors from category **V** to the category **S** (for SOIL) called L_C and L_N respectively. A natural transformation between these functors η compares the mapping from source to target of the two functors and enables us to derive equations which must hold if stability is to be achieved. If the natural transformation holds, there is a balance in nature.

6   Concluding Remarks

The constructions in category theory are very similar to those of the new theoretical models based on the natural system approach. Category theory, with its rigorous mathematical basis, therefore offers a promising route as a basis for underpinning multi–level eco–models and as a tool for constructing universal diagrams representing the various balances and adaptions. Work is continuing at Newcastle on developing a generalized categorical model which can be adapted to handle a variety of complex ecosystems including chaotic ones.

References

1. L.Press, Personal Computing: Emerging Dynabase Tools, *CACM* **37**(3) 11–19, 1984.

2. R.M.Sibly & R.H.Smith, Behavioural Ecology : ecological consequences of adaptive behaviour, 25th Symp British Ecological Society, Oxford, 1985 (620pp).

3. F.Cramer, *Chaos and Order : the complex structure of living systems*, Weinheim, New York, 1993 (249pp).

4. B.N.Rossiter & M.A.Heather, *Database Architecture and Functional Dependencies Expressed with Formal Categories and Functors*, Comp. Sci. Tech. Rep. no.432, University of Newcastle (30pp), 1993.

5. M.Barr & C.Wells, *Category Theory for Computing Science* , Prentice–Hall, 1990.

6. P.J.Freyd & A.Scedrov, *Categories, Allegories* , North Holland Mathematical Library **39**, 1990.

7. P.Berge, Y.Pomeau & C.Vidal, *Order within chaos : towards a deterministic approach to turbulence*, Wiley, New York, 1986.

8. M.A.Heather & B.N.Rossiter, An Interoperable Theory of Higher Level Logic and Integrable Law for the Environment, in: *Towards a Global Expert Systems in Law*, Florence 1993, Bargellini, G, & Binazzi, S, (edd.), Cedam, Padua **II** 1995.

9. M.A.Heather & B.N.Rossiter, Global Theories for Global Problems, *IV International Conference towards the World Governing of the Environment* , Venice, Italy, June 1994.

10. P.Ormerod, *The Death of Economics*, Faber, London, 1994.

11. Information technology – *Reference Model of Data Management*, Standard ISO/IEC 10032 (1993).

12. Nelson, D A, & Rossiter, B N, Prototyping a Categorical Database in P/FDM, in: *ACM Workshop ADBIS*, Moscow, June 1995, **I** 247–258 (1995).

13. C.Rosenzweig & M.L.Parry, Potential Impact of Climate Change on World Food Supply, *Nature* **367**(6459) 133–138, 1994.

14. C.R.Chapman & D.Morrison, Impacts on the Earth by asteroids ans comets: assessing the hazard, *Nature* **367**(6458) 33–40, 1994.

15. J.M.Mclillo, A.D.McGuire, D.W.Kicklighter, B.Moore III, C.J.Vorosmarty & A.L.Schloss, Global Climate Change and Terrestrial Net Primary Production, *Nature* **363**(6426) 234–240, 1994.

Experiences building a pure object-oriented modelling system architecture

O. David

Department of Geography/Geoinformatics, Friedrich-Schiller-University Jena, Loebdergraben 32, D-07743 Jena, Germany

Abstract

Modelling and simulation of complex environmental problems becomes more interesting by using modern software construction techniques for building large scale models. A closely problem-oriented and transparent software design and it's efficient implementation seems to be orthogonal in the past. The *pure object-oriented approach* in modern software construction leads to a very homogenous system view for developing software in general. In modelling and simulation object-oriented system view becomes more and more interesting because of problem-oriented representation of enviromental research topics. The paper presents the design of the *Object Kernel* as a part of the *Object Modelling System* developed at the Department as an integrated system for building modells on the top of existing library components using. Emphasis is given on the design of the modelling system architecture.

1 Introduction

Computer simulation and modelling becomes more and more interesting for problems in environmental science since more computing power has migrate from the computing centre to the desktop and comfortable software tools are available for using this power most effecent.

In reality there is a lot of software code available for simulation in environmental science. But most of them are very closed applications written for special purposes with special data in- and output, graphical representation, etc. Because of handling complex environmental problems in simulations the software architecture was conveniently optimized for special operating systems and hardware features like for example multithreading, vector processing and pipelining. Therefore programming languages like FORTRAN are prefered because they are suilable for intensive and efficient numerical computations with no main relevancy to software engineering and design aspects. There are two main separate groups in closed modelling systems, *model developer* are designing and implementing the whole modell and the *modell user* are applying the modells. A lot of functions and libraries based on individual work of environmental scientists where written over the years. There are for example a lot of closed models in geohydrological solute transport modelling on PC and UNIX platforms. To control all information about models and simulation software like system requirements, interfaces, parameters, etc. *meta information systems* are used to manage the correct usage of models.

To avoid individual solution for modelling software, software (modell) development systems where developed like for example the *Modular Modelling System MMS* [13]. They are used as a tool set by a modeller to build a modell based on a set of modules implementing modelling algorithms, data import and export, graphical representation. With this knowledge the modelling process can now be subdivided into three main classes:

- *Modell application* A model user is applying an existing executable modell for simulation. He is only varying parameters to observe the behaviour of the modell in his work. That's the way how to apply an executable modell in environmental analysis. The user has to know about the parameter for fitting the modell. He knows mathematical basics but noting about programm implementation.

- *Modul combination* A model user/developer is building a new modell based on a set of existing modules. He is replacing one module with another implementing a different computing approach. This work can merged into the modell application. Not only parameters has to be varied but algorithms too. The developer/user knows about modell parameters and behaviour and interfaces of modules.

- *Modul development* A developer is implementing modules which can be used for Modul combination. Thats the way how to transfer human knowledge representing in mathematical methods into algorithms. This work often is relatively independ from real application of modells. Than comprehensive the available module library than more powerful the the system benefit. A developer detailed knows a programming language (C,C++,FORTRAN,..)

Using a modelling tool supporting these three classes of module development, the modelling application is represented not only by a strict separation of modell development and application. The modell application is characterized by the variation of *Parameters/Data* and *Algorithms*. The modelling application is now becoming a character as a kind of *programming*. Modell application and a following new modell combination can be performed alternating. The development and application process is merging. But this work is only possible by using a well supporting tool in this process. It has to give a non programmer the power to combine existing modules in a right order to fit the modelling scheme without any knowledge about programming paradigms, data structures and formats and so on.

This paper will point out how can a architecture for a modelling system based on the pure object-oriented approach be constructed. This architecture is used within the *Object Modelling System.* the modelling engine of the Object Modelling System. Section 2 gives an overview about our system requirements and object definitions in modelling, section 3 will introduce the design and implementation of the *Object Kernel* and section 4 will show the embedding of the kernel in a modelling system.

2 Some Modelling Development System Design Aspects

The development of the Object Modelling System was derived from a project called "Object-Shell" [5, 4]. The goal of this project was to provide an object-oriented user interface on the top of the UNIX operating system to enable an object-oriented interaction with UNIX. Such a tool is very usable for prototyping-oriented software development in general [2] whereas user want testing only aspects or a particular behaviour of software prototypes. These requirements in general software develpment are identical to the modelling process. Modeller sometimes wants to build an executable modell only for some application cases. Thats a typical procedure in scientific computing.

There where defined some design requirements in building such a system shell.

- The system should only satisfy the pure object-oriented approach in system design. Operations allowed are the *object creation and deletion* and a *type save interaction* of object via messages.

- The prototyping (modelling) system should be work on the top of C++ libraries. Because this language is becomming more and more popular, a lot of code is available. C++ allows to integrate C and FORTRAN routines as well.

- To achieve best performance balance in program (model) generation and execution, pure interpretative or compiling working methods are rejected because of many disadvantages in prototyping/modelling [5].

An incremental compiling and linking method should be applied saving both (i) short program preparation time for building an executable and (ii) a time efficient run time to handle complex simulation.

- Platform independence should be reached as far as possible to support a range of different architectures. RISC architectures are prefered because of relatively simple machine code structure. This feature is important for the efficient mapping of object-oriented commands into an executable without any system compiler or linker.

- Operating system support should be given to systems supporting a common file format for executables. This is complied by using the Executable and Linkable Format (ELF) [9] of some SVR4 UNIX derivates. Solaris, Irix and newer versions of Linux are supporting this format. The ELF defines a common file structure for executables under different operatring systems and hardware architectures.

- All software for developing such a system should be based on public domain products. Therefore the GNU tools are used for software developing exclusively.

Matching all these topics at the same time available prototyping oriented development platforms are not suitable enought for such a system. They only satifying particular aspects as noted above. A good orientation for developing such a system is given by the operating system research. *Object-oriented Microkernel Operating Systems* seems to be qualified for an interactive programming/modelling. They deal with some interesting architecture characteristics like for example a non complex kernel responsible for object management, migration and message passing and some additional but limited features like inter-process communication. In order to this architecture paradigm a kernel for prototyping/modelling was designed to fulfit a real application in modelling.

Next section will point out the design and implemantation of the *Object Kernel*.

3   The *Object Kernel*

Kernel architectures in computing science reserach in general are characterizes by a basic and limited functionality to realize a well defined amount of tasks. This design is supporting the maintenace, flexibity, and robustness of the sourounding software environments due to strong separated resposibilities. Kernels are dealing with features like ressource management, communication, strong hardware related operations. In operating system (research) such examples are for example the *MACH* kernel [1] the architecture of the *iAPX 432 System* [10] and the *Spring Operating System* [8]

Hence it is recommended to model the architecture of the *Object Modeling System (OMS)* closely to object-oriented (operating) systems. The message controled interaction of object-oriented operating system components and the direct support of object-oriented user processes are main properties of such systems as well [12, 3].

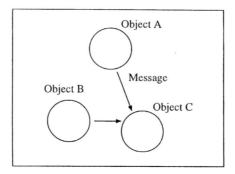

Figure 1: Object Interaction

Kernels in terms of object-orientation are only responsible for *object creation* and *communication*. They are realizing a full object management which consist of:

1. Ressource management for object spaces and

2. controlling the type save message interaction between objects with message identifiers and parameters (Figure 1).

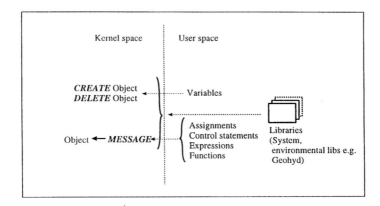

Figure 2: kernel-approach

If such an approach wants to be applied for a prototyping/modelling system a C++ system library is required to implement programming basics like *assignments, control statements, expressions, etc.* The implemented *Kernel System Library* provides ca. 20 C++ classes in a smalltalk like order for realizing basic programming features. Now an unique system view of all nessesary elements of a programm/modell can be applied in object-oriented terms. It doesn't matter how complex are algoritms inside a message implementation of an object. The object kernel is only handling the interface of the message to realize the C++ method call (C++ terminology). The principle is illustrated in Figure 2.

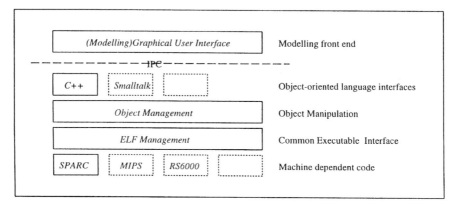

Figure 3: The Object Kernel Layered Architecture

The implemented *Object Kernel* structure is shown in Figure 3. This is a layered architecture whereas the different boxes are partial implemented as C++ classes. There are well defined interfaces for using features of the layers but only the nearest underlaying class can be accessed. The direct access from for instance the object management to the hardware related layers on the bottom of the Figure is prohibited. The layer responsibilities can be notet as follows:

Language Interface Layer This Layer implements the language interface of the kernel. It defines the syntax of the accepted object-oriented language. Currently there is a C++ style syntax available with some enhancements to manage instruction objects a la smalltalk. The definitions for lexical and syntax analyses are implemented using the GNU *flex* and *bison* tool. The Language Interface Layer is transfering statements of a concrete incomming (modelling) source script into general calls to the underlaying object management layer. This layer can only perform syntax checks of a source script.

Object Management Layer It is realizing the object management in terms of storing meta information about all related (i) *static information* of C++ classes like inheritance relations, class size, message (method) list, parameter lists of messages, etc. and (ii) *dynamic information* like objects and their relations. The layer is storing all these information into several hashtables. It can verify the semantic correctness of a command.

Elf Management Layer This layer/class is managing the analysis and manipulation of an executable. There are routines for extracting headers, sections (e.g. text, data, rodata) and additional informations like symbol or string tables from an existing executable. Using this layer a fast access to particular region inside the executable for object management is possible. These routines are applicable for all ELF supporting operating systems.

OPCode Layer The OPCode Layer implements the processor related parts of an executable. This means real opcode instruction for some basics like (i) *parameter* pushing into stack or registers and (ii) *calls* to a library function inside an executable. These instructions are usualy placed inside the text-section of an executable to enable a processor-specific execution. The number of needed opcodes for object management is only about 14 on Sparc architecture example.

The layers are direct corresponding to C++ classes in structure. and is reflected by the class and module dependencies. The OPCode Layer has an additional so called abstract interface class, which deals with interface methods to all derived classes. The amount to port the software to other processor architectures like MIPS or RS6000 is limited by adding a new C++ class which realizes the Opcode. If the operating system is not supporting the ELF Format, the ELF Layer has to be replaced by a COFF (Common Object File Format) for instance. Then the porting amount is increasing.

The total size of source code nessesary to build the *Object Kernel* for the SPARC/Solaris architecture with C++ interface is only about 4500 lines of C++ Code, the executable *Object Kernel* takes only 70 KB. The complexity of the system is minimal. Maintenance and error localization becomes easy. In comparison to known modell development environments, the programm (modell) preparation time using our incremental compiling and linking scheme takes up to 1000 times faster than conventional methods.

4   Embedding the Object Kernel into a Modelling System Architecture

The *Object Kernel* is currently representing only one element in modelling system development. Beside the *Object Kernel* there are components like *Databases DB, Decision Support Systems DSS* or *Graphical User Interfaces*

GUI (Figure 4). The GUI component in particular is separated from the modelling kernel. Graphical user interface and the *Object Kernel* are communicating using an Inter Process Communication IPC facility (Figure 3). Currently we are using a TCP socket communication and we defined a protocoll for information exchange.

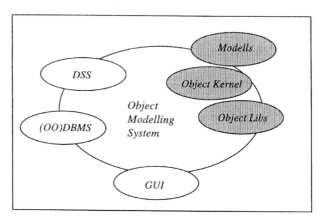

Figure 4: Interacting Subcomponents in OMS Development

Using the GUI the modeller is constructing his software modell with icons, pictograms, etc. He is designing the interaction of objects using a graphical editor. All static informations are provided by the kernel.

The GUI is generating a script corresponding to the *Object Kernel* language syntax (e.g. C++). The information exchange protocol transmits the source to the kernel, which is permanently running on a *modelling server*. The kernel is listening on defined ports and is expecting modelling commands. Inside the protocol there are commands for executable generation, execution, tracing, etc. The generation and execution of a model is located on the modelling server. Results are transmitted to the GUI.

This architecture of a pool of loosely coupled tools leads to more flexibility in modelling. Both the GUI and the kernel can run on the same machine but it is not nessesary. Also PC's can work as an modelling front end running MS-Windows. They tcp network support is the only precondition to integrate it. As GUI we are prefering public domain software products like the Tk Toolkit [11], which is available on many platforms.

5   Conclusion and Outlook

Because of the very suitable application of object-oriented techniques for software design and implementation also in environmental domains, this approach in software construction is used for building a modelling system.

A closely design of a modelling system related to microkernel based object-oriented operating systems was realized. Such a system is characterized by a maximum support of interactivity due to these simple object operations and a low complex structure.

Building such development systems in object-oriented structure too, the whole modelling system architecture becomes much easier because of the integrative system architecture for modelling system, modelling libraries and models. Every simulation within the system is viewed as an interaction between objects. Objects can be all resources in problem description and solution.

The previous work on the *Object Kernel* shows the suitability of the pure object-oriented approach building a system for experimental software development in general and modelling software in special. Due to the homogenous system view in library design and modelling system development architecture the complexity of the modelling system decreases to a minimum. Maintenace, realbility, robustness and type security are some feautures caused by the pure object-oriented approach and micro-kernel implementation.

The flexibility of the system application is determined by an direct support of several Operating Systems using the SVR4 Executable and Linkable Format and RISC Architectures. These configurations seems to be the most popular HW/SW architectures in scientific applications because they are scalable from desktop to supercomputers in one architecture, cost efficient and universal usable. The usage of a distibuted modelling architecture based on the UNIX IPC facility and the separation into (i) a graphical modelling front end for interactive modell construction and (ii) a modelling enginge (*Object Kernel*) for the modell generation and execution gives more interaction power to the PC/windows and moves the modell execution onto the computing-oriented RISC/UNIX world.

Our future work is mainly focused on the integration of more HW platforms like the new Pentium Architecture and the additional *Object Kernel* support of Windows NT. On the same time we are also active in adopting existing environmental software to the object modelling system. We started with different libraries in groundwater and solute transport modelling in conjunction with the approach of Hydrological Response Units (HRU) [7] in geohydrolocical river basin modelling.

References

[1] M. J. Accetta, R. V. Baron, W. Bolosky, and D. B. Golub. Mach: A New Kernel Foundation for UNIX Development. In *Summer Conference Proceedings*, pages 93–112. USENIX-Association, June 1986.

[2] W. R. Bischofberger and G. Pomberger. *Prototyping-Oriented Software Development, Concepts and Tools.* Springer-Verlag, 1992.

[3] R. H. Campbell. Principles of Object-Oriented Operating System Design. Technical Report UIUCDCS-R-89-1510, University of Illinois at Urbana-Champaign, April 1989.

[4] O. David. Entwurf und Implementation einer objektorientierten SHELL. Internationales Wissenschaftliches Kolloquium, September 1994.

[5] O. David. *Entwurf und Implementation einer objektorientierten UNIX-Shell, ein prototyping-orientierter Ansatz.* PhD thesis, Technische University Ilmenau, February 1995.

[6] O. David. Interactive Environmental Modeling using an Object-Shell. In *Proc. International Congress on Modelling and Simulation MODSIM 95*, volume 4, pages 185–190. MSSA, MSSA, November 1995.

[7] W.-A. Fluegel. Delineating Hydrological Response Units (HRU's) by GIS analysis for regional hydrological modelling using PRMS/MMS in the drainage basin of the River Broel, Germany. *Hydrological processes*, 9:423–436, 1995.

[8] G. Hamilton and P. Kougiouris. The Spring Nucleus: A Microkernel for Objects. In *Proc. 1993 Summer USENIX Conference*, pages 147–160, June 1993.

[9] Intel Corp. *Excutable and Linkable Format (ELF), Portable Formats Spec.*

[10] Intel Corp. *System 432/600 System Reference Manual*, 1981.

[11] J. K. Ousterhout. *An Introduction to Tcl and Tk.* Addison Wesley, 1992.

[12] V. F. Russo. *An Object-Oriented Operating System.* PhD thesis, University of Illinois at Urbana-Champaign, 1991.

[13] University Colorado at Boulder. *Modular Modeling System Users's Manual*, January 1993.

The atmospheric pollution monitoring virtual network of Tarragon (Spain): virtual reality applied to environmental education and science diffusion

P.J. Celma, I.W. Chiarabini, R. Pérez

Ecotechnology Division, Engineering Department, CETS Institut Químic de Sarrià, Ramon Lull University, Via Augusta 390, 08017 Barcelona, Spain

Abstract

The purpose of this paper is the introductory presentation of the Atmospheric Pollution Monitoring Virtual Network of Tarragon, a set of Virtual Reality worlds for low cost PC computer based systems. The software developed detailedly reproduces the physical pollution monitoring network implanted in the industrialized area of Tarragon, Spain. This automatic network collects on-line pollution data and meteorological information that are processed by the SODCAT system, a comprehensive air quality surveillance model that carries out on-line 6h prognostic simulations. The educational aspects of this system are supported by the presented Virtual Reality based software, intended to present and diffuse the efforts of the Autonomic Government in air quality protection. This work present the current situation and introduces the expected final results.

1. Introduction

The SODCAT system is a comprehensive PC computer system for pollution on-line monitoring and forecasting for the Tarragon coastal area, developed at the

Institut Químic de Sarrià of the Ramon Llull University under request and with the support of the Autonomic Government of Catalonia.

The Tarragon area is one of the most industrilized areas in Spain, with several industrial emissions and multiple interaction of chemical species. In 1987 an automatic Atmospheric Pollution Monitoring Network was designed to continuously surveil the air quality levels around the area and detect hazard episodes (BISA[1]). Presently, the automatic network is composed by eighteen automatic pollution measuring cabins, a Doppler Sodar and a 50m high meteorological tower (GDEQ[2][3])

In 1994 started the development of the SODCAT system as a complementary tool for atmospheric pollution control and an indispensable instrument for efficient air quality level forecasting. It makes use of the monitored data and relies its 6 hours anticipated prognostics on a set of mathematical models that take into account multiple aspects of the local pollution dispersion. A Virtual Reality Scientific Visualizer has been developed to integrate environmental 3D real-time scientific presentation as well as 2D classical scientific charting (Celma et al. [4]).

The Atmospheric Pollution Monitoring Network conforms the experimental measurements support of the SODCAT system, nevertheless the Autonomic Government's effort in air quality surveillance is also being assisted in his task of educational diffusion and presentation of results.

To support this educational and diffusion spirit started at the end of 1995 the modeling and designing of the Atmospheric Pollution Monitoring Virtual Network of Tarragon, representing in a Virtual Reality world the integrated monitoring network of the local area

The purpose of this project framed in the SODCAT system is to develop a fully interactive Virtual Reality duplicate of the topographical area assembled to detailed interactive and animated models of each of the measuring equipment. Educational visiting and exploration of the implanted network is made possible to the main public through PC based Virtual Reality software. Operation and instrumentation of each of the components of the network is detailedly reproduced allowing free user interaction and queries. Numerical simulated results from the SODCAT system are planned to be imported in the generated Virtual Reality environment, being a formidable support to atmospheric science diffusion.

The developed set of virtual worlds will be freely and massively distributed in the next future on floppy media, although the study of offering Virtual Reality navigation through Internet is currently in an advanced stage.

This first paper presents a general overview of the physical Atmospheric Pollution Monitoring Network of Tarragon and details the current and future situation of the Virtual Reality based replica.

2. Virtual Reality in science and education

Virtual Reality is of primal interest in scientific and educational applications as it's bringing an encouraging new perspective in educational software development. Scientific and technical information are highly difficult to analyze and furthermore to assimilate and comprehend for the nonskilled main public. Therefore, an attractive graphical interface for establishing communication is essential. On the other hand, educational software applications need to induce user curiosity and interest to satisfy their goal, that is information assimilation and retention.

Virtual Reality is basically a three-dimensional computer graphic environment reproducing the physical real world under study. It establishes a real-time interaction between the user commands and the three-dimensional graphics being displayed. The user may be guided or left to observe and interact with this virtual world.

Educational applications find in Virtual Reality a perfect media to connect to the main public, dramatically lowering the barrier to communicate complex and technical information. Virtual Reality worlds induce human connatural exploration skills, therefore they generate highly interest rates and fully assure their goal in message retention and assimilation.

3. Atmospheric Pollution Monitoring Network's domain

The SODCAT system covers an area of $196km^2$, constituting a 14km sided square centered at the 41°9' latitude and 1°12' longitude coordinates that may be considered almost flat terrain (Figure 1 & Figure 2). The area is highly industrialized with multiple chemical plants, ranging from fine chemicals production to refineries. The industries are grouped in three main industrial parks. The Northern park contains approximately ten industries, while the Tarragon City and Southern parks comprise around twenty chemical plants each (Figure 1).

The Atmospheric Pollution Monitoring Network is an instrument ensemble that collects the necessary data to carry out the pollutants' dispersion simulation of the SODCAT system. The network is distributed geographically around the area and is formed by eight pollution receptor monitoring cabins, ten pollution emission monitoring cabins, a Doppler Sodar and a meteorological tower (Figure 1). The minute averaged collected information is sent via 1.200 BPS radio modems to the Control Center that is located at the Environmental Government Delegation of Tarragon. The data is then processed by the software specifically developed.

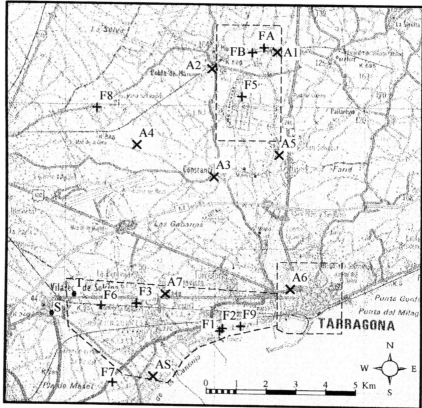

Figure 1: Geographical location of the Atmospheric Pollution Monitoring Network and industrial parks (S: Sodar, T: Tower, A: Receptor cabins, F: Emission cabins)

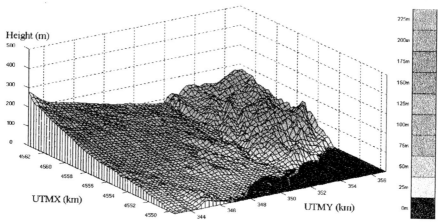

Figure 2: Orographic reproduction of the area under study.

a) Receptor cabins

The pollutant concentrations at receptor site and some surface meteorological parameters are collected by the receptor cabins. The atmospheric pollutants that are presently automatically monitored are SO_2, NO_x, Total Organic Compounds, CO, O_3, H_2S and HCl. Particulate matter, VOCs and lead concentrations are manually measured. The measured meteorological parameters comprise wind speed and direction, air temperature, relative humidity, barometric pressure, solar radiation and total rainfall.

b) Emission cabins

The pollutant emission control is carried out at selected stack release locations. Currently, ten sources of six different plants are monitored. The measured pollutants are CO, SO_2, NO_x, Total Organic Compounds and Opacity. On the other hand, some physical parameters as speed, temperature and volumetric flow of emission gases are also continuously monitored.

c) Doppler Sodar

A Doppler Sodar is located in the area. This instrument monitors the vertical and horizontal wind speed, thermic structure and height of the mixing layer up to 500m.

d) Meteorological tower

A 50m high tower retrieves the necessary meteorological information concerning the surface layer. The meteorological instruments are distributed at different height levels, measuring radiation, rainfall, temperature, relative humidity and three-dimensional wind.

4. Atmospheric Pollution Monitoring Virtual Network

The "Atmospheric Pollution Monitoring Virtual Network of Tarragon" is a set of fully interactive and immersing three-dimensional virtual worlds describing in detail the instrumental network implanted in the Tarragon area.

The Virtual Reality environment has been developed under Superscape® Virtual Reality Toolkit™ (VRT) and can be viewed with any of the Visualiser™ for DOS, Windows 3.x or Windows 95 freely distributed by Superscape®.

A Netscape 2.0 plug-in of the Visualiser™ is currently in its beta stage, but it will allow in the future to interact with any virtual world directly through the World Wide Web (WWW). Nevertheless, utilities for conversion of the developed worlds to Virtual Reality Modeling Language (VRML) consent to extend the target group to any existing computer platform.

The Visualiser™ requires a 486DX at 100MHz for a satisfactory performance, a video graphics accelerator is recommended. Navigation is accessible through keyboard or standard mouse, but a spaceball is suggested for demonstration purposes.

The currently first stage of development of the software covers a fully developed Tarragon Topographical Virtual Model of the area and a first approximation to the Atmospheric Pollution Receptor Virtual Cabin, still in its refining stage.

4.1 Tarragon Topographical Virtual Model

The Topographical Virtual Model is a three-dimensional projection of the topographical and cartographical data made available through the Atmospheric Environment Protection Service of the General Directorate of Environmental Quality, Environmental Department, Generalitat de Catalunya.

Iconic objects allow user queries about the location and identification of each measuring entity, as well as accessing to a fully detailed virtual model with a single mouse click. Therefore, this virtual world allow the user to have a comprehensive view of the network distribution around the area and serves as the main selecting menu to visit each equipment. The iconic objects clearly state the type of measuring instrument and the location is fully identified through the projected cartographical information of rivers, roads and urban buildings. The altimetrical scaling of the model dramatically enhance the attractiveness of its interaction, allowing fly and walk through (Figure 3).

Specialized import/export modules have been developed under Matlab®v4.2 to import different file formats as Arc/Info and Drawing Interchange File (DXF) as well as altimetrical and land use matrix information. The imported data is then manipulated, projected and exported to the VRT format. Matlab® offers a perfect environment for vectorial and two-dimensional data handling, therefore it has been chosen to support all data processing for the SODCAT system. A fully graphical low cost GIS (Geographic Information System) application based on Matlab®, currently in its final stage, has been used to process the topographical data (Urdiales & Chiarabini[5]).

The altimetrical data of the area (Figure 2) is interpolated onto a 32x32 triangled surface representing the terrain orography. The same data is used to create an altimetrical 800x800 pixel texture colored accordingly to height above the sea level. Cartographical information of rivers, roads and urban building is projected over the altimetrical texture producing the full set of combinations. The iconic objects are located accordingly to their exact geographical emplacement and made interactive to mouse click and movement. Mouse movement on selected objects allows on screen identification while mouse clicking swaps to each detailed virtual world.

4.2 Atmospheric Pollution Receptor Virtual Cabin

The Atmospheric Pollution Receptor Virtual Cabin is a fully interactive and detailed three-dimensional replica of one of the eight identical receptor cabins located in the area. External measuring devices are exactly reproduced and user queries about functional purposes are allowed. Through an interactive opening

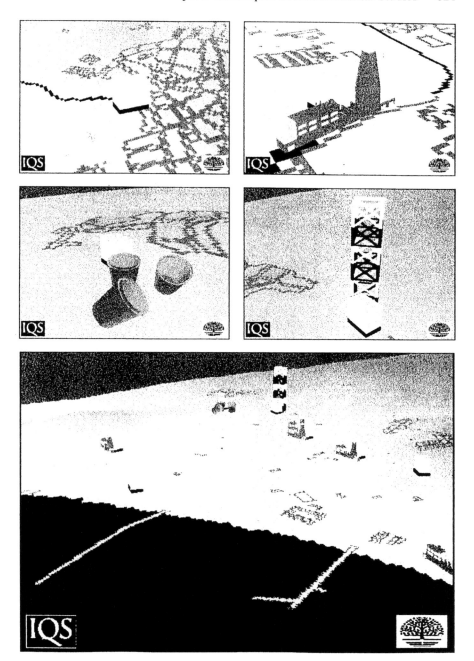

Figure 3: Topographical Virtual Model and iconic objects.

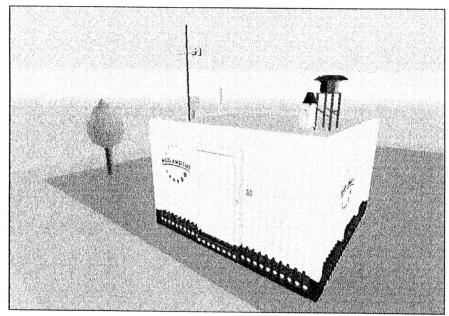

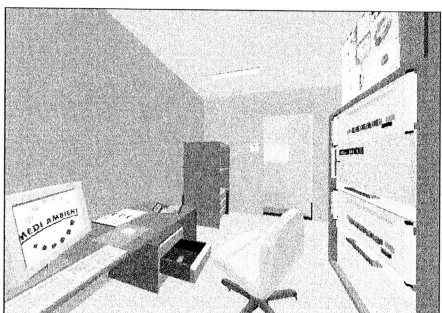

Figure 4: Pollution Receptor Virtual Cabin.

door the interior instruments are made visible and can be explored and their function investigated.

This world allow the user to virtually visit and explore even to minimal details one of the implanted cabins (Figure 4).

The exterior shell has been constructed using a mathematical algorithm assembled in Matlab® and then exported to Superscape® VRT format. Textures of digitized real decorations of the cabin are applied to enhance realism. The exterior instruments reproduce exactly the location and dimensions of the real equipment. Light illumination has been set to reproduce the corresponding daylight of the computer time and date, a lamp post is used to illuminate the environment at evening and night hours.

The interior currently reproduces primary details regarding the automatic measurement equipment and items distribution and dimensions. Each of the real devices for pollution automatic measurement is clearly identified and exactly located. A light switch controls the internal illumination of the cabin, while interactive office equipment and intelligent objects are properly placed to enhance interaction and exploration attractiveness. The monitor screen will present real simulated outputs of the SODCAT system when activated. The desk drawer's contain real abbreviated reports of the cabin functioning and instrument revision.

Collision is detected and run through walls, floor, ceiling and objects is forbidden, hence highly extending the realism.

5. Future steps and Internet distribution

The implemented virtual worlds will be extended further in detail level in the next future. Refined automatic animated instrument and objects will be located to provide full realism and attractiveness. Animated demonstration of sample taking from the manual pollution measurement instruments will be added. On the other hand, sound will be extensively used.

Modified derivations of the Pollution Receptor Virtual Cabin will be joined to the Virtual Emission Cabin, Sodar and Meteorological Tower Models currently under initial construction.

Inclusion of the Lagrangian particle dispersion simulation outputs from the SODCAT system as well as pollutant isoconcentration surfaces, as animated polygonal clouds, is currently under its preliminary study.

The virtual worlds are made experimentally available to the public through Internet, full interactive WWW access and download is expected in the near future.

6. Summary and conclusions

This paper has presented a general overview to the Atmospheric Pollution Monitoring Virtual Network of Tarragon, a Virtual Reality interactive and full sensing immersive world replica of the real Atmospheric Pollution Monitoring Network implanted in the Tarragon area. Its background and educational intention have been detailed. Both the current situation and the expected final stage have been presented.

Special emphasis is given to the innovative aspects of applying Virtual Reality on low cost PC platform for educational purposes, making technical and scientific information assimilation attractive to the main public. Internet application has been detailed for a massive distribution of the software as well as for Virtual Reality interaction with the developed worlds through the WWW.

7. Acknowledgments

The authors wish to thank the support of the Atmospheric Environment Protection Service, General Directorate of Environmental Quality, Environmental Department, Generalitat de Catalunya.

The technical assistance of the staff of RTZ[6] regarding the Superscape® VRT™ software is acknowledged.

8. Bibliography

1. Brain Ingenieros S.A., "Automatic monitoring network project for atmospheric pollution surveillance and prevision in Catalonia", elaborated under requirement of the Health and Social Security Department, Generalitat de Catalunya, Spain, 1987, (Catalan).
2. General Directorate of Environmental Quality, "Surveillance and prevision network of atmospheric pollution in Catalonia. Manual and automatic measurements: 91/92 term", Environmental Department, Generalitat de Catalunya, Spain,1993, (Catalan).
3. General Directorate of Environmental Quality, "Air quality in Catalonia. Manual and automatic measurements: 92/93 term", Environmental Department, Generalitat de Catalunya, 1994, (Catalan).
4. Celma, P.J., Chiarabini, I.W., Agraz, V., "The SODCAT System: Pollution on-line monitoring and forecasting in the heavily industrialized area of Tarragon, Spain", *Submitted to Air Pollution 96*, CMP Publications, Southampton, 1996.
5. Urdiales, E. & Chiarabini, I., "Implementation of a GIS graphical system under Matlab® for topographical data processing", internal report N°U1, Atmospheric Modeling Branch, Ecotechnology Division, Institut Químic de Sarrià, Ramon Llull University, Barcelona, 1996, (Spanish).
6. RTZ, "II Symposium of Superscape® Virtual Reality Toolkit™ users", Barcelona., March 1996.

An integrated modelling system for real-time control of atmospheric pollution (SICEG)

J.I. Ibarra,[a] E. Piernagorda,[b] J. Perena[b]

[a]IBERDROLA, S.A., Santiago de Compostela 100, 28035 Madrid, Spain

[b]IBERSAIC, Alcalá 265, 28027 Madrid, Spain

Abstract

An integrated modelling system has been developed by IBERDROLA and IBERSAIC for the real-time control of routine releases of passive pollutants into the atmosphere. The system has been called SICEG and it has been installed at the Guardo power station in September 1995. SICEG has been designed to provide guidance to normal and especially planned operations at any industrial facility, in the control and minimization of the environmental impact. The system integrates a communication module, a meteorological preprocessor, a modelling module and a display module. All these processes have been implemented in a PC Pentium based computer under a 16 bits WINDOWS environment.

1. Introduction

In April 1990, IBERDROLA received the approval to carry out an R+D research project under the generic reference PIE-134.036. The aim of this project was the assessment of meteorological and atmospheric dispersion models by the use of tracer techniques, in particular SF_6. In November 1990, an intensive field tracer campaign took place in the surroundings of the Guardo Power Station (GPS), north of Spain. The power plant is located within a deep canyon, 300m depth, with a large plateau extending at the outlet of the valley. A total of 14 tracer experiments were carried out, combining surface and elevated releases of SF_6 through a 165m stack. A number of spanish and international institutions participated in the experiments. In 1993 the project ended with the calibration and testing of a set of atmospheric models ranging from classical gaussian models to 3D non-hydrostatic models. In 1994 the Power Division of IBERDROLA takes the decision to support the implementation of an integrated modelling system for the real-time control of gaseous emissions from the GPS. In 1995 the new system, SICEG, is developed taking advantage of the lessons learned during the R+D project and installed at Guardo by the end of September 1995. Since then, the system is running normally and producing the expected advise to the staff operating the power plant.

2.   SICEG's structure

SICEG is integrated by four main modules: (1) a communication module called MIR, (2) a meteorological preprocessor called METPRO, (3) a modelling module called MODAR and (4) a GUI display module. The modelling module is integrated by a 3D mass-consistent meteorological model (MATHEW) and a 3D particle-in-cell dispersion model (ADPIC). A description of these modules is presented in the paragraphs to come.

- **MIR**. The communication module performs remote and periodic calls via modem to monitors and sensors located at the GPS stack, the onsite meteorological station and to a local environmental network. MIR collects the SO_2 release rate, SO_2 concentrations around the power station and meteorological data measured at 10m. All the data gathered are subjected to a quality control process and stored in a historical data base managed by EXCEL 5.0.

- **METPRO**. The meteorological preprocessor implements a regression model which is the result of applying a statistical stepwise regression approach over each of the horizontal components of vector wind. The regression technique was applied during the intensive tracer campaign to 9 meteorological stations deployed over a domain of 40x40km. Then, the resulting regression model has the capacity to estimate wind speed and wind direction at 9 different positions based on the information collected at a reference station (the GPS meteorological station). It is also feasible, the extrapolation of a vertical profile drawn from experimental information collected by an accoustic system installed during the experimental campaign. METPRO also provides turbulence information, variable mixing height values and characteristics of the releases (temperature and exit velocity of gases).

- **MATHEW**. It is a well-known diagnostic, mass-consistent meteorological model. MATHEW adjusts a non-divergent 3D wind field based on a variational technique. MATHEW was the model which provided the best approach during the calibration exercise of different numerical platforms. This model is particularly suited to work in complex terrain and a version for PC has been developed as a result of the project. At the GPS, MATHEW has been configured to work with a horizontal resolution of 500x500m and 14 vertical levels, 50m in depth each. The domain at work is 40x40km.

- **ADPIC.** It is a 3D atmospheric dispersion model based on the particle-in-cell technique. ADPIC has been modified to work on a PC based platform and a new turbulence scheme has been implemented in the model. The new turbulence scheme was developed during the R+D project and it is particularly suited to characterize the atmospheric turbulence features at the GPS site. ADPIC provides 3D concentrations at each cell and also surface concentrations at specific points of interest (i.e. urban areas, local monitors, etc.).

- The GUI has been designed to facilitate an interactive dialog between a user and the different modules implemented in SICEG. It is based on a set of small windows where wind fields, isocontours of surface concentrations, plume trajectory paths and point specific concentrations are shown (Fig.1). A quick exploration of what has happened in the last 48 hours is also available since this information is permanently stored in memory and updated every operational cycle. It is also possible to explore all the data stored in the historical data base along with the printing of monthly reports of releases, based on the specifications of the spanish legislation.

3.   Functions

SICEG V1.0 implements the following calculational and management configuration:

- Remote calls and capture of meteorological data, data of releases from a stack and surface concentration data collected at a local automatic network.
- Calculation of 3D wind fields and map of concentrations in near real-time.
- Display of alarms in case of overpassing maximum release levels or maximum aceptable surface concentrations.
- Checking of historical data bases and results for the last 48 hours.
- Manager of errors and quality control of input data.
- Manager of monthly reports for releases.
- Interactive configuration of File and Directory systems, updating of

geographical coordinates and point of interest, layout of local network monitors, updating of the communication protocol and the setting up of security passwords (Fig.3).

4.   Cycle of operation

The standard cycle of operation implemented in SICEG has been divided into four sequential phases, as shown below:

- Phase of remote capture of data
- Phase of calculation (data preprocessing + modelling)
- Phase of graphical display
- Phase of user navigation

Under normal circumstances, the first phase of remote calls takes about 5 minutes whereas the calculation and representation periods extend over ≈ 15 minutes. Thus, the standard cycle of operation takes about 20 minutes under normal conditions up to a maximum stated in approximately 30 minutes. SICEG has been programed to produce outputs (wind fields and maps of surface concentrations) every hour, leaving ≈ 30 minutes for user operations, such as data base browse or report printing.

In case of problems in the communication phase, SICEG retries those failed communications, to capture lacking data, for no more than 15 minutes. When the programed time period of communication is over, it resolves the situation by deciding if prime information, to go ahead with the modelling module, has been collected or not. If negative, SICEG gives free hands to the user and waits for the following hourly based operational cycle. By prime information is intended a set of critical data necessary to perform calculations with a minimum of reliability, such as wind conditions and/or information about the emissions through the GPS chimney.

5.   Consultations

In the Menu bar of the Application, a bottom has been designed to allow the user a set of consultations or checkings. Version 1.0 of SICEG has the capacity to explore the historical data base, where observed and predicted information is hourly stored; the on-line 48 hours of predicted wind fields and map of surface concentrations; the configuration of the entire system and the printing of emissions' reports (Fig.2).

By using this bottom, the user can check observed meteorological data and surface concentrations, along with the monitored releases at the stack. It is also possible to see the observed surface concentrations at the selected points of interest. To do so, the user must identify the desired time interval, the station and the parameter or parameters which must be shown on the screen.

Consultations to the on-line 48 hours data base provide sequential plots of predicted wind fields, contours of surface concentrations, X-Y view of the cloud of released particles and predicted surface concentrations at selected points of interest.

6.   Utilities

The main utilities implemented in version 1.0 of SICEG can be summarized in the following points:

-     Hard-copy of screen plots.
-     Configuration of the system (Fig.3).
-     Automatic safety copies of data bases into a backup hard disk.
-     Data exports in EXCEL format over a magnetic device (i.e. a $3^{1/2"}$ disk).
-     Manager of security passwords.
-     Input data quality control.
-     Errors' manager.

A manager to handle security passwords has been implemented. The control of passwords is applied to the normal operation options in the system, along with to the configuration of the system. Every time the user gets into the different options programed in the Menu bar, the system requires a password to accept or denied access to the interior of the Application.

A quality control of data is applied during and at the end of the communication phase looking for consistency in the meteorological, emission and surface concentrations data. The process is focused on controling validity of the operational range of the data and the analysis of time series for the repetitivity/replication of generic data.

The errors' manager is primarily based on a pre-set codification scheme and a separate typification of the potential errors. The typification scheme discriminates among different type of errors, such as file errors, memory errors, overflow and mathematical errors, comunication errors and I/O errors. Every time an error occurs, it is stored in a .log file to ease the analysis and further diagnostic, and a message is also displayed on the screen. If the error level is not critical, SICEG goes ahead with the operational cycle, but if it becomes a fatal error, the Application stops and advise the user to analyze and resolve the error before continuation.

7.   Potential of the design

The current design of version 1.0 of SICEG includes a module for the assimilation of the topography at a given site. Version installed at the GPS has been programed to handle a domain of 40x40km, although it can be easily changed and updated to a different domain by using the topo module. The main feature to be retained is the desired spatial resolution for a given domain. This is an aspect to be carefully analyzed case by case since any standardization on this respect might have strong influence on the final predicted wind field and on the map of surface concentrations.

SICEG can simulate up to a maximum of three release points within the study domain and up to 5 different species. The version installed at the GPS only simulates one single release point and one single pollutant, the SO_2 case.

SICEG has been designed to assimilate a maximum of 7 meteorological variables at the station of reference, and it has capacity to simulate up to 9 additional meteorological points distributed all over the study domain. It is also feasible to handle a local network integrated by up to 14 monitoring stations fully instrumented (i.e. SO_2, NOx, O_3, etc.).

The current version installed at the GPS includes a channel to monitor the operational output of the plant in MW as an alternative to the temperature of gases, not monitored at GPS. A regression model, calibrated with emission data from the GPS and also from a second power station owned by IBERDROLA, estimates the temperature of exit gases based on the readings of the megawatts produced by the power plant.

8.   Computer requirements

SICEG requires the following minimum hardware:

- A personal computer, IBM AT compatible.
- CPU i-Pentium 90 MHz
- 8 MB RAM (16 MB recommended)
- SVGA graphic card/monitor
- 20 MB free in disk
- Microsoft compatible mouse
- A second hard-disk for backups (optional)

The required software is as follows:

- MS-DOS, v.5.0
- MS-Windows, v.3.1
- MS-Excel, V.5.0
- MS-Visual C++, v.1.5
- MS-Fortran Power Station

The Application is delivered with a Technical Manual, a User's Guide and the installation disks.

9.   On-coming new features

Version 1.0 of SICEG has been designed to deal with releases of low-reactive pollutants, such as SO_2 and NOx. During 1996 new features are going to be implemented. The main added features will be the following:

- A meteorological prognostic module.
- The capacity to simulate photochemical processes by using simplified schemes compatible with Pentium based microprocessors.
- A separate module for radiological calculations (for nuclear applications).
- A manager to provide direct guidance during planned and unplanned plant operations.
- Migration into a graphical environment compatible with WINDOWS 95 or WINDOWS NT.

The meteorological prognostic module would help to identify the expected average life-time of the current meteorological situation and the probability of change into a set of different meteorological scenarios in the short-term (4-8 hours).

The implementation of photochemical and radiological modules is intended to extend the application of SICEG to other set of scenarios within the environmental impact context and to become a support tool in emergency situations.

The operational manager is intended to become a kind of an interactive option which would provide guidance and advise important for the decision making in both operational strategies and especial operations. The entire set of guidance will be oriented to minimize the environmental impact and to reach the operational objectives in terms of control and environmental management at a power station.

Migration into WINDOWS'95 or/and WINDOWS NT is the natural evolution of SICEG towards 32-bits based computer platforms. These operating systems would allow the implementation of a multiprocess design of the Application. In doing so, the time of the standard operational cycle will be reduced making it possible a user-machine interaction free from limitations in the handling and control of the different processes involved in the actual version of SICEG.

References

1. Ibarra, J.I., 1991: "Atmospheric dispersion experiments over complex terrain in a Spanish valley site (Guardo-90)", Proceedings of the Specialist's Meeting on Advanced Modeling and Computer Codes for Calculating Local Scale and Meso-Scale Atmospheric Dispersion of Radionuclides and their Applications, OECD/NEA Data Bank, Saclay, 6-8 March 1991.

2. Ibarra, J.I., (1992): "The SF_6 tracer during the Guardo-90 atmospheric dispersion study", Revista Energía, XVIII, May-June 1992, pp. 73-81 (in spanish).

3. Ibarra, J.I., (1992): "Turbulence and boundary layer parameterisation during valley wind flows", Fourth International Conference ENVIROSOFT92, 7-9 September 1992, Portsmouth, UK.

4. Ibarra, J.I., (1992): "Valley winds observations at the Guardo valley site", Sixth Conference on Mountain Meteorology, September 29 - October 2, 1992, Portland, Oregon (USA).

5. Lyons, W.A. and. J.I.Ibarra, (1993): "Evaluation of complex terrain dispersion predictions using a fine mesh prognostic mesoscale model", 86th Annual Meeting & Exhibition Air and Waste Management Association, Denver, 14-18 June, CO.

6. Ibarra, J.I., (1993): "Evaluation of Atmospheric Dispersion Models during the Guardo Experiment", 20th International Technical Meeting on Air Pollution and its Applications, November 29-December 3, Valencia (Spain).

7. Ibarra, J.I. and Encarnación R. Hurtado, (1995): "An E-ϵ turbulence closure scheme for modelling nighttime shallow drainage flows", 10th World Clean Air Congress, May 28 - June 2, 1995, Helsinki.

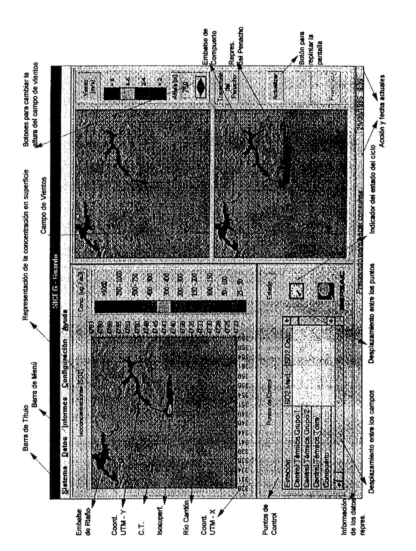

Fig. 1.- Principal frame of the SICEG's GUI where windows for wind fields, contours of SO$_2$ concentrations, plume path and observed *vs* predicted data are shown.

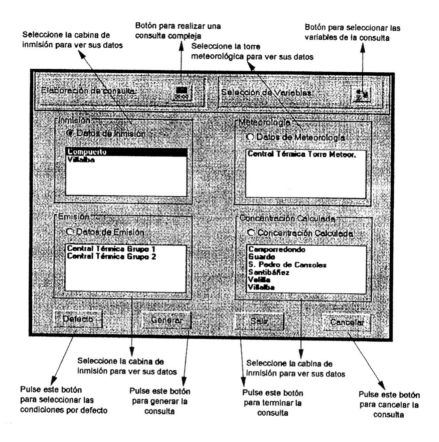

Fig. 2.- Window designed for consultations to the historial data base.

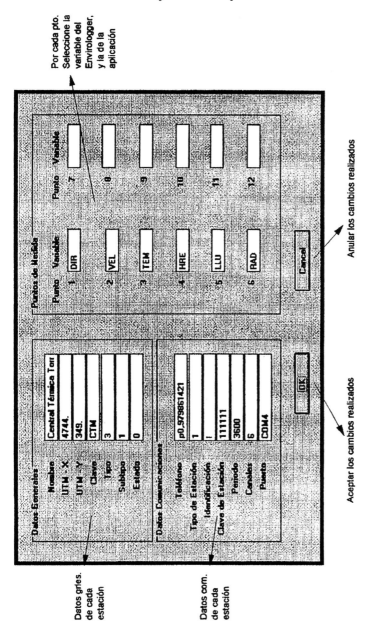

Fig. 3.- Window designed for dialogs of configuration applied to meteorological stations.

Air quality management: an effects-based approach - (development of a sophisticated environmental management technique based upon the integration of air quality modelling within a GIS/DSS framework)

K. Fedra,[a] R. Pemberton,[b] J. Elgy,[b] A. Mannis[c]
aEnvironmental Software and Services GmbH, P.O.Box 100, A-2352 Gumpoldskirchen, Austria
bCivil Engineering Department, Aston University, Birmingham B4 7ET, UK
cSchool of the Built Environment, University of Ulster, Newtownabbey BT37 0QB, Northern Ireland, UK

Abstract

In order to develop sustainable and scientifically defensible environmental strategies, development planners, regulators, and policy-makers require reliable information relating to the current and future state of the environment, and to the options for minimising levels of pollution *and the effects of pollution* in a cost-effective way. Techniques for environmental management may be greatly enhanced by combining rigorous administrative protocols with appropriate software tools (based on state-of-the-art data analysis and simulation techniques) to create a "hybrid" environmental management system.

This paper describes the AIDAIR R&D project to develop an effects-based air quality management system by the seamless integration of such tools.

The software architecture is based on a geographical information system (GIS). Embedded within this structure are a suite of predictive models, and it is backed by a decision support system (DSS), including multi-factorial, multi-objective cost/benefit optimisation.

The GIS basis of the system offers the potential for the user to characterise the relationships between the spatial and temporal distribution of the sources of pollution, the ambient air quality and the nature of affected receptors.

Having incorporated the spatial and temporal distribution of receptors with the GIS, and cost/benefit factors in the DSS, it becomes possible to formulate an effects-based and cost-effective air quality management strategy, which enables the policy-maker to maximise improvements in air quality whilst minimising any economic disbenefits.

1 Introduction

In order to develop sustainable and scientifically defensible environmental strategies, development planners, regulators, and policy-makers require reliable information relating to the current and future state of the environment, and to the options for minimising levels of pollution *and the effects of pollution* in a cost-effective way. The opportunity exists to greatly enhance techniques for environmental management by combining the use of administrative protocols with appropriate software tools (based on state-of-the-art data analysis and simulation techniques) to create a "hybrid" environmental management system.

A software structure based on a GIS, but with the inclusion of predictive models, and a DSS (including multi-factorial, multi-objective cost/benefit optimisation), should satisfy the need for appropriate tools to assist environmental problem-solving in those areas shown in Figure 1.

Figure 1:   Areas of environmental problem-solving for which a sophisticated effects-based air quality management system may be used.

Such a structure can provide, for example, the tools required to **both quantify and evaluate** the factors governing ambient air quality in any defined area. A GIS-based system offers the potential for the user to characterise the relationships between the spatial and temporal distribution of the sources of pollution, the ambient air quality and the nature of affected receptors. Incorporating the spatial and temporal distribution of disease within the GIS, and incorporating cost/benefit factors in the DSS, make it possible to formulate a health effects-based and cost-effective air quality management strategy which may enable the policy-maker to maximise improvements in air quality and health whilst minimising any social disbenefits.

The technical and administrative problems which must be overcome in order to develop a seamlessly integrated system are considerable, but the potential rewards are equally great.

2 The Policy Context

Air quality considerations are a major feature in health, environment and sustainable development documents, including those of the World Health Organisation,[25] United Nations,[21] European Environment Agency (Stanners & Bourdeau[20]), European Union (CEC[13,14]) and the UK (DoE[4,5]).

In Western Europe declining urban air quality is a particularly prominent feature of the environmental agenda, and has promoted a number of political commitments to address the problem. Two key initiatives are the European Union's Draft Framework Directive on Air Quality Assessment and Management (CEC[12]), and the UK Government's intention to establish air quality management as a statutory function of local authorities (DoE[7]). Both of these proposals acknowledge the need for innovative, effects-based approaches to environmental risk assessment.

2.1 EU Initiatives in Air Quality Management

A Framework Directive on Air Quality Assessment & Management (CEC[12]) has been proposed. Amongst its objectives, the Directive is intended to redress problems identified with the operation of the EU's existing four air quality standards (CEC[8,9,10,11]).

Its aims include:

a) Continuation of the process of greater public access to environmental information within the EU

b) A more central role for **Air Quality Standards** (AQS) based on protection of environmental quality and health. Current AQSs are to be revised by 1996, and further standards devised by 1999 for an initial total of 14 pollutants. Standards will be expressed as a series of values as follows:

- *Alert Thresholds* would require the public to be informed of an air pollution episode

- *Long Term Limit Value* (LTLVs) represent the AQSs to be achieved within 10 to 15 years

- *Current Permitted Values* (CPVs) represent incremental stages in achieving the reduction from actual ambient concentrations to the LTLVs over each year of the transition period

c) Harmonisation of air quality assessment and procedures in order to secure reliable and comparable data across the EU, with monitoring based on population size, density and air quality categories

d) Maintenance and improvement of air quality, based on the following defined categories:

- *Areas of Poor Air Quality*, where the air quality exceeds the CPV, an improvement plan must be devised and implemented in order to achieve the CPV as soon as possible and the LTLV within the specified deadline

- *Areas of Improving Air Quality*, where air quality exceeds the LTLV but is below the CPV, no deterioration in air quality is permitted and the LTLV must be achieved within the specified deadline

- *Areas of Good Air Quality*, where air quality is below the LTLV, no action is required.

The Directive has been described as opening a new phase in the EU's air quality policy, in that it represents a departure from the traditional practice of the parallel application of product or emission standards and air quality standards, towards an *"effects-based approach supported by source-based controls"* (DoE[6]).

This new approach brings with it a need for the development of true management (decision-making) tools, which represent a considerable advance over the many simple data collection and modelling systems available (Aria[1], Paul[18], Wahlsted[t24]).

One innovative, effects-based methodology for environmental risk assessment and source management is exemplified by the AIDAIR Research and Development project. This project aims to develop a "hybrid" toolkit in which administrative protocols are combined with state-of-the-art hardware and software, in order to develop a spatial decision-support environment for air quality management strategy development and implementation.

3 The Development of an Effects-Based Air Quality Assessment and Management System

3.1 Technical and Administrative Approach

Systems analysis techniques are applied in the design of a multi-variate, multi-objective DSS (Fedra[17]), which assists the formulation of efficient and cost-effective planning strategies, and this optimisation software is run interactively with air quality simulation software. An expert system using matrix scaling techniques (Fedra et al[16]) aids in the identification of areas in which polluting activities should be avoided and, given a set of alternative scenarios, and a set of pollution control and social cost/benefit functions, will both calculate a minimum cost strategy and calculate the maximum environmental benefit for any given level of financial investment.

The technical aspects of the system are supported by a range of administrative protocols, each detailed in a "methodology" report (Pemberton[19]). Titles currently include:

- Criteria for Accepting a Model
- Evaluating the Performance of a Model
- Preparing a QA Plan for an Ozone Precursor Emission Inventory
- Photochemical Modelling and Performance Evaluation
- Control Strategy, Design and Evaluation
- Preparing a Modelling Protocol
- Model Application
- Application of Modelling Results by Senior Management
- Preparation of Emission Inventories for Use in Photochemical Modelling
- Introduction to Local Air Quality Management
- Introduction to Integrated Environmental Information and Decision Support Systems

The environmental management methodology is based on the experience and results derived from a number of precursor projects undertaken by the Advanced Computer Applications (ACA) Group of the International Institute for Applied Systems Analysis (IIASA). These include:

- **EARSS (Agenda 21)**
  A State-of-Environment (SoE) assessment & reporting system for UNEP

- *AirWare*
  An environmental assessment and management system for the cities of Geneva and Vienna

- **XENVIS**

  An environmental information and risk management system for the Dutch Government

- **MEXSES**

  Environmental information and impact assessment system (Fedra & Winkelbauer[15])

- *WaterWare* **(EUREKA EU 487)**

  A river basin information and management system for Thames Water International

These systems provide an integrated information system framework, which combines databases, a GIS, simulation and optimisation models, and expert systems technology within an easily accessible, easy-to-use interface, based on multi-media technology.

3.2 *AirWare*, the Precursor of AIDAIR

The *AirWare* system brings together the following components:

- Databases (emission inventories, meteorological data, and model scenarios)

- A rule-based expert system for the estimation of point and area source emissions and their location.

- A geographical information system, based on GRASS (CERL[3])

- Simulation models; currently ISC2 (USEPA[23]) & PBM (USEPA[22]), but optionally UK-ADMS (Carruthers et al[2]) & CALGRID (Yamaratino[26]).

- An optimisation model

3.2.1 Databases

The emission inventories are available either through the map display by selecting sources for display and editing of their characteristics, or from a parallel listing of named sources. Basic source characteristics applicable for a given source are stored in the inventories. Examples of such source characteristics are:

- Location
- Emission rates for various pollutants
- Stack parameters
- Cost functions for alternative pollution abatement technologies

The meteorological database allows the display of weather data and the selection of either a particular set of observations as the basis for a short-term model run, or the definition of a longer period (usually an entire year) for a long-term simulation. In the latter case, a pre-processor generates the frequency distributions required by the long-term model from the time series of observations selected by the user.

3.2.2 Rule-Based Expert System

An embedded expert system can be used to derive emission estimates from basic technological data, such as fuel consumption and characteristics, or production technologies and volumes.

3.2.3 Geographical Information System

The geographical information system provides tools to display, access, and manipulate spatially defined data which is used in 4 ways:

a) To supply the models directly (e.g. a digital elevation model, or land cover used to estimate surface roughness and surface temperature differentials)

b) To derive emission estimates (e.g. urban areas and major traffic arteries used to estimate point, area and line source emissions)

c) To assess environmental impact (e.g. different land use of vulnerability to various pollutants)

d) As geographical background data for the spatial orientation of the user, the location of sources, and as spatial reference for model results.

3.2.4 Simulation Models

The simulation models of the current *AirWare* system include an implementation of EPA's Industrial Source Complex, ISC2, model (USEPA[23]), a Gaussian model which may be run both for short episodes and with long-term frequency data. These models predict ground level concentrations of pollutants, such as SO_2, NO_x or particulates.

For summer (photochemical) smog and ground level ozone, a photochemical box model, PBM, simulating daily episodes based on EPA's PBM code (USEPA[22]) may be used. It is driven by the same weather scenarios and shares the emission inventories (for point and area sources of NO_x).

Emissions of volatile organic compounds are estimated with the embedded expert system, using emission coefficients and a set of rules. Alternatively, a three layer finite element model, used in conjunction with a spatially distributed wind field generator, may be used for dynamic short-term runs over a few days.

3.2.5 Optimisation Model

The output of the long-term Gaussian model may be used as an emission and impact scenario for the optimisation module. Using a source-receptor matrix and a spatially distributed, non-linear environmental impact function which assigns different weights to different land use or population zones, this component finds cost-effective strategies for pollution abatement.

Each controlled source has a number of alternative control technologies available, including the option, in some cases, to reduce emissions to zero. Each option is associated with costs, and for a given overall budget the model finds the most effective (in terms of environmental impact) investment strategy. Varying either the budget, time-scale, or discount rate for the cost estimate results in a large number of scenarios, which may be further analysed by a discrete multi-factorial, multi-objective optimisation tool (Zhao et al[27]).

An "optimal" emission scenario can then be used again at the level of the simulation model, and tested with a broad range of individual weather scenarios (rather than the frequency data used for the long-term model) to test the abatement strategy under specific (e.g. worst case) assumptions.

3.2.6 Model Output

Model results may be displayed as:

- Colour coded overlays over the background maps

- 3-D displays over the digital elevation model

- Set of symbols representing emission reductions at the source locations in the optimisation model

- A set of time series diagrams in the case of the photochemical box model

3.2.7 Other Features

As a management tool for local and regional governments, regulatory agencies, and major industries, such a system is sufficiently flexible to adapt to existing and emerging EU and international environmental standards and regulations. Links with established models of industrial and domestic energy use and traffic flow patterns allow the analysis of a uniquely broad range of environmental planning and management scenarios.

The basic client/server implementation facilitates integration with existing components, such as existing databases or monitoring networks. It also provides the means for direct, but distributed access to the main information resources through widely available, low-cost terminal equipment such as PCs, using Internet-based WWW browser software, or dedicated client software. The latter feature not only supports efficient and shared use in spatially distributed institutions, but also provides an attractive means for public access to environmental data.

3.3 New Developments in AIDAIR

Applicability is currently being further enhanced by the incorporation of on-line links to traffic flow and air quality telematic systems. In conjunction with short-term emission and dispersion models, this will provide a real-time air quality modelling capability, and hence a near real-time management system.

The system under development incorporates facilities for:

- The collection, storage, analysis and display of data from real-time data acquisition systems, including air quality data, meteorological data and traffic flowdata, and includes tools for their statistical and spatial analysis.

- The compilation of point, area and mobile source emission inventories in data-poor situations with the assistance of rule-based expert system estimation methods, and includes databases of pollutants, pollution control technologies and the cost of control.

- The operation of a range of **recognised and validated** emission, advection and diffusion models for the simulation and prediction of air quality for abroad range of reactive pollutants, in three dimensions, and over a variety of time scales for comparison with the appropriate standards, for real-time air quality forecasting and for "what-if" analysis in support of planning, epidemiological or regulatory functions.

- Multi-factorial optimisation of possible outcomes using tools which allow the user to design, optimise and evaluate strategies and policies for pollution control or technology investment, after taking account of environmental, technological, and socio-economic variables.

- The display and spatial analysis of any geographical (visual or numeric) data in three dimensions, including all data supplied to or generated by the system, and incorporating satellite image overlays.

- The dissemination of information via a user-friendly multi-media interface permitting remote access through WAN technology (Internet, WWW).

4 Conclusions

The combination of administrative and technical tools, and the adoption of an effects-based approach distinguish this project from the purely emission-based, hardware/software approaches of the various modelling systems on offer or under development. It is perhaps these features, together with the project's novel use of both transportation demand models and real-time telematics, combined interactively with photochemical air quality models and epidemiological data, all within an established GIS/DSS cost/benefit analysis framework, which makes the resulting "hybrid" unique.

References

1. Aria, *ADSO: System for Analysing Atmospheric Dispersion*, Aria Technologies, Le Charlebourg, Colomes, France, 1994.

2. Carruthers, D.J., et al *UK Atmospheric Dispersion Modelling System*, Cambridge Environmental Research Consultants, Cambridge, UK, 1993.

3. CERL, *GRASS 4.1*, USA Construction Engineering Research Laboratories, Champaign, IL, USA, 1994.

4. Department of the Environment, *This Common Inheritance: Britain's Environmental Strategy*, Cmnd 1200, HMSO, London, 1990.

5. Department of the Environment, *Sustainable Development: The UK Strategy*, Cmnd 2426, HMSO, London, 1994.

6. Department of the Environment, *Improving Air Quality*, Department of the Environment, London, 1994.

7. Department of the Environment, *Air Quality: Meeting the Challenge*, Department of the Environment, London, 1995.

8. CEC, Council Directive on Air Quality Limit Values and Guide Values for SO_2 and Suspended Particulates (80/779/EEC), *Official Journal of the European Communities*, 1980 (30 Aug), **L229**.

9. CEC, Council Directive on A Limit Value for Lead in the Air (82/884/EEC), *Official Journal of the European Communities*, 1982 (31 Dec), **L378**.

10. CEC, Council Directive on Air Quality Standards for NO_2 (85/203/EEC), *Official Journal of the European Communities*, 1985 (27 Mar), **L087**.

11. CEC, Council Directive on Air Pollution by Ozone (92/72/EEC), *Official Journal of the European Communities*, 1992 (13 Oct), **L297**.

12. CEC, Proposal for a Council Directive on Air Quality Assessment and Management, COM(94)109, *Official Journal of the European Communities*, 1991, **C216**.

13. CEC, Council Regulation on Substances that Deplete the Ozone Layer (594/91/EEC), *Official Journal of the European Communities*, 1991 (14 Mar), **L67**.

14. CEC, Resolution of the Council on the Fifth Community Policy and Action Programme in Relation to the Environment and Sustainable Development (1993-2000), *Official Journal of the European Communities*, 1993 (17 May), **C138**.

15. Fedra, K. & Winkelbauer, L.   MEXSES: An Expert System for Environmental Screening, *Proceedings of the Conference on Artificial Intelligence Applications*, pp 294-298, IEEE, Piscataway, NJ, USA, 1990.

16. Fedra, K., Winkelbauer, L. & Pantulu, V.   *Expert Systems for Environmental Screening: An Application in the Lower Mekong Basin*, RR-91-19, International Institute for Applied Systems Analysis, Laxenburg, Austria, 1991.

17. Fedra, K. Integrated Environmental Information and Decision Support, *Computer Support for Environmental Impact Assessment*, eds. G. Guariso and B. Page, IFIP Transactions B-16, pp 269-288, Elsevier BV, Amsterdam, Holland, 1994.

18. Paul, P. *Le Project Climatologique Regional Transfrontalier REKLIP - Objectifs et Premiers Resultats*, CEREG, Institut de Geographie, Universite Louis Pasteur, Strasbourg, 1995.

19. Pemberton, R.   *Phoenix EMS: List of Background Papers*, Civil Engineering Department, Aston University, Birmingham, UK, 1995.

20. Stanners, D. & Bourdeau, P. (eds) *Europe's Environment: The Dobris Assessment*, European Environment Agency, Copenhagen, Denmark, 1994.

21. United Nations, *The Earth Summit: Agenda 21 - United Nations Programme of Action from Rio*, United Nations, New York, USA, 1992.

22. USEPA, *User's Guide for the Photochemical Box Model, PBM*, US Environmental Protection Agency, NC, USA, 1984.

23. USEPA, *User's Guide for the Industrial Source Complex, ISC2, Dispersion Models - Volume 2: Description of Model Algorithms*, US Environmental Protection Agency, NC, USA, 1992.

24. Wahlstedt, B. *The Airviro System: System Specification*, Indic AB, Norrkoping, Sweden, 1992.

25. World Health Organisation, *Environmental Health: The European Charter and Commentary*, World Health Organisation, Geneva, Switzerland, 1990.

26. Yamaratino, R.J. *CALGRID: A Mesoscale Photochemical Grid Model*, California Air Resources Board, Sacramento, CA, USA, 1990.

27. Zhao, Ch., Winkelbauer, L. & Fedra, K. *Advanced Decision-Oriented Software for the Management of Hazardous Substances - Part IV: The Interactive Decision-Support Module*, International Institute for Applied Systems Analysis, Laxenburg, Austria, 1985.

SECTION 11:
SOFTWARE IMPLEMENTATION

A numerical simulation of channel flow augmentation upon different pumping schemes

A.C. Neto,[a] A.C. Lock[b]

[a]*Researcher of the CNPq at the Fundação Centro Tecnológico de Hidráulica, 120 Av. Professor Lúcio Martins Rodrigues, 05508-900 São Paulo, SP, Brazil*

[b]*Lecturer of the Department of Civil Engineering, Faculty of Engineering and Applied Sciences, University of Southampton, Highfield, Southampton, Hampshire SO9 5NH, UK*

Abstract

A serious problem which happens very often in several torrid regions of the globe is the intermittence of natural channels. An usual scheme to attenuate this is pumping water from the underground to the channel. However, in spite of being a feasible policy, it does not avoid water from circulating back to the aquifer before being used for social and environmental purposes, due to the interaction between the aquifer and the channel. It is improbable that optimal strategies can be reached by downright simulation, but numerical simulation can be used to estimate the dependence of pumping rates, pump positions and their combination upon the performance of the scheme. Furthermore, optimization of a certain set of nonlinear objective functions subject to inequality nonlinear constraints of the state variables makes use of optimization algorithms coupled with simulation models. This article presents a methodology based on the Finite Element Method to simulate and somehow to predict the performance of the many possible pumping schemes.

1 Basic Assumptions

Two-dimensional transient saturated flow through an homogeneous isotropic unconfined aquifer without lateral recharge can be described by the well-known Boussinesq equation derived from Darcy's law and Dupuit's assumption as follows:

$$\sigma \frac{\partial h}{\partial t} - K \left[\frac{\partial \left(h \frac{\partial h}{\partial x} \right)}{\partial x} + \frac{\partial \left(h \frac{\partial h}{\partial y} \right)}{\partial y} \right] + p + \sum_{i=1}^{np} [Q_i . \delta(x - x_i, y - y_i)] = 0 \tag{1}$$

where σ is the storage capacity, K is the hydraulic conductivity, p is the flow that leaks from the aquifer to the channel per unit length of the channel, h is the hydraulic head, np is the number of pumps, Q_i is the discharge from the aquifer into the channel of the ith pump positioned at coordinates (x_i, y_i), and δ is the Dirac Delta function.

As for the channel, its flow can be computed by means of the equations due to Saint-Venant which are the two following expressions:

$$\frac{\partial A_w}{\partial t} + \frac{\partial q}{\partial t} - p - \sum_{i=1}^{np}\left[Q_i . \delta(s - s_i)\right] = 0 \tag{2}$$

$$\frac{\partial q}{\partial t} + \frac{\partial\left[\dfrac{q^2}{A_w}\right]}{\partial s} + gA_w\left(\frac{\partial\zeta}{\partial s} + \frac{\partial z}{\partial s} + \frac{n^2 q|q|P_w^{\frac{4}{3}}}{A_w^{\frac{10}{3}}}\right) + \sum_{i=1}^{np}\left[Q_i\delta(s - s_i)\frac{q}{A_w}\right] + \frac{pq}{A_w} = 0 \tag{3}$$

where q is the channel flow, A_w is its cross-sectional wetted area, P_w is its respective wetted perimeter, s is the channel length dimension, g is the acceleration due to gravity, ζ is the channel depth of flow, z is the channel bottom height relative to a reference datum, n is Manning's roughness coefficient, and s_i is the point where the ith pump discharges into the channel. The link between the aquifer and the channel is performed by both the pumping rate terms Q_i and the leaking term p.

Although it is normal that a real channel only partially penetrates the aquifer, the supposition of penetration to the base of the aquifer can be assumed for some problems. Oakes and Pontin[5] used this assumption based on the observation that "chalk transmissivity has higher than average values in the vicinity of the river tracks". Then average values of transmissivity can be selected for the river neighbouring in order to compensate for the increasing discharge values which result from the assumption of no-vertical flow. Analytical models of stream-aquifer systems may be used in simplified formations and relate the channel leakage to the aquifer drawdown, according to $Q_i=\Gamma D$, where Γ is the reach transmissivity which is defined as the volume flow rate out of the channel per unit length of the channel per unit drawdown, and D is the drawdown at a fixed distance from the channel. This kind of model was developed by Morel-Seytoux and Daly[4]. A theoretical formulation describing leakage from clogged channels that partially penetrate aquifers was reported by Chin[1]. He developed an expression which related the reach transmissivity to the mean channel width, the distance of drawdown measurement from the channel centerline and the ratio of drawdowns on both sides of the channel. Many authors used a linear relationship between the channel free surface and the groundwater table levels. Heckele[2] assumed that the discharge per unit length that leaks from the aquifer to a static channel is directly proportional to the product of the wetted perimeter by the difference of the hydraulic heads of the aquifer and the channel. Maddock[3] presented a conjunctive stream-aquifer simulation model using finite differences to develop an algebraic technological function. The interactions between the stream and the aquifer were assumed to be such that the stream acted as a constant head boundary to the aquifer. In other words, his supposition was that there were sufficient flow in the stream so that losses from the stream to the aquifer did not affect the head levels in the stream. The saturated thickness of the aquifer was supposed to be large compared to that of any drawdown, so the transmissivity was considered constant with respect to depth, and the aquifer was then linear. Willis[6] developed a model for a linear aquifer where the interaction between the stream and the groundwater system was represented by a time varying Dirichlet boundary condition. In the present work, it was chosen a model of the leakage from the aquifer to the channel represented by the following equation:

$$p = \lambda.(h - \gamma) \tag{4}$$

where λ is a resistance factor, and γ is equal to the channel free surface level if h is above the channel bed bottom, and is equal to Z otherwise.

After discretizing equation (1) using finite elements over the space, and by finite differences over time, under fixed prescribed water table level boundary conditions, the following system is obtained:

$$\mathbf{A}.\left(\mathbf{h}^{n+1} - \mathbf{h}^n\right) + \theta_1.\Delta t_1.\left[\mathbf{b}^{n+1} + \mathbf{C}.\mathbf{q}^{n+1} + \mathbf{H}.\mathbf{p}^{n+1}\right] + (1 - \theta_1).\Delta t_1.\left[\mathbf{b}^n + \mathbf{C}.\mathbf{q}^n + \mathbf{H}.\mathbf{p}^n\right] = 0 \tag{5}$$

where $\mathbf{A}$ is the capacitance matrix, $\mathbf{h}$ is the vector of the water table levels, $\mathbf{b}$ is a time varying vector which depends upon the present value of $\mathbf{h}$, $\mathbf{q}$ is the vector of the pumping rates, $\mathbf{C}$ is a connective matrix whose elements are all nil apart from some diagonal elements whose values are equal to one, which refer to the aquifer nodal positions of the pumps, $\mathbf{p}$ is the leakage vector which is dependent on the values of all the unknowns h, ζ and q, $\mathbf{H}$ is the matrix which positions the sink/source nodes represented by the channel line over the aquifer domain, θ_1 is a time discretization parameter which varies from zero to unity, Δt_1 is the groundwater time step, and n refers to the time stage. Using the same methodology for equations (2) and (3), it is obtained the discretized system bellow:

$$\mathbf{g}^{n+1} - \mathbf{g}^n + \theta_2.\Delta t_2.\left(\mathbf{d}^{n+1} + \mathbf{E}.\mathbf{q}^{n+1} + \mathbf{f}^{n+1} - \mathbf{p}^{n+1}\right) + (1-\theta_2).\Delta t_2.\left(\mathbf{d}^n + \mathbf{E}.\mathbf{q}^n + \mathbf{f}^n - \mathbf{p}^n\right) + \Delta t_2.\mathbf{v} = 0 \tag{6}$$

where $\mathbf{g}$ is the vector where the odd elements are the channel depths of flow, and the even elements refer to the channel discharge values, $\mathbf{d}$ is a time varying vector which depends on both g and h, $\mathbf{E}$ is the matrix that connects the pumps to the channel in the same fashion as the already defined matrix $\mathbf{C}$, $\mathbf{f}$ is a transient vector dependent on all the unknowns, $\mathbf{v}$ is a constant vector due to the momentum caused by the weight, and finally θ_2 and Δt_2 have the same meaning as for θ_1 and Δt_1. Both systems (5) and (6) are non-linear and can be solved by Newton-Raphson method.

2 Numerical Experiments

In order to test the validity of the assumptions made for the combination of channel and aquifer movements, the mesh of figure 1 was generated. The mesh was composed of 779 nodes and 1440 triangular elements forming a rectangular area of $2,000 \times 1,800$ m^2. A channel crossed the rectangle from top to bottom positioned in its central "north-south" line. The channel had a constant slope of 1%, Manning's roughness coefficient of 0.028, and its cross-section was a non-symmetrical trapezium with a basis 2.00 meters wide and lateral slopes of 10^0 and 5^0. Four pumps namely A, B, C and D were then located at the points shown in figure 1 with the capabilities of pumping from the aquifer to the upstream boundary node of the channel. The channel was composed of 19 nodes and 18 elements 100 m long. Pump A was in the right bank 200 m away from channel node 13, pump B was in the left bank 300 m away from channel node 10, pump C was also in the left bank 200 m away from channel node 7, whereas pump D was in the right bank 500 m away from channel node 8.

The net gain was defined as the ratio between the net flow augmentation and the pumping flow. Several schemes of channel flow augmentation were then carried out to measure the net gain of the different strategies.

The initial condition of the channel was the empty bed, and then a gradually increasing input discharge was applied at the upstream boundary until reaching 12.00 m^3/s after 4 hours, and then maintained constant. The time increment for the channel was 1 minute and the parameter θ was 0.75. The aquifer system was run using $\theta=1.00$ and $\Delta t= 1$ hour. The leakage parameter λ was arbitrarily chosen as 5×10^{-6} m/s. To analyze the situation of the aquifer at the steady state, it was supposed that the channel bottom at the upstream boundary was 48 metres above the reference datum, while the downstream boundary was 30 metres over this reference horizontal plane. Water table level at 48 metres high (relative to the same reference datum) was the initial condition of the aquifer, and it was prescribed as this value in all its external boundaries. Then the flow was simulated for approximately 2.5 years when the system eventually reached the steady state. This was done in order to obtain a baseline situation for the system which could then be used as a reference. The baseline system was then compared with the conditions when using a pumping scheme. The channel in this case is a line sink across the centre of the aquifer. In other words the channel was considered as a gorge ripping the aquifer.

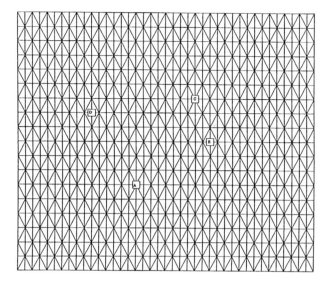

Figure 1   The finite element mesh.

2.1. Dependence of the net gain upon the pumping rate

Water was pumped from pump A under several different pumping rates to analyze the net gain of the use of this pump for channel flow augmentation. The net gain is defined as the ratio of the outflow discharge augmentation to the pumping rate. Table 1 shows the output flow of the channel, the flow augmentation and the net gain for 10 different pumping rates at pump A.

It can be seen that the net gain decreases as the pumping rate increases. The relationship between pumping rate and output flow augmentation is non-linear, and it is expected to be dependent upon the channel position, the channel flow, the aquifer transmissivity, the pump position, the pumping rate, the leakage parameter λ and the channel flow parameters. Figure 2 shows the situations of the aquifer for different pumping rates of A at the steady state. It can be observed that the greater the pumping rate is, the less elongated and the more concentric the contour lines become. This is due to the increasing influence of the pumping on the definition of the depths of flow and the consequently decreasing dependence upon the leakage to the channel.

Pumping rate (m³/s)	Output Discharge (m³/s)	Flow augmentation (m³/s)	Net gain (%)
0.000	12.053886195	***********	*************
0.001	12.054717324	0.000831129	83.112870000
0.002	12.055548373	0.001662177	83.108865000
0.005	12.058041037	0.004154841	83.096824000
0.010	12.062193861	0.008307665	83.076652000
0.020	12.070493383	0.016607188	83.035940500
0.040	12.087067390	0.033181194	82.952985750
0.060	12.103606928	0.049720737	82.867887000
0.080	12.120110600	0.066224404	82.780505250
0.100	12.136576875	0.082690879	82.690679300
0.120	12.153004053	0.099117858	82.598214833

Table 1 Flow augmentation and net gain for several pumping rates at pump A.

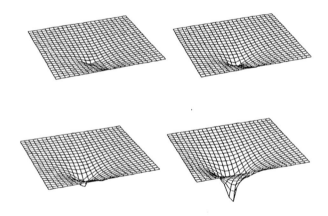

Figure 2 Three-dimensional visualization of the aquifer flowdepths at the steady state for Q_A=0.02, 0.04, 0.06 and 0.12 m³/s, respectively.

2.2 Dependence of the net gain upon the pump combination

Sometimes, when dealing with river flow regularization schemes subject to constraints upon either a minimum water table level (constrained state variable) or a maximum individual pump capacity (upper limited control variable), it is not possible to regularize the discharge of the channel using a single pump. A common procedure is to combine the effect of several pumps located in different points of the aquifer to distribute the drawdown over the whole domain.

Although it is possible to obtain a table like table 1 for several pump positions, it would not be feasible to get a net gain table for a combined pump scheme by summing up the rows of the tables of each individual pump scheme, as the problem is nonlinear and superimposition of separate solutions would not lead to the solution obtained when using associated pumps, except in the case of confined aquifers. Therefore each combined scheme must be analyzed by simulating the flow using the actual pump distribution.

A discharge of 0.120 m^3/s was siphoned from pumps B, C and D to compare the net gains with the net gain of the already shown scheme of pumping from A. Table 2 shows that a better performance was obtained when pumping from B than from A, mainly due to its greater distance from the channel. When pumping from D the performance was even better, because of its greater distance from the channel. Two combinations of the four pumps were used to pump a total discharge of 0.120 m^3/s from the aquifer to the river. The first scheme used the pumps A and C with pumping rates of 0.060 m^3/s each. When using a discharge of 0.030 m^3/s for each one of the four pumps a better result was obtained; but the scheme which used pump D alone was still the best choice.

Pumping scheme	Output flow (m^3/s)	Flow augmentation (m^3/s)	Net gain (%)
No pumping	12.053886195	*********	**********
0.120 m^3/s in A.	12.153004053	0.099117858	82.598214833
0.120 m^3/s in B.	12.155018147	0.101131951	84.276626250
0.120 m^3/s in D.	12.162316293	0.108430098	90.358414917
0.060 m^3/s in A and C.	12.153171401	0.099285206	82.737671750
0.030 m^3/s in A, B, C and D.	12.155943831	0.102057635	85.048029417

Table 2 Efficiency of channel flow augmentation pumping schemes.

Other combinations could be tried, but if it were necessary to choose one of the five schemes shown here, one would be inclined to decide for the scheme which uses only pump D.

Although it is unlikely that an optimal solution could be achieved using exhaustive repeated simulation alone, due to the countless alternatives which could be supposed, the model described here could furnish data to more complex nonlinear optimization schemes coupled with the finite element method. The consequences of the pumping strategies upon the groundwater can be better observed by examination of the figures 3 and 4.

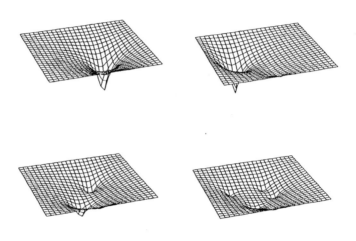

Figure 3 Visualization of the different pump position schemes.

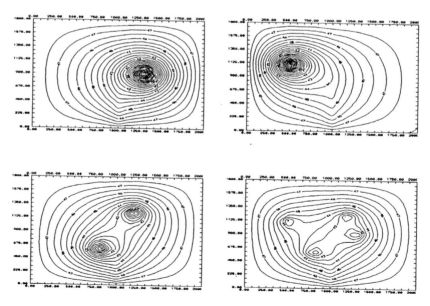

Figure 4 Contour lines of the water table level for different pumping position schemes.

The figures show that when using any one of the single pump schemes the contour lines tend to acquire an elongated form as the distance from the pump increases. This is due to the fact that sink points (like the pumps) produce concentric contour lines, whereas sink lines (like the channel which can be thought of as a line composed of a sequence of point sinks with increasing discharge rate) generate these elongated contour lines. Since the pumping rate is roughly twice the magnitude of the sink line, i.e. the sequence of all the sink channel nodes combined, the influence of the leakage to the channel is more accentuated far from the pump. When using more than one pump the effect of this line sink is less important despite the use of smaller individual pumping rates. This is due to the spatial distribution of the pumps; two pumps somehow form a line sink, and in a certain sense four pumps can be viewed as a poor discretization of a quadrilateral sink.

4 Discussion

The present model seems to be a usefull tool on the choice of channel regularization schemes. However, a more realistic model should include other constraints which exist in the real world such as the following.

1)- The pumping rate is not constant, but a function of the vertical distance between the topographic surface and the water table level. Moreover, as there is energy loss within the pipes, the distance from the pump to the point of the channel where the water will be discharged must be taken into account.

2)- Including the price of the pipes and pumps, not to mention maintenance costs could lead to another choice, if costs are either one of the multiple objective functions or one of the upper limited constraints.

3)- If the choice is constrained to excessive drawdowns that could cause damage to agriculture, the natural environment and the aquifer itself, schemes which produce less water table level drop would tend to be chosen.

In spite of the coherence of the results obtained here, if Neumann and Cauchy boundary conditions were imposed for the aquifer, certainly more realistic results would be obtained. Although the program has the capability of prescribing both transient depths of flow and time varying discharges in the boundaries, prescribed depths of flow were imposed in the boundaries as a first approach to test the validity of the connection between the aquifer and the channel, because initial conditions could be more easily implemented.

In any case, this study can offer information to researches aimed at optimizing channel flow augmentation, regularizing river discharges, managing water table levels subject to constraints upon the channel flows and controlling the use of pumps.

5 Acknowledgements

The authors would like to thank the efforts of the two following Brazilian Agencies: FAPESP (Fundação de Amparo à Pesquisa do Estado de São Paulo) and CNPq (Conselho Nacional para o Desenvolvimento Científico e Tecnológico), without whose help the consecution of this research would have been impossible.

6 References

1. Chin, D. A. Leakage of clogged channels that partially penetrate surficial aquifers, *Journal of Hydraulic Engineering*, 1991, **117(4)**.

2. Heckele, A. *Numerische Optimierungsverfahren zur Bewirtschaftung flächig ausgedehnter Gundwaβer Systeme.* Ph.D. thesis, Fridiriciana University of Karlsruhe, Germany, 1987.

3. Maddock, T. The operation of a stream-aquifer system under stochastic demands, *Water Resources Research*, 1974, **10(1)**, 1-10.

4. Morel-Seytoux,H. J. and C. J. Daly. A discrete kernel generator for stream-aquifer studies, *Water Resources Research*, 1975, **11(2)**, 253-260.

5. Oakes, D. B. and J. M. Pontin. Mathematical modeling of a chalk aquifer, Water Research Centre, Stevenage, 1976.

6. Willis, R. A stochastic planning model for conjunctive ground and surface water resources managements, pp 473-481, *Proceedings of the International Conference on Water Resources Development*, Taipei, Taiwan, May 1980.

Development of water quality models within the TELEMAC system and recent applications

C. Moulin,[a] A. Petitjean,[a] J. Gailhard[b]

[a]Laboratoire National d'Hydraulique, 6 Quai Watier, 78400 Chatou, France
[b]Département Environnement, 6 Quai Watier, 78400 Chatou, France

Abstract

The impact of industrial activity on water quality is currentely a major concern at EDF. That is why mathematical and computational tools have been developed and validated to support decision makers in managing the environment. This paper presents recent developments in that field within the TELEMAC system. Specific attention has been devoted to deal with tidal flats and mass conservation problems. The application of the system to water quality modeling is illustrated with 2 examples : a maritime application concerning the impact of sewage on the Morbihan Gulf, and a fluvial application dealing with heavy metals in the River Seine.

1 Introduction

1.1 Importance of water quality at EDF

Water is of the utmost importance to EDF as it is the source of energy for hydroelectricity and because it is a convenient means for cooling thermal power plants. EDF, who manages 75% of surface water in France, has learnt to share water with other users and to feel concerned about the impact of its activities on water quality. For example, important research efforts are devoted to the study of the impact of power plant outfalls and to the impact of reservoir emptying on downstream water quality.

1.2 General considerations on water quality

Obviously the definition of the scales of a water quality problem is of major importance because the temporal and spatial variations of processes and parameters cover a wide range of scales. For example the temperature of a river has daily, seasonal and yearly variations and it can also be measured at a local scale or more globally across the area of a catchment.

The choice of the physical, chemical or biological variables to be considered in the model, and the description of their interaction, is often difficult and must be adapted for the relevant scales of the problem.

Generally, the variation in the concentration of a given substance with time is the result of advective and dispersive transport, of internal reactions, and of the action of external sources and sinks. A good knowledge of hydrodynamic variables is required when travel time directly influences physical or chemical processes. In particulate transport the behaviour of suspended sediments has to be represented carefully since suspended sediments act as the first link to the ecosystem during the propagation of a pollutant.

2 Water quality models within the TELEMAC system

2.1 TELEMAC-2D and SUBIEF

Most of the processes of sub-aqueous systems can be expressed mathematically by a set of differential equations. These equations can be treated within the TELEMAC system, which is a set of programs developed by the Laboratoire National d'Hydraulique (LNH) to deal with free surface flows. These programs share the same finite element library and the same pre-and post-processors.

TELEMAC-2D computes the hydrodynamics of the flow by solving the shallow water equations [1]. Different turbulence models are available including the Elder model and a k-epsilon model.

SUBIEF computes the transport of one or several tracers within a 2D free surface flow with the " homogeneous concentration over the depth " approximation. The model also assumes that the tracers are carried passively in the flow and do not affect the hydrodynamics as SUBIEF uses the hydrodynamic data produced by TELEMAC-2D. This data then allows the creation of the convective field used to solve the transport equations. This approach significantly reduces the computational time.

SUBIEF has an open architecture in which the number of variables, and all their physical constants and interactions are not part of the model but are only part of the data.

As a result of recent developments, SUBIEF can now be used as the basis for water quality models and several pre-programmed modules of water quality are available :

- heavy metals model,
- biomass model,
- dissolved oxygen model,
- thermal model.

2.2 Conservative or non-conservative formulation ?

In this section we discuss the choice of the form of the tracer transport equation as it can be written in a conservative or non-conservative form.

The conservative variable is the tracer mass :

$$M = h \, c$$

where :   h is the water depth

c is the tracer concentration

In the conservative formulation, the equation to be solved is :

$$\frac{\partial M}{\partial t} + \text{div}(M \, \vec{u}) = \text{div}(\overline{\overline{K}}.h \, \overrightarrow{\text{grad}} c) + S_M \qquad (1)$$

with $\vec{u}$: depth-integrated velocity (m/s)

$\overline{\overline{K}}$: dispersion tensor

S_M : source terms of tracer mass

Obviously, eqn. (1) provides the best basis for building a mass-conservative discretization of the transport equation. However, one is more often interested in concentrations than in masses, e.g. the final results of a thermal application are usually in temperature rather than heat.

When M is obtained, we then have to compute :

$$c = \frac{M}{h} \qquad \text{where } c \to \infty \text{ when } h \approx 0.$$

An immediate consequence of such an operation ($c \to \infty$) would be the loss of the monotonicity of the concentration computation. If M has a linear spatial discretization, then c does not belong to the finite element approximation subspace. Also giving a minimum value for h in order to achieve monotonicity would not give a general method. Hence this formulation has not been retained in SUBIEF.

Starting from eqn (1) :

$$\frac{\partial(hc)}{\partial t} + div(hc\,\vec{u}) = div(\overline{\overline{K}}.h\,\overrightarrow{gradc}) + S_M$$

the non-conservative formulation is obtained by splitting the equation into :

$$c\frac{\partial h}{\partial t} + h\frac{\partial c}{\partial t} + h\vec{u}.\overrightarrow{gradc} + c\,div(h\vec{u})$$
$$= div(\overline{\overline{K}}.h\,\overrightarrow{gradc}) + S_M$$

As (h,u) obeys the continuity equation : $\dfrac{\partial h}{\partial t} + div(h\vec{u}) = S_h$ (2)

(assumed to be correctly computed) where S_h are source terms for h, we get :

$$h\left[\frac{\partial c}{\partial t} + \vec{u}.\overrightarrow{gradc}\right] = div(\overline{\overline{K}}.h\,\overrightarrow{gradc}) + S_M - cS_h$$

S_M is given by an imposed concentration c_i of tracer within the water discharge to be diluted. Hence, we have $S_M = c_i\,S_h$ and :

$$\frac{\partial c}{\partial t} + \vec{u}.\overrightarrow{\text{grad}c} = \frac{1}{h}\text{div}(\overline{\overline{K}}.h\ \overrightarrow{\text{grad}c}) + \frac{c_i - c}{h}S_h \qquad (3)$$

A *finite element formulation* of eq.(3) can be summarised as :

$$\int_\Omega [\frac{\partial c}{\partial t} + \vec{u}.\overrightarrow{\text{grad}c}\ ]\ \varphi_i\ d\Omega$$

$$= \int_\Omega [\frac{1}{h}\text{div}(\overline{\overline{K}}.h\ \overrightarrow{\text{grad}c}) + \frac{S_M - cS_h}{h}]\varphi_i\ d\Omega \qquad (4)$$

Specific developments have been made to ensure a correct treatment of tidal flats and mass conservation.

2.3 Mass conservation and choice of a numerical scheme

At a continuous level, eqn.(3) is theorically mass conservative. However, we have to obtain a discrete formulation of eqn.(3), starting from a discretisation of eqn.(1), in order to get a numerical proof of the tracer mass conservation.

The first advection scheme implemented in SUBIEF was the method of characteristics with a linear interpolation. Its main advantage was monotonicity but it tended to be highly diffusive and no theoretical proof of the tracer mass conservation could be obtained.

A theoretically mass conservative method may also be obtained with a pure finite element method, without fractional steps. Good mass conservation has been obtained with a SUPG weighing function (Hervouet & Moulin, [2]), but with the down-side of this technique being a decrease on the monotonicity of the solution. PSI advection schemes seem to be a good answer to this monotonicity problem (Janin& al., [3]).

2.4 Tidal flats

Equation (3) includes some terms where there is division by h. However, their effect is minimal when applied on the dispersion and on the source terms. There would be no problems with division by h if we considered that eqn.(3) had only to be solved where there is water ($h \neq 0$). However, in order to acieve this in the numerical solution choices have to be made as to how dry zones are defined.

A first approach to dealing with dry elements is to solve eq.(4) over the whole finite element domain, making sure that the 1/h terms are correctly computed, using some criteria such as :

$$\text{assuming } \frac{1}{h} = 0 \text{ if } h < \varepsilon$$

$$\text{computing } \frac{1}{h} \quad \text{if } h > \varepsilon$$

This method raises two questions :

- Defining ε,

- Solving eqn.(4) over the whole domain will not prevent the tracer concentration from reaching dry areas, especially in the case of a highly diffusive advection scheme.

A second way of dealing with dry elements is to solve eqn.(4) over a sub-domain defined by the wet elements only. This solution has physical backing as eqn.(4) does not have to be solved where there is no water. As a result this method will obviously not allow tracers to appear in dry areas. However the problem is still how to determine whether an element is dry or wet.

One solution is simply to define a "mask" term which is calculated for each element :

$$\text{mask(iel)} = 0 \text{ if the element is dry}$$
$$\text{mask(iel)} = 1 \text{ if the element is wet}$$

and solving :

$$\sum_{iel} \int_{iel} \text{mask(iel)} \left[\frac{\partial c}{\partial t} + \vec{u}.\overrightarrow{\text{grad}}c \right] \varphi_i \, d\Omega$$

$$= \sum_{iel} \int_{iel} \text{mask(iel)} \left[\frac{1}{h} \text{div}(\overline{\overline{K}}.h \, \overrightarrow{\text{grad}}c) + \frac{S_M - cS_h}{h} \right] \varphi_i \, d\Omega$$

This solution works well with element by element techniques in which the system matrices are built for each element (Hervouet, [3]). A criterion has been found at LNH to define the dry elements without any "$h < \varepsilon$" test.

3 User interface and quality assurance

3.1 User interface for water quality models

An interface has been developed which allows the easy formulation of water quality models. The water quality model can be simply described in a single file which defines variables, parameters and equations (fig.1). A filter then interprets this file and automatically translates the water quality model into SUBIEF format, creating a number of data files which include some internal source code. For post-processing, RUBENS is a graphics visualization software package that offers many types of graphics representations and allows the manipulation and processing of visual data as well as the superimposition of measurements onto the visual data.

3.2 Quality Assurance procedures

The TELEMAC software is developed under EDF Quality Assurance procedures. The model validation stage is of prime (if costly) importance. Validation of SUBIEF was performed by Laboratoire d'Hydraulique de France (LHF Grenoble) so that the validation team was independent from the development one. The validation document follows the recommendation of the Group of European Hydraulics Laboratories and provides the user with a data base of test cases representative of phenomena encountered in practical studies [4,5]. As a complement to the validation stage, feedback from users is taken into account through a Users Club.

4 Examples of application

4.1 Study of heavy metals in Seine river

Although the impact on the downstream ecosystem is very low, the Environment Department of EDF-DER and LNH are involved in the study of the influence of copper and zinc downstream of nuclear power plants. Numerical modelling is a part of the study which also includes accurate chemical and toxicity analyses.

The first two steps in the modeling proceedure consists of setting up the hydrodynamics and the sediment transport model. A 10 km long reach of the Seine river downstream of Nogent sur Seine was modelled with TELEMAC-2D. The erosion and deposition of suspended sediments were taken into account in SUBIEF using relevant source and sink terms,

$$S = Q_e - Q_d$$

where Q_d and Q_e are the deposition and erosion rate per unit area per second $(kg/m^2/s)$.

The deposition rate is given by Krone's formulation [7].

$$Q_d = W_s\, C\left[1 - \left(\frac{u_*}{u_{*d}}\right)^2\right] \qquad \text{for} \qquad u_* < u_{*d}$$

where u_* is the shear velocity, u_{*d} the critical shear velocity under which deposition occurs, C the concentration and W_s the settling velocity.

The erosion rate is given by Partheniades [6].

$$Q_e = M\left[\left(\frac{u_*}{u_{*e}}\right)^2 - 1\right] \qquad \text{for} \qquad u_* > u_{*e}$$

where u_{*e} is the critical shear velocity above which erosion occurs. It is dependant on soil concentration C_s. M is determined experimentally.

The computations predicted the regions of the river where deposition was most likely to occur. These corresponded to areas of low shear stresses. Deposition was found to be concentrated in a section of the river which is 300 m long. This region should be a sediment trap and field studies are to be intensified in this area.

The third step in the modelling proceedure is the development of the heavy metal transport model itself. Heavy metals are represented by dissolved and particulate phases. Adsorption and desorption of metal on the sediment are represented by first order reversible reactions which lead to an equilibrium characterized by the distribution coefficient Kd. The particulate metal can also be deposited on the bottom depending on the flow conditions.

Before the final model can be used to simulate resuspension of heavy metals during floods, calibration measurements have to be made. Experience has shown that the quality of the final results directly depends on the calibration of the model, although the calibration proceedure may be time consuming. The hydrodynamic model requires data from field surveys which calibrate the roughness coefficients of the river bed. The sediment submodel needs measurements and constraints on the settling velocity, the erosion rate, and the critical deposition and erosion shear stress for each granulometric class. The ion-exchange model is calibrated with both in situ and laboratory experiments which determine Kd for each metal.

4.2 Pollution of the Morbihan Gulf

A second application, supported by the administration of the Morbihan district, deals with the Morbihan Gulf, in south Brittany. It is a complex tidal system, with a large surface area (125 km2) compared to the width of its entrance (less than 1 km). The Gulf also has many islands and tidal flats illustrated in fig.2. Due to the location of the Gulf and its pleasant climate, tourist development and water quality has become a major concern for the Morbihan district. Hence a numerical model was set up to simulate the impact of pollution discharge at points around the Gulf. The aim of the model was to assist decision makers in choosing solutions which improved the protection of the environment [8].

The first step was to reproduce the complex flow patterns within the Gulf as a validation test for the model.

The flow was calculated with TELEMAC-2D on a mesh including more than 9000 nodes. The flow is dominated by the tide whose amplitude at the entrance reaches 4.3 meters. In the narrow straits strong currents can reach 9 knots, while, in contrast, many eddies appearing in the lee of islands have low velocities. The results from TELEMAC-2D are in close agreement with the available measurements and show many of the details of the complex flow pattern, e.g. correctly reproducing the secondary flow that exists in transversal channels.

Pollution due to faecal bacteria, mainly produced by sewage discharged by 18 outfalls, was modelled with SUBIEF. The faecal bacteria are non-conservative pollutants which survive for 1 or 2 days in this salty environment. T90 was calibrated by comparison with IFREMER measurements on oysters.

Concentrations that were lower than 10 bacteria per 100 ml were obtained almost everywhere in the Gulf and in the downstream part of the rivers (fig.2).

Nevertheless water quality might be improved in the upper reaches of some rivers. Refurbishment of water filtering systems in these area should improve the present situation.

In the near future the model will be extended to larger time scales to deal with other pollutants such as inorganic particles containing nitrogen or phosphorus.

5 Conclusion

In this paper, we have shown how a finite element system, initially devoted to hydrodynamics and sediment transport, can be used to deal with water quality problems. Work is still in progress which is attempting to extend the modelling to long term simulations. The choice of the ecological model and its calibration nevertheless remain a difficult problem.

The examples have shown that a mixed approach combining experimental and numerical models is required to produce results within a reasonable time frame. The broadening of the project team, which should include hydraulic engineers, environmentalists and economists, is another way to progress on applied studies.

Acknowledgements

We are gratefully indebted to:

- J. Gailhard for providing computational and experimental results of heavy metals on Seine river,

- J.M. Janin for providing computational and experimental results on Morbihan Gulf.

References

[1]   Hervouet J.M., Van haren L.
TELEMAC-2D release 3.0. Principle note. EDF-DER HE-43/94/052/B.

[2]   Hervouet J.M., Moulin C.
New advection schemes in TELEMAC-2D release 2.3. Assesment of the SUPG method. EDF-DER HE-43/94/052/B.

[3]   HERVOUET J.M.
Element by element methods for solving shallow water equations with F.E.M.
IXth International Conference on Computational Methods in Water Resources, Denver, USA, 1992.

[4]   MOULIN C.
SUBIEF release 3.1 : validation document. EDF-DER, to be published.

[5]   IAHR
Guidelines for documenting the validity of computational modelling software.

[6]   PARTHENIADES E.
Erosion of Cohesive Soils.*Journal of the Hydraulics Division*, ASCE, Vol. 91, 1965

[7]   KRONE R.B.
Flume Studies of the Transport of Sediment in Estuarine Shoaling Process. Technical Report Hydraulics Engineering Laboratory, Univ. of California, Berkeley, 1962.

[8]   JANIN J.M., MARCOS F.
Qualité des eaux du Golfe du Morbihan. Utilisation d'un modèle hydrodynamique. IVth Journées nationales Génie Côtier, Génie Civil, Dinard, France, 1996.

Water Quality file

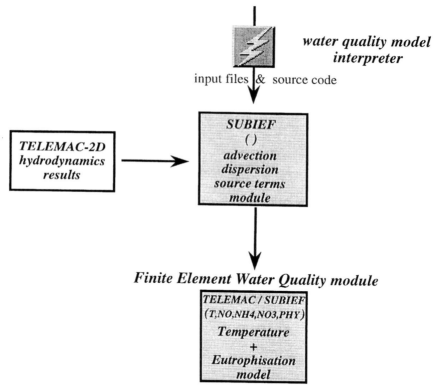

```
[T]   = Temperature (°C)
[NO]  = organic nitrogen (mgN/l)
[NH4] = mineral nitrogen NH4 (mgN/l)
[NO3] = mineral nitrogen NO3 (mgN/l)
[PHY] = phytoplancton (µgchla/l)
...
d/dt[T]   = 1/{ρ}{c}{H} ({RS}+{RA}-{RE}-{CV}-{CE})
{M} = {Mineralisation}*[T]/20*[NO]
{N} = {Nitrification}*[T]/20*[NH4]
d/dt[NO]  = {f}*((1-{d})*{DP})*[PHY] - {M}
d/dt[NH4] = {f}*({d}*{DP}-{e}{CP})*[PHY] - {N} + {M}
d/dt[NO3] = -{f}*((1-{e})*{CP})*[PHY] + {N}
d/dt[PHY] = ({CP}-{DP})*[PHY]
...
```

water quality model
interpreter

input files & source code

TELEMAC-2D
hydrodynamics
results

SUBIEF
()
advection
dispersion
source terms
module

Finite Element Water Quality module

TELEMAC / SUBIEF
(T,NO,NH4,NO3,PHY)
Temperature
+
Eutrophisation
model

Figure 1 : generation of water quality models within SUBIEF

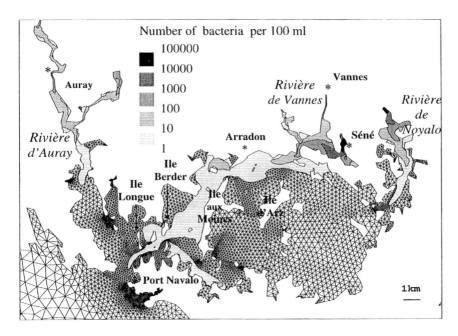

Figure 2 : Finite Element mesh of the Morbihan Gulf and
computed bacteria concentrations

The SIMULCO software: description of modelling and examples of application

J.-P. Roumégoux

Energy-Air Pollution Laboratory, The French National Institute for Transport and Safety Research (INRETS-LEN), 69675 Bron, France

Abstract

To assess the pollutant unit emissions for road vehicles, we combine engine emissions maps with engine operating conditions as calculated by a vehicle simulation model. The SIMULCO software, using Windows on a PC microcomputer, takes into account: the vehicles characteristics (engine, transmission, mass, aerodynamics ...); various operating parameters, such as the vehicle load or the use of accessories (ligthing, air conditioning ...); the trips characteristics (road gradients and curves, speed limits, stops); and the drivers characteristics (throttle opening, engine revolution speeds ...). This model calculates instantaneous operating conditions, in particular the engine torque and speed, wich in turn enables the calculation of pollutant emissions (CO_2, CO, HC, NOx, particulate) and of fuel consumption. It offers three possibilities: firstly driver simulation, where operating conditions and the vehicle dynamics are dependent on driving practice; secondly the kinematic following, as the case of a driving cycle, where the vehicle dynamic is imposed; and thirdly the calculations at steady speed. Results show the great influence on fuel consumption and pollutant emissions of vehicle mass, vehicle speed, road gradient, air conditioning use ..., for different kinds of vehicles travelling on various trips or driving cycles.

1 Introduction

Global concerns in terms of preservation of the environment and energy savings led to an ever increasing attention being paid to such requirements and a more accurate consideration of these phenomena in all the fields related to road freight and passenger transport. In France it is estimated that road transport account for about 70% of gas pollutant emissions such as carbone monoxide,

CO, unburnt hydrocarbons, HC, and nitrogen oxides, NOx, and for 25% of carbone dioxide, CO_2, emissions, which are all contributors to the greenhouse effect. There is thus no need to demonstrate the adverse effects of vehicle-related pollution on people health. Reduction in CO, HC and NOx emissions requires exhaust gas treatment, and reduction in CO_2 emissions requires reduction in fuel consumption. The SIMULCO computation model, which is a technical assessment software package, has been developed considering all these conditions. The first version was designed and developed with the financial support of the French Agence de l'Environnement et de la Maîtrise de l'Energie (ADEME: the French Agency of the Environment and Energy Control).

2 Software overview

The SIMULCO software package concern various assessments relating to road vehicles (passenger cars, buses and coaches, duty vehicles and heavy vehicles).It enables to calculate vehicle operating conditions, fuel consumption and pollutant emissions using simulation methods. This software is operated using a PC type microcomputer under Windows ™.

2.1 Operating principles

Four computation modes are available:
- the driving simulation mode: vehicle dynamics is determined by system input data including vehicle, road and driver parameters; at each computation step, corresponding to a second time interval, the vehicle instant speed is compared to a "target" speed which allows determining the driver's decision: accelerating, maintaining the speed or slowing down depending on the cases, gear changing ;
- the cycle monitoring mode: speed and engaged gear are imposed as a function of time; this corresponds to regulatory test cycles such as European cycles ECE15 and EUDC;
- the kinematic monitoring mode: only speed is imposed as a function of time; this can include for example data recording under real-world vehicle operating conditions;
- the steady speed mode: the vehicle is driven at steady speed on the road considering a prescribed road gradient and gear ratio.

2.2 Input data

Input data includes parameters related to the vehicle, the driver and the trip to be performed. Relevant files are created using software integrated modules and are stored in corresponding libraries. Vehicle-related data includes various vehicle technical characteristics (mass, aerodynamics, tires, engine and drive train), engine fuel consumption and pollutant emissions maps (CO_2, CO, HC, NOx and PM) and control and energy consumption features of the auxiliaries (alternator, engine cooling fan, lighting and air conditioning systems, ...).

Driver-related data, mainly used in the driving simulation mode, includes vehicle dynamic behaviour (centrifugal and longitudinal accelerations experienced, maximum speed), the gearbox use and the engine operating range (rotation speeds). Trip-related data differs depending on whether the trip is to be performed in the driving simulation mode, or considering the cycle or the kinematic to be used. In the first case, data must be collected as a function of the distance travelled. This includes road types (city, road and motorway) to which correspond speed limits, road gradients, curve positions and radii, stop positions and durations. In the second case, data is collected as a function of time: vehicle speed and if required the gear ratio selected. In addition, a number of data, considered as parameters, is entered when running the simulation programme: vehicle load (constant or variable), auxiliary uses, engine operating conditions at stop (idling or switched off engine).

2.3 Available results

In the most general case, i.e the driving simulation mode, the model calculates the vehicle position, speed, the gear ratio engaged, the engine torque and rotation speed, the energy of external forces applied to the vehicle, fuel and pollutant mass flows at second-time intervals over a prescribed trip. From the results obtained which can be stored optionally in a specific file, the integrated values over the whole trip are calculated:
- average vehicle speed (in km/h),
- average fuel consumption (in liter per 100 km),
- average pollutant emissions (in g/km and in g/kWh),
- energy balances of the power train (efficiencies in %) and of the vehicle (energies of applied forces in MJ/100 km),
- statistical data relating to engine operating conditions versus engine torques and rotation speeds (in % of time).
The software package enables to graph the results obtained (values calculated at second-time intervals), or a number of input data (for example road longitudinal profile), using a graph plotter after file formatting.

3 Description of modelling

3.1 Formulation of the physical phenomena

Simulation includes programmation logics; computation as such is carried out using different formulations such as power transfer equation and motion equation.

3.1.1 Power transmission equation
The power available at the vehicle driving wheels for the vehicle drive equals the engine supplied power less the transmission mechanical loss power, and less the auxiliaries mechanical power. During an acceleration, the engine supplied power is reduced by its angular acceleration power and the power available at

the driving wheels is also reduced by the angular acceleration power of all the rotating parts (transmission and wheels). The application of the kinetic moment and of energy conservation theorems allows drawing up the following equation:

$$C_{rm}.V/r = \eta.k_b.k_p.(C_m - C_{aux}).V/r - I.\gamma.V/r^2 \qquad (1)$$

where C_{rm} is the torque available at the driving wheels for the vehicle drive, C_m is the engine torque, C_{aux} is the auxiliaries torque, V is the road linear speed, r the driving wheels radius, η is the transmission mechanical efficiency, k_b and k_p are the gear box and driving axle ratios, I is the moment of inertia of all the rotating parts (engine, transmission, wheels), reduced to the driving wheel axle and γ is the vehicle acceleration. η, k_b, k_p and I are function of the gear selected.

3.1.2 Motion equation

Using the power transmission equation and the balance of the forces applied to the vehicle, the motion equation which determines the γ longitudinal acceleration from the C_m engine shaft torque and the F_r running resistance can be written as follows:

$$\gamma.(M + I/r^2) = \eta.k_b.k_p.(C_m - C_{aux})/r - F_r \qquad (2)$$

$$\text{with} \quad F_r = R.M.g.\cos\alpha + 1/2.\rho.S.C_d.V^2 + M.g.\sin\alpha \qquad (3)$$

where α is the road horizontal inclination angle, R is the tire rolling resistance factor (function of vehicle speed V), M is the vehicle mass, g the acceleration due to gravity, ρ the air density, S the vehicle frontal area, and C_d the aerodynamic drag coefficient. Acceleration γ is assumed to be constant for the computation step duration.

3.2 Engine torque

The calculation of engine torque depend on the computation mode selected (see 2.1). In the case of driving simulation mode, this torque is determined by the driver's decision; in particular, if the driver decide to accelerate, the engine torque is a function of the full load torque:

$$C_m = dtr.C_{mf} \qquad (4)$$

where dtr is the driver's torque ratio (generally between 0.5 and 1.0) and C_{mf} is the full load torque depending on engine speed. The driver's torque ratio is also dependent on the fuel injection control system: with an engine speed control system (case of lorries diesel engines) in the place of an engine load control system (case of cars petrol engines), dtr is always equal to 1.0.
In the others cases (cycle or kinematic monitoring and steady speed modes), the engine torque is calculated using eqn (2), after auxiliaries torque determining.

3.3 Auxiliaries torque

Modelling the power absorbed by auxiliaries is performed in a very simple way: input data includes the mean load at which each auxiliary is used on the travel to be follow (using ratio). The auxiliaries torque is thus a function of engine speed only. Modelling the auxiliaries features is also very simplified.

3.3.1 Alternator

Alternator input characteristics are the voltage U_a, the speed's ratio (alternator/engine) and two points of the full load curve (speed, maximum current intensity, and efficiency). These two points allows us determining the coefficients of expressions giving us the maximum current intensity I_{max} as a function of alternator speed N_a, and the electrical and mechanical power loss P_{a_loss} as a function of speed and current intensity I delivered:

$$I_{max} = i_0 + i_1.N_a \qquad (5)$$
$$P_{a\_loss} = p_{0a}.N_a + p_{2a}.I^2 \qquad (6)$$

Giving us the full load electrical power $P_{e\_aux}(i)$, and thus the total I intensity, and a using ratio $r_{e\_aux}(i)$ of each (i) electrical auxiliary (fans, ligthing, ...), we calculate the whole electrical power and finally the total power P_a absorbed by the alternator:

$$P_a = \Sigma r_{e\_aux}(i).P_{e\_aux}(i) + P_{a\_loss} \qquad (7)$$

3.3.2 Air conditioning

Air conditioning (cold air) input characteristics are the fan's electrical power, the speed's ratio (compressor/engine) and the mechanical compressor rated speed N_{c0} and power P_{c0}. These last values allows us determining the mechanical power absorbed by the compressor P_c as a function of compressor speed N_c and of the using ratio r_{cond}:

$$P_c = r_{cond}.P_{c0} . N_c/N_{c0} \qquad (8)$$

3.3.3 Mechanical fan and other mechanical drived auxiliaries

Mechanical drived auxiliaries input characteristics are two points of the full load curve (engine speed, power absorbed, and mechanical efficiency). These two points allows us determining the coefficients of expressions giving us the maximum available mechanical power P_{m_aux} and the power loss P_{m_loss} as a function of engine speed N_e:

$$P_{m\_aux} = m_0 + m_1.N_e \qquad (9)$$
$$P_{m\_loss} = p_{0m} + p_{1m}.N_e \qquad (10)$$

The total power P_m absorbed by the mechanical drived auxiliaries is also determined by the using ratios $r_{m\_aux}(i)$ of each (i) mechanical auxiliary :

$$P_m = \Sigma r_{m\_aux}(i).P_{m\_aux}(i) + P_{m\_loss}(i) \qquad (11)$$

3.4 Fuel consumption and pollutant emissions

Input data for each vehicle comprises massics flows of fuel consumption and pollutant emissions at idle, and fuel and pollutants maps (massics flows too) as a function of engine torque and speed. These maps have been established on engine test bench, at steady state under stationnary thermal conditions (warm engine). For a given engine operating point, defined by torque and speed values as calculated by the model, fuel consumption and pollutant emissions flows are determined by a double interpolation (or extrapolation) following the torque and speed on test bench measured values. If the so calculated fuel consumption is zero or less, then the fuel consumption and pollutant emissions are set to zero. With an engine speed control system (acting on fuel injection), the fuel consumption is set to zero when the vehicle is slowing down with a negative engine torque.

4 Examples of results

Four examples demonstrating the influence of the driver, the vehicle use and the road infrastructure were selected. They include two vehicles: a passenger car of Renault Clio type, equipped with a catalyst petrol engine, and a heavy vehicle of tractor-semi trailer type, eg. Renault R340, equipped with a diesel direct injection turbocharged engine.

4.1 Driving mode influence: Speed on motorway

Simulated driving modes for a 65 km motorway trip correspond to the following maximum speeds; gentle driving style : 110 km/h for the passenger car and 70 km/h for the heavy vehicle; average driving style : 140 km/h for the passenger car and 100 km/h for the heavy vehicle; very fast driving style : 180 km/h for the passenger car and 140 km/h for the heavy vehicle (heavy vehicle speed did not actually exceed 127 km/h).
The results shows the significant influence of maximum speed. For example, as regards the passenger car (table 1), fuel consumption can vary by a factor of 2 and CO emission by a factor of 40. It should be noted that NOx emission from the passenger car reached the maximum value for an average driving style and nearly equals zero for a very fast driving style.
With respect to heavy vehicle (table 2), pollutant emissions and fuel consumption vary to a lesser extent as compared to passenger cars. The mass effect is all the more significant since the vehicle weight is high.

Table 1: Fuel consumption and pollutant emissions for a passenger car on motorway, as a function of the driving style. Calculation parameters: half load operated vehicle (total weight 1175 kg) , no auxiliary use.

Driving style	gentle	average	very fast	fast-average (%)
average speed (km/h)	107.8	134.8	155.5	+ 15 %
consumption (l/100 km)	6.2	8.5	12.2	+ 44
CO2 (g/km)	140	181	204	+ 13
CO (g/km)	1.25	8.33	47.5	+ 470
HC (g/km)	0.06	0.21	1.04	+ 395
NOx (g/km)	0.41	1.02	0.07	- 93

Table 2: Fuel consumption and pollutant emissions for a heavy vehicle on motorway, as a functionof the driving style. Calculation parameters: half load operated heavy vehicle (total weight: 26750 kg), no auxiliary use.

Driving style	gentle	average	very fast	fast-average (%)
average speed (km/h)	66.1	89.5	97.3	+ 9 %
consumption (l/100 km)	28.4	35.0	38.4	+ 10
CO2 (g/km)	642	790	853	+ 8
CO (g/km)	2.47	1.77	1.64	- 7
HC (g/km)	1.05	0.91	0.96	+ 5
NOx (g/km)	7.76	8.66	8.96	+ 3
particles (g/km)	1.08	0.85	0.73	- 14

4.2 Influence of auxiliary use: The air conditioning unit case

About ten years ago, very few upper range European passenger cars were equipped with air conditioning units. At present, air conditioning units are available as a common optional extra, even for lower range vehicles such as Citroen AX or Renault Clio. The conditioning unit operating conditions have a significant impact on fuel consumption and pollutant emissions (table 3), both under road or motorway driving conditions and under urban conditions (European cycle ECE15). It should be noted that the conditioning compressor rated power amounts to about 4 kW, which is considerable. The results recorded were used to compare continuous conditioning unit operating conditions at full load (actually it is usually operated under intermittent conditions), and no use conditions. The car was equipped with other auxiliaries which were not operated (no load auxiliary conditions) thus generating no power consumption, except for the alternator which derived power from the engine shaft, even when it was not supplying any current. The significant impact of the air conditioning unit in urban conditions must be highlighted: fuel consumption and pollutant emissions are increased by 30%, and NOx emission by a factor of 3.

Table 3: Influence of the air conditioning unit operating conditions (no or with and variation in %) on fuel consumption and pollutant emissions for a passenger car, as a function of the route type. Calculation parameters: half load operated passenger car (total weight 1175 kg), for an average driving style.

	on the motorway			on the road			urban cycle		
	no	with	%	no	with	%	no	with	%
average speed (km/h)	135.1	133.3	-1%	82.6	81.0	-2%	18.7	18.7	0%
consumption (l/100 km)	8.7	9.7	+11	5.8	6.5	+12	8.2	10.6	+29
CO_2 (g/km)	185	196	+ 6	131	148	+13	186	243	+30
CO (g/km)	9.42	16.6	+76	0.91	1.14	+25	0.59	0.78	+32
HC (g/km)	0.24	0.42	+75	0.06	0.06	+ 0	0.10	0.13	+30
NOx (g/km)	1.04	0.92	- 12	0.21	0.37	+76	0.03	0.10	+200

4.3 Influence of engine operation at stop: idling or switched-off engine

Engine switching-off under interrupted traffic conditions is contemplated in the aim of reducing fuel consumption and pollutant emissions. The impact of such a measure depends on stop duration and on the system used for engine restarting. Potential gains are assessed (tables 4&5) for a passenger car over the ECE15 cycle and for a bus over an INRETS cycle. The gains recorded are significant, in particular with respect to the passenger car: 13% for fuel consumption and 10 to 20% for pollutant emissions.

Table 4: Influence of engine operating conditions at stop on fuel consumption and pollutant emissions for a passenger car in urban conditions. Calculation parameters: equivalent inertia 1130 kg, no auxiliary use.

Operation	idling engine	switched-off engine	variation %
consumption (l/100 km)	8.2	7.1	-13 %
CO_2 (g/km)	186	161	-13
CO (g/km)	0.59	0.47	-20
HC (g/km)	0.10	0.09	-10
NOx (g/km)	0.03	0.02	- ? (*)

(*) insufficient accuracy

The gains to be obtained under real-world conditions are lower than the indicated values. Pollutant emissions and fuel consumption under engine restarting conditions must be taken into account, except if the energy required for engine starting has been previously regenerated (under deceleration conditions for example) and stored.

Table 5: Influence of engine operating conditions at stop on fuel consumption and pollutant emissions for a bus. Calculation parameters: 1/3 load operated bus (total weight 14100 kg), no auxiliary use.

Operation	idling engine	switched-off engine	variation %
consumption (l/100 km)	39.7	38.2	- 4 %
CO_2 (g/km)	1100	1065	- 3
CO (g/km)	6.36	5.92	- 7
HC (g/km)	1.66	1.50	- 10
NOx (g/km)	25.3	24.6	- 3
particles (g/km)	0.75	0.67	- 11

4.4 Road gradient influence under steady speed conditions

Fuel consumption and NOx emission shows the following variations as a function of the road gradient (figure 1). With respect to the passenger car, at 130 km/h, fuel consumption increases with the road gradient by about 1 litre/100 km per gradient percent. NOx emission which is very low under high gradient conditions (- 4% and + 4%) reaches its maximum value for a gradient close to 0.5%. With respect to the heavy vehicle, relative fuel consumption variations are similar to those observed for NOx emission. At 100 km/h, fuel consumption and pollutant emissions equal zero for gradients lower than or equal to - 2%. Between - 2% and + 4%, fuel consumption increases with the road gradient at a rate of about 24 litres/100 km per gradient percent. The 42 l/100km value, on a horizontal road at 100 km/h, using the 18th gear, increases to 145 litres/100 for a 4% gradient travelled at 30 km/h using the 10th gear.

5 Conclusions

5.1 Anticipated applications

SIMULCO has a twofold application: computer-aided design and evaluation. The model is used to assess the impact of a number of variations in the road-vehicle-driver system, in energy consumption, pollutant emissions and vehicle performance. In terms of computer-aided design these applications can be of interest to car makers and motor vehicle equipment manufacturers (impact of technological changes), and to developers of road infrastructures (studies into alternative road projects), etc...In the evaluation field, these applications are aimed at organizations such as ADEME and a number of R&D departments: comparison of various routes, vehicle selection aid, influence of the driving mode, etc.

5.2 Development prospects

The first SIMULCO version must be considered temporary; studies must be continued to test good software operating conditions under all possible cases, and to validate the results recorded on the vehicles using a comparative method. Then a number of enhancements will be contemplated: considering vehicle operation under cold engine conditions, leading to excess fuel consumption and pollutant emissions with respect to stabilized thermal conditions; this will require developing the modelling of thermal exchanges to predict temperature changes (cooling water, exhaust gases, etc.). In addition, software adaptations can be anticipated considering specific applications such as assessing the energy-air pollution balance for a road infrastructure, which requires the development of a number of interfaces in terms of software input data and additional calculation in terms of output data.

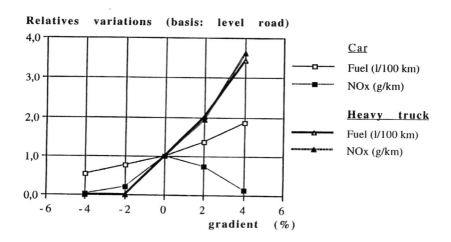

Figure 1: Relative variations in fuel consumption and NOx emission at steady speed as a function of the road gradient (base: horizontal road). Calculation parameters: half load operated passenger car (total weight 1175 kg), no auxiliary use; full load operated heavy vehicle (total weight 40000 kg), no auxiliary use.

Reference

Roumégoux, J-P. *Modèle de simulation Simulco, Version 1.0 - Organisation, fonctionnalités, utilisation,* rapport LEN 9501, INRETS-ADEME, 1995.

A parallel cellular simulator for bioremediation of contaminated soils

S. Di Gregorio,[a] R. Rongo,[a] W. Spataro,[a] G. Spezzano,[b] D. Talia[b]
[a]Department of Mathematics, University of Calabria, 87036 Rende (CS), Italy
[b]CRAI, Località S. Stefano, 87030 Rende (CS), Italy

Abstract

The bioremediation of contaminated soils is one of main strategies for site clean-up. The most important principle of bioremediation is that micro-organisms (mainly bacteria) can be used to destroy hazardous contaminants or transform them into a less harmful form. Currently, we are facing this problem in the CABOTO project within the PCI ESPRIT framework. The CABOTO objective concerns the design and implementation of a parallel simulator for the bioremediation of contaminated soils by using models based on the cellular automata (CA) theory. For the parallel implementing of the simulator has been used the CAMEL system, a parallel environment for the simulation and modelling of complex systems based on CA.
This paper describes the model used to simulate the contamination and the bioremediation of the soil, the main features of the CAMEL system and the parallel implementation of the simulator by CAMEL. Finally, experimental results are described.

1 Introduction

For the implementation of tools that allow the prediction of the behaviour of complex dynamic phenomena it is necessary to have models that can describe with accuracy the real systems. The implementation of these models often requires massive amounts of computation that can be only offered by parallel processing. Massively parallel computers allow to simulate scientific and engineering problems exploiting high performance. They represent the natural computational support for modeling and simulation of large complex phenomena and systems. The adoption of high performance computing simulation in a wide variety of sectors and activities demonstrates that this technology can be used to improve complex industrial processes in terms of efficiency, safety and security of service.

Soil decontamination is a major environmental problem in all industrial countries: while different decontamination methods are known, the evaluation of their effectiveness requires a number of laboratory or pilot plant tests. However, appropriate forecasting of real field results from these tests could be achieved only by resorting to mathematical models, which are fairly complicated and require high computing power. An important goal in order to face soil pollution problems is the ability to simulate contaminant transport, diffusion and transformation processes for reliably estimating their impact and forecasting the effects of different decontamination strategies.

Several factors are involved when the full cycle is considered: contamination by liquid transport, adsorption, precipitation, chemical reactions, biological degradation kinetics and so on. If the classical approach of differential calculus is adopted, particular complexities could arise because of heterogeneous soil composition and because of the intricate physical-chemical-biological interactions. In fact, such a way is very hard to be followed because it is extremely difficult to solve analytically the equations for such phenomena without making simplifications that neglect in many cases substantial characteristics of the phenomenon.

Here we propose a CA approach like Karapiperis & Blankleider [1], where the fluid flow with diffusion-transport of contamination-decontamination agents inside the soil and their mutual influences are viewed as a dynamic system based on local interactions with discrete time and space, where the space is represented by cubic cells. Each cell is characterized by specific values (the state) of selected physical parameters, representing physical-chemical specifications, relevant to the evolution of the phenomenon. A transition function is defined in order to describe state variation and chemical and biological interactions. This model has been implemented using CAMEL a parallel software environment based on cellular automata theory developed by Cannataro et al. [2].

2. Bioremediation model

In the specification of a CA for soil contamination-decontamination, we refer to the contamination of a porous soil caused by a phenol rich water solution. This problem involves two phases only (water and air) while of course in other cases it might be necessary to consider additional phases (e.g. an oil phase in the case of hydrocarbon contamination).

The CA model that is used to simulate the phenomenon is composed by a large number of cells, where each cell being of mesoscopic size does not describe the phenomena which take place inside a single pore (typical pore dimensions being 10-100 μm) but those inside a portion of soil. Our model is three-dimensional and is necessary for real scale simulations, in order to describe heterogeneous regions.

The cellular automaton specification is given by:

$$SOIL = (R_3, A_1, A_{6a}, A_{6b}, X, Q, \sigma, \gamma_1, \gamma_{6a}, \gamma_{6b})$$

where

- $R_3 = \{(x, y, z) | x, y, z \in N, 0 \leq x \leq l_x, 0 \leq y \leq l_y, 0 \leq z \leq l_z\}$ is the set of points with integer co-ordinates in the finite region, where the phenomenon evolves. N is the set of natural numbers.

- $A_1 \subset R_3$, specifies the water source cells at air-soil contact, where contamination ends; such cells obey to the particular transition function γ_1 substitutive, but compatible with σ. In cells of A_1 the contaminated water appears.

- $A_{6a}, A_{6b} \subset R_3$, specify the two last down layers of cells, which obey respectively the particular transition functions γ_{6a}, γ_{6b} different from σ.

- The set X identifies the geometrical pattern of cells which influence the cell state change. They are the cell itself and the "up", "north", "east", "west", "south" and "down" neighbouring cells, which are individuated by the indexes 0, 1, 2, 3, 4, 5, 6

$$X = \{(0, 0, 0), (0, 0, 1), (0, 1, 0), (1, 0, 0), (-1, 0, 0), (0, -1, 0), (0, 0, -1)\}$$

The finite set Q of states of the elementary automaton:

$$Q = Q_w \times Q_p \times Q_{outf}^6 \times Q_{K\_sat} \times Q_{cap\_thr} \times Q_{c\_ph\_s} \times Q_{c\_ph\_w} \times Q_b$$

where the substates are:

* Q_w is the water content in the cell.

* Q_p is the *effective porosity* parameter. It depends on physical-chemical characteristics of the soil in the cell and individuates the quantity of water drainable from the cell in condition of full saturation, i.e. the porous volume, which can be filled by water. For example, a cell with completely impermeable soil inside has the *effective porosity* parameter equal to zero, while the *effective porosity* of a cell is greater if the soil is more permeable.

* Q_{outf} represents the water flow toward the six neighbourhood directions from the central cell. Note that the inflows are not explicitly considered, they are obtained trivially by the outflows.

* Q_{K_sat} is the hydraulic conductivity at the saturation, considering that, inside the cell, the soil is homogeneous.

* Q_{cap_thr} is the *capillary water threshold* substate, i.e. the maximum cell water which is effected by the capillary forces.

* $Q_{c\_ph\_w}$, accounts for the phenol concentration in the cell water.

* $Q_{c\_ph\_s}$, accounts for the phenol mass adsorbed in the cell per unit mass of dry solid.

* Qb, is the biological mass inside the cell.

• $\sigma:Q^7 \rightarrow Q$ is the deterministic state transition function for the cells in R_3, a large outline of its specification will be given in the next sections.

• $\gamma_1:Q^2 \times N \rightarrow Q$ is the transition function of source cells, where water and solutes have *origin* at each SOIL time interval $t \in N$; according to the assigned feeding, the outflow to the inferior cell only is calculated.

• $\gamma_{6a}:Q^2 \rightarrow Q$ is the deterministic state transition function for the cells in the first down layer, it specifies the behaviour of ideal cells, which receive water only from the superior cell and transmit it to the inferior one without retention.

• $\gamma_{6b}:Q^2 \rightarrow Q$ is the deterministic state transition function for the cells in the second down layer, it specifies the behaviour of ideal cells, which receive illimitable water without transmit it at all.

Such ideal cells with strange transition functions γ_{6a}, γ_{6b} are considered in order to describe forced conditions in laboratory experimental devices.

At the CA step 0 the initial configuration is defined, specifying all the starting values of the cell substates. Usually the water outflows may vanish. At each next step, the function σ is applied to all cells in R_3 {$A_1 \cup A_{6a} \cup A_{6b}$}, while at the same time step the functions γ_1, γ_{6a}, γ_{6b} are respectively applied to A_1, A_{6a}, A_{6b}, so the configuration is changed and an evolution step of SOIL is obtained.

Let us remark that contamination is a non stationary process, so that some usual simplifications cannot be used and really dynamic models are required.

3 CAMEL overview

CAMEL has been implemented on a parallel computer composed of a mesh of 32 Transputers connected to a host node. The current implementation of CAMEL does not limit the number of Transputers which can compose the parallel computer, so no changes should be necessary in the software of CAMEL whether a very large number of Transputer should be used.

The system is composed by a set of *macrocell* processes each running on a single processing element of the parallel machine and by a *controller* process running on a master processor. Each *macrocell* process implements several elementary cells and makes use of a communication system which handles the data exchange among cells.

CAMEL offers the computing power of a parallel computer although hiding the architecture issues to a user. It provides an interface which allows a user to dynamically change the parameters of the simulation during its running, and a graphical interface which shows the output of the simulation. The approach taken in CAMEL is to make parallel computers available to application-oriented users hiding the implementation issues coming from their architectural complexity. CAMEL allows the solution of complex problems as described in Di Gregorio et al. [3], which may be represented as discrete across a square or hexagonal grid or lattice.

CAMEL implements a cellular automaton as a SPMD (*Single Program Multiple Data*) program. CA are implemented as a number of processes each one mapped on a distinct PE that executes the same code on different data. According to this approach a user must specify only the transition function of a single cell of the system he wants to simulate by CARPET, a cellular language defined to program CA applications in CAMEL. CARPET (*CellulAR Programming EnvironmenT*) is a high-level language that extends C with some additional constructs to describe the rules of the transition function of a single cell of a cellular automaton. The main features of CARPET are the possibility to describe the state of a cell as a set of typed substates each one by a user-defined type, and the simple definition of complex neighbourhoods (e.g., hexagonal), that can be also time dependent, in a n-dimensional discrete cartesian space.

The CAMEL system allows, by the Graphical Interface (GI), to rapidly and interactively explore and analyse very large amounts of scientific data gathered during the execution of computer simulations.

A tool called *IVT (Interactive Visualization Tool)* has been added to CAMEL to improve data visualization. Utilizing data computed by simulation, *IVT* provides a variety of functions and services, including 2 and 3-dimensional graphical displays of data, hard copy of graphical displays and text, interactive colour manipulation, animation creation and display, rotation of the images, saving of data in files according to different data formats. Further, the tool allows quick access to the software for the novice as well as the advanced user and provides an intuitive easy-to-learn-and-use interface.

One of the most well-known visualization software package is certainly MATLAB. It offers outstanding numerical computation functionality together with sophisticated graphical presentation. By MATLAB we designed IVT provides a window-oriented graphical interface for an interactive manipulation of graphical displays.

The integration of IVT with the CAMEL system allows to extract more information from raw data gathered during the simulation using the MATLAB's graphics capabilities (i.e. 3D visualization of 2D images) and the numerical transformation functions. A software interface controls the communications between CAMEL and IVT. The communication occurs connecting dynamically, under Windows, the two applications. Further, this approach will provide the remote visualization of data and the possibility to use IVT with other parallel applications.

4 Model implementation

The bioremediation model has been implemented using the CARPET language. The transition function σ is composed of three *sequential steps* one for each of phenomena that constitute the model. An *evolution step*, performed by each cell, of the transition function concerns:

1. computing the water outflows from a cell toward the adjacent cells;
2. computing the solute adsorption/desorption inside a cell and the solute dispersion from a cell toward the adjacent cells;
3. updating the cell substates considering the chemical-biological effects of biomass and its growth/reduction.

During the first step the new water substates values are computed according to the following calculations, which must be executed sequentially in each cell:

- first of all, the new water content in the cell is calculated, considering the water balancing between current content *cell_w*, inflows and outflows; inflows and outflows are determined by the contributions of capillary (*infc* and *outfc*) and gravitational (*infg* and *outfg*) inflows and outflows;
- the new outflows determined by the capillary forces (*outfc*) are calculated;
- the outflows determined by the gravitational forces (*outfg*) and the new global outflows (*outf*) are calculated.

The water outflow from a cell (or alternatively part of the inflows into the adjacent cell) depends on the capillary and gravity forces; we will account on the main features of the transition function for the water flow.

Capillary forces involve a suction effect, which covers the porous surface with a water layer. The maximum possible thickness of the water layer depends on the attractive inter-molecular forces between water, air and soil.

We can characterise the equilibrium by a suction threshold (*cell_cap_thr*), which is the minimum water quantity inside the cell above, when the suction effects are practically negligible. Part of outflows (*outfc*) from the central cell to its neighbours are originated because of balancing the capillary forces, when the value of *cell_cap_thr* is not reached in the neighbours.

The capillary forces can be considered approximately proportional to the suction index *suct_ind*, here defined as ((*cell_cap_thr* - *cell_w*) /*cell_cap_thr*), when   *cell_w* < *cell_cap_thr* (0 otherwise). So a balance can be given by outflows, which minimise the total difference of the suction index between a cell (the central cell) and its six neighbours after each time interval.

Computation involves two sequential actions :

- Selection of the neighbouring cells (the so-called "eliminated cells"), toward which capillary water outflow is not possible, because their suction index is larger than the suction index of the central cell or because other neighbouring cells have by comparison so low suction index that the capillary water flows necessarily only toward such cells.

- Computation of the values of the outflows toward the not eliminated cells, in order that the same value of the suction index is assumed by the central cell and the not eliminated cells.

The gravitational water in a cell is the part of water which is practically not sensible to capillary forces. Its flow is ruled by the hydraulic conductivity, which depends upon the hydraulic conductivity at the saturation K_sat, the porosity p and the water content w according to a function depending on the type of soil.

In order to calculate the flow between the cell and its neighbour down, the conductivity value, which is considered, is the minimum of the conductivities of the two cells.

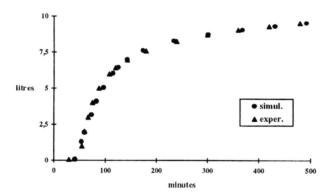

Figure 1. The simulation data of water flow.

The simulation of the first hours of an experiment is reported in figure 1. The model captures a fluid-dynamical behaviour, difficult to be simulated. The experimental results reported here for comparison have been performed in a pilot plant, under carefully controlled conditions by Banti et al. [4]. The model involves a number of parameters, which can only partly be directly determined. The remaining parameter values have been found either by tuning them at hand, or by applying the novel optimization technique of genetic algorithms.

4.1 The solute flow through the soil

A dynamic description of the adsorption/desorption of solute onto the pore wall was found necessary, and two different techniques to achieve such dynamic description have been tested:

1. The solute distribution between bulk water and pore walls described by the *Freundlich* isotherm was considered as the goal, and a linear dynamics towards that goal was defined: at each time step, only a fraction of the difference between the actual and the goal values was allowed to move from the bulk to the wall or vice versa.

2. A mass balance equation for the solute was used, with a term, describing the flow from bulk to walls, proportional to the solute concentration in the bulk, and a term, describing the opposite flow, proportional to the concentration of the adsorbed solute.

Moreover, a further aspect that has been taken into account is the water moves from one cell to another, carrying with it its solute content, a mixing takes place with the immobile water (i.e. residual + irreducible) which is present in the target cell. Therefore the solute concentration (at least initially) decreases, as it is diluted in a larger amount of water. Since the immobile water may take fairly high values, this dilution phenomenon may be quite important. Moreover, it would be unrealistic to suppose that an instantaneous dilution in all the immobile water may take place, so it could be limited to a fraction of the overall immobile water. Here again a set of models has been considered: in all simulations the mixing with the immobile water has been introduced in the model, but in to two different ways, explicit or implicit. In the former case a parameter has been used, which describes the fraction of immobile water involved in mixing, while in the latter case the phenomenon has been implicitly taken into account by assuming a lower value of immobile water than the experimental one..

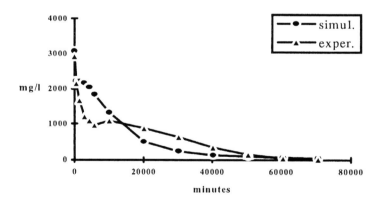

Figure 2. The simulation data of drained water contamination.

Figure 2 compares the experimental data of the phenol concentration in solution with the simulation results

4.2 The biomass transition function

Strictly biological effects are here considered, i.e. metabolic processes, which alter the contaminant or the nutrient as well as bacterial reproduction processes. Bacterial colonies may be either mobile or substantially fixed. The motion of bacteria could be described by the same techniques useful for solute transport, adding possibly chemiotropism.

So far not enough data are available to decide about the necessity to take into account bacteria transport in the case of interest here. We will therefore limit ourselves to describe the algorithm for bacterial metabolism and growth within each cell: this will suffice if bacterial transport can be neglected, while, if it is not the case, the appropriate transport equations should be also included.

Looking to the Monod formulae, we propose to adopt any solution method suitable for a finite cubic closed volume, considering a time interval Δt, equal to the computation step of the CA.

5 Experimental results

We measured the water flow and contamination model execution times (seconds) on our Transputer-based machine for different number of cells per processor, iterations and processor configurations.

Figure 3 shows the speed-up results for a 32x17x13 (7072 cells) grid with 10 substates for the water flow model and 12 substates for the contamination event model. Comparing the results of the simulation with the experimental data we deduce that 100 steps of the simulation correspond to about 12 minutes of the real phenomenon.

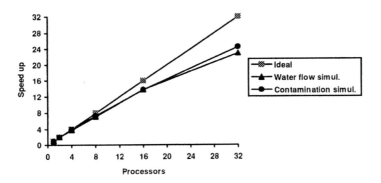

Figure 3. Speed-up measures.

Furthermore, we point out that the execution time of a single step is about 0,5 seconds. This result is very important because many iterations are necessary to complete our simulation on a long period basis (64 days). Consequently, it is worth noting that 24 hours of the real phenomenon can be simulated in about 1 hour and 40 minutes.

Figure 4 shows a snapshot of the water flow simulation by IVT tool described in section 3.

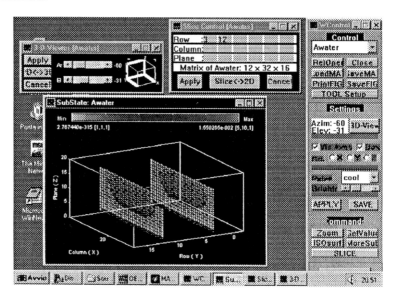

Figure 4. A snapshot of the water flow simulation by the IVT tool.

References

1. Karapiperis T & Blankleider B. Cellular Automaton Model of Mass Transport with Chemical Reactions, *Bericht Nr.93-06 Paul Scherrer Institut*, Villingen, Schweitz,1993.

2. Cannataro M., Di Gregorio S., Rongo R., Spataro W., Spezzano G. & Talia D. A Parallel Cellular Automata Environment on Multicomputers for Computational Science, *Parallel Computing,* 1995, **21**, 803-824.

3. Di Gregorio S., Rongo R., Spataro W., Spezzano G. & Talia D. A Parallel Cellular Tool for Interactive Modeling and Simulation, *IEEE Computational Science & Engineering,*1996, **4**.

4. Banti A., Mazzullo S., Giorgini M. & Serra R. A Detailed Analysis of in-situ Bioremediation: Pilot Plant Experiments with a Phenol-Contaminated Soils, *in Proc. of the Conf. Water Resources and the Environment*, Como, 1996.

A generative computer tool to model shadings and openings that achieve sunlighting properties in architectural design

D. Siret
Laboratoire CERMA Ura CNRS, École d'Architecture de Nantes, Rue Massenet, 44300 Nantes, France

Abstract

We present a generative tool based on a new simulation method for making sunlighting an actual formgiver in architectural design. Our system reverses the common simulation process. It works with intuitive sunlighting properties such as " this area must be sunless the afternon in summer ". Given a property, it computes a complex geometrical volume figuring the sunlighting phenomenon in time and space. This volume provides a visualization of the sunlighting constraint and it enables the designer to model the different solutions — shadings or openings — that exactly check on the given property. We illustrate the system with two demonstrative examples in architectural design and we introduce some new developments under consideration at the present time.

1 Generative computer tools purpose

1.1 Generative vs evaluative tools

In architectural and urban design context, many simulation tools have been developped for appraising interactions between constructions and their environment. Most of these tools are based on an evaluative approach. They use a simplified or a physical model of a phenomenon to compute the resulting state of the phenomenon in the future construction : thermal response, lighting intensity, noise level, and so on.

These simulation tools generally need a full description of buildings (topology, geometry, materials). In architectural and urban design processes, they cannot be used before the main schemes are completed. As they work on definite plans, these evaluative tools do not suit design practice. They come at the latest stage of design, whereas key decisions are already taken. Whenever the computing results do not match the architect's wishes or the program cons-

traints, the design has to be started again or, more often when possible, corrective solutions must be found. This is the well-known iterative process :

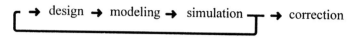

As opposed to this evaluative process, generative computer tools emphasize a constructive point of view. A definition of these generative tools has
been recently proposed by L. Khemlani [1] : « These are tools which are able to
generate alternative solutions for various limited aspects of the design that satisfy some given well-definable specification. A generative process is essentially one of searching through all possible solutions to a given problem to find
those that meet specific goals. The idea of generative tools is *not* to produce *the*
solution to a problem (...). Instead, a generative tool aims to produce alternative
solutions meeting *some* objective criteria, possibly in the form of ' tests '.
Using more subjective, non-computable criteria, the architect can then make a
selection from among the solutions offered by the computer, with modifications and refinements if necessary. »

Following this definition, generative approach reverses the common simulation process in environmental design. Its purpose is not to compute the state
of a phenomenon in a given construction, but rather to suggest ' all possible '
constructions that achieve a given state of a phenomenon for a given environment. Thus, for generative computer tools, the question to answer is not : *what
kind of properties can I observe for this construction in this environment ?* but
rather : *what kind of construction(s) must I build in this environment to achieve
these properties ?* In that way, generative tools suit design practice, as suggested by R. Woodbury [2] since 1991.

1.2 The reverse simulation approaches

Generative tool starts from result — the property to achieve — to produce alternative solutions that make this property real. Two ways of realizing this reverse simulation process can be found : the ' Generate and test ' method which
is used by Khemlani to design windows configurations from an energy couns-
cious point of view, and the ' Inverse simulation ' method that we advocate in
this paper, following Schoeneman & al. [3]

1.2.1 Generate and test method

For a specified environmental property, this method consists of :
  • firstly, defining and generating a set of solutions available to design,
  • then, evaluating all these solutions to find those that meet the given
    property.

Khemlani used this method for window design. Different windows configurations available are first generated (within important constraints in order to
reduce the combinatorial explosion and to enable the simulation) and then, they
are tested with a simple simulation model. The subset of windows that achieve

the desired property, from which the designer may choose one, is considered as the solution to the problem.

1.2.2 Inverse simulation method

The second method considers that generative tools reverse the classical process of simulation, that is in some way, they inverse the simulation model. Given a property to achieve, this method consists of :

- firstly, computing an inverse model of the simulation process for the given property,
- then, using this model to define alternative solutions.

All solutions that the inverse model provides achieve the property. As the set of solutions may be infinite, new kinds of properties can be brought in to explore them from different points of view. This method has been proposed by Schoeneman & al. to determine the light intensities that most closely match a desired lighting effect drawn in an architectural scene. It was there introduced within the larger field of ' inverse problems ' studied in physical Sciences.

Inverse simulation method is opposed to direct one as shown on figure 1 below. The diagrams illustrate these opposite processes for a given phenomenon (the arrow) in a given context (the profile of a simple building). Direct simulation methodology consists of computing the state of the phenomenon (the gray shape) for the given context. On the contrary, inverse simulation starts from the result. Given a state of the phenomenon, it consists of computing the ' solutions ' which enable the construction to generate this state. For instance, these solutions may describe the way of transforming the building in order to achieve the given state, if necessary.

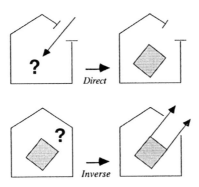

Figure 1. Direct and inverse simulation processes

2 A method for the inverse simulation of sunlighting

Direct sunlighting simulation is a well-known process based on geometrical methods. It consits of computing all sunless or sunlit shapes in a given geometrical construction, for a given environment and a given time period [4].

The inverse problem is less known. It aims at determining the solutions that achieve a given sunless or sunlit geometrical spot in a given construction, for a given time period. These solutions are members of two wide families of architectural objects : the openings (holes) that achieve ' sunlit ' properties, and the shadings that achieve ' sunless ' properties. It is clear that, at least for shadings, many alternative solutions can be found to check on a given sunlighting property (a characteristic of inverse problems).

2.1 The sunlighting property to achieve

Datas and results are reversed in both direct and inverse simulation processes. For the inverse sunlighting problem, datas are the parameters of the sunlighting property to achieve. Such a property can be formalized as a set of three parameters (P, S, T) where :
  • P is the geometrical spot that receives sunlighting (a convex polygon),
  • S is the sunlighting qualification (sunlit or sunless),
  • T is the time period during which P must check on the qualification S.
Thus, a sunlighting property (P, S, T) is written : " P must be S during T ". It can be composed at any stage of the design process, to define the massings, to make windows, to draw sun visors, and so on.

2.2 The complex sunlighting volume associated with a property

The target is to make a given sunlighting property true : " the spot P *must be* sunlit (or sunless) during the time period T ". Our inverse simulation method is based on the computation of the complex sunlighting volume defined by P and T. We denote $\Pi(P, T)$ this volume (an artefact connecting time and space). Any sunlighting property (P, S, T) generates such a volume which embodies the exact set of points that may shade the polygon P for all the instants within the given time period T.

The three diagrams on figure 2 (next page) illustrate the method we use to build the sunlighting volume Π associated with a given property (P, S, T). Firstly, we compute the simple volume π defined for any point p of the plane of the polygon P, and for the time period T figured as a geometrical shape on the sky vault. The volume π embodies all the shading points for the point p during T. It has its vertex on p and its edges follow the solar directions p-t_i defined by the time interval T. As described in a previous paper [5], we use an original meshing of solar trajectories to make the shape T square with some intuitive lived time periods such as ' the end of the afternoon in spring ', ' early morning in May ', and so on.

Geometrically speaking, the complex volume $\Pi(P, T)$ results from the boolean union of all the simple volumes π set in each point of the polygon P. Assuming that P is convex, the result of this union squares with the displacement of π round the perimeter of P (figure 2, at the left bottom). This displacement produces the boundaries of the sunlighting volume $\Pi(P, T)$. It is computable using common boolean algorithms between polygons.

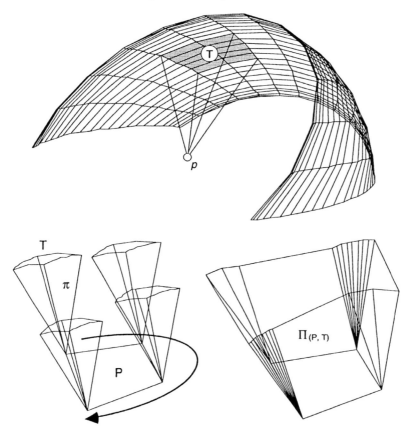

Figure 2. Three steps in the construction of $\Pi(P, T)$.

2.3 Shading and opening solutions

Given the volume $\Pi(P, T)$ associated with the property (P, S, T), one can define the solutions to the inverse sunlighting problem, that is : the openings or the
shadings that enable the polygon P to be S during the time period T. In other
words, these solutions make the property (P, S, T) become true.

An opening is any hole that lets the sunbeams reach the spot P during T. A
shading is any object that makes a shadow on P during T. There always exists
an opening or a set of openings (a shading or a set of shadings) that checks on a
given sunlighting property.

Therefore, a sufficient condition for P to be sunless during T is that there
exists at least one object intersecting the sunlighting volume $\Pi(P, T)$. Recipro-
cally, a necessary condition for P to be sunlit during T is that no object in the
scene intersects, even partially, the volume $\Pi(P, T)$.

Given these simple rules, there are two ways of achieving a sunlighting property when necessary :
- to create the opening(s) resulting from the intersection between $\Pi(P, T)$ and all the objects in the scene, for type (P, Sunlit, T) properties,
- to model at least one shading object that intersects the volume $\Pi(P, T)$, for type (P, Sunless, T) properties.

Examples introduced in section 3.2 below illustrate both methods.

3 A generative tool for sunlighting design

3.1 Overview

We have developped a generative tool based on this inverse simulation method. This tool works in association with a common CAD system. It provides interactive implements to define a sunlighting property in any complex architectural or urban scene. Given such a property :
- a rectangle that figures the sunlighting zone P,
- the qualification *sunlit* or *sunless* S applied to this zone,
- an intuitive time period T,

... the system computes the associated sunlighting volume. Then, according to the qualification S, it generates new openings or shadings that achieve the property as explained below. In addition, our system provides direct simulation tools for appraising real sunlighting states before composing hypothetical properties.

3.1.1 Openings modeling

Openings that check on a ' sunlit ' property are computed as exact geometrical intersections between the sunlighting volume associated with the property and all the objects of the scene. If an object cannot be drilled, the property cannot be achieved and the process fails. Otherwise, all the intersections are created. These openings must be then designed to get an architectural or urban relevance in the project context : windows, bays, places, etc. Facing the ' raw holes ' computed by the system, a designer is free to give his own architectural interpretation. Of course, the initial spot P is slightly altered as exact openings are transformed into realistic windows.

3.1.2 Shadings modeling

Most of the time, the number of shadings that achieve a ' sunless ' property is infinite. The system uses geometrical planes, that define polygons when intersecting the sunlighting volume, to figure shadings. It first suggests a default plane resulting in any polygon. Then, it enables the designer to graphically modify this plane and to explore different solutions. The shading polygon is automatically transformed as and when the plane is. Any kind of ' raw shading ' can be outlined in that way : tree, façade, porche, sun visor, and so on. Such as openings, the exact shading polygons computed by the system have to be designed to get an architectural relevance in the project context.

3.2 Demonstrative examples in architectural design

Graphics on the next pages illustrate both methods. The project concerns a small building in France (latitude 47° N), which volumetry has been fisrt outlined following the program. Yet, two openings had to be made.

For the first one (figure 3 a, b, c, d), the architect decided to enable the area figured as the rectangle P to be sunlit during the time period T ' the middle of the morning in winter '. Given the property (P, sunlit, T), the system computed the sunlighting volume shown on (b) and then, created the exact opening drawn on (c). The architect followed his own interpretation to design the window figured over the hole in graphic (d). As this window recovers the largest part of the exact opening, the resulting sunlighting spot for the time period T closely matches the rectangle P.

The form of the second opening (figure 4 a, b, c, d) has been decided under the shape of the rectangle P on the South-West façade. For such a large bay, the problem is to avoid summer-time overheating. Thus, the property (P, sunless, the beginning of afternoon in summer) has been first composed and the system computed the associated sunlighting volume shown on graphic (b). The process the designer used to explore alternative shading planes that achieve the property cannot be figured here. The result is an incline plane intersecting the sunlighting volume through the exact shading polygon shown on (c). Following his interpretation, the architect finally turned this polygon into an architectural sun-visor and designed the opening as shown on (d).

4 Future developments

These concern extensions to the existing tool which all converge to a best integration of the generative methodology into the design practice. Three main objectives are outlined below.

4.1 Management of numerous properties

Our system deals with properties by turns. A more suitable method would be to manage all properties together, that is to detect and to solve incompatible or redundant sunlighting effects during the design process. A solution to this problem consists of computing the boolean intersection and difference between the sunlighting volumes associated with the properties. Openings or shadings that achieve both properties must check on some rules related to the resulting partition : intersecting one part and not the others, for instance. Such rules can be formalized in a logical way. An intelligent tool working on these logic rules may infer relevant advices to manage properties together.

4.2 Achievement of fuzzy properties

Another limit of our system is the restriction to the binary qualification sunlit or sunless. A designer would prefer to deal with fuzzy sunlighting qualifications such as ' enough sunlit ', ' not too sunless ', and so on. That means he may accept that properties are roughly, rather than perfectly, achieved. Some of the fuzzy logic methods are under consideration to solve this problem.

4.3 Modeling of realistic openings and shadings

At the present stage of development, our system generates exact geometrical shapes figuring openings and shadings that achieve a given property. It is the designer's responsability to interpret and to transform these shapes into realistic shadings and openings that make sense according to the project context. However, an architect may prefer obtaining directly some relevant realistic objects, especially when these objects are difficult to model. An interesting example is the one of landscape design where ' tree-shadings ' can be used for achieving sunlighting properties as well as visual ones. In that case, an architect would prefer working on alternative realistic trees, rather than on geometrical shapes. Our system could satisfy such a request by exploring a knowledge base of tree and plant features. For a given sunlighthing property, such an expert system would suggest alternative realistic tree-shadings solutions, according to their height, their volume, their leaves in winter or summer, and so on. Of course, this methodology could be extended to many kinds of relevant openings and shadings by using different architectural and urban knowledge bases.

5 Concluding remark

Advantages of generative approaches are numerous and quite still unexplored. Their scope could be extended to various environmental properties as well as architectural ones. The ' declarative methods ' emphasized by Lucas & al. [6] and mainly applied to geometrical modeling, constitute an alternative way for such issue.

6 References

1. Khemlani. L., Genwin : A generative computer tool for window design in energy conscious architecture, *Building and Environment*, Vol. 30 n° 1, 1995
2. Woodbury R.F., Searching for designs : Paradigm and practice, *Building and Environment,* Vol. 26 n° 1, 1991
3. Schoeneman C., Dorsey J., Smits B., Arvo J., Greenberg D., Painting with Light, *Computer Graphics Proceedings*, Anaheim, California, 1993
4. Groleau D., Marenne C., Gadilhe A., Climatic simulation tools : an application for a building project in an urban space, in *Proc. of the 3th Inter. Conf. on Solar Energy in Architecture and Urban Planning*, Florence, Italy, 1993
5. Siret D., Declarative modeling methods related to ambient environment in architectural and urban design, *Revue internationale de CFAO et d'Informatique graphique*, Vol. 10 n° 5, 1995
6. Lucas M., Desmontils E., Declarative modelers, *Revue internationale de CFAO et d'Informatique graphique*, Vol. 10 n° 6, 1996

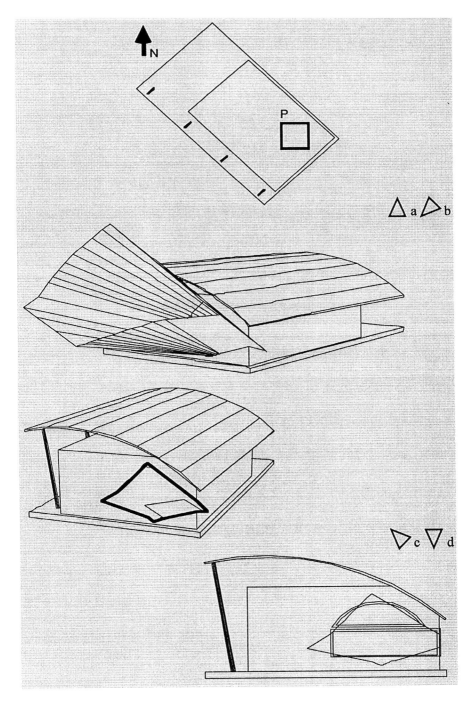

Figure 3. ' P, Sunlit, the Middle of the morning in winter '

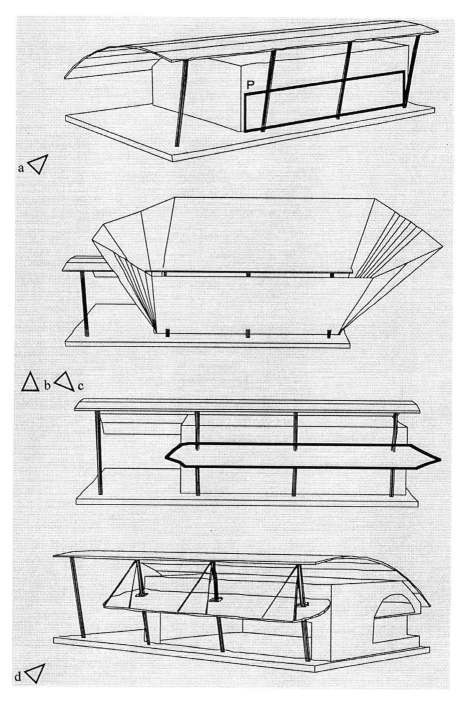

Figure 4. ' P, Sunless, the Beginning of afternoon in summer '

OilTrack™: a computational model to simulate oil spills trajectories in water

R. García-Martínez,[a,c] J.J. Rodríguez-Molina[b]
aInstituto de Mecánica de Fluidos, Universidad Central de Venezuela, Caracas, Venezuela
bInstituto de Urbanismo, Universidad Central de Venezuela, Caracas, Venezuela
cVisiting Professor at Wessex Institute of Technology, Ashurst Lodge, Ashurst, Southampton, UK

Abstract

This article describes the *OilTrack*™ mathematical model to simulate oil spills trajectories in water. The model calculates the 2-D velocity field in the spill zone and simulates the oil spreading on the water surface. A Graphical User Interface (GUI) that runs under MS-Windows™ makes the model accessible to professionals with little computer training. Comparisons with analytical solution, laboratory experiments and various spreading formulations show that *OilTrack*™ is an accurate and useful tool to study the oil trajectory from instantaneous or continuous spill accidents in water necessary for contingency planning, and environmental impact assessments.

1 Introduction

Oil spill simulation is a very complex task where a large number of interacting factors participate. The trajectory of a particular spill is of fundamental importance since it will determine the oil impact on coastal areas and other sensitive ecosystems. Also, the thickness distribution over the spill area will influence the evaporation of the oil and consequently its volume decrease rate.

Most oil spills models available today use the Mackay [10] formulation to simulate oil spreading. This formulation was recently corrected and modified by García-Martínez et al. [8]. However, these formulations assume circular spreading that seldom occur in reality.

This article resumes the fundamentals and applications of the *OilTrack*™ mathematical model to simulate oil spills in water.

The *OilTrack*™ computer program consist of two models. The first one is a two-dimensional vertically-averaged model that calculates the unsteady velocity field and water levels in the zone of interest (*water current model*). The second one, is a Lagrangian spill model that calculates the trajectory and oil thickness using the results from the velocity field model. The oil spill model may be also applied using field velocity data.

To simplify the application of the models, a graphical user interface was developed in MS-Windows™allowing the interactive discretization of the spill zone. Once data is introduced, models are executed and results may be printed or plotted in a very flexible manner. The graphical user interface makes the model accessible to professionals with little computer training.

OilTrack™ is a useful tool in the design of contingency plans, spill response and environmental impact assessments.

2   Fundamentals of the Mathematical Model

The mathematical model is based on hypothesis that follow the behaviour of oil observed in the laboratory and field spills. Section 2.1 describes the water current model. Section 2.2 presents the spill model and section 2.3 treats with the determination of diffusion coefficients.

2.1   Water Current Model

To determine the water velocity field, *OilTrack*™ implements a hydrodynamic numerical model [4] that simulate wind and tidal currents under the shallow-water approximation. To solve the equations, it uses an explicit time discretization based on the finite difference MacCormack time-split scheme [9]. This model has been extensively applied and validated solving different problems in large rivers and estuaries [4, 5, 6, 13]. The model calculates the vertically averaged velocity and the water elevation at each computational cell. The velocities are later used by the spill model to compute advective particle trajectories.

The water current model is optimized to work in microcomputers and may be applied in practically any geographical configuration.

2.2   Oil Spill Model

Many oil spill models available today use Eulerian schemes that are often accompanied by serious difficulties like spurious oscillations leading to unphysical representation of the phenomena. Instead of using such Eulerian approach where the pollutant concentration is calculated at certain fixed points, a Lagrangian approach was chosen to model oil spreading and trajectory in water based on experimental results and data reports from many oil spills. In this method the spilled mass is represented by a predetermined (and large) number N of tracer particles. This transport model, based on tracer or Lagrangian concepts, calculates the particles trajectories in a turbulent flow field considering advection by the flow and dispersion of the oil in water column. At desired time intervals the model may calculate the oil thickness using a local Eulerian balance explained later.

The main advantage of this technique compared to the traditional Eulerian method is the elimination of *numerical* diffusion.

In turbulence theory the concentration of a substance may be represented by a finite number of particles moving with the velocity at their instantaneous position [11]. Usually, it is assumed that the velocity can be expressed as a mean flow *advective* velocity u_a plus a smaller scale random fluctuation due to turbulent diffusion u_d.

To calculate the particle trajectories the model solves the following ordinary differential equations

$$\frac{dx_i}{dt} = u_{ai} + u_{di}, \tag{1}$$

$$\frac{dy_i}{dt} = v_{ai} + v_{di}, \tag{2}$$

$$\frac{dz_i}{dt} = w_{ai} + w_{di}, \tag{3}$$

where x_i, y_i, and z_i are the coordinates of the i particle, t is the time, u_{ai}, v_{ai}, and w_{ai} are the water velocities that advect particle i in x, y, and z directions respectively and u_{di}, v_{di}, and w_{di} are the particle i velocities due to diffusion in the x, y, and z directions respectively.

Equations 1 – 3 may be solved using Euler method as follows

$$x_i^{n+1} = x_i^n + \Delta x_i^n + \Delta x_i^{n'} \tag{4}$$

$$y_i^{n+1} = y_i^n + \Delta y_i^n + \Delta y_i^{n'} \tag{5}$$

$$z_i^{n+1} = z_i^n + \Delta z_i^n + \Delta z_i^{n'} \tag{6}$$

where Δt is the time interval, x_i^{n+1}, y_i^{n+1}, and z_i^{n+1} are the i particle coordinates for time $(n+1)\Delta t$, x_i^n, y_i^n, and z_i^n are the i particle coordinates for time $n\Delta t$, $\Delta x_i^n, \Delta y_i^n$, and Δz_i^n are the advective particle displacements defined by

$$\Delta x_i^n = u_{ai}\Delta t, \tag{7}$$

$$\Delta y_i^n = v_{ai}\Delta t, \tag{8}$$

$$\Delta z_i^n = w_{ai}\Delta t \tag{9}$$

and Δt is the time interval, $\Delta x_i^{n'}, \Delta y_i^{n'},$ and $\Delta z_i^{n'}$ are the i particle displacements due to the random velocity fluctuations defined as

$$\Delta x_i^{n'} = u_{di}\Delta t, \tag{10}$$
$$\Delta y_i^{n'} = v_{di}\Delta t, \tag{11}$$
$$\Delta z_i^{n'} = w_{di}\Delta t \tag{12}$$

The horizontal advective field (u_a, v_a) is calculated by the water current mathematical model and the vertical velocity w_d considers the buoyancy forces due to differences in densities between water and oil through the Stokes law.

The random velocities (u_d, v_d, w_d) due to diffusion are obtained using Montecarlo sampling [7] in a range of velocities $[-U_r, U_r]$ proportional to the diffusion coefficient. U_r is calculated as follows. Assuming large-scale Brownian motion [11] the isotropic diffusion coefficient ϵ may be expressed as [1]

$$\epsilon = \left(\frac{1}{2\tau}\int_{-\infty}^{+\infty} x^2 f(x)dx\right) \Big/ \left(\int_{-\infty}^{+\infty} f(x)dx\right) \tag{13}$$

where $f(x)$ is a probability distribution that determines the particle displacement x due to the random fluctuations in time τ. It may be proved that equation 13 depends mainly on the variance of $f(x)$ and not on its specific form. Therefore, this model adopts a *tophat* distribution that, introduced in (13) determine the relationship between ϵ and the diffusion velocity fluctuation range $[-U_r, U_r]$ in the following way

$$U_r = \sqrt{\frac{6\epsilon_x}{\Delta t}} \tag{14}$$

where ϵ_x is the diffusion coefficient in x direction.

An analogue deduction may be performed to obtain the random velocity fluctuation ranges in y and z directions $([-V_r, V_r], [-W_r, W_r])$.

To calculate oil concentration at a particular point (x, y, z), the model counts the number of particles inside a control cubic volume around the point. The concentration corresponding to each particle is the initial concentration divided by the number of particles used to represent the spill.

To calculate oil thickness Z, the following expression is used:

$$Z = N(\Delta A)Z_p \tag{15}$$

where, $N(\Delta A)$ is the number of particles contained in the reference surface area ΔA, Z_p is the *particle thickness* defined as:

$$Z_p = \frac{V}{\Delta A N_p} \tag{16}$$

V is the initial spill volume, and N_p is the total number of particles used to represent the oil.

2.3   Diffusion Coefficients

Careful estimation of the diffusion coefficients is one of the basic factors affecting oil spills, since they determine the random velocity ranges and consequently the oil trajectories.

The complexity of the dynamic process that govern turbulent dispersion in water and the spreading dynamics justify an empirical approach to estimate diffusion coefficients. In this work it is proposed to use horizontal diffusion coefficients such that the particle spreading calculated with the Lagrangian scheme is equivalent to the solution given by the modified Mackay formulation [8].

The resulting expression gives a relationship between the oil spreading law and the diffusion coefficient D as follows:

$$D = \frac{kt^m}{2} \qquad (17)$$

where k depends on the oil properties including density and spreading coefficient and m depends on the spreading regime (see details in [8]).

3   Comparison Experiments

3.1   Analytical Solution Comparison

The oil spill model was compared with a 3-D analytical solution. The simplest solution corresponds to the instantaneous release of mass M of a solute in an unconfined static fluid at $t = 0$. The resulting concentration distribution is given by [2]

$$C = \frac{M}{(4\pi t)^{3/2}(\epsilon_x\epsilon_y\epsilon_z)^{1/2}} \exp\left(-\frac{x^2}{4\epsilon_x t} - \frac{y^2}{4\epsilon_y t} - \frac{z^2}{4\epsilon_z t}\right). \qquad (18)$$

Figure 1 shows the concentration at $(0,0,0)$ calculated by *OilTrack*™ and from the analytical solution (18). Note that both solutions agree from about $t = 400$ seconds. For previous times, the numerical solution gives smaller concentration values than the analytical solution because (18) tends to infinity as t tends to zero, while the *OilTrack*™ model assumes a large but finite initial concentration.

The concentration fluctuations in the numerical solution are typical of stochastic methods like the one used in the model.

Concentration (ppm)

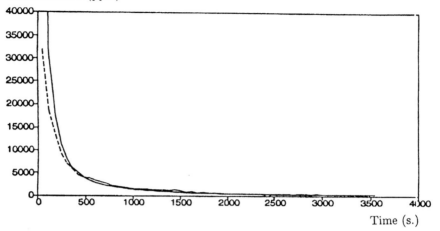

Time (s.)

Figure 1: Comparison of *OilTrack*™ results with analytical solution. —— Analytical solution, - - - model results

3.2   Comparison with Experiments

To test the validity of the model, a set of oil spreading experiments in quiescent water were done in the laboratory. A total of thirty four oil spills were tested in a 14 m. long, 9 m. wide and 0.60 m. deep tank. Three Venezuelan oils, Victoria, Mesa, and Lago Medio with different physical properties were used in the tests. The oil-water and oil-air interfacial tensions were measured in the laboratory for each oil sample using a Fisher ring tensiometer. Densities were also measured before each test. Details of these tests are reported in references [12] and [3].

Figure 2 presents the time variation of the radius of the circular slick measured with Mesa oil as well as the calculations performed with various formulations. It may be seen that the proposed mode (Rf2) compares well the experimental data. Mackay thick and thin formulas differ considerably with the experiments and with the other formulations.

It is important to remark that Mackay's thick slick formulation greatly underestimate the spreading radius in all test performed in this study.

4   Use of the Model for Simulating Oil Spills

OilTrack™ runs under the MS-Windows™ ver. 3.1/Win95/Win NT operating systems on a 486 microcomputer with at least 8 Mb RAM, and requires 4 Mb of hard disk space for installation. A VGA color monitor

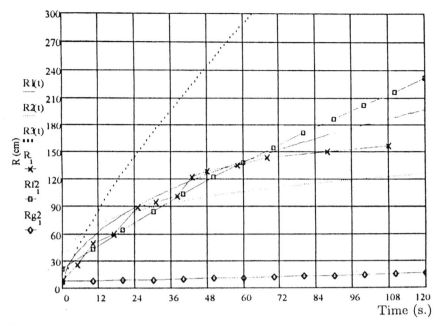

Figure 2: Comparison between Mesa oil spreading experiments X, Mackay formulation (Rg2) and the present model (Rf2)

and a graphical output device are highly recommended. Geographical data may be input with digitizing table. The *OilTrack*™ graphical user interface allows the interactive discretization of the spill zone.

The water current model requires the following basic site and oceanographic data

- Bathymetry (bottom elevations). This information may be input digitizing the map of the zone,

- Water elevation time series at the open boundary when currents are generated by tides or

- Water velocities at the model open boundary,

- Wind velocity at 10 m height, and

- Bottom roughness coefficient (Manning).

The oil spill model requires the following data

- Oil physical properties such as density, spreading coefficient, etc.,

- Total spill volume,

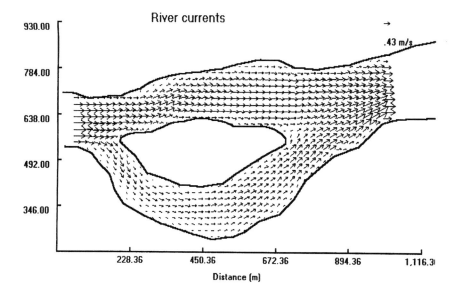

Figure 3: Water currents calculated with *OilTrack*™.

- Duration of spill,

- Spill coordinates (x_o, y_o, z_o),

- Water density,

- Wave conditions (amplitude, direction, etc.) and

- Total simulation time.

This information is introduced through windowed dialogue boxes and may be stored in hard disk files for later retrieval. User input is validated to avoid out of range values. Default or recommended values are presented when dialogues opens.

Once the user has entered all necessary data, models may be executed and results presented in a variety of graphic options. As figure 3 shows, 2-D water currents are represented as vector fields for any chosen time. Graphs may be printed in the MS-Windows™ printer or plotter connected to the microcomputer.

Figure 4 presents a 2-D graph for oil particles at a specific time. These graph may be animated to observe dynamically the oil trajectory. Vertical concentration contours may be also observed. This graphical options allow to have a 3-D idea of the oil cloud position at any time.

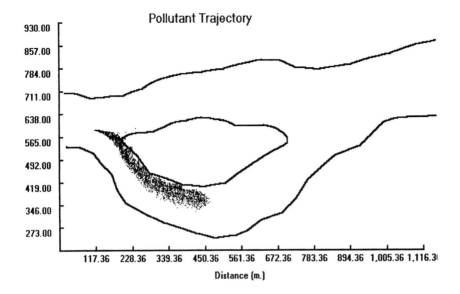

Figure 4: Oil particle positions.

The user may also select, using the mouse or the digitizer, any coordinate inside the model domain to graph velocity, water elevation, oil concentrations and thickness variations with time.

5 Conclusions

This article resumes the development and application of the *OilTrack*™ mathematical model to simulate oil spills in water.

Results show that the 3-D Lagrangian approach used, correctly model the dispersive behaviour of oil in water.

Validation with analytical solutions and experiments show that the model is able to simulate spills predicting oil spreading and trajectory.

The *OilTrack*™ graphical user interface makes it accessible to professionals with little computer training. The model is an useful tool to study the oil trajectory from instantaneous or continuous spill accidents in water necessary for contingency planning, and environmental impact assessments.

References

[1] Einstein, A. Über die von der molekularkinettischen Theorie der Wärme geforderte Bewegung von in ruhenden Flüssigkeiten suspendierten Teilchen. *Ann. Physics.* 4. 1905.

[2] Fischer, H. B., List, E. J., Koh, R. C. Y., Imberger, J. and Brooks, N.H. *Mixing in Inland and Coastal Waters*. Academic Press, New York, 1979.

[3] Garban, J.R., and Tinoco, J.C. (1995). *Interaction of Oil Slick in the Water Surface* (in Spanish). Engineer Thesis at Faculty of Engineering, Universidad Central de Venezuela, Caracas, Venezuela. 180 pages.

[4] García, R. and Kahawita R. Numerical Solution of the Saint Venant Equations by the MacCormack Finite-Difference Scheme. *Int. J. for Num. Meth. in Fluids.* **6** (5), 259–274, 1986.

[5] García, R., Febres, B. and Villoria, C. *Simulating the Evolution of Dredged Trenches with 2-D Numerical Models*. 5th International Conference of Computational Methods and Experimental Measurements. Mtl. Canada, Jul. 1991.

[6] García, R. *Eulerian-Lagrangian Model for Simulating Flow Fields and 3-D Pollutant Transport in Rivers, Lakes, and Coastal Waters*. SIAM Pan-American Workshop on Applied and Computational Mathematics. Caracas, Venezuela. Jan. 10–15, 1993.

[7] Hunter, J.R., Craig, P.D., and Phillips, H.E. On the Use of Random Walk Models with Spatially Variable Diffusivity. *J. Comput. Physics* **106**, 366–376. 1993.

[8] García-Martínez, R. *A Correction to the MAckay Oil Spreading Formulation*. 19th Artic and Marine Oilspill Program Technical Seminar AMOP. June 12-14, Calgary, Canada. 1996.

[9] MacCormack, R. W. The Effect of Viscosity in Hypervelocity Impact Cratering. *AIAA paper* 69-354. 1969

[10] Mackay, D., Buist, I., Mascarenhas, R., Paterson, S. (1980): Oils spills processes and models. *Environment Canada*, Ottawa, Ontario. 86 pages.

[11] Maier-Reimer, E. On Tracer Methods in Computational Hydrodynamics. Chap. 9 of *Engineering Applications of Computational Hydraulics* Vol. 1. Ed. by Abbott, M.B. and Cunge, J.A. Pitman 1982.

[12] Quintero, Y., Rivero, G. and Sorondo, M. (1994). *Simulation of Oil Spreading in the Water Surface* (in Spanish). Engineer Thesis at Faculty of Engineering, Universidad Central de Venezuela, Caracas, Venezuela. 143 pages.

[13] Villoria, C. and García, R. *Prediction of Current and Sediment Deposition Patterns in Puerto Miranda Oil Terminal Using 2-D Mathematical Models.* 23th International Conference on Coastal Engineering Oct. 4–9, Venice, Italy. 1992.

A graphical software interface for traffic emissions analysis

L. Della Ragione,[a] F. D'Aniello,[a] M. Rapone,[a] V. Shatsov[b]

[a]*Instituto Motori, Consiglio Nazionale delle Ricerche, V.le Marconi 8, 80125 Napoli, Italy*
[b]*Academy of Sciences, Institute for System Analysis, Moscow, Russia*

Abstract

To assure a satisfactory air quality in congested urban areas traffic managers have to consider vehicle emissions among the effects of traffic planning and management. In this paper the general structure of an information system capable to support traffic management with traffic and emissions analysis is presented. The information system is made of two modules: TRANSNET and EMISSION. TRANSNET implements a programming tool for the modelling of traffic flows in the transportation network of a city. EMISSION links the modelling of traffic with the evaluation of vehicle emissions corresponding to vehicle operating conditions determined by traffic. In particular, in this paper the current implemented version of EMISSION is presented, which represent information relative to an individual vehicle in the traffic flow of Naples city. EMISSION uses driving cycles to characterise vehicle operating conditions and traffic image to characterise traffic conditions relative to determined driving cycles. It displays fuel consumption and emissions measured in the laboratory using determined driving cycles. This application is useful to compare traffic conditions, driving cycles, fuel consumption and exhaust emissions connected with different streets and traffic levels in the urban network.

1 Introduction

In highly congested urban areas, traffic planners have to consider vehicle exhaust emissions among the effects of traffic plan and traffic management policy. This is more emphasised in Italy, where a specific law obliges City

Major to stop vehicle circulation in the case pollution level exceeds limits established by law.

Thus, since the first phase of traffic plan conception, planner should evaluate expected pollution levels related to predictable critical traffic conditions for different zones. Pollution levels depend on a number of factors, traffic emissions are one of them, which can be up to a certain extent controlled.

Traffic emissions are a function of flow vehicle composition and of individual vehicle exhaust emissions. Vehicle composition can be assumed on the basis of statistical surveys. Thus the problem is solved by the determination of vehicle operating conditions and related exhaust emissions for different traffic conditions.

In this paper, the structure of a decision support system for traffic planners capable to support traffic management with traffic and pollutant analysis is presented. The development of DSS is in progress with the co-operation of Institute for System Analysis of the Academy of Science of Moscow and Istituto Motori of Research Council of Naples. The information system is made of two modules: TRANSNET and EMISSION. TRANSNET implements a programming tool for the modelling of traffic flows in the transportation network of a city. EMISSION links the modelling of traffic with the evaluation of vehicle emissions corresponding to vehicle operating conditions determined by traffic.

TRANSNET is based on some well known models and algorithms [1] [2] and some specific methods developed [3] [4] for the evaluation of traffic flows.

EMISSION input information is based on the experimental information detected on road, on results of a complex procedure of data analysis and classification founded on multivariate statistical analysis [5], on emission laboratory measurements [6]. An experimental approach was followed to determine vehicle operating conditions representative of different traffic condition for each road. Vehicle operating conditions are determined by driving cycles performed by each vehicle through the traffic. Typical driving cycles are determined by statistical analysis of vehicle kinematics recorded on the road by instrumented vehicles. Cluster analysis is used to determine driving cycles by velocity profile detected on the road and to classify them into groups of driving cycles representative of different traffic conditions [5]. Driving cycles determined by this way are used to measure emissions in the laboratory.

EMISSION is intended to evaluate emissions of the vehicle flow on the basis of a given vehicle composition of flow. Vehicle emissions are measured in laboratory using driving cycles determined for each vehicle category. At the current status of research, experimental results achieved so far are relative to just one medium size vehicle. Actually EMISSION uses driving cycle to characterise vehicle operating conditions, an image of traffic recorded while that cycle was detected to characterise traffic conditions, on-road   and laboratory measured fuel consumption to compare the two different situations, exhaust emissions measured in the laboratory with the medium size gasoline car.

Further developments will be the extension of emission data base to many different vehicle category to allow the emission analysis to consider different traffic vehicle composition for emission evaluation.

2 General structure of DSS.

In fig. 1 the general structure of the information system is presented. The shaded area refers to Emission module. Transportation Network and Traffic Flows is common information of the two modules. In fig.2 the structure of graphical interface for emission analysis is shown.

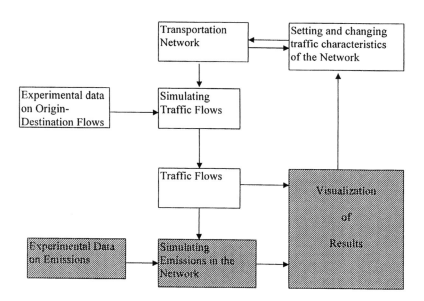

Figure 1: The general structure of the information system

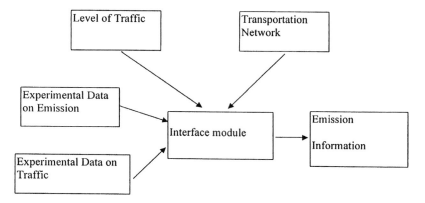

Figure 2: Graphical interface for emission analysis

3 Program TRANSNET.

The program TRANSNET implements a programming tool for the modelling of traffic flows in the transportation network of a real city. The program is designed as a Windows application for IBM-compatible PC with the self-explanatory "user-kind" interface, which provides the following functionality:

- Input and edit an information about the transportation network. This information includes both the geographical data (coordinates of nodes and arcs) and the traffic characteristics, i.e. the set of characteristics, that are necessary to perform the modelling of traffic flows in the network

- Modelling of traffic in the network: evaluation of an origindestination flows, assignment of OD flows on the network, evaluation of shortest paths etc.

3.1 General structure of information

From the perspective of modelling traffic flows, transportation network is considered as an oriented graph, which consists of nodes (intersections) and arcs (links). Each node and arc has a set of characteristics, that define the conditions and the cost of moving along the arc, and the cost of any manoeuvre at the intersection.

3.1.1 Hierarchical structure

The following hierarchical structure of information is used:

- Basic element of information is a "line", as an ordered sequence of "points". Two neighbour points in a line are connected by a pair of arcs (namely, the "forward" and the "backward" arc). Points on a line correspond to an intersections (in a case of streets) or stations of the "railway-like" lines (see below).

- Lines of the transportation network are combined into "sets of lines". It is preferable to combine lines with similar characteristics in a single set (highways, main streets, small streets etc.). This allows the use of the "default" arc characteristics for each line and simplifies the description of the network.

- On the top of hierarchy we distinguish two main "categories" of lines: "street" and "railway-like" lines. The latter include the lines of transport types, that use their own separate way, and do not occupy street space in the city (railway, underground, subway lines etc.).

3.1.2 Traffic characteristics of arcs
The traffic characteristics of arcs are the following:

- Class of the arc. This characteristic defines the cost (the travel time)-flow function of an arc.
- Number of lanes and the parking permission flag.
- Length of the arc
- Flags, which define the permission of tracks and the presence of public facilities (bus, trolley-bus etc.) in the arc.

3.1.3 Traffic characteristics of intersections
The complete description of an intersection is given by a matrix of costs of manoeuvres at intersection. The specification of the individual matrix for each intersection in large networks seems to be rather difficult task for the user. However the majority of intersections in real networks can be described with the use of predefined "standard" cost characteristics.

The program distinguish following standard manoeuvres at intersections:
- forward move - move along the current line (it is not necessary a geometrical straight line);
- overturn - move back along the same line;
- right turn and left turn - relative to the direction of the forward move.

The additional information, that can be specified for the standard intersection, is the "scheme of permitted turns", i.e. matrix of flags, which define the permission of each manoeuvre at intersection.

If an intersection has non-standard characteristics (costs or geometry) it is declared as "user-defined", and the matrices of costs is inputted by user.

3.1.4 Structure of information
The structure of information provided for the system of "default setting" of traffic characteristics of network objects. Namely, the default arc characteristics, specified for a set of line (single line), are automatically applied to each line in this set (each arc of this line). These defaults can be overridden for the individual lines (arcs), if necessary. Similarly the default scheme of turns, specified for both directions of moving along a line, is automatically applied to each point in this line.

This principle of "default setting" provides the essential decrease of user input.

3.1.5 Classes of arcs

The main characteristic of an arc is a cost-flow function, which defines the dependence of cost of moving along the arc on the current flow through the arc. The structure of information contains a database of predefined cost-flow functions, which are applied to arcs of different classes. The specification of class in the description of arc referred this arc to a particular function from this database.

Cost-flow function can be defined either in the analytical (polynomial or exponential approximation) or in the table form. In a first case the coefficient of an analytical functions are inputted, otherwise the name of the file, that contains a table is specified.

3.1.6 Origin-destination zones

The origin-destination zones are presented by a set of points of special category. These points are linked to the normal points with the so called "virtual" arcs. Main characteristics of OD-points are:
- total out-flow;
- total in-flow;
- cost-characteristics of virtual arcs, which link OD-point to the normal points of the network.

3.2 Models and algorithms

The main modelling tasks, performed by the program are:
- Finding the shortest paths (i.e. paths with a minimal cost). The effective shortest paths algorithm [1] was adopted to take into account the costs of manoeuvres at intersections (turn penalties).
- Evaluation of an origin-destination flows. Evaluation of an origin-destination flows is based on the entropy maximising procedure. The "weighted" entropy function is constructed with the use of the empirical "travel cost distribution" function [3] [4].
- Solving the traffic assignment problem. The linear approximation method of the solution of equilibrium (capacity constrained) traffic assignment problem was implemented. The classical Frank and Wolfe algorithm [2] was adopted to incorporate several classes of travellers (those using the public facilities, private cars and tracks).

3.3 The interface of the program

Main elements of the user interface are:
- Menu and a control bar for selecting the appropriate functions and commands.
- "Map window", which contains a graphical representation of the network.

- Specialised dialogs and pop up windows, that are created at a run-time for input of data, specification of parameters, choosing different options, visualisation of requested information etc.

4 Program EMISSION

Program EMISSION is designed on the base of the program TRANSNET and is intended (as a perspective goal) to link the modelling of traffic with the evaluation of the emissions and pollution in the city. The current version of a program serves as a programming tool for the visualisation of an empirical data on the emissions and pollution, and, in particular, the graphical representation of data for the transportation network under issue.

4.1 General structure of emissions data

4.1.1 Traffic level data
The empirical study shows that the variety of different traffic conditions in the city can be divided into groups with similar characteristics with respect to emissions and a fuel consumption the so called" traffic levels". Each traffic level is characterised by the following set of data:
- emissions of different types (CO_2, CO, HC, NOX, and particulate)(g/km);
- fuel consumption (g/km);
- reference to a typical "traffic cycle";

Traffic cycle describes the typical travel rate of a vehicle: the dependence of the velocity and gear on time. Traffic cycle data is stored in a file and is linked to the traffic level data via reference to a file.

An additional visual characteristic of traffic level is an image of the typical traffic conditions in a street. These images (if available) are stored in files and linked to the traffic level data via reference to file.

4.1.2 Emissions data
The traffic level data sets for all basic traffic levels are combined in a container; this container constitutes the complete description of emissions data. The program handles the general emissions data (i.e. the emissions data, which characterise the average emissions for the whole city), as well as the individual emissions data sets, which can be specified for each street (or arc) in the city.

The special database contains the description of routes, for which the measured emissions data is available.

4.2 The program interface

4.2.1 Main menu
Main menu of the program consists of the basic command groups (see the description of the program TRANSNET) and an additional group Emissions. Group Emissions include 2 command items:

- General data;
- Street data.

These commands implement the requests on displaying the general emissions data, and the emissions data for the individual streets, correspondingly.

4.2.2 Main window

Main window contains:
- the Map window;
- the "Route view" dialog.

Map window contains the graphical representation of the network. The routes, for which the measured data is available, are shown on the map. "Route View" dialog is used to select an appropriate route, and to track route street-by-street or point-by-point. The selected route/point on a route is highlighted on the map. (fig. 3)

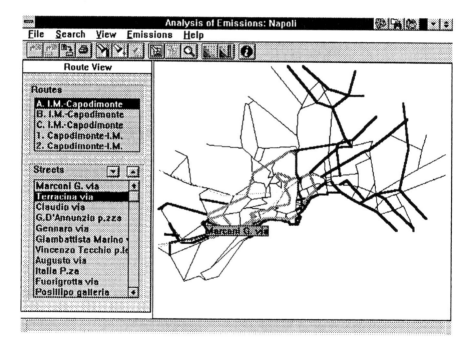

Figure 3: Main window

4.2.3 Emissions data

The general emissions data and the individual emissions data for the currently selected street is displayed and edited via the "Emissions" dialog. The "Emissions" dialog contains:

- a list of traffic levels;
- histogram of emissions and   fuel consumption for the traffic level, currently selected in the list;
- histogram of the traffic cycle;
- image of typical traffic conditions.

Edit button calls the special "Edit" dialog   for editing the data of the selected traffic level.

This dialog is used for changing the values of emission/consumption, and setting the references to the traffic cycle data file (fig. 4) , and image file (fig.5).

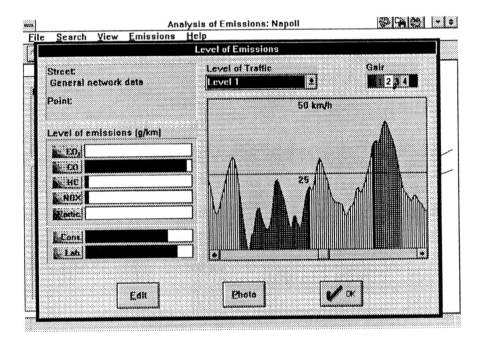

Figure 4: Emission window representing driving cycle

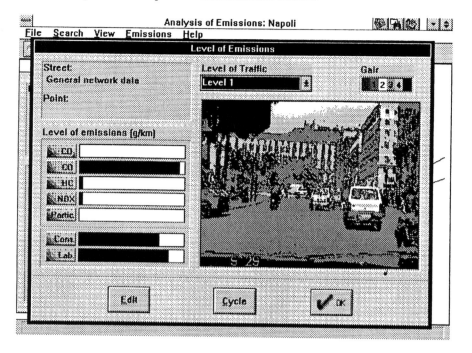

Figure5: Emission window representing traffic image.

Key Words: Traffic management - information system - emissions

References

1. Diniz, E.A., Efficient algorithms of finding shortest paths in a network"
Transport Systems, Institute for Systems Studies, Moscow, 1978.
2. LeBlank, L.J., Morlok, E.K., & Pierskalla, W.P., An efficient approach to
solving the road network equilibrium traffic assignment problem,
Transportation Research,. 1975, **9** pp 309-318.
3. Shmulyan, B.L., Entropy model of Urban System, *Automatics and
Telemechanics,* N10 (in russian), 1979.
4. Wilson,A.G., Entropy in Regional and Urban Development, Pion, London,
1974.
5. Della Ragione, L., Luzar, V. & Rapone, M., Determination of Driving
Cycles for Emissions Modelling in Urban Areas: a case study, *Proceedings of
the UTE 95 First International Conference on Urban Transport and the
Environment*, Southampton, UK, 1995.

6. Della Ragione, L., D'Aniello, F., Luzar, V. & Rapone, M., Experimental Evaluation of Fuel Economy and Emissions in Congested Urban Traffic, *Proceedings of the 1995 SAE International Fuel & Lubricants Meeting & Exposition*, Developments and Advances in Emission Control Technology SP-1120, Toronto, Canada, 1995.

INDEX OF AUTHORS

Computational Mechanics Publications

oastal Environment

ited by: **A.J. FERRANTE** and
A. BREBBIA, *Wessex Institute of Technology,*
uthampton, UK

is book will contain the proceedings of the First ernational Conference on Environmental Problems in astal Regions, to be held during August 1996 in Rio de neiro, Brazil. The importance of accurately modelling ese regions is emphasized by the need for a better derstanding of their normal behaviour and response to treme conditions. Densely populated areas or sites of ajor industrial developments have become a major ernational environmental concern.

rovisional Partial Contents: Hazard Mitigation; Risk nalysis; Environmental Impact-Assessment and tigation; Harbours, Ports and Marinas; Littoral Drift; astal Erosion; Siltation and Dredging; Oil Slicks and ills; Sewage and Chemical Pollution.

;BN: 1853124362   1996   360pp   £96.00/$144.00

ydrotrak

veloped by: **R. GARCIA-MARTINEZ** and **J.J.)DRIGUEZ,** *Universidad Central de Venezuela*

YDROTRAK is a sophisticated hydrodynamic and llutant transport software, which was developed to assist ofessionals and researchers working to solve water llution problems. It consists of two models. The first one a two-dimensional vertically-averaged model that lculates the unsteady velocity field and water levels in the ne of interest (water current model). The second one is three-dimensional pollutant transport model that lculates the trajectory and concentration of the pollutant ing the results from the velocity field model or field data.

simplify the application of the models, the Windows aphical user interface allows the interactive scretization of the zone where the pollutant is discharged. nce data is introduced, models are executed and results ay be printed or plotted in a very flexible manner. The aphical user interface makes the model accessible to ofessionals with little computer training. Hydrotrak runs der the Windows ver. 3.1. operating system on a 386 icrocomputer with at least 8 Mb RAM, and requires 4MB f hard disk space for installation.

demonstration disk is available on request.
SBN: 1853124869   1996   £950.00/$1490.00

See our 1996 catalogue on the Internet at
http://www.cmp.co.uk

Environmental Engineering Education and Training

Edited by: **T.V. DUGGAN** and
C.A. BREBBIA, *Wessex Institute of Technology,*
Southampton, UK

This book contains the proceedings of the First International Conference on Environmental Engineering Education and Training, held during April 1996 in Southampton. There are five main environmental problems requiring urgent engineering solutions: efficient use of energy sources; safe disposal of waste; recovery and recycling of materials; repair of existing damage and environmentally sound design of future technology. These factors have produced pressure on industry, both by legislation and public concern, and on engineering teachers and other educators to develop course curricula to include environmental issues in programmes of study.
ISBN: 1853123935   1996   288pp   £89.00/$134.00

Air Pollution IV

Edited by **C.A. BREBBIA** and **H. POWER,** *Wessex Institute of Technology, UK* and **B. CAUSSADE,** *Institute de Mecanique des Fluides de Toulouse, CNRS, France*

This book contains the proceedings of the Fourth International Conference on Air Pollution, being held during August 1996 in Toulouse, France. Development of experimental and computational techniques can be used as a tool for the solution and understanding of practical air pollution problems. It is then possible to evaluate proposed emission control techniques and strategies.
Partial Provisional Contents: Process Studies; Chemical Transformation Modelling; Emission Inventories; Regulatory Bodies; Indoor Pollution; Urban and Suburban Transport Emissions; Data Analysis and Observation; Global Studies.
ISBN: 1853124222   July 1996   888pp
£245.00/$368.00

All prices correct at time of going to press. All books are available from your bookseller or in case of difficulty direct from the Publisher.

Computational Mechanics Publications,
Ashurst Lodge, Ashurst, Southampton,
SO40 7AA, UK.
Tel: 44 (0)1703 293223   Fax: 44 (0)1703 292853
E-Mail: cmp@cmp.co.uk
Internet: http://www.cmp.co.uk

Computational Mechanics Publications

Air Pollution III

Edited by: **H. POWER** and **C.A. BREBBIA**,
Wessex Institute of Technology, UK and
N. MOUSSIOPOULOS, *Aristotle University of Thessaloniki, Greece*
These four volumes contain the proceedings of the Third International Conference on Air Pollution, held in Greece during September 1995.
SET ISBN: 1853123110;1562522353 (US, Canada, Mexico) 1995 1360pp £316.00/$485.00

Air Pollution Theory and Simulation
Vol 1
Partial Contents: Air Pollution Models and Environmental Modelling; Turbulence and Diffusion; Wind Flow and Dispersion Modelling; Chemical Transformation Modelling.
ISBN: 1853124419;1562523511 (US, Canada, Mexico) 1995 456pp £116.00/$178.00

Air Pollution Engineering and Management
Vol 2
Partial Contents: Pollution Engineering; Deposition of Air Pollutants; Aerosols in the Atmosphere; Monitoring and Laboratory Studies; Data Analysis and Processing; Emission Inventories and Modelling; Transport Emissions.
ISBN: 1853124427;156252352X (US, Canada, Mexico) 1995 536pp £135.00/$207.00

Urban Pollution
Vol 3
Partial Contents: Features of the Athens Basin Wind Flow in View of Recent Experimental Work; Simulation of the Paris Heat Island During Two Strong Pollution Events; Nested Dispersion Simulation over the Lisbon Region.
ISBN: 1853124435;1562523538 (US, Canada, Mexico) 1995 216pp £60.00/$92.00

Observation and Simulation of Air Pollution:
RESULTS FROM SANA AND EUMAC (EUROTRAC)
Volume 4
Edited by: **N. MOUSSIOPOULOS,** *Aristotle University of Thessaloniki, Greece* and **A. EBEL,** *Universitat zu Koln, Germany*
Partial Contents: Meteorological Effects on Air Pollution Variability on Regional Scales; Changes of some Components of Wet Deposition in East Germany.
ISBN: 1853124443;1562523546 (US, Canada, Mexico) 1995 152pp Paperback £40.00/$61.00

Environmental Engineering Education

Edited by: **T.V. DUGGAN,** *Wessex Institute of Technology, Southampton, UK* and
C.A. MITCHELL, *University of Queensland, Australia*
This book covers the field of environmental engineering in such areas as water, energy, nuclear engineering and the educational aspects.
Partial Contents: Introduction; Environmental Engineering: Where Do We Go From Here?; Environmental Engineering and its Impact on Engineering Education.
ISBN: 1853124796 1996 approx 300pp approx £89.00/$137.00

Computer Techniques in Environmental Studies V
Edited by: **P. ZANNETTI,** *Failure Analysis Associates Inc., California, USA*
These two volumes contain the proceedings of the Fifth International Conference on the Development and Application of Computer Techniques to Environmental Studies, held in San Francisco Bay, USA in November 1994.
SET ISBN: 1853122726; 1562521969 (US, Canada, Mexico) 1994 736pp £174.00/$267.00

Pollution Modeling
Vol I
Partial Contents: Atmospheric Studies/Air Pollution; Oil Pollution of Sea Surface/Surface Water Quality; Groundwater Quality; Hazardous Waste.
ISBN: 1853123706; 1562522949 (US, Canada, Mexico) 1994 320pp £84.00/$129.00

Environmental Systems
Vol II
Partial Contents: Global Issues; Meteorology and Meteorological Modeling; Modeling; Ecological Modeling; Adverse Effects; Regulatory Software; Environmental Systems; Decision Support, Expert Systems and Information Systems.
ISBN:1853123714;1562522957 (US, Canada, Mexico) 1994 416pp £109.00/$167.00

All prices correct at time of going to press. All books are available from your bookseller or in case of difficulty direct from the Publisher.

Computational Mechanics Publications
Ashurst Lodge, Ashurst, Southampton,
SO40 7AA, UK.
Tel: 44 (0)1703 293223 Fax: 44 (0)1703 292853
E-Mail: cmp@cmp.co.uk
Internet: http://www.cmp.co.uk